Intermediary Organization of the Skeleton

Volume II

Author
Harold M. Frost, M.D.
Department of Orthopedic Surgery
Southern Colorado Clinic
Pueblo, Colorado

CRC Press, Inc.
Boca Raton, Florida

Library of Congress Cataloging-in-Publication Data

Frost, Harold M., 1921-
 Intermediary organization of the skeleton.

 Includes bibliographies and index.
 1. Human skeleton—Diseases. 2. Human skeleton.
3. Orthopedia. I. Title. [DNLM: 1. Bone and Bones.
WE 200 P938i]
RD680.F76   1986   612'.75   85-16595
ISBN 0-8493-5948-1 (v. 1)
ISBN 0-8493-5949-X (v. 2)

This book represents information obtained from authentic and highly regarded sources. Reprinted material is quoted with permission, and sources are indicated. A wide variety of references are listed. Every reasonable effort has been made to give reliable data and information, but the author and the publisher cannot assume responsibility for the validity of all materials or for the consequences of their use.

All rights reserved. This book, or any parts thereof, may not be reproduced in any form without written consent from the publisher.

Direct all inquiries to CRC Press, Inc., 2000 Corporate Blvd., N.W., Boca Raton, Florida, 33431.

© 1986 by CRC Press, Inc.

International Standard Book Number 0-8493-5948-1 (Volume I)
International Standard Book Number 0-8493-5949-X (Volume II)

Library of Congress Card Number 85-16595
Printed in the United States

PREFACE

" . . . the great obstacle to progress is not ignorance but the illusion of knowledge."

(Daniel Boorstin, 1984)

As I wrote this a squirrel raiding the bird feeders (again) took off on local telephone wires, loaded to the gunwales with sunflower seeds, no doubt to rejoin his family. He ran easily on a wind-swayed wire some 3 mm in diameter, although a mathematical analysis once showed that such animals cannot balance on such unstable supports (and another, that humming birds cannot hover in midair). But neither squirrels nor humming birds seem to know that, for they do their respective "impossibles" effortlessly and surely.

The above paragraph and Boorstin's one-liner contain three messages that apply to the content and objectives of this book.

First, many young people in skeletal science tend to assume its most important knowledge and discoveries already exist, so future progress will depend more on exploiting them than on new kinds of information and understanding. Many established authorities have the same view of the long range prospects of the field and those observations would apply also to scholars in the times of Herodotus and to contemporary biochemists tinkering with the genome of a mouse.

Such factors partly explain a tendency of some authorities to conclude that if existing knowledge cannot explain something then it does not exist, even though an unbiassed eye can see it does. Exactly such receptions met the original descriptions of the bone remodeling BMU, the "On-Off" and the "kσ" properties described in Chapter 4, and the flexure-drift relation, the flexural neutralization mechanism, microdamage and the MES property described in Chapter 7, when they were first described between 1960 and 1964.

Yet history offers this lesson that has not changed since the age of Sumer: *always* more knowledge than already crowds any library, more "X", awaits perception.

Second, physiological systems play by their game rules rather than any conceived by man. For perhaps 300 years a few men spent parts of their lives trying to discern and understand those rules. Nevertheless, nature spent over a billion years and did billions of experiments in every one of them to devise, test, and perfect her rules, and to fit them one to another in intricate ways, as the life evolved that encircles this planet. Therefore it is an unwitting intellectual arrogance to assume that natural systems play only by those rules our brief and often naive research and thinking have already uncovered.

In that regard, the squirrel-humming bird analyses may have contained truth and avoided errors of comission but they also omitted other truths which would explain rather than deny the facts. It follows in the long run that it might be better to seek what such analyses lacked, rather than to defend them as sufficient or to reject them completely because they do not account for everything known about the matter at hand. But that is seldom done for humans tend to insist that if *a* truth is genuine it must be *the* truth.

Third, to understand the human body better requires observing and thinking about it in ways that use what is known but are not constrained by it or by orthodox concepts. In different words, a novel idea could be taken as at least a possible part of the whole, rather than as automatically suspect. In point of fact, fundamental progress in any field of science always came from fundamental changes in its concepts and kinds of knowledge, and blocking such changes always postponed that progress. To repeat, squirrels and humming birds reveal that something beyond once orthodox concepts of mechanical and physiological dynamics, some "X", also existed and made their "impossibles" real.

In my judgment those observations apply with special force to the skeletal field. For diverse reasons that stem from its nature, personalities and history, many contemporary authorities automatically suspect real innovation in theory, concepts or thought. They react to such innovation by seeking its flaws, and if any are found then they reject it entirely without exploring it further (yet, rare indeed the fundamental innovation that lacks flaws in its initial conception or description). Their basic reaction to true innovation is "attack and destroy", rather than try to build on it, and human nature provides many more such destroyers than builders. When those guardians of the *status quo* and conceptual stasis borrow the knowledge and ideas of other fields — such as mechanics, biochemistry, statistics, physiology, biology, physics, genetics — they use them only within orthodox limits and they discourage innovative uses of such knowledge and ideas by others.

They do not usually do that out of ill intent, for they were taught that negativism by the effects of earlier, limited and often flawed basic knowledge and understanding of this field, which caused most proposed basic theoretical innovations before recent times to vanish promptly under avalanches of contradicting facts.

But in the 1980s that negativism is becoming inappropriate, for knowledge and understanding have progressed to the point that the attack and destroy reaction holds this field back in significant ways. That must and will change, although other lessons of history suggest that change may meet strong resistance.

This monograph organizes some core knowledge of skeletal physiology and disease in a new way which allows both prediction and further development. It could not have been done until a missing link in the basic chain of skeletal knowledge, the skeletal intermediary organization (IO), became known and understood at least provisionally.

The evidence that supports that new way is not really new. It comes from the laboratory and the clinic, from current journals and dead archives, from the knowledge and concepts of many fields of science, and from many minds past and present, and the many references in this book only sample them. To repeat, the innovation in this book lies in how it organizes the evidence. While novel indeed, that way does not contradict the evidence, and it also brings out previously unperceived meanings that apply to the definition, study, prevention, diagnosis and treatment of all skeletal disease. It has value to the degree that it helps to understand the skeleton and to manage its malfunctions, using the term, skeleton, in its broadest sense. It has already been tested in the laboratory and the clinic and it betters its antecedents, and many of the new meanings it exposed have been verified. One example: the results of the human ADFR osteoporosis treatment set up and coordinated by Dr. Colin Anderson in London, Ontario.[23,24] Conceived solely from the properties of this new way, its success portends a shift in the collective perception of the IO, from the status of esoteric knowledge and visionary therapeutics, to the status of something practical and useful.

In metaphor, this book provides a pathfinding sojourn into a fascinating, rich and complex new territory rather than a trip on a worn highway to some settled and staid destination. It provides more truth if not yet all the truth, and it marks considerable progress achieved by the legions of men who followed Hippocrates, Galen,[270] Vesalius,[744] Mendel,[502] Harvey,[313] Hunter,[348] Virchow,[752] Lister,[461] Wolff,[782] Albright,[9-11] and Jaffe.[361-363]

Part of that progress is marked by some facts which will be new to most who read this work. It is now possible to predict major features of bone, joint, tendon, ligament and fascial architecture from first causes and initial conditions, and to predict the clinical types, pathogenesis and anatomical-pathological features of most osteoporoses, osteomalacias, aseptic necroses, dwarfism, stress fractures, chondrodystrophies, rickets, children's acquired deformities, and malfunctions of the skeletal healing and homeostatic functions. Also, it is now possible to understand certain modes of failure of artificial bones, joints, and ligaments.

But if we have come far we must go farther still, and builders or heurists, not destroyers, will lead the way.

This is a book for builders.

Harold M. Frost, M.D.

THE AUTHOR

Harold M. Frost, B.A., M.D., is currently an active orthopedic surgeon in the Southern Colorado Clinic in Pueblo, Colorado. He received his BA degree at Dartmouth College, and his M.D. degree at Northwestern University in 1945. A 2-year period in the Navy and orthopedic residencies in Worcester, Mass. and Buffalo, N.Y. followed. He entered orthopedic practice in Buffalo after finishing training in 1953, became assistant professor of orthopedic surgery at Yale University in 1955, joined the orthopedic department of Henry Ford Hospital in Detroit in 1957, and became its chairman in 1966 as well as clinical professor of surgery at the University of Michigan. He resigned and went to Pueblo in 1973. He belongs to many clinical orthopedic and skeletal research societies and is currently adjunct professor of anatomy at Purdue University.

Dr. Frost's research began at Buffalo, and continued at Yale and Henry Ford Hospital, where he founded its present Calcified Tissue Laboratory. His work focussed on what is now known as the intermediary organization of the skeleton. In its course he developed the current standard dynamic histomorphometry systems for compact and spongy bone. That research involved collaborations with more than 100 fellows and colleagues and led to 409 publications, including 12 texts, symposia and monographs, and contributions to over 20 other symposia. His personal research ceased in 1973 but he continues to lecture on clinical and basic science matters abroad as well as in the U.S., to publish, and to plan, evaluate and interpret the research of other individuals and institutions. He has become recognized and respected as one of the most productive basic science heurists in the field of orthopedic surgery. The present volumes synthesize 37 years of work in clinical orthopedics and 30 years in skeletal research. That work touches on and integrates information conventionally assigned to the fields of orthopedic surgery and oral surgery, and skeletal biomechanics, pathology, anatomy, histology, cellular biology, physiology, growth and development, metabolic bone disease, joint disease, trauma, the healing processes, children's deformities and the mechanical determinants of the architecture of hard and soft skeletal tissues on both micro- and macroscopic scales.

TABLE OF CONTENTS

Volume I

Chapter 1
Historical Background
Abstract 1
I. Introduction 1
    A. The Roots of Scientific Holism 2
    B. The Skeletal Problem 3
        1. A Renal Thought Experiment 3
        2. A Skeletal Transform 3
        3. The Scenario Contains a Thorn 3
    C. The Roots of Skeletal Atomism 5
        1. The Kidney 6
        2. The Skeleton 7

Note 1
Misextrapolation from Laboratory to Patient 11

Note 2
Default Judgments (or Unwitting Assumptions) 17

Chapter 2
Biological Organization
Abstract 19
I. Introduction 19
II. Some General Properties of Biological IOs 19
    A. The Ladder Property 19
    B. The Association-Organization Property 20
    C. The Function-Property Ratio 21
    D. Functions Act Upwards in the Ladder 23
    E. The Function-Control Relation 23
    F. Space-Time Domain Scaling 24
    G. The Whole is More Than and Different From the Sum of its Parts 25
    H. The Andistributive Property 25
    I. The Losing Betting Odds Property (LBO) 27
    J. The Downwards Analysis 27
    K. The Dendritic Property 27
III. Final Comments 28

Note 1
New Tricks and Old Dogs 31

Note 2
Biological Continua, Switches, and Saturation 39

Note 3
Definition of Disease 45

Note 4
Proof of a Function 47

Chapter 3
The Skeletal IO and the Lower IO (L_1)

Abstract 49
I. Introduction 49
    A. An Aborted Career 49
    B. The Basic General Functions Provided by the IO 50
        1. Histogenesis 50
        2. Growth 52
        3. Modeling 52
        4. Maintenance 52
        5. Repair 54
        6. Homeostasis 54
    C. Skeletal Organization 54
II. The Lower IO (L_1) 55
    A. The Common Plan 55
    B. Programming Modes and Variations 63
    C. Some Basic Properties of the L_1-Level System 64
        1. The MC Mode 65
        2. The N Mode 66
III. Functions of L_1 Level Entities 74
    A. Histogenesis 74
    B. Growth 75
    C. Micromodeling 76
    D. Rigidity 79
    E. Strength 79
    F. Elastic Compliance 79
    G. The LOBO Effect 79
    H. Destruction 80
    I. Molecular-Level Turnover 80
IV. Illustrative Questions 80

Note 1
On the Former Controversy Surrounding the Skeletal IO 83

Note 2
Cities and Skeletons: An Analogy 89

Chapter 4
The Middle IO (L_2)

Abstract 93
I. Introduction 93
II. Growth 95
    A. Proliferation 95
    B. Architecture 96
    C. Growth 97
    D. Modeling 97
    E. Programming Effects 99
        1. Articular Cartilage 100
        2. Epiphyseal Plate 100
        3. Perichondral Ring 103
        4. Fibrous Tissue 103

III.	Macromodeling		104
	A.	General Properties	104
	B.	Mechanical Determinants	105
	C.	"S"- and "L"-Mode Mechanical Effects	108
	D.	The Primal Role of Precursor Cells ("P") in Modeling	110
	E.	The "krσ" Phenomenon	110
IV.	Remodeling		111
	A.	Basic Lamellar Bone Remodeling Properties	112
	B.	The State Equations of Remodeling	121
	C.	The MCN Analog for Remodeling	121
V.	Illustrative Questions		125

Note 1
Dynamic Histomorphometry 129

Note 2
Some of the Histomorphometric Proof of the Bone Remodeling BMU and
Coupling 133

Chapter 5
The Upper IO (L_3)
Abstract 147

I.	Introduction		147
	A.	Summation in Bulk: Part of Nature's Camouflage	147
	B.	General Properties of the Upper IO	148
II.	The Skeletal Envelope System		150
	A.	Bone Envelopes and Envelope-Specific Behavior	150
	B.	Mechanisms Underlying the Envelope-Specific Bone Behavior	151
	C.	Locally Mediated Regulation of the IO	153
	D.	Series and Parallel Gating	162
III.	Temporal Coherence		162
	A.	The Nature of Temporal Incoherence	162
	B.	Inducing Temporal Coherence as a Treatment Stratagem	163
	C.	Temporal Coherence as a Research Tool	166
	D.	Some Requirements of Coherence Treatment/Research	169
IV.	The Epiphyseal System		172
	A.	Structural Features	173
	B.	Functions	173
V.	Comments on Basic L_3-Level Functions		178
	A.	The Basic Roles of the Recruitment Function	178
	B.	Modeling	179
	C.	Remodeling	179
	D.	The IO and Disease	180
	E.	Repair	180
	F.	Homeostasis	180
	G.	Future Research	181
	H.	Intrinsic and Extrinsic IO Disease	181
VI.	Illustrative Questions		182

Note 1
A Chronology of Some Basic IO Facts, the Discovery of which Awaited Dynamic
Histomorphometry 185

Note 2
Mechanical Usage Effects (MU) on Bone Remodeling 193

Interface 197

Chapter 6
The IO, Osteoporoses, and Other Matters
Abstract 199
I. Introduction 199
    A. Definitions 200
II. Comments on OPn and OPs 203
    A. Growth Modeling-Dependent and Remodeling-Dependent OPn and OPs 203
    B. Naturally Reversible and Irreversible OPn and OPs 203
    C. Abnormal Bone Fragility 205
    D. Envelope Sizes 212
    E. Bone Balances and Turnover 212
    F. Regional Differences in BMU-Based Remodeling 213
    G. BMU Activation Effects 213
    H. ΔB·BMU Effects 215
    I. The LOBO Property 216
    J. Microdamage 216
    K. Steady-State Turnover and Activation Function 216
    L. Accelerated Gains and Losses of Bone in Adults 217
    M. Fluoride and OPn 217
        1. Background to Fluoride Treatment 218
        2. Theoretical Possibilities 219
        3. Histomorphometric and Other Clues to the Above Matters 221
        4. Prevalent Notions about Fluoride Mechanisms 222
        5. Existing Evidence 222
        6. Provisional Synthesis 223
III. Comments on Osteomalacias 224
    A. Increased Amounts of Osteoid 224
    B. Osteoid Thickness 225
    C. Osteoid Thickness and Surface Extent Should be Independent 230
    D. The On-Off Phenomenon 230
    E. Ratio of Fractional Resorption Surface to Fractional Formation Surface 230
    F. Other Mineralization "Defects" 231
IV. Comments on Modeling 232
    A. Micromodeling and Macromodeling are Different Functions 232
    B. Macromodeling and Remodeling are Different Functions 232
    C. Bone Malunions Cannot Correct Spontaneously in Adults 232
    D. Periosteal and Cortical-Endosteal Modeling can be Independent 232
    E. Fibrous Tissue Length and Thickness are Independent 233
V. Comments on Growth 233
    A. L_1-Mode and L_2-Mode Growth can Vary Independently 233
    B. The Common Plan of the IO Provides a Basis for Creating Different Kinds of Tissues with Different Properties and Functions 233
    C. Growth and Modeling are Separate Functions 234

		D.	Modeling Requires Growth but the Converse is not True 234
		E.	The Framework of Skeletal Growth and Maturity 234
VI.	Illustrative Questions.. 236

Note 1
The Referent 241

Note 2
Recent Examples of Anholistic Skeletal Thought.. 251

Chapter 7
Mechanical Determinants of Lamellar Bone Macromodeling
Abstract 267
I. Introduction... 267
  A. Orthopedic Biomechanics 267
    1. Its Subdivisions 267
    2. Properties of Materials...................................... 267
    3. Kinematics... 267
    4. Pathophysiology of Trauma and Healing......................... 268
    5. Structural Design of the Adult Skeleton 268
    6. Design Principles of the Growing Skeleton 268
  B. General Remarks 268
  C. The Design and Construction Strategy in Nature 268
  D. Laws: Wolff's and Others 269
  E. Review of Basic IO Functions... 270
II. The IO Mechanisms of Lamellar Bone Modeling 271
  A. Bone Macromodeling.. 271
    1. The Resorption Drift.. 272
    2. The Formation Drift 273
    3. Drift Systems 274
    4. Further Properties of Lamellar Bone Macromodeling 274
III. Mechanical Controls of Lamellar Bone Modeling........................... 275
  A. The Growth-Modeling Axiom 276
  B. The Strain/Stress Axiom... 276
  C. The Minimum Effective Strain (MES)................................ 276
  D. The Dynamic Strain Axiom 279
  E. The Strain-Averaging Mechanism 279
  F. The Flexure-Drift Axiom... 281
  G. Drift Barriers 283
  H. The Magnitude Axiom 284
  I. Exclusions.. 284
  J. The Modeling Relations 285
IV. Organ-Level Examples of Bone Macromodeling 286
  A. Vertebral and Metaphyseal Inwaisting................................. 286
    1. Flexural Neutralization...................................... 286
  B. Natural Long Bone Curves.. 289
  C. Abnormal Long Bone Curves 290
    1. The Compression-Formation and Tension-Resorption Thesis... 293
  D. Modeling of Young or Recently Fractured Bone 293
  E. Neuromotor Effects... 293

V. Discussion 295
  A. The Relative Modeling Rate..295
  B. The Bone Modeling Transducer: An Hypothesis 297
  C. Bone Microdamage...300
  D. Biomechanical Competence of Fracture Callus 302
  E. Conclusion 303
VI. Illustrative Questions...305

Note 1
The MES Concept: Some Analytical Meanings 307

Note 2
A Chronology of Post-1960 Understanding of the Mechanical Determinants of
Skeletal Modeling 321

Note 3
Structural Adaptations to Mechanical Usage..325

Index 353

Volume II

Chapter 8
Mechanical Determinants of Hyaline Cartilage Modeling
Abstract 1
I. Introduction... 1
  A. The Antecedents... 1
    1. Nervous System Problems 2
    2. Joint and Limb Alignment Problems............................. 3
  B. General Remarks... 3
II. The IO Chondral Growth and Macromodeling Mechanisms 6
  A. Growth 6
  B. Macromodeling... 7
III. The Chondral Modeling Axioms 9
  A. The Postnatal Domain 9
  B. Strain Governs Chondral Modeling..................................... 9
  C. The Minimum Effective Signal (MES) 9
  D. The Dynamic Averaging Property 10
  E. Chondral Modeling Occurs During Growth...........................10
  F. The Chondral Growth/Force Response Curve (CGFR Curve) 10
  G. Chondral Modeling State Equations 12
IV. Clinical Examples of L_o-Level Chondral Modeling 13
  A. Epiphyseal Plate...13
    1. Knee Alignment 13
      a. The Negative Feedback Mode............................14
      b. The Positive Feedback Mode.............................16
      c. Epiphyseal Plate Diameter................................17
  B. Articular Cartilage..20
  C. Neuromotor Relationships to Chondral Modeling..................25
V. Comment 26
  A. The Relative Growth and Modeling Rates............................26
  B. Chondral Dominance of Skeletal Architecture......................26
  C. Chondral Creep..27

VI. Illustrative Questions..32

Note 1
The Composition and Certain Other Properties of Bone, Cartilage,
and Fibrous Tissue 39

Chapter 9
Mechanical Determinants of Fibrous Tissue Modeling
Abstract 43
I. Introduction..43
    A. Preamble..43
    B. General Remarks...44
II. The IO and Growth, Modeling, and Turnover 45
    A. Growth 45
    B. Modeling 49
        1. Micromodeling..49
        2. Macromodeling 49
    C. Turnover 49
        1. Fibrous Tissue Remodeling BMU..............................49
        2. Molecular-Level Turnover 51
III. Fibrous Tissue Modeling Phenomena 51
    A. General Facts..51
        1. Control of Tendon Length 51
        2. Control of Bulk Strength.......................................51
    B. The Dynamic Peak Strain Axiom.....................................52
    C. The Time-Averaging Axiom..52
    D. Tension-Creep Compensation..52
    E. The Minimum Effective Strain (MES)...............................52
    F. The Stretch Hypertrophy Axiom 53
    G. State Equations for Fibrous Tissue Modeling.......................55
    H. Rules of Thumb 55
IV. Discussion 56
    A. Fatigue and Microdamage 56
    B. Biomechanical Competence of Fibrous Tissues......................57
    C. Conclusion 59
V. Illustrative Questions...60

Note 1
The Attuned Perceptions 63

Chapter 10
The IO and Homeostasis
Abstract 73
I. Introduction..73
    A. Preamble..73
    B. Pre-IO Concepts 74
II. The Skeletal Role in Homeostasis 79
    A. The Histological System..79
        1. Osteoclasts..79
        2. Osteoblasts...82
        3. The Remodeling BMU...83

	B.	The Bone Surface-Lining Cell System 84
	C.	The Percolation System 85
		1. Interpretation 91
III.	Synthesis 95
IV.	Illustrative Questions 96

Note 1
Some Properties of Physiological Balances 99

Chapter 11
The Regional Acceleratory Phenomenon (RAP)

Abstract 109
I.   Introduction 109
     A.   Preamble 109
     B.   General Remarks 110
II. The RAP 110
     A.   Characteristics of the RAP 110
          1.   The Causes 110
          2.   The Nature of the RAP 110
          3.   Its Anatomical Distribution 112
          4.   Its Duration 115
     B.   Clinical Examples of RAP Effects 116
          1.   Potentiated Bone Healing 116
          2.   The Pathological RAP 117
          3.   Arthrofibrosis 118
          4.   Neuropathic Soft Tissue Problems 118
          5.   The Charcot Joint 118
          6.   Rheumatoid Phenomena 121
          7.   Obtunded RAP of Unknown Cause 121
     C.   The RAP in Research 123
          1.   The RAP in Histomorphometry 123
          2.   Longitudinal Bone Growth 123
          3.   Bone Mechanical Loading Experiments 124
III. Comment 125
     A.   Some System Properties of the RAP 125
     B.   Experimental Design and the RAP 125
     C.   Experimental Studies of the RAP 125
     D.   Causative Mechanisms of the RAP 127
     E.   A Final Common Path? 128
     F.   The RAP: It is Ubiquitous 128
IV. Illustrative Questions 128

Note 1
On Drug and Endocrine Effects on the Skeletal IO 131

Chapter 12
The Bone Repair Process

Abstract 135
I.   Introduction 135
     A.   A True Ministry 135

	B.	Background 135
II.	Biologic Processes 137	
	A.	The Processes 137
	B.	The Biology.. 138
		1. Temporal Overlap and Smearing............................... 144
		2. Biomechanical Competence 145
		3. Normal Variations of Traumatically Induced Bone Repair 147
		4. The General Biologic Functions Involved....................... 147
III.	Clinical Problems Related to the Bone Repair Mechanism..................... 147	
	A.	Clinical Problems in the Repair of Bone Trauma........................ 147
		1. Rate Problems 148
		2. Problems of Quantity and Kind 150
		a. Insufficient Callus.. 150
		b. Impaired Mineralization 152
		c. A Remodeling Disorder 153
	B.	The Bone Repair Mechanism in Disease 153
IV.	Clinical-Biological Synthesis... 153	
	A.	The Skeletal IO 153
	B.	The Abnormal RAP 154
	C.	Abnormal "P" Response.. 155
	D.	Abnormal "D"... 155
	E.	Abnormal "A".. 155
	F.	Defective "M"... 156
	G.	Abnormal "O" and "C" 156
	H.	Temporal Incoherence 156
V.	Therapeutic Prospects.. 157	
	A.	Electrical Stimulation (ES) 157
	B.	Bone Grafting 158
	C.	By-Passing the Callus Phase... 158
	D.	Direct Biological Stimulation 159
	E.	Future Research 159
	F.	Summary 160
VI.	Illustrative Questions... 160	

Note 1
On the Acceptance of IO-Oriented Skeletal Research 163

Chapter 13
Choosing Appropriate Model Systems for Skeletal Research
Abstract 165
I.	Introduction... 165	
	A.	Background to the Terminology 165
II.	The Skeletal IO (and Growth, Macromodeling, and Remodeling) 166	
	A.	General Remarks 166
	B.	Growth 168
	C.	Lamellar Bone Macromodeling... 168
		1. Control... 168
		2. Age Incidence 169
		3. Distribution 169
		4. Operational Features 169

		5.	More on Distribution and Control............................171
		6.	Distributive Properties 171
		7.	Functions of Macromodeling...................................172
	D.	Lamellar Bone Remodeling 173
		1.	Age Incidence, Distribution, and Control 173
		2.	Operational Composition.......................................173
		3.	Distributive Properties 173
		4.	The ΔB·BMU..174
		5.	Remodeling Variants in the Spongiosa of Growing Animals...174
		6.	The Functions of Remodeling 175
III.	Concerning Intraskeletal Mechanisms of Bone Disease 178
	A.	Growth Modeling-Dependent Lamellar Bone Disease 178
	B.	Remodeling-Dependent Bone Disease 183
	C.	In Vitro Systems: Their IO-Oriented Properties 184
	D.	Envelope-Specific Phenomena..186
	E.	Skeletal Turnover Mechanisms 187
		1.	Turnover in the Growing Skeleton 189
		2.	Turnover in the Adult Skeleton 190
IV.	Propositions and Postulates 191
V.	Illustrative Questions..193

Note 1
On Tissue and Organ Culture Techniques 195

Chapter 14
An Optimizing Strategy for Skeletal Research
Abstract 197
I. Introduction...197
II. The Algorithm 198
    A. Some Goals of Medical Research 198
    B. Two Ways to Find Cures...199
        1. The Random Path...199
        2. The Downward Search...199
    C. A Basis for an Optimizing Algorithm 199
        1. The Function/Property Ratio 200
        2. The Function/Disease Relation 200
        3. The Odds of Random Choices 200
        4. Functions Evolve Upwards..200
        5. Identifying the Causative Malfunctions 201
        6. The Upwards Search..202
        7. Recapitulation..202
        8. The Control of the Causative Malfunction 203
III. Comment 205

Note 1
Theory: Some of Its Value and Uses 211

Note 2
Some Effects of an Investment in a Scientific Posture.................................217

Chapter 15
Some Design Problems in Hip Replacement Prostheses: An Analysis
Abstract 221
I. Introduction 221
II. The Problem 222
    A. Brief History 222
    B. Subsequent Course 224
III. The Natural Solution 225
    A. The Problem is Not 225
    B. What the Normal Hip Achieves 226
    C. How the Normal Hip Achieves it 227
IV. Man's Solution 232
    A. Base Plate Fit 232
    B. Effects of Local Unit Load Multiplication 232
    C. Differential Longitudinal Stiffness 235
    D. The Cone Piston Effect 235
    E. The Shear-Shunt and Skid Mechanisms 237
    F. The Outflaring Cortical Flexure 241
    G. Load Focusing can Multiply Trivial Total Loads into Catastrophic Local Ones 241
V. Possible Solutions 242
    A. Transmission of the Vertical Load 242
    B. The Shear Skid and Shunt 242
    C. The Inwaisting Factor 243
    D. Toggling 244
    E. Mechanical Fit and Tolerances 244
    F. The Cement 244
    G. Biologic Fixation 244
    H. The Biology 244
    I. Generalization to Other Situations 245
VI. Illustrative Questions 245

Note 1
Communication between Laboratory and Clinic 247

Chapter 16
A Summation
I. A Perspective 251
II. A Metamorphosis 251
III. Some Conclusions 254
    A. Some Lessons of the Recent Past 254
    B. The Future 256

Note 1
Pre-IO Concepts of Skeletal Physiology 257

Note 2
Science and Recipes: A Postface 265

Appendix I
I. Introduction 267
II. Definitions of Symbols in the State Equations 268
III. Glossary 268

Acknowledgments 285

References
I. Symposia and Multiauthor Texts...287
II. Specific Citations 288
III. Group Citations 314

Index 319

Chapter 8

MECHANICAL DETERMINANTS OF HYALINE CARTILAGE MODELING

"Science, like life, feeds on its own decay. New facts burst old rules; then newly divined conceptions bind old and new together into a reconciling law."

(William James)

ABSTRACT

Growing hyaline cartilage responds predictably to time-averaged dynamic mechanical unit loads. Under increasing unit tension loading its growth increases, and within a low range of unit compression loading it responds likewise, but when the unit compression loads exceed some limit, further increases of compression retard growth. Those mechanical responses, at times modified by mechanical creep of already existing cartilage, underlie the mechanical control during growth of the gross architecture of all load-bearing chondral structures in the body, including articular cartilage, epiphyseal and apophyseal plates, and the growths of the bony attachments of tendons, ligaments, and fascia.

I. INTRODUCTION

This field now has a reasonably detailed theory of the mechanical determinants of chondral macromodeling.[267] It is still qualitative rather than quantitative and it is largely phenomenological, meaning it describes what actually happens, but it has also worked well in the predictive as well as the retrospective senses, and it identifies a number of mechanisms, questions, and phenomena that would not otherwise be perceived, and therefore would not likely be studied experimentally either. Equally, the discovery of the coupling between R and F in the bone remodeling BMU identified a previously unknown mechanism that has since become an active and intriguing subfield of research.[822] That macromodeling theory is developed in this chapter, beginning with some of its background.

A. The Antecedents
Anterior poliomyelitis used to be common in American children; its paralytic residuals could cause serious skeletal deformities and maldevelopment during subsequent growth.[638-640] By 1960, orthopedic surgeons around the world had learned how to prevent many of those deformities by combinations of bracing, joint arthrodeses (e.g., permanently fusing or stiffening joints), physical therapy, and transfers of functioning tendons and muscles to new locations to compensate for the adverse effects on growth — as well as on function — of paralyzed muscles.[804] That expertise developed by trial and error, empirically, for no formal rules existed then for the effects of mechanical usage on the macromodeling of growing bone or cartilage, so suitable preventions and corrections could not be devised from theory. Accordingly, they had to be found by experience. For example, given nonfunctioning peroneal muscles, the insertion of the anterior tibial would be transferred to the base of the fourth metatarsal to replace the missing pronator function of the peroneals while continuing to provide ankle dorsiflexion. A posterior tibial could be transferred to the front of the foot to replace paralyzed ankle dorsiflexors, and at a suitable age the subtalar joint would be arthrodesed to prevent a subsequent valgus deformity of the foot and to free the child of a brace.

During the 1950s the polio vaccines virtually eradicated polio in this country. As one result, in many children's orthopedic clinics the legions of polio victims shrank and were replaced by a smaller group of children with various forms of cerebral palsy, particularly spasticity. Consequently, orthopedic surgeons who had become experts at managing postpolio problems now had to confront those of the spastic child. That change exposed baffling problems for, and as an example, when the surgical procedures that managed a postpolio equinus satisfactorily (e.g., drop foot) were done for a spastic equinus they could fail, while the braces that managed such an equinus quite well after polio could cause severe additional deformities in the spastic child without correcting the original problem. By about 1970 such problems afflicted the surgical and orthotic management of spastic deformities with considerable controversy.

The author supplied part of the rationale that improved that situation in the form of two separate insights concerning the mechanisms that underlay both the genesis of the spastic deformities, and the failures of proven postpolio procedures to work well in spastic children. One of those insights related to the nervous system, the other to the principles of action that govern the mechanical control of chondral modeling. The essences of the former insights were published and are abstracted next, because they bear on the matters discussed in this chapter and the previous one.

1. Nervous System Problems

It was realized that much of the surgical success with postpolio problems stemmed from the fact that the undamaged central nervous system (CNS) still retained the ability to learn new patterns of muscle contraction, both individually and as groups, during gait and hand function. For that reason, a cooperative child could learn to make a normal muscle contract during a different and unnatural phase of gait, and thereby substitute for other, paralyzed muscles. That retrainability in the postpolio child applied to all muscles of both upper and lower extremities because the disease damaged the motor communication of the CNS with the peripheral muscles but it did not damage the internal data processing and learning functions of the CNS.[267] Also, when braces were used to contain some paralytic deformity during growth the involved joints could be moved without resistance from uncontrollable muscle contraction, and the paralysis greatly reduced the total and peak loads on such joints, by factors ranging from two- to over fivefold.

In the spastic child the real damage affected the information processing inside the CNS itself and left largely intact the communication between the CNS and the peripheral muscles. Because of that central damage, the spastic CNS could not learn to make a given spastic muscle or muscle group contract in a new pattern.[267] The author summed that up as a rule of thumb in an article around 1966 that is still valid, thus: *one cannot retrain a spastic muscle*.[340] It followed that such muscles usually actively and uncontrollably fought the restraints of braces, which in turn increased the typical loads on the joints, but now in patterns that were made abnormal by the braces themselves. It also followed that a really spastic muscle continues to contract in its original and predetermined pattern with respect to gait or hand function no matter where a surgeon moves its insertion or origin, and no matter what efforts a physical therapist focuses on it. That knowledge suddenly made it possible to predict the effects of particular muscle or tendon transfers on gait and hand function (and also of any bracing) and also the mechanical loads such muscles would apply to various joints.[267] With the aid of that insight, beginning about 1962 the author found himself doing many of the same operations that had been done in postpolio children, but now for quite different purposes. While not perfect, that new predictability has proven usefully accurate and consistent since and it changed the surgical treatment of the spastic child by making it far more predictable than in 1960.[340] To repeat, those matters were summarized and presented in monograph form in 1972.

2. Joint and Limb Alignment Problems

Here it was realized that most deformities of the older spastic child represent the effects of chondral modeling on growing epiphyseal plates and articular cartilage, effects dictated by the abnormal muscle contractural patterns that were in turn dictated by the "unteachable" spastic CNS, and often also by the effects of braces that might conceal spastic muscle dynamics but failed utterly to correct them.

The recognition between 1958 and 1968 that the limb deformities of spastic as well as postpolio youngsters reflected perfectly normal adaptations of normal growing hyaline cartilage to abnormal mechanical force environments focused attention on the then unknown game rules of chondral modeling. By 1967 most of the insights found in this chapter were worked out. They were published in 1972[236] and a correction for certain tension effects was published in 1979.[244]

Those spastic and postpolio children provided the chief laboratory and the natural experiments that, to paraphrase William James' words, overturned old ideas with newly perceived facts, which in their turn revealed at least some of the principles that govern how growing hyaline cartilage responds to mechanical usage. Their description will begin with some general functions of cartilage, observing that Note 1 provides a synopsis of chondral composition, histology, and other properties pertinent to the domains of this chapter and book.

B. General Remarks

In skeletal growth and development and both before and after birth, cartilage plays a dominant role for it determines the original number, location, and shape of the fetal models of most bones, as well as the number, locations, orientations, alignment, size, and shape of the articulations.[108,650] Postnatal cartilage functions establish bone and limb length, body height and proportions, facial and joint size and shape, the alignment of the segments of the spine and the appendages, the locations, size, and growth of tendon, ligament, and fascial attachments to bone, and the properties of growing and mature joints.[H,190,521,650] Cartilage malfunctions of varied kinds cause most children's skeletal deformities, and many other skeletal diseases, as well as all joint disease. They have significant effects on bone macromodeling by the mechanism described in Section IV.A of Note 3 in Chapter 7. It follows that chondral physiology is complex and that its IO should have important effects on many structures, functions, and diseases that arise from or otherwise involve cartilage.

This chapter concerns some effects of mechanical forces on postnatal chondral growth, modeling, and macroarchitecture in L_o-space time, and Figure 1 illustrates several examples of those effects. That chondral modeling capability arises in the special mechanisms the IO provides to the skeleton. As with bone, the IO can cause IO-extrinsic disease, IO-intrinsic disease, and a diverse but less frequently occurring combined group. This chapter emphasizes IO-extrinsic problems.

Like the previous chapter, this one will describe first the basic intraskeletal mechanisms that underlie chondral growth and modeling, then some of the rules that seem to govern their behavior, and finally some of the functions they seem to provide to the intact subject. Because structures evolve as one ascends the skeletal organization ladder, and because so far no other authority has related conventional anatomical terminology to the various levels in that ladder in a systematic way, Table 1 lists some chondral structures according to their level in that ladder; Note 1 briefly describes the composition of cartilage. The remainder of this chapter will deal mainly with L_3-domain mechanisms, their functions and regulation, and their effects on L_o-domain anatomy. First, however, a digression merits a few comments.

In effect, this chapter views cartilage as a basic brick used by a mason to construct higher-order structures. The mason here would be those L_2- and higher-domain factors

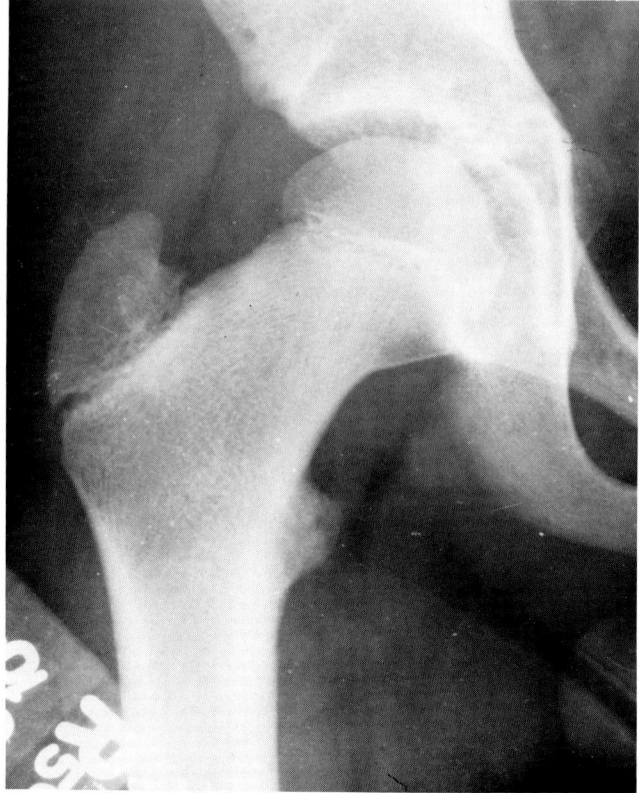

FIGURE 1. An AP X-ray of a normal left hip of a 10-year-old girl. It shows the capital epiphyseal plate, the apophyseal plate of the greater trochanter, the lesser tuberosity of the femur, the acetabulum or socket, and the trajectories of the spongiosa in the femoral neck. Four species of hyaline cartilage are present in this structure, although being radiolucent the cartilage itself casts no X-ray shadow. Articular cartilage occupies the space between the head and the neck, and it spends its life loaded only in compression. An apophyseal plate separates the greater trochanter from the underlying femur and it carries only tension loads throughout its life from the hip abductor muscles. Another layer of hyaline cartilage separates the bone itself from the tendon attachments to the greater and lesser trochanters, and at this age the periosteal surface of the upper cortex of the neck is also covered by cartilage, while both surfaces of the lower cortex and the cortical-endosteal surface of the upper cortex are not. Note that the plane of the epiphyseal plate of the caput or head is perpendicular to the major trabecular orientation beneath it. The structure shown here represents a combination of coordinated and patterned chondral and bone macromodeling effects.

that govern macromodeling. The IO magnifying glass reveals some of the sequences of the events, enablement, and regulation that make lower-level IO entities organize and assemble, brick-like, to form higher-level ones. All such entities begin with the basic stuff, L_1-domain hyaline cartilage, that, as Note 1 records, contains characteristic kinds and proportions of Type II collagen, proteoglycans, noncollagenous proteins, chondroblasts, chondrocytes, and chondral precursor cells. As a relation then:

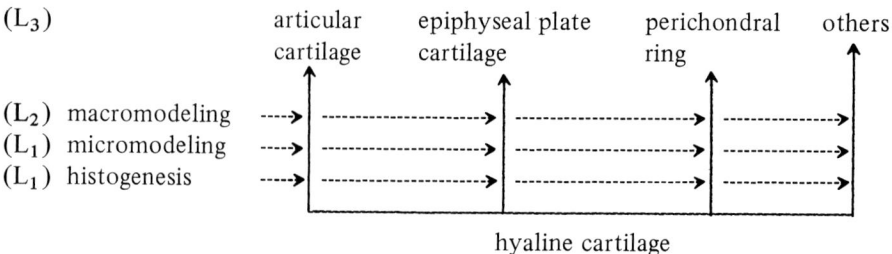

Table 1
CHONDRAL STRUCTURES BY LEVEL OF ORGANIZATION

Level of organization	Structure
(L_{sk}) Articulated skeleton	Whole skeleton, including bones, joints, tendons, muscles, innervation
(L_o) Organ	Intact joint, including both articular cartilages, subchondral bone, synovia, ligaments, menisci
	Intact intervertebral disc, including end plates, disc, annulus, and subchondral bone
	Larynx, trachea
(L_3) Upper IO	Entire articular cartilage, entire epiphyseal plate, entire perichondral ring, complete ligament, articular cartilage-subchondral bone complex, epiphyseal plate-metaphyseal spongiosa complex, complete epiphysis, complete synovia, a meniscus, individual laryngeal cartilage plates
(L_2) Middle IO	Chondral material or tissue of articular cartilage, epiphyseal plate, menisci, perichondral ring; material of subchondral bone, synovia, metaphyseal spongiosa
(L_1) Lower IO	Simple tissues: hyaline cartilage, elastic cartilage, fibrocartilage, fibrous tissue, lamellar bone, woven bone, synovia
(L_c) Cell level	Chondroblasts, chondrocytes, precursor cells, synovial cells, endothelial cells

Table 2
CLINICAL EXAMPLES OF CHONDRAL GROWTH ERRORS

Gigantism	Malnutrition-induced growth retardation
Postparalytic limb shortening	Postparetic limb shortening
Limb overgrowth, post-fracture	Limb overgrowth, AV fistula
Limb overgrowth, neurofibromatosis	Ollier's disease[a]
Limb overgrowth due to other causes of a RAP	Stickler's disease[a]
Morquio's disease[a]	Turner's syndrome
Hemiatrophy	Corticosteroid-induced growth retardation
Pituitary dwarf	Marfan's syndrome
Achondroplasia[a]	Jansen's disease[a]
Rickets, nutritional	Pseudoachondroplasia[a]
	Rickets, congenital-familial[a]

[a] These are IO-intrinsic disease; the other examples are IO-extrinsic

Here, "others" signifies additional chondral structures such as the attachments to bone of tendon, ligament, and fascia, plus apophyseal plates and the tracheal, laryngeal, and costal cartilages. For each of those structures the hyaline cartilage provides a basic and IO-intrinsically determined kind of clay and straw for making bricks. For each structure, extrinsic factors cause micromodeling during its L_1-mode growth to create a particular fiber grain and spatial cellular organization that a competent histologist usually can recognize at a glance under a high power as characteristic of the tissue.[149] For each structure, extrinsic L_2- and higher-domain factors, perhaps different in some respects form those that control L_1-level micromodeling, also direct macromodeling to create the different forms recognized at a glance by the competent gross anatomist.

Little is known about the details of those micro- and macromodeling activities, partly due to lack of a formal synthesis that would suggest specific and potentially productive questions to investigate, meaning to put to suitable experimental model systems.

Table 2 lists some chondral growth disorders, including the predominant IO-extrinsic forms and the less common IO-intrinsic ones.

II. THE IO CHONDRAL GROWTH AND MACROMODELING MECHANISMS

A. Growth

This occurs at least in two modes. Both involve the proliferation of precursor cells to increase the total number of chondral cells, which then add new chondral matrix to increase the size of the tissue or structure. Endocrine agents exert a systemic control on such growth in the general sense that must act concurrently with local influences.[72,86,87,139,685,695] Much has been learned about the systemic control of chondral growth and its dependence on somatotrophic hormone and one or more somatomedins, and also on the ability of other endocrines (adrenal corticosteroids, estrogens) to depress it. For the needs of this chapter, that systemic control of general chondral growth can be assumed, for the discussion focuses on how mechanical factors can modify that growth to affect the size and configuration of a growing structure. As for the aforesaid modes of growth, in the L_1-mode growth the "P:A:M" processes merely enlarge the size of the clone, and the L_2-mode growth adds to that an increase in or multiplication of the number of clones.

Cartilage grows differently from bone, for most chondral precursor cells lie inside of a cartilage matrix rather than on its surface. That makes the molecular sieve effect of the matrix potentially important in how extrinsic endocrine, chemical, and physical agents can reach and affect the cells within the tissue. That effect, plus other local anatomical factors, can gate the access of circulating agents to the cells inside growing cartilage, and thereby modify in various ways their responses to common mechanical factors. Chapter 4 provided three L_2-level examples of that phenomenon that are reviewed briefly next. Figure 5 in Chapter 4 explained the anatomical terminology used in this chapter.

In an epiphyseal plate such as that in Figures 3 and 4, the circulating agents that affect its growth-dependent precursor cell proliferation initially pass through an epiphyseal capillary that is extrinsic to the chondral tissue, then across "L" or the thin local space between the epiphyseal capillary and the cartilage, to enter the latter or "M", and finally reach the proliferating or "P" cells inside that matrix. The molecular sieve property of the matrix somewhat impedes the diffusion of the large molecules of the polypeptide and steroid hormones that affect its growth. Therefore, the sequence is "S" → capillary → "L" → "M" → "P", and the programmed relation could look like this:

$$F(G_c) = \begin{array}{c} S \rightarrow C \\ \downarrow \\ L \\ \downarrow \\ P:D:O:A:M \end{array} \qquad (1)$$

Epiphyseal plate growth determines bone length[765] and limb and joint alignment[244] as described later.

In the perichondral ring described in the legends of Figures 3 and 7, the germinal cells lie mostly outside a well-established chondral matrix so no or little molecular sieve effect applies to the case, while the capillary probably forms an intrinsic part of the L_2-level entity. The sequence then is "S" → capillary → "L" → "P", and its state equation might look like this:

$$F(G_{cp}) = \begin{array}{c} S \\ \downarrow \\ \boxed{\xrightarrow{} L} \\ C:P:D:O:A:M \\ \uparrow \end{array} \Rightarrow \quad (2)$$

Perichondral ring growth determines the diameters and cross-section areas of epiphyseal plates and joint surfaces,[765] and therefore it affects their unit loadings under superimposed loads.[267]

In articular cartilage, as shown in Figures 3 and 7, the agents that regulate growth pass through the synovial capillary, then the synovia itself (L_s), then the synovial fluid (L_f), and finally the cartilage matrix to reach the precursor cells within it. The sequence then is "S" → "C" → "L_s" → "L_f" → "M" → "P", so the state equation could look like this:

$$f(G_{ac}) = \begin{array}{c} S \rightarrow C \rightarrow L_s \rightarrow L_f \\ \downarrow \\ \boxed{} \\ P:D:O:A:M \\ \uparrow \end{array} \Rightarrow \quad (3)$$

Articular cartilage growth determines the shapes and thus the types of free motion of joints.

Such gating features should account for some of the characteristic differences in the growth rates of the epiphyseal plate, articular cartilage, and perichondral ring in the same epiphysis supplied by the same blood at the same time.[263, 267] Growth of an epiphyseal plate typically exceeds that of perichondral cartilage, which exceeds that of articular cartilage, excepting a few anatomical locations where special circumstances modify those relationships. Those differences underlie many of the characteristic differences in gross dimension and proportions of the structures in question.

Some malfunctions of the chondral growth process known to clinicians were listed in Table 2, observing that in many of them chondral and bone modeling errors also exist.

B. Macromodeling

Local mechanical loads plus other local anatomical, physical, and biochemical factors that are extrinsic to a chondral tissue can restrain its growth in some places and directions and potentiate it in others to mold visible anatomical features. That molding represents chondral macromodeling.[244] Both micro- and macromodeling occur in all three kinds of cartilage, but the former only establishes the intrinsic architecture of the tissue during its production regardless of where and when it is produced, and over domains of 0.01 to 0.1 mm. As a relation:

$$L_1\text{-mode growth} \xrightarrow{\uparrow} \text{micromodeling} \quad (4)$$
$$L_1\text{-domain extrinsic factors}$$

The macromodeling processes form the naked eye level or extrinsic macroarchitecture of growing chondral structures, and they will be referred to henceforth as simply "modeling". The programmed state equation of chondral modeling under the control of mechanical forces (S_m) could look like this:

$$F(M) = \boxed{\begin{array}{c} S_m \\ \downarrow L \\ \hline C:P:D:O:A:M \end{array}} \Rightarrow \qquad (5)$$

In words, here "S_m" signifies an "S"-mode mechanical load carried by a growing layer of cartilage or "M", say the body weight on the knee articular cartilage, so the latter strains as a result, which somehow creates within it a special signal that modifies in appropriate ways the proliferation and differentiation of the precursor cells or "P" within the chondral matrix, so it also modifies the synthesis of new matrix by the daughter cells, and thus local chondral growth. Accordingly, mechanically controlled chondral modeling depends on the properties of two sequential gates: "M" and "P". As a relation, one could also write more concisely:

$$L_2\text{-mode growth} \longrightarrow \text{macromodeling} \qquad (6)$$
$$\uparrow$$
$$L_2\text{-domain extrinsic factors}$$

An "S"-mode or domain mechanical load would represent the total load applied to, say, the proximal tibial articular surface by the femoral articular surface above it. It could include body weight, the pulls of the various muscles that cross the knee joint, and any forces due to accelerations of the superimposed body mass by muscles, including those that move the hip and ankle during running and jumping. On the other hand, an "L"-mode or domain mechanical load would represent the load on some small part or L_2-domain region of that joint or other kind of chondral surface. The load would be applied by the corresponding small region of the opposite joint surface in contact with it at the moment. As another example, the groove in the posteroinferior lateral malleolus is caused by the pressure of the peroneal tendons that lie in that groove and that use it as a kind of pulley that changes the direction of the tension loads transferred from the peroneal muscles to the midfoot. Therefore, such a small regional load is only a part of the "S"-mode load on the whole joint surface, and the loads on many small L_2-domain regions from all over the joint surface sum up to form that "S"-mode load. The distinction may seem subtle but it is important and the legend to Figure 4 will present one reason why it is important. In this respect, chondral and bone modeling differ greatly, for the "S"-mode loads that control bone modeling do not directly affect the bone modeling cells at all. Rather, they cause special things to happen on the bone surfaces that adjacent cells lying on, but not in, that surface can perceive and react to, provided they lie close enough to the surface.

In different words, bone is shaped as a sculptor shapes plaster, by removing some at one surface and adding some at another. Cartilage models like a puff of smoke, by patterned and spatially oriented internal expansions.[236,238]

Mechanical influences on L_3-mode and higher-level chondral growth behavior follow a few axioms aptly expressed in words, plus another conveyed better by a graph than by language.

III. THE CHONDRAL MODELING AXIOMS

Six currently recognized axioms originally proposed and described in 1972 to 1979 have clinically visible effects on skeletal architecture.[236,821] Their relationship to earlier analyses of these matters is somewhat obscure. A number of authors refer without specific citation to a "Heuter-Volkmann" rule that holds that chondral growth decreases with increasing compression. This author and several colleagues have been unable to locate the source of that rule (which in its proper context is correct). Creuss[822] also refers to a rule by Delpech, but that too has not been located. However, the following rules were all conceived by the author after 1965, and if it should prove that some of them were only rediscoveries of earlier proposals by others, that will gladly be acknowledged.

A. The Postnatal Domain

These axioms operate primarily after birth, for chondral modeling in utero involves some special added features that this text will not discuss.[267] In effect, however, and as noted in the previous chapter, one set of principles constructs the preformed miniature models of bones and joints in the embryo and fetus, and after birth another set of principles inherent in the chondral tissues of newborn infants adapts their architecture to their mechanical usage while they grow. One could consider the fetal anatomy as the initial conditions of an equation, the aforesaid principles of action as the way in which the various terms in that equation respond to mechanical usage under the constraints imposed by the initial conditions, mechanical usage as the independent variable, and each mature skeletal structure as one solution to that equation and the relationships and interactions that it represents.

B. Strain Governs Chondral Modeling

As applies to bone as well, probably load-induced dynamic strain most directly controls cartilage modeling, and strain in shear seems like a convenient candidate.[267] However, the evidence that is available at present that bears on that matter is more tenuous than that for the role of fluid streaming effects in controlling bone macromodeling. Still, that implies that dynamic strain gradients and the associated internal hydrostatic pressure gradients in a chondral structure could affect the flow of the interstitial fluid in it and thereby contribute to the mechanism that transduces and amplifies the strain into other kinds of signals that cells can perceive directly and respond to. The signal generated by that dynamic strain might be electrical, chemical, mechanical, something else, or some combination but, in effect, where "S" signifies a mechanical load and "M" signifies solid cartilage, the basic sequence for the mechanical control of chondral macromodeling is S→M→P, as in Relation 5.

The somewhat different kinds of effects that can be exerted simultaneously on "P", the precursor cells, by the hormones and biochemical factors that this system can respond to could then modify how it responds to strain. That would allow the same kind of cartilage to respond differently to identical dynamic strains but in different locations and environments. The three major kinds of cartilage — hyaline, elastic, and fibrocartilage — also have different chemical compositions and kinds of cells which could also make them respond differently to identical stimuli. Those factors provide attractive explanations for at least some of the observed differences in chondral growth and modeling in different parts of the skeleton.

C. The Minimum Effective Signal (MES)

As in all feedback systems, and as shown in Relation 5 in the previous chapter, a minimum strain-generated signal strength should arise before this modeling system can detect and respond to it.[260] The system would not respond to signals below that limit but would to any above it up to the point of saturation, so trivial loads would generate trivial signals, meaning signals that have no clinically apparent effects on chrondral macromodeling. Nontrivial loads would represent those larger ones that do affect that modeling, and the MES range would represent the transition between the two extremes. Those statements describe a gating property of "M" that would affect the responses of "P" in Relation 5. The magnitude and physical nature of the chondral MES both remain obscure at present, but the extent of the trough on the horizontal axis of the CGFR curve in Figure 2 represents it. The MES should represent an L_2-level phenomenon and the transducer that converts strain into signals that chondral cells can perceive and respond to may well also be or be related to streaming potentials and their associated physical-chemical and chemical effects. Certainly, electrical potentials do occur in strained fresh hyaline cartilage.[v-y]

As applies to the bone case, probably some chondral disorders exist in which the basic malfunction is an MES for strain that has too high a set-point, and others that have too low a set-point.[267]

D. The Dynamic Averaging Property

Like bone in this respect, from clinical observations the author deduced by 1972 that cartilage adapts its architecture to some history of its mechanical usage, meaning it responds to an integration of many nontrivial dynamic loads averaged over some period of time.[236,265,267] A single brief load, even if large, has no known clinical effect on subsequent chondral architecture unless it injures the tissue. Unlike bone, however, constant loads may and in fact probably do have significant effects on chondral macromodeling, although some of those effects may relate to the creep phenomenon described in the Discussion[526] as much as to any effects on precursor cell activity. For example, a constant valgus force of correct magnitude applied by a brace or cast across a growing knee or foot can change its alignment predictably by modifying the rates of longitudinal growth on opposite sides of those structures, thereby correcting a bow leg or knock knee. The dynamic averaging property may also arise at the L_2-level of the IO. It is not a known property of unassociated skeletal cells of any kind. As a relation, the above three axioms might be written thus, where $F(M)$ signifies macromodeling:

$$F(M) = F(\text{growth}):F(\text{dynamic, time-averaged strains} \geqslant \text{MES}) \qquad (7)$$

E. Chondral Modeling Occurs During Growth

Chondral modeling changes and guides local growth to control local macroarchitecture, so modeling ceases when growth ceases, meaning at maturity in species that mature in that sense, as man does.[244,267] The osteophytes evoked by degenerative and other joint diseases in adults represent a local reactivation of the growth process in the marginal articular cartilage, usually in L_2-mode, and usually in response to or accompanying a local RAP, and under those circumstances some local chondral macromodeling could and does occur too, as shown in figures in Chapter 11.

F. The Chondral Growth/Force Response Curve (CGFR Curve)

Figure 2 shows the corrected CGFR curve as described in 1979.[244] Its vertical axis plots the absolute growth rate, say in millimeters per year, of any given hyaline chondral plane such as a growth plate or an articular cartilage. As a continuum, the horizontal axis plots the unit tension and compression loads carried by that plane, with maximum tension on the left, maximum compression on the right, and the zero load point in between them. (The Glossary, Appendix I, defines unit loads.) To reiterate, the matter of how those unit loads should be best expressed — in terms of magnitude, frequency, range, rate, and duration — remains unknown at present.

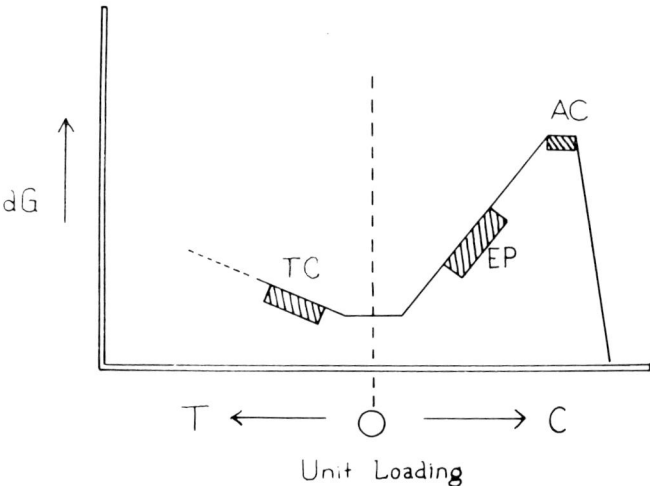

FIGURE 2. The chondral growth/force response (CGFRC). The horizontal axis of this graph plots the time-averaged dynamic unit tension (T) and compression (C) loads on a plane of growing hyaline cartilage as a continuum, increasing tension lying to the left of the zero load point and increasing compression to the right of it. The rate of growth (dG) plotted on the vertical axis could be expressed in units such as millimeters per year. The curve plots the growth responses of a given chondral plane to changes in its time-averaged unit loading. The hatched regions show where on this curve the typical unit loadings lie for tendon and ligament attachement to bone (TC), epiphyseal and apophyseal plates (EP), and articular cartilage (AC). Perichondral ring would presumably lie in the trough of the curve. Note that the tension limb of the curve has the smallest slope and the growth-descending compression limb has the greatest. To repeat, this curve applies to hyaline cartilage. It is not known at present if it applies also to fibrocartilage and elastic cartilage.

That curve describes how the growth of any particular chondral plane responds to changes in its own typical loading, so it compares the plane to itself but at different times.[244] For example, it does not compare a given articular to a given intervertebral disc cartilage, nor the proximal humeral epiphyseal plate to the overlying articular cartilage, nor the distal radial epiphyseal plate to the epiphyseal plate of the proximal phalanx of the thumb in the same limb. The obvious differences in longitudinal growth rates of such chondral planes should reflect in part the influences of IO factors already mentioned, and possibly also inherent differences in the responsiveness of their cartilage cells,[68,267] a matter that tissue culture studies could shed some light on.

The CGFR curve has the following three major features.

1. It has two inflections or major changes in slope. A lower *trough* or minimum lies at the point of zero loading, and the horizontal extent of that trough represents the MES values for cartilage macromodeling, noting that a different and possibly smaller MES may apply to the micromodeling process.[267] The second and upper inflection of the curve forms a *peak* on the compression loading side.
2. The slope of the growth-descending compression limb (the Heuter-Volkmann range to the right of the peak) exceeds the slope of the growth-ascending compression limb to the left of the peak, which in turn exceeds the slope of the

Table 3
PROVISIONAL CHONDRAL MODELING LAWS

1. Those laws apply to postnatal growth.
2. Modeling acts by modifying local growth speed.
3. Load-induced strain generates a signal in cartilage that affects its cells.
4. Increasing unit tension loads on a chondral plane increase its growth.
5. Increasing unit compression loads also increase growth up to the CGFR peak.
6. Beyond the CGFR peak, increasing unit compression loads retard chondral growth.
7. Those growth responses apply to time-averaged, dynamic, unit loads.
8. Articular cartilage unit loading normally lies near the peak of the CGFR curve.
    Epiphyseal plate unit loading normally lies on the growth-ascending limb.
    Tendon and ligament attachments to bone normally load on the tension limb.

tension limb to the left of the trough. Thus a given change in the unit loading on a growing domain of cartilage would cause the largest change in growth on the descending compression limb and the least on the tension limb. Those properties have some clinically obvious effects on skeletal architecture, and some of them will be described later.

3. Normally articular cartilage (AC) loads on or near the peak of the curve,[267] and never in tension, so both increases and decreases in its loading would retard its growth; increases would have greater effects because of the greater slope of the curve to the right of the peak.[267] Epiphyseal and compression-loaded apophyseal plates (EP) normally load on the growth-ascending compression limb,[244,267] so increasing loads would increase their growth, although only up to the peak of the curve, and conversely. In tension (TC), growth increases as tension increases,[244,267] which occurs in the cartilage layers that anchor tendons, ligaments, and fascia to bone, as well as in traction apophyses such as the tibial tubercle, the lesser and greater trochanters of the femur, and the olecranon.

Much remains unknown about the probably complex histological, biochemical, and physical determinants of the above general phenomena. They represent potentially fruitful subjects of research in the immediate future, and the legend of Figure 23 later will suggest some pertinent ideas. Table 3 summarizes the above axioms, which may be taken as the provisional chondral modeling laws.

G. Chondral Modeling State Equations

To recapitulate, in mechanically governed chondral modeling or $F(M_c)$, an "S"-mode load on the cartilage causes the cartilage to create a signal(s) that affects the resident precursor cells, and thence the rest of the system, so the state equations could look like this, where the dotted arrow now indicates the feedback loop that allows feedback modes of behavior to occur:

$$F(M_c) = (S):(M:P:D:O:A:L) \qquad (8)$$

The programmed relation was shown earlier as Relation 5. For this activity, the "L" term represents a passive and wide-open gate,[267] although it is an active gate for circulating endocrine and biochemical agents. That means that the parallel gating mode described in Chapters 3 to 5 should operate here.

Table 4
CLINICAL EXAMPLES OF CHONDRAL MODELING ERRORS
(INTRINSICALLY NORMAL CHONDRAL IO ASSUMED)

Epiphyseal Plate Errors

Coxa valga	Genu valgum	Cubitus varum
Coxa vara	Genu varum	Cubitus valgum
Blount's disease	Postparalytic ankle valgus	Postparetic limb malalignments
Humerus varus	Spastic back-knee	Ehrenfried's disease
Tibial torsion	Femoral torsion	Scoliosis

Articular Cartilage Errors

Subluxing patella	Metatarus Adductus	Tibial torsions
Congenital hallux varus	Club foot	Femoral torsions
Pes planus	Rockerbottom foot	Coxa valga
Ball-socket ankle	CDH	Paralytic hip dislocation
Paralytic back-knee	Congenital subluxing shoulder	Club hand
Enlarged radial head, postfracture		Spastic hindfoot varus
Upper extremity torsions (Erb's, spastic, postparalytic)		
Congenital radial head subluxation		

Note: Most IO-intrinsic chondral disorders exhibit both growth and modeling malfunctions. See Table 2 in Chapter 7 and Figures 11 and 12.

IV. CLINICAL EXAMPLES OF L_o-LEVEL CHONDRAL MODELING

The above material appears in action in children's clinical problems, and it can help the clinician prospectively and therapeutically, as well as retrospectively and diagnostically, for it allows predictions of how mechanical forces will affect chondral growth, so corrections can be devised for deformities that depend on further growth. It can also point to the loading abnormality that caused a given deformity of IO-extrinsic origin, thereby aiding the diagnosis of mechanical load imbalances in growing limbs;[267] and Section I referred to some clinical successes that stemmed directly from that use of the new knowledge. Table 4 lists some clinically known examples of chondral modeling errors, and separately for those of articular cartilage and those of epiphyseal plates.

Some epiphyseal plate phenomena will be discussed first and articular cartilage ones afterwards. As noted elsewhere in this book, epiphyseal plates have two major organ-level functions: their growth increases the length of a bone, and their macromodeling controls limb alignment across joints such as the elbow, knee, and digits. The latter function is described first and the human knee will provide the illustrative typical case.

A. Epiphyseal Plate
1. Knee Alignment

Figure 3 shows X-rays of a normal, growing human knee. Figure 4 diagrams the epiphyseal plate of the upper tibia as seen in the AP view, meaning from the front. The plate lies essentially perpendicular to the long axis of the bone, and between the metaphyseal spongiosa below and the epiphysis above. The epiphysis takes loads from the joint and transfers them to the plate and thence to the underlying metaphysis and finally to the compacta. In doing so, the epiphysis spreads or *defocuses* the more concentrated joint loads more evenly over the whole plate,[244,267] so while one side — e. g., medial or lateral — of the plate may carry higher unit loads than the other, the epiphy-

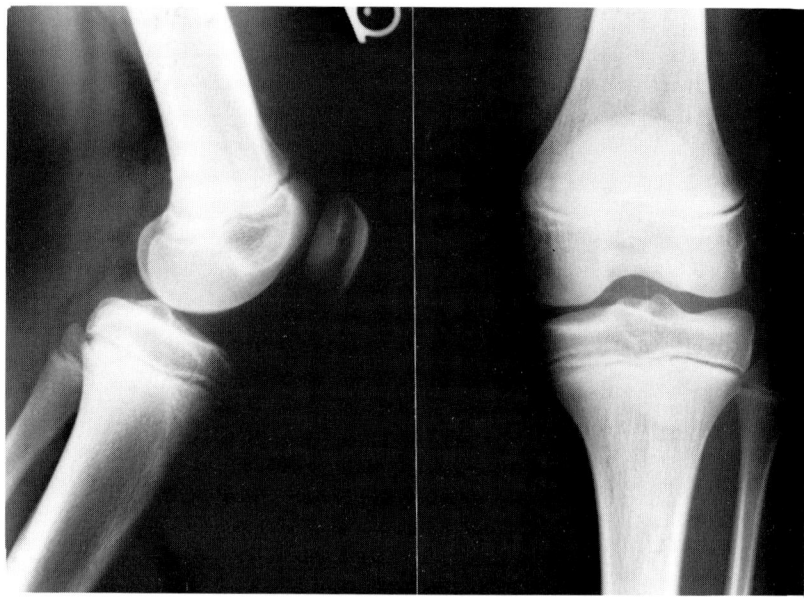

FIGURE 3. Lateral and AP X-rays of the normal knee of a colleague's 12-year-old daughter. Note the contours of the articular surface, the transverse epiphyseal plates, the metaphyseal flairs, and the slight posterior placement of the proximal tibial epiphysis relative to the central axis of the underlying shaft (compare to Figure 17 in Chapter 7). The space between the bones is occupied by the articular cartilage, and in part too by the menisci. A shell of cartilage named the perichondral ring also surrounds the sides of the bony epiphysis, but it is also radiolucent and therefore not seen here. The load-focusing and -defocusing effects of the ossification centers in the epiphyses can also be appreciated (X-ray courtesy Dr. C. Bartecchi, Southern Colorado Clinic).

seal "pressure pad" prevents large loading inequalities and load concentrations from localizing to small regions of the plate.[267] Figure 4 illustrates that defocusing mechanism. Since they normally load on the growth-ascending compression climb of the CGFR curve (as shown in the graph in Figure 2), epiphyseal plates can control tibiofemoral alignment in two important and clinically long known modes: a negative feedback physiologic one[267] and a positive feedback pathologic one.[267] It is of some interest here that the rough approximate unit loading on adult human articular cartilage is on the order of 20 kg/cm². One may infer that epiphyseal plate loading is somewhat similar, not forgetting in the process that the loading property that is important in life should also include strain, strain rate and frequency, range, and some period of time (Poss[822]).

a. The Negative Feedback Mode

In a normally aligned knee,[603] the time-averaged nontrivial *unit* (not total) loading should be the same on both medial and lateral sides, as the drawing on the upper left as well as inset A of Figure 5 implies.[308] If a small abnormal varus, e.g., bow legs, arises in the knee (Figure 5B), the vertical loads on it then add cantilever-origin flexural forces to the essentially uniaxial vertical compression loads of body weight and muscle pull. That cantilever effect adds flexural compression to the vertical compression load already on the right side of the plates, so their loading moves to the right on the CGFR curve below (circles) and accordingly their growth slightly increases. On the left sides of the plates the flexural tension reduces the already present vertical uniaxial compres-

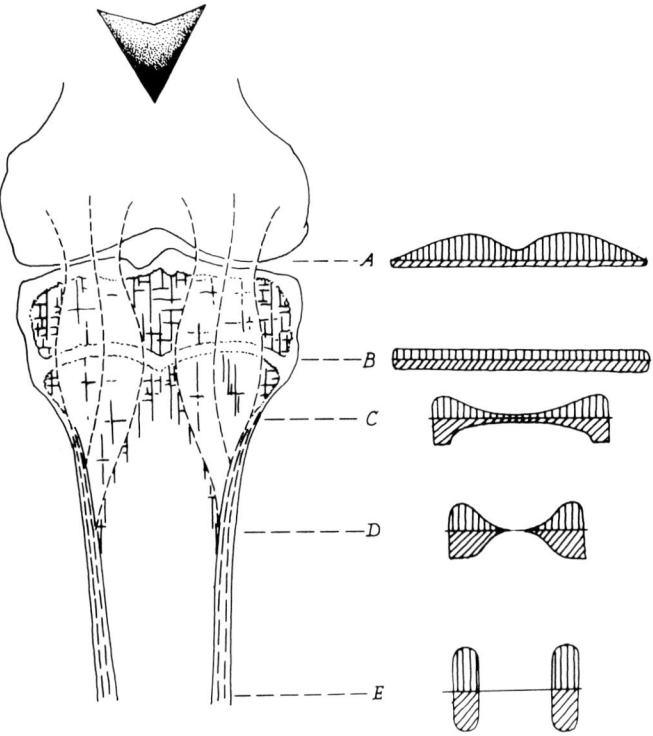

FIGURE 4. Diagram of an AP view of a child's knee, showing certain relationships between articular cartilage, epiphysis, epiphyseal plates, metaphyseal spongiosa, and diaphysis. The dotted vertical lines plot vertical compression load trajectories. They tend to concentrate or focus in the middle of the condyles at the joint surface. The stiff epiphysis defocuses or spreads those trajectories more evenly over the epiphyseal plate, and the underlying metaphyseal spongiosa then transfers them gradually to the diaphyseal cortex. The diagrams on the right show the distribution of load across given diameters (upper vertically hatched regions) and the distribution of structural support along the same diameters (lower, slanted hatched regions). The greater the vertical height, the greater the local unit loading or support. Where the load distribution is not matched by a mirror-image support distribution, significant differences in unit loading of the tissue occur, e.g., load focusing, and as indicated the tissue most affected by that load-focusing mechanism is the articular cartilage.[267]

sion, so their loading moves to the left on the CGFR curve as in the curve below (squares); that causes their growth to decrease slightly. Those two changes in growth speed proceed to correct the varus angulation. They stop when the unit loadings on the two sides equalize again,[244] which happens when the varus becomes corrected (Figure 5C).

That feedback control of joint and limb alignment occurs in the knee, elbow, all IP joints, the ankle, the vertebral centrae and the epiphyseal plates of the heads of the humerus, and the lateral four metacarpals and metatarsals, all as seen from a direction lying within the planes of their hinge-like motions.[267] It is a phenomenological observation rather than an hypothesis and represents a major organ-level or L_o-domain function of L_3-mode macromodeling of epiphyseal plates.[267] That negative feedback modeling response also explains how a 5-year-old youngster restores normal alignment and

16 *Intermediary Organization of the Skeleton*

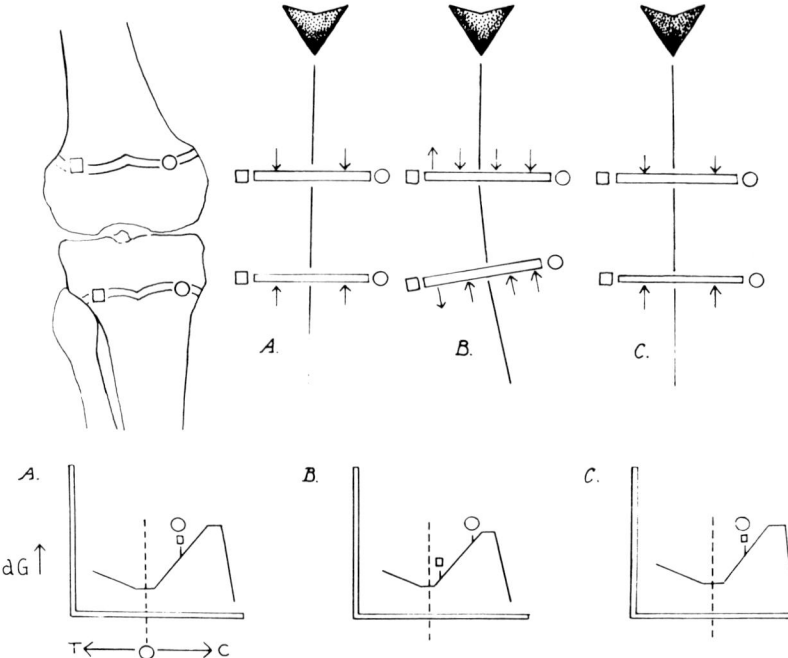

FIGURE 5. *Upper left:* a child's knee viewed from the front with the distal femoral and proximal tibial epiphyseal plates shown. Squares identify the left sides of the plates on the CGFR graphs below, and circles the right sides. The normal unit loading situation on those plates is shown at A. Each side of each plate carries equal time-averaged unit compression loads, as indicated by the short arrows, so their locations on the CGFR curve at inset A below coincide. When a varus deformity arises as in B, the unit loadings on the two sides become unequal for reasons given in the text, and as shown by the square and circles on the corresponding CGFR curve below. The resulting changes in growth rate then correct the varus, as in C, and the correction ceases then because growth on the two sides becomes equal again. This represents a *negative feedback* mode of control of the alignment of the femur and tibia at the knee and in this view.[267]

direction of growth to the distal radial epiphysis after a malunion of the radius following a fracture, as well as analogous phenomena in all other long bones (see also the correction in the orientation of the capital epiphyseal plate in Figure 2 in Chapter 7).

b. The Positive Feedback Mode

As in Figure 6, a greater amount of varus can move the loading on the right sides of the plates onto the steep descending limb of the CGFR curve, which would markedly retard or even arrest growth on those sides. Yet the reduced compression on the left sides only modestly retards growth there, so that combination causes this severe deformity to progress with further growth.[244,267]

Of course, when genu valgum (e.g. knock-knee) arises, the same mechanisms operate as above but the sides are simply reversed, and when the above alignments arise from a malunion of a fracture, the same mechanisms operate.[267] Those too are observed phenomena rather than hypotheses. If the malunion causes a small malalignment then the resulting negative feedback mode of response of its subsequent growth will realign it normally. On the other hand, too great a malalignment will invoke the positive feedback mode of response and corrective surgical osteotomy will be required to salvage the appearance, alignment, and function of the adjacent joint.

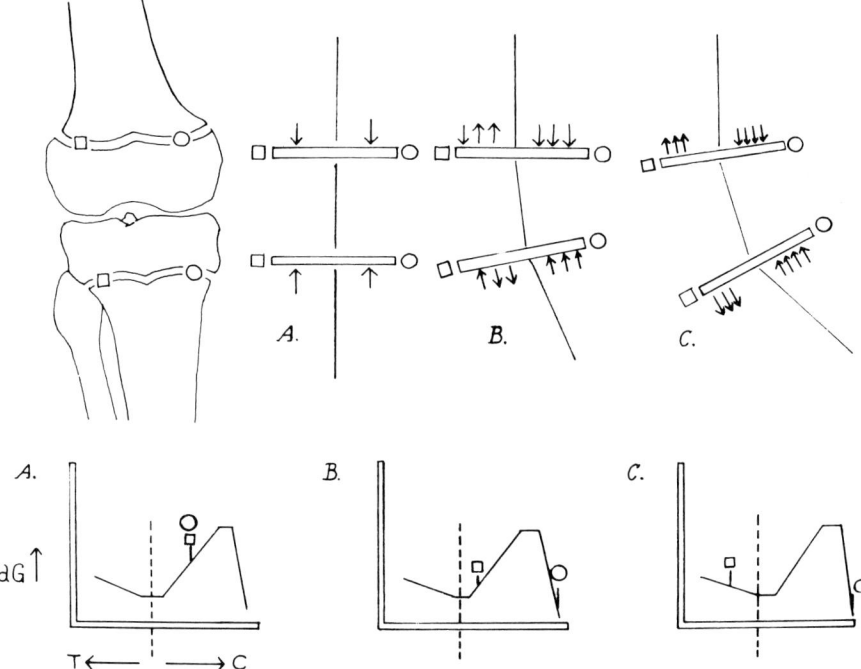

FIGURE 6. Similar to Figure 5, but illustrating a genu varum so great that it changes the response of the system from the negative to the positive feedback mode, as a result of which the deformity becomes worse with time rather than better. Clinical examples of such positive feedback response modes occur in Blount's disease, congenital coxa vara, and Madelung's deformity as well as after certain kinds of injury. Note: The square in the middle CGFR curve locates the loading characteristic of one part of the epiphyseal plates in the drawing above it so it represents an L_2-domain "L"-mode effect as mentioned earlier in the text. When one adds to that the circle representing the other side of those plates the mode analyzed escalates to "S", the domain escalates to L_3, and a new property emerges: the relationship of the growth rates on the two sides of the plate that now controls limb alignment during growth.

The above material explains why epiphyseal plates tend to align essentially perpendicular to their time-averaged nontrivial dynamic compression loads.

c. Epiphyseal Plate Diameter

Excepting the femoral head, where special circumstances not discussed here modify the response, the total cross-section area of an epiphyseal plate fits the magnitude of its major loads in the sense that when its peak unit loads exceed some natural limit the diameter of the plate then increases to provide more area to carry the total loads on the joint.[267,658] The growth of the perichondral ring controls that diametric growth by increasing the diameter and the total cross-section area of the plate.[379,765] The ring itself does not carry a significant fraction of the vertical load on the whole plate, and it will continue to grow — although more slowly — in the absence of nontrivial loads, as in the lower extremity bones of a high-level meningomyelocele (for example see Figure 17 in Chapter 7.) Figure 7 illustrates this mechanism, which so far has not attracted the attention of other biomechanical authorities.[267] Accordingly, when the loading on the plate exceeds some as yet unquantitated tolerable limit, then perichondral ring growth increases to enlarge the area of the plate and thus the total number of square millimeters of plate material that share the total load in it. That happens by adding more cartilage columns at the periphery of the plate, e.g., by clone multiplication. It pro-

18   *Intermediary Organization of the Skeleton*

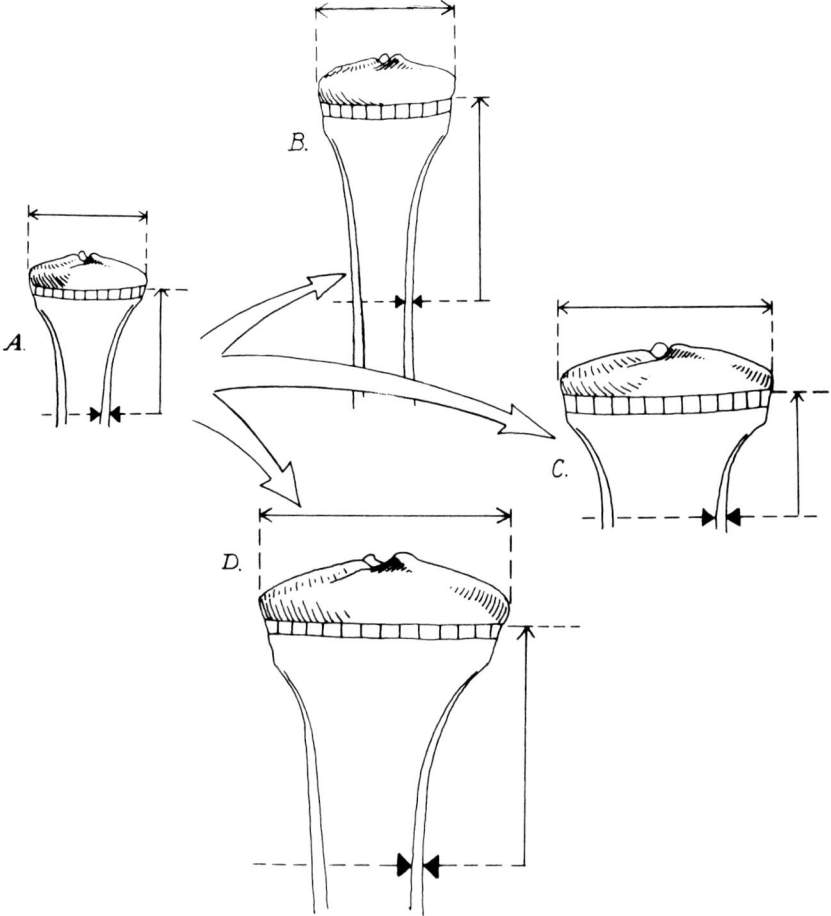

FIGURE 7. The growth of the child's proximal tibia at A involves clone elongation or growth of the chondral columns in the epiphyseal plate. That L_1-mode longitudinal growth by itself only adds new length to the bone without affecting the diameter of the epiphysis. Figure 7B shows the effect on bone configuration of pure L_1-mode growth. In Figure 7C the clone multiplication that perichondral ring growth adds to the above mode to form L_2-mode growth would, if it occurred by itself, only increase the diameter of the epiphysis without affecting the length of the bone. When the two phenomena combine, as they normally do, then growth in both length and diameter occur, and in typical proportions as in Figure 7D. The long-slender and short-slender proportions shown at B do occur in life, both in normal children and in a variety of congenital disorders, as do also the short-stocky and long-stocky proportions shown at C. The horizontal arrows indicate perichondral ring growth which adds further columns and width to the epiphyseal plate, and more area to the articular cartilage. The vertical arrows indicate the longitudinal growth of the chondral cell columns in the epiphyseal plate. To repeat, each vertical line in the epiphyseal plates represents a group of chondral cell columns. The short dark arrows spanning the lower right cortexes of each figure mark a constant reference level from which one may determine the amount of longitudinal growth above it.

vides a special feedback mechanism that can keep the typical unit loading of the structure on the appropriate part of the CFGR curve during growth, and it functions similarly for the articular cartilage.[267] Figure 8 also diagrams that effect for the human knee. On the left, five groups of columns of chondral cells in each plate, the distal

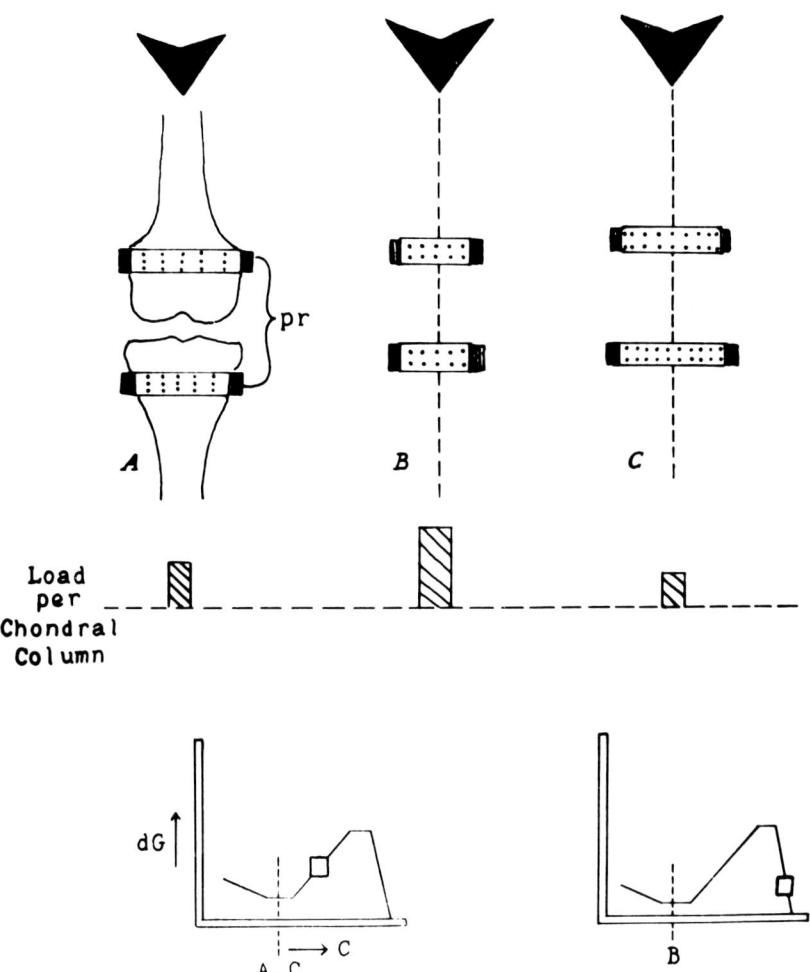

FIGURE 8. On the left, five groups of chondral columns in the epiphyseal plates of the distal femur and proximal tibia carry the vertical compression load applied to this knee joint. The bar graph and the CGFR graph below indicate the normal unit loading on the typical chondral column, and on the precursor cells that directly control its rate of longitudinal growth. At B a large increase has occurred in the total load on the knee, which has proportionally increased the unit loading on the typical chondral column and perhaps even moved it over the peak of the CGFR curve and down onto the growth-descending limb. If so, that would then invoke the positive feedback mode of response to any malalignment. Such overloads would also change the articular cartilage relationships illustrated in Figure 13, with deleterious effects on articular geometry and modeling. When such overloads occur in a normal body the perichondral ring (PR) reacts by adding more cartilage columns at the periphery of the epiphyseal plate and articular cartilage. As in C, that provides more columns to carry the total load, so it reduces the load on the typical individual column back to normal, as shown by the bar graph below and the CGFR curve at A, C. It also places articular cartilage loading back on or near the peak of the curve.

femoral and the proximal tibial, share the superimposed compression load. In the middle, a large increase has occurred in the superimposed load, which increases the loading on each column of cells in the epiphyseal plates. If uncorrected, that could move their loading well over and down the growth-descending compression limb of the CGFR curve, which would change their modeling responses from the above-described negative feedback mode to the positive feedback mode.[267,658] In real life, such an overload normally enables a growth response of the perichondral ring (PR) that adds more chondral columns alongside the preexisting ones. The resulting increased cross-section area of the plate lowers the unit loading on each column of cells back to the range determined as optimal by nature and by the response characteristics of the system.[267,658] This system also has a saturation threshold such that pathologically large overloads can lead to damage to the local cells and tissues and enable a repair response that is both different from the normal macromodeling and that dominates it.

B. Articular Cartilage

This tissue grows about one fifth as rapidly as the epiphyseal plate in the same epiphysis,[378,765] and it normally loads on or near the peak of the CGFR curve; while it grows more slowly than epiphyseal plate (probably in part due to the synovial "L" effects described earlier as well as in Chapter 4), it normally grows about as fast as it can.[244,267] Consequently, both increases and decreases in its time-averaged unit loading will retard its growth but increases will have the larger effect (because of the steeper slope to the right of the peak of the CGFR curve).[267] Also, the stiff epiphyseal ossification center protects epiphyseal plates from large loading inequalities concentrated on small parts of the plate, as shown earlier in Figure 4, but articular cartilage directly contacts the opposite joint surface so a small incongruity or misfit in their surfaces can cause large loading increases on small parts of those surfaces.[267] It follows that some parts of a joint surface can readily load far down on the growth-descending compression limb of the CGFR curve. Finally, at a typical moment in a hinge joint or ginglymus such as the knee, the epiphyseal pad distributes or defocuses its joint surface loads over the whole growth plate, but a smaller area of articular surface usually carries it.[267] Thus, the momentary unit loadings on parts of an articular cartilage can readily exceed those on the adjacent epiphyseal plate. That simple process combines biomechanically important load-focusing and -defocusing mechanisms that at least partly should explain the different normal loadings of those two types of cartilage on the CGFR curve.[267] Figure 4 diagrammed those load-focusing and -defocusing effects. They are important in the modeling of growing bones and joints, and when they malfunction, whether as the result of disease or as the consequences of trauma or surgery, they can cause clinically serious joint problems. The following examples illustrate some of the effects of the above properties on joint configuration.

The ball and socket ankle joint — The normal ankle allows free flexion-extension because, from the lateral view, its shape resembles a ball within a socket, as in Figure 9, top. Normal activities also impose valgus-varus or side-to-side loads across the ankle, but subtalar joint mobility normally stress-relieves (more accurately, load-relieves) the ankle of such loads, which the strap muscles — posterior and anterior tibial (PT) and the peroneals (P) — absorb by performing what Radin calls "negative work" or energy dissipation, as in the middle row of insets in that figure which now show the AP view of the ankle.[804]

If the subtalar joint (STJ) is or becomes rigid early in life, as in the lower row of insets, then the valgus-varus loads crossing the ankle will correspondingly tilt the talus in the ankle mortise, making its corners impinge at times on the malleolar cartilage and at others on the inferior tibial articular cartilage. Those concentrated unit loads lie well down on the growth-descending limb of the CGFR curve, which greatly retards the growth of the impinging parts relative to their load-relieved neighbors. Over time, and spread out over the whole joint surface by the varied motions of normal daily activities, that phenomenon makes the growing joint try to adopt the one configuration that allows such motions without incongruity: a true ball and socket.[244,267] Those phenom-

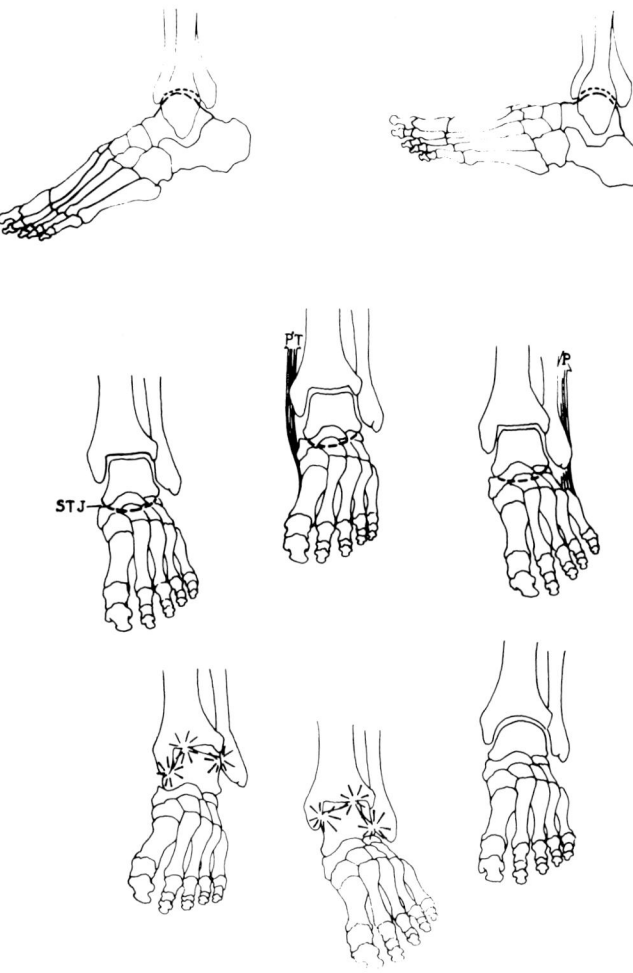

FIGURE 9. Diagram of the genesis of a ball and socket ankle joint due to an early or congenital synostosis of the subtalar joint, as described in the text. The final ball and socket configuration is shown at the lower right.

ena also retard the total growth of the articular cartilage so the tibia becomes slightly shorter than normal and the talus lacks its normal vertical height.[267] Both of those features regularly occur in this deformity and Figure 10 illustrates an actual example. It also illustrates a negative feedback mode of response of articular cartilage modeling to its loading situation.[267] From it and innumerable other but similar phenomena, one may abstract the following clinically useful rule of thumb: a growing joint adopts a surface configuration that eliminates incongruity under the motions imposed on it by its typical useage.[267] Do not forget here the first of the six basic chondral modeling axioms, which states that the remaining ones apply to postnatal matters, not embryonic ones.

The above phenomena can also operate in a positive feedback mode, and that mode underlies some of the features of congenital hip dysplasia[334,335] and dislocation, as well as other joints that have been reviewed elsewhere,[244,704] and the hip joint in Figure 2 in Chapter 7 illustrated an example.

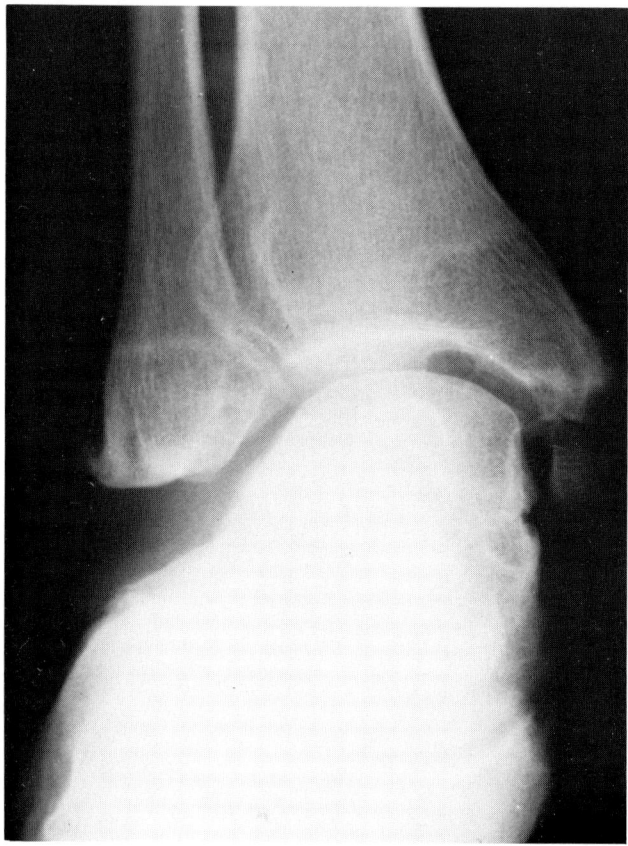

FIGURE 10. An AP X-ray of a man with a ball-and-socket ankle joint. The subtalar joint on this side was congenitally synostosed or stiffened, so during growth articular chondral modeling shaped the talotibial joint into a true ball and socket as described in this text as well as in 1972 and 1979, and as illustrated in the previous figure. The man's other ankle and subtalar joint were normal. The writer respectfully disagrees with those who postulate that this disorder arises from IO-intrinsic disorders in the local ankle tissues.[562] (Case courtesy Dr. W. B. Johnson, Detroit.) (From Frost, H. M., *Calcif. Tiss. Int.*, 28, 181, 1979. With permission.)

The time-averaging property — An epiphyseal plate forms a natural time marker for the longitudinal growth of the overlying articular cartilage, because the distance between the plate and the joint surface grows only at the articular cartilage (the growth of the epiphyseal plate only adds length to the metaphysis).[379,765] The knee X-rays in Figure 11 show that along the longitudinal axis, the height of the distal femoral epiphysis nearly equals that of the proximal tibial epiphysis in infants, but exceeds it in the adolescent.

This may seem inexplicable, since at any moment the unit loads on the load-bearing parts of the femoral and tibial articular surfaces must be identical and likewise their total loads, while the same synovial fluid nourishes both growing surfaces.

In large part at least, that reflects another default judgment: the unwitting assumption that momentary and time-integrated effects on this system are the same. Yet to repeat, and as the time-averaging axiom pointed out earlier, they are not.

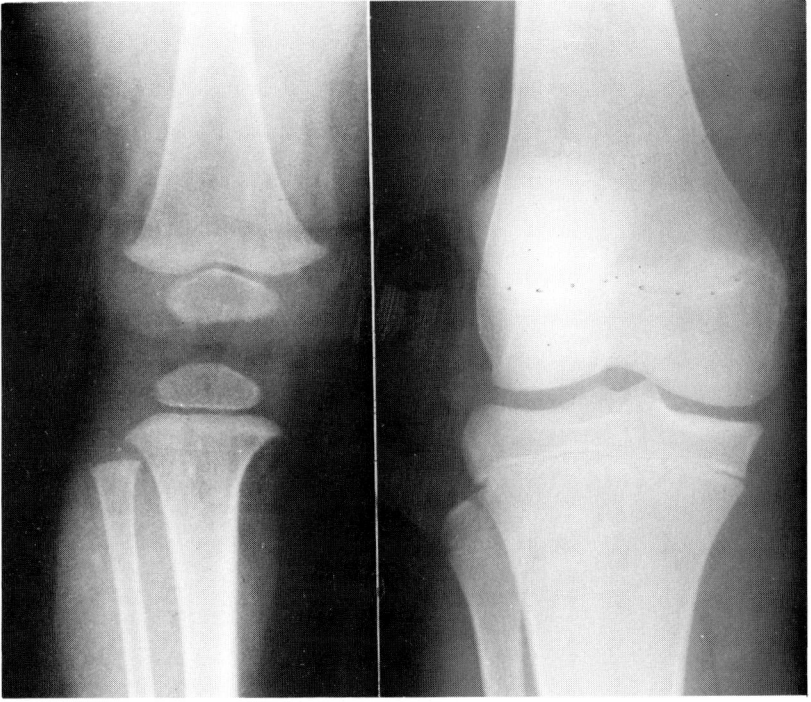

FIGURE 11. AP X-rays of the right knee of a 3-year-old child on the left and a 15-year-old on the right. The heights of the ossification centers of the femur and tibia are the same on the left, but on the right the femoral center is about twice the height of the tibial. The text explains the mechanical and chondral factors that appear to account for that difference. On the right, note that the epiphyseal plates are slightly convex towards and beneath the focused loads on the joint surface (X-rays courtesy of Dr. Charles Hanson, Pueblo).

The geometry of this joint, and of all other analogous ones, causes essentially the same part of the tibial joint surface to always accept any downward load from the femur, but quite different parts of the femoral articular surface transmit it when the knee is extended and flexed. Thus, the time-averaged load on a given unit area of the central tibial articular cartilage (and thus its time-averaged strains) will exceed that on the femoral articular cartilage even though their instantaneous loads do remain equal.[267,658] Other authors recently have begun to refer to this feature as a dynamic strain history (Carter[822]). Figure 12 diagrams the situation. As the strain-averaging axiom indicated earlier, macromodeling responds to time-averaged loads and strains rather than to single ones. Accordingly, in this situation the time-averaged tibial articular cartilage unit loading moves to the right from the peak down onto the steep, growth-descending limb of the CGFR curve, while the femoral articular cartilage unit loading moves from the peak down to the left and onto the less steep growth-ascending limb of the curve.[267] That combination allows the latter cartilage to grow slightly faster than the former so its epiphysis becomes taller with time.[244] One could say that that loading combination retards the potential growth of the femoral cartilage less than it retards the tibial one.

Parenthetically, that phenomenon also causes the tibial articular surface to remain slightly concave and the femoral surface to remain convex during growth, a feature that also applies to other analogous joints such as the elbow, shoulder, ankle, radi-

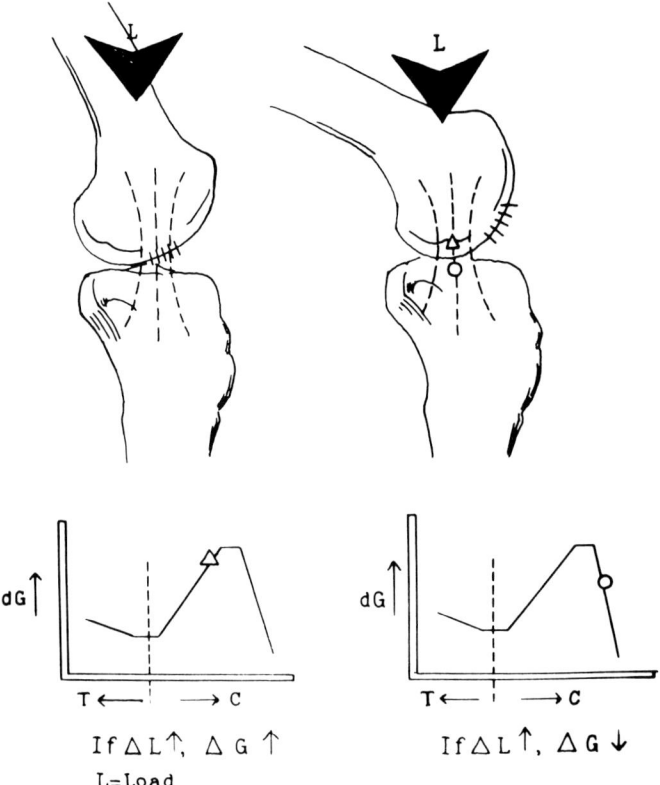

FIGURE 12. A human knee seen in lateral view. The two drawings at the top show that the normally somewhat focused or concentrated loads that cross the joint are always carried by essentially the same parts of the tibial articular cartilage but by different regions of the femoral articular cartilage when the knee is extended and flexed (we will avoid complicating the discussion by adding to it the menisci). Accordingly, the time-averaged unit loading on the tibial cartilage exceeds that on the femoral cartilage so they lie on different parts of the CGFR curve, as shown in the graphs below for the femoral (triangle) and tibial (circle) joint surfaces. To repeat an earlier-made point, that curve plots the chondral growth responses to time-averaged unit loadings, not to momentary ones. Accordingly, and as the CGFR curves below indicate, the femoral articular cartilage (triangle) grows faster than the tibial (circle), so the epiphysis of the former becomes taller. Also, as the relations below the graph indicate, the growth of the femoral cartilage (dG) responds to a slight increase in its time-averaged unit loading (L) by increasing because it lies on the upslope of the compression curve, while the tibial growth responds by decreasing because it lies on the downslope. Those responses keep the femoral contour convex during growth, and the tibial contour concave; the same factors apply to those contours as seen in the AP view in Figures 1, 3, and 11.

ocarpal, MP, talonavicular, atlanto-occipital, and IP joints. Using the knee as the illustrative case again, it does so because the above-described shifts in their typical loading on the CGFR curve cause the femoral articular cartilage to respond to any slight increase in local loading by growing slightly faster.[267] As many recent studies have shown, the central parts of the femoral surface as seen in both the lateral and AP views carry somewhat larger unit loads than the peripheral parts, and for that reason they should grow slightly faster than the peripheral parts, which would make the contour of the joint surface in this view convex.[267,658] On the other hand, the time-averaged loading of the central part of the tibial articular cartilage lies to the right of the peak of the CGFR curve as just noted, so it would respond to any slight increase in its unit loading by growing even more slowly than before, and in fact more slowly than its surrounding less heavily loaded parts, so it remains slightly concave.[267,658] Figure 13 diagrams those effects also.

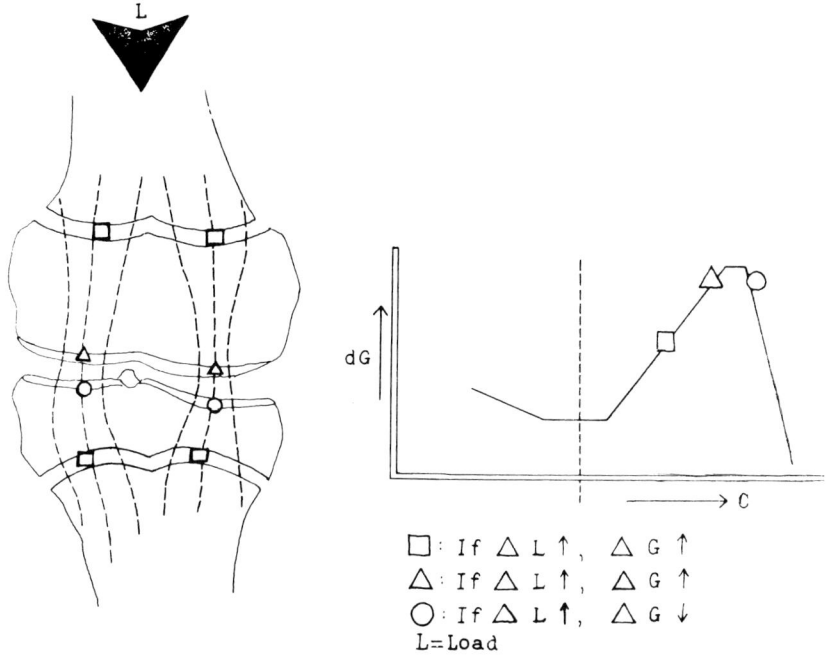

FIGURE 13. A child's knee seen from the front (AP view), showing the epiphyseal plate and articular cartilages. The loads tend to focus or concentrate at the joint line and in the centers of the two condyles of the femur and of the tibia each. The ossification centers in the epiphyses defocus those loads so they distribute more evenly over the epiphyseal plates than over the articular cartilage, but the defocusing is not complete, which leads to the curvature shown in those plates, and which are not present in the infant. The more highly loaded regions under the centers of the condyles grow slightly faster than the regions in the center of the plates and at their margins, so the contours of these plates, which are essentially flat in the newborn, later develop the curvatures shown here and in Figure 12. The squares, circles, and triangles locate the chondral loadings on the CGFR curve in the graph at the right, and the relations below the graph reveal how the growth of each chondral layer (dG) responds to a slight increase in its time-averaged unit loading (L). The reasons for those differences in loading on the femoral and tibial articular cartilages were explained in the previous figure.

Note Bene: Failure to account for some of the effects of time underlies much of the confusion and many of the misunderstandings, apparent paradoxes and enigmas, and even controversies that have characterized earlier efforts to understand the role of mechanical factors in controlling all skeletal architecture. It has been widely assumed, but usually as a default judgment, that instantaneous and integrated effects were the same, yet clearly they are not, as first pointed out in 1963 to 1964.[808,823]

To return to the matter of epiphyseal height, similar phenomena appear at all hinge joints with comparable epiphyseal plate markers, including the radiohumeral and shoulder joints, and they can be inferred from other anatomical markers for the ankle, MP, DIP, PIP, ulnohumeral, and patellofemoral joints.[804] Figure 13 summarizes some of the above material, which bears repetition because it has proven difficult for many people to absorb at first. It is different and thus novel; nonetheless, it is part of the truth of this system.

C. Neuromotor Relationships to Chondral Modeling

The previous chapter noted that even on weight-bearing joints the major loads come

Table 5
SOME BIOMECHANICAL FUNCTIONS OF NORMAL HYALINE CARTILAGE

1. Epiphyseal plate macromodeling determines limb valgus-varus relationships, as well as part of limb torsions and most of limb length.
2. Articular cartilage macromodeling determines joint configuration, and the spatial orientation and ranges of free and restricted motion, as well as part of limb torsions.
3. To occur in significant amounts, chondral modeling requires active chondral growth, so it acts primarily during childhood.
4. Micromodeling determines the internal preferred fiber orientation in all kinds of young cartilage, and shear-creep phenomena add to it in older cartilage.
5. Local mechanical strain presumably most directly controls local chondral macromodeling and micromodeling.
6. A RAP can accelerate the above processes.

from muscle forces rather than body weight, and they can readily exceed body weight by factors of 2, 5, or even on occasion, 10×.[27,241,474] It follows that chondral modeling should respond primarily to the dictates of muscle loads, including their lines of action in tissue space and their points of application to the skeleton.[244,267,821] Furthermore, since the nervous system coordinates muscle contractural strengths and patterns, it follows that the chondral modeling contribution to joint configuration and limb alignment must reflect certain properties of that neurologic control as well.[244,267,293,639,640,794,821] Those matters have been explained in some detail elsewhere and they are predictive with useful accuracy in dealing with those children's skeletal deformities of neuromotor origin. While obvious in retrospect, they remained concealed from earlier analyses of such matters by another default judgment. Analysts rather naturally assumed that since the architecture of the finished, mature skeleton is highly stereotyped, meaning essentially the same from one person to another, its construction must faithfully and blindly follow some predetermined architectural blueprint regardless of postnatal usage. The realization that the postnatal skeleton is architecturally fluid to a significant degree, and that the sterotypism of the mature skeleton reflects the sterotypism of its mechanical usage during growth coupled to the inital conditions established by the blind execution of the embryonic and fetal structural blueprint,[267] is too new yet to have been digested by a majority of the workers in the field.

V. COMMENT

Table 5 lists some functions of chondral modeling as it relates to the intact organism. They merit inspection.

A. The Relative Growth and Modeling Rates

Like bone modeling, the maximum potential effects of chondral modeling on skeletal architecture relate directly to the relative growth rate (the inverse of the time needed at the given absolute growth rate to double the dimensions of a structure), and to what fraction of a child's total modeling potential remains uncompleted. Only that uncompleted fraction can be modified or molded for therapeutic purposes.[267] Accordingly, chondral modeling is a game of actively growing vertebrate skeletons. Figure 14 diagrams this effect for equal linear growths in diameter of an infant and an adolescent femoral head above, and of an infant and an adolescent bone diaphysis below.

B. Chondral Dominance of Skeletal Architecture

A useful rule of thumb holds that cartilage conducts and bone follows in the skeletal

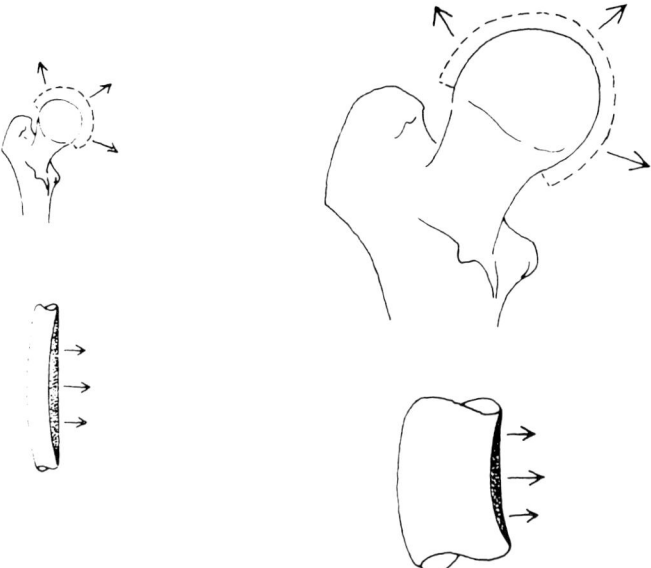

FIGURE 14. Relative and absolute growth and modeling rates. *Top row:* both femoral heads sketched here add the same number of millimeters to their diameters annually, but that causes a greater percentage or relative increase in the diameter of the infant's caput on the left than of the 12-year-old one on the right. Thus, their absolute linear growth rates are equal but their relative rates are unequal. *Bottom row:* a 4-mm formation drift arises on both diaphyseal segments shown here so their absolute linear modeling rates are equal. That causes a larger percentage or relative increase in the diameter of the infant's bone on the left than of the adolescent's bone on the right. Again, the absolute linear modeling drifts are equal, but the infant's relative modeling exceeds the adolescent's.

structural orchestra.[267] As examples, chondral features determine bone length and metaphyseal diameter via growing articular cartilage and epiphyseal and apophyseal plates. They also control the configuration and size of joints, their rotatory and angular alignments and rages of motion, and the locations of ligament, tendon, and fascial attachments to bone. They affect the flexural loads, moments, and strains in bone as described in Section IV of Note 3 in Chapter 7. During growth, bone replaces previously elaborated cartilage by the endochondral ossification mechanism, and as it does so it simply copies the configuration of the overlying cartilage on a different scale from the histogenesis of bones in the embryo, but otherwise resembling it.[267]

Because of that chondral dominance, IO-intrinsic chondral disorders can cause a chondral modeling abnormality that an intrinsically normal bone modeling process will react to by fitting the bone architecture to the mechanical problems and other situations caused by the chondral abnormality.[293] Hence, many of the skeletal anatomical features of, e.g., rickets, Morquio's disease, achondroplasia, and multiple hereditary exostoses, in which chondral modeling malfunctions due to IO-intrinsic or IO-extrinsic problems arise, but the bone modeling modality remains normal. Figures 15 and 16 illustrate such phenomena.

C. Chondral Creep

In mechanics, creep means a slowly progressing permanent deformation of rigid

28 Intermediary Organization of the Skeleton

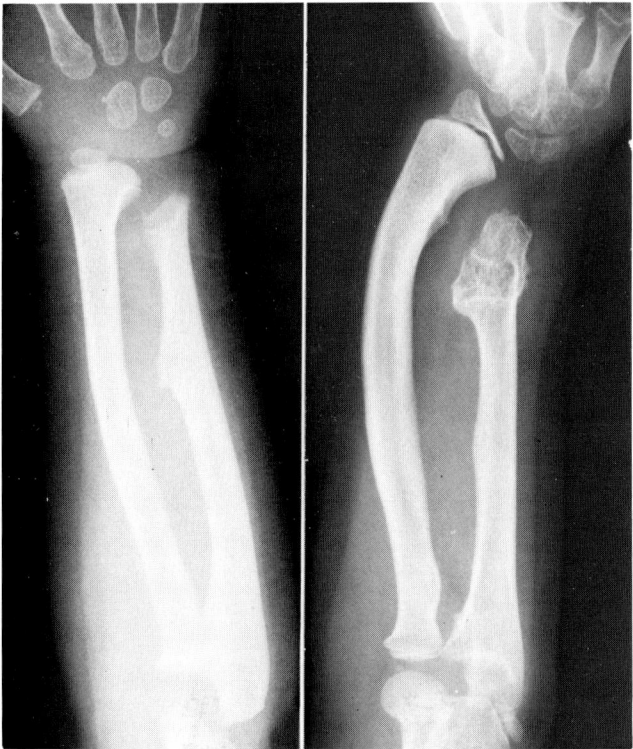

FIGURE 15. AP X-rays of the forearm of a youngster born with multiple hereditary exostoses or Ehrenfried's disease, an IO-intrinsic disorder of hyaline cartilage in which hyaline cartilage islands occur on the metaphyseal periosteal surfaces, often at the attachments of fascia, ligament, or tendons to bone. They grow centrifugally rather than longitudinally with respect to the long axis of the bone, causing irregular transverse expansions or masses at the ends of growing bones. The growing cartilage is itself normal and is replaced by bone by the same normal processes that replace growing epiphyseal plates with bone, so while unmineralized cartilage does not show on ordinary X-rays, the masses do become visible on X-rays after bone has replaced them. Radius on the left, ulna on the right. *Left:* the appearance of this forearm at about age 3 years. *Right:* several years later a marked deformity of the radius had developed. The bone and chondral macromodeling laws suggested to the author that the bony deformity was IO-extrinsic and reflected the mechanical demands of the abnormal local chondral growth and modeling. The malorientation of the distal radial epiphyseal plate implied a restraint or tether holding back its longitudinal growth only on its ulnar side. The foreshortened ulna meant that longitudinal growth at the distal ulna was retarded, and the ulnar collateral ligaments that connect the distal ulna to the ulnar side of the carpus and the ulnar side metacarpals should represent the aforesaid tether. The fibrous tissue modeling laws described in the next chapter also suggest those ligaments would be increased in diameter. At surgery, the author found the ulnar collateral ligaments over twice the normal thickness. They were resected and an osteotomy was done on the distal radius to correct its bowing and to restore normal alignment of the distal radial epiphysis relative to the longitudinal axis of the radius. The cosmetic deformity was corrected by that means and it did not recur during subsequent growth. The surgery was simple technically but sophisticated in its conception and its success lent support to the reasoning that suggested it. Here then an experiment from the laboratory of the clinic that revealed the principles of action that relate bone, chondral, and fibrous tissue modeling to mechanical factors.

matter under a load.[241,272,384] The beams of a bridge can strain or deflect elastically in flexure in response to the traffic loads crossing it, thousands of times a day for decades. However, over 10 years or so those beams will also develop a gradual permanent sag named creep that occurs at varying rates in all known rigid structural materials, including the structural materials of the body. Normally, however, it is clear that cartilage exhibits much larger amounts of creep than fibrous tissues, bone, and the hard tissues of teeth.[x]

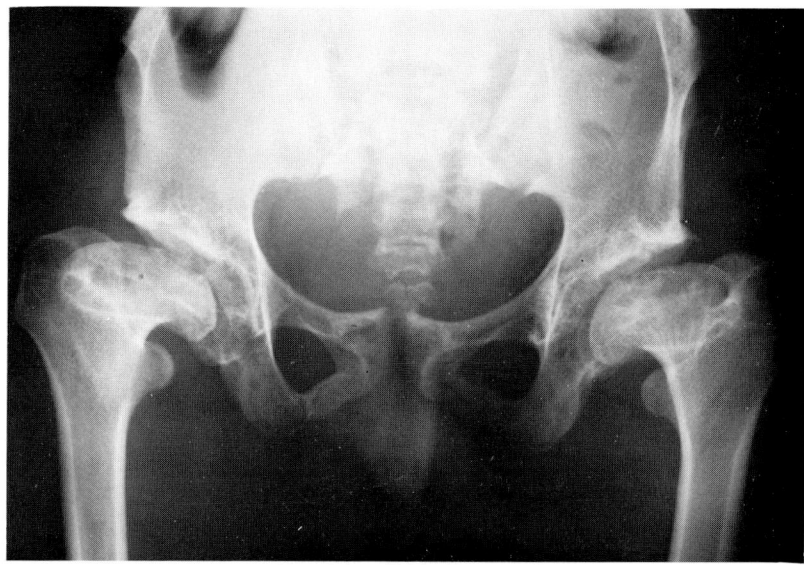

FIGURE 16. This late adolescent has Morquio's disease, a form of dwarfism due to an IO-intrinsic disorder in hyaline cartilage. As a result of that disorder, chondral growth occurs too slowly and its mineralization occurs even more slowly. During growth, that causes increased thickness of articular cartilage and an increased mean tissue age of both articular and epiphyseal plate cartilage. That increased mean age, plus a qualitative abnormality in the materials properties of the cartilage, also allowed greatly increased amounts of both unit and total chondral creep to occur in heavily loaded joints and epiphyses. Partly from that creep, epiphyses tend to become squat, meaning wide in relation to their height, as these femoral heads show. That creep has also caused the articular surface to deform from the ideal spherical shape to an ovoid or egg shape.

Other things being equal, the longer the time the more creep will accumulate in a unit length of the material, and the longer the beam the greater its total sag. Therefore, *unit creep* exists in a unit length of a structure, and *total creep* adds up over the whole length and/or thickness of the structure.

L_2-Domain hyaline cartilage shows creep in addition to its transient viscoelastic deformability, and to repeat, far more so than mature bone. Under pressure, as when a compression load descends on a joint surface or an epiphyseal plate, cartilage tends to deform or flow, somewhat as butter will under the pressure of one's thumb, but much more slowly. It can also creep in torque, like a bar of butter when twisted.[526] However, the architectural effects of that creep differ according to the anatomical location and type of cartilage involved.

Little total creep occurs in normal epiphyseal plates, partly because their great width relative to their thickness provides great resistance to any creep in shear, and partly because the endochondral ossification process constantly replaces old with new epiphyseal plate material, so its *mean tissue age* is only a few weeks, during which time little creep can accumulate.[267] A form of creep in torque at epiphyseal plates[526] does underlie some limb torsions, however, as well as the effectiveness of certain orthotic and surgical muscle-balancing methods of correcting them.[267]

Significant total or L_3-domain chondral creep appears to occur in life under four conditions. An abnormally thick cartilage can increase its creep in proportion to its thickness,[267] just as a long beam sags more than a short one, other things staying equal (e.g., vitamin D-resistant rickets, secondary hyperparathyroidism, hypothyroidism).

When the quality of the cartilage becomes abnormal, its unit creep can increase too,[267] just as a wooden beam creeps more than a steel one of similar dimensions (e.g., rickets, Morquio's disease, slipped capital femoral epiphysis, scoliosis). Also, the older a given volume of cartilage, meaning the greater its mean tissue age, the more creep it can accumulate[267] (e.g., conditions of growth retardation, as in rickets and malnutrition). For example, the growth-related turnover of epiphyseal plate cartilage is normally much more rapid than of articular cartilage, so under pathological loading conditions and/or when the cartilage has an abnormal quality that increases its susceptibility to unit creep, much larger creep-induced distortions will appear in the articular cartilage than in the epiphyseal plate. Figure 16 provides an example in the hip joint. Pathological loading situations, by increasing the local shearing loads, can cause increased creep in perfectly normal cartilage, as in Figure 17,[769,771] where the anatomy and mechanics of the situation now focus the pathology on the epiphyseal plate and adjacent bone, and spare the articular cartilage. Finally, combinations of the above effects can and often do appear in some diseases (e.g., all forms of rickets, Morquio's disease, congenital coxa vara, achondroplasia, Jansen's disease, hypothyroidism, Blount's disease). Increased creep in shear can occur under the pathologically large shearing loads caused by severe genu varum or valgum in otherwise normal children, as well as in coxa vara, in Ehrenfried's disease, in Madelung's deformity, and as the result of forces applied by certain splints or corrective casts in growing children. Figure 17 provides a clinical example.

Consequently, some limb alignment and joint abnormalities reflect the effects of pathologic chondral creep, usually in shear,[267,658] to which the regional chondral and bone modeling processes then try to respond in biomechanically perfectly appropriate ways.[267] While those combinations occur they have not received systematic study.

Chondral creep also underlies the apparently disorganized cartilage columns seen by histological examination of the epiphyseal plates in many forms of rickets, Morquio's disease, and the disordered cartilage in the medial proximal tibial epiphyseal plate in Blount's disease.[267] To visualize this, imagine two plates separated by a couple of centimeters and held together under a tension load with numerous parallel threads. The tension will hold the threads in regular and parallel alignment; a longitudinal cut through that set-up would reveal that alignment. Then bring the plates together and rotate and shear them back and forth a few times. The threads will curl and interwine in various ways and a longitudinal section through them would now show exactly the same kind of disarray known to histologists in the growth plates of diseases such as the above. Figure 18 diagrams the phenomenon.

Creep phenomena should also occur in normal adult articular cartilage and they could explain some of the differences in orientation of its superficial and deep collagen bundles. Since articular cartilage grows more slowly than epiphyseal plate, and not at all in adults, it is not rapidly replaced by the growth and endochondral ossification mechanisms; a given unit volume of it has a much greater mean tissue age, meaning it has had a much longer time to accumulate total creep and its effects on the organization of the tissue.[267]

The life-long synthesis of the molecular components of the organic matrix by the resident chondrocytes in adult articular cartilage[H,336,471] may compensate for some structural effects of that creep. For example, the writer infers that the alignment of its surface collagen fibers parallel to both the surface and the direction of motion of the opposite joint surface probably reflects the spatial orientation of that creep, as some of the organic material synthesized deep inside the cartilage by the resident chondrocytes gradually moves or "flows" towards the surface in the direction of motion of the joint surface, dragging with it the more slowly turning over collagen, somewhat as bottom-anchored sea weed bends with an outgoing tide.[267] The lower right inset in

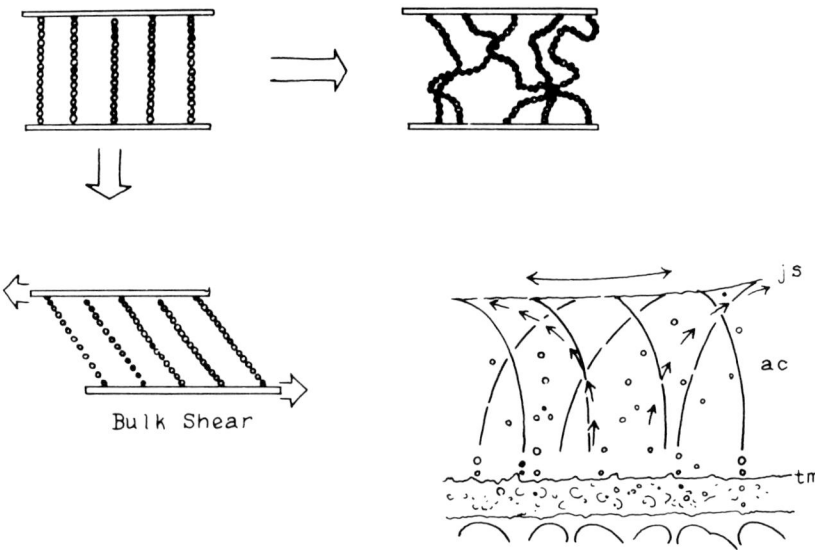

FIGURE 18. *Left:* a diagram of a longitudinal section through a normal epiphyseal plate, showing the regular, parallel columns of chondral cells. Below is seen the effect of a continuous shearing load on the plate. Its substance deforms or creeps slowly and plastically in response to that load, somewhat as butter would. To the right is the appearance observed in rickets and in many chondrodystrophies. It reflects a chaotic creep in shear that intertwines the chondral columns in intricate ways, as described in the text. *Lower right:* a diagram of a vertical section through an articular cartilage (ac), with the joint surface (js) above, the direction of motion of the opposite joint surface shown by the double-ended arrow, tide mark (tm) and bone below. The trajectories of the major collagen bundles are shown by the heavy lines, and they follow the pattern diagrammed here. The chondrocytes lying within the lacunae continually synthesize new proteoglycans and release them into the matrix. The author believes that over time, and because articular cartilage has a relatively long mean tissue age, the proteoglycans gradually move towards the joint surface as the short arrows suggest, tending to follow the directions of motion of the surface of the opposing articular cartilage, and dragging after them the bundles of collagen to form the arcades that are shown.

the chondrocytes residing in the tissue. That flow would drag along with it the much more slowly turned over collagen fibers.

VI. ILLUSTRATIVE QUESTIONS

While chondral growth and modeling pathophysiology underlie numerous problems of children's orthopedics and pediatrics, they have received disproportionately little study so far. Most published work has focused on cell- and molecular-level matters and on chemical composition and immunologic properties,[336] and very little work has been published that studies the behavior of various kinds of cartilage as integrated and organized systems, i.e., from the special points of view that apply to the IO of the tissue and in its L_2- to L_3-space-time domains. Here again one finds the unwitting assumption that to understand clinically apparent malfunctions of cartilage one needs only to learn enough about macromolecules and unassociated cells, whereupon the answers to clinical enigmas would come forth automatically. As a result, unanswered basic questions abound, and that continues into the problems of joints in adults, where attention has also focused so far on mostly molecular- and cell-level problems, although that begins to change. For example, Radin, Vignon, Courpron, and some oth-

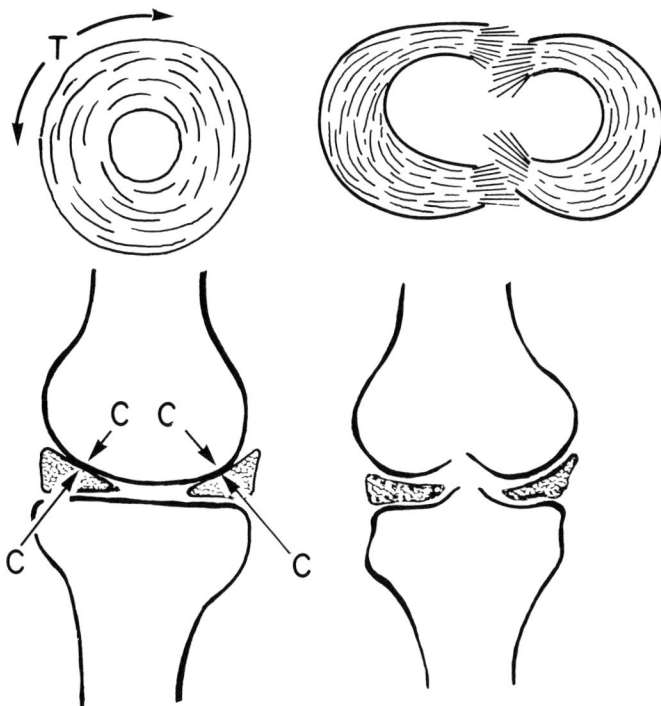

FIGURE 19. The menisci in the knee are made of fibrocartilage, and they act as flexible pressure pads that have typical collagen rigidity under tension loads alone. They carry about half or a bit more of the vertical compression loads that cross the knee. To do that they must retain their "hoop" stiffness, shown in diagrammatic form in the left drawing. The real-life situation is shown on the left, where each end of the meniscus attaches firmly to the tibia in the intercondylar region. While the infant's knee grows to adult size, as do the menisci, what mechanisms adjust the growth of the hoop length of the menisci so they continue to bear some of the weight? If they grew too little they would become too short and the femoral condyles would come to ride completely on them, deloading the central condylar regions. If they grew too much or for some reason stretched too much, then they would not carry part of the load and the central region of the joint, by carrying too much of it, would overload, and in fact the author suspects that actually causes many or most of the cases of osteochondritis dissecans that occur during growth in the central part of the femoral condyles.[267] Clearly some effective feedback mechanisms operate here that relate vertical load and the hoop tension strains it must cause in the menisci, to their bulk stiffness, length, and strength, and that compensate for their natural creep, But what are they? (From Frost, H. M., *Orthopaedic Biomechanics*, Charles C Thomas, Springfield, Ill., 1973. With permission.)

ers have done a few more holistically conceived studies aimed at the causes of degenerative adult joint disease. Some sample questions follow.

Exactly what agents, both "S" and "L" and of physical, biological, and biochemical nature, enable chondral growth and then regulate its continuum phase? How? Do different articular cartilages and epiphyseal plates have inherently different magnitudes of response to the same enabling and regulating agents? How do hyaline cartilage, fibrocartilage, and elastic cartilage differ in that respect? Why? Do different agents control L_1- and L_2-mode growth? How? How do age, sex, diet, race, climate, species, and activity affect such things?

Exactly what kinds of load-induced signals modify chondral growth to control its modeling? Are streaming effects analogous to those in bone involved? If so, how? What other agents can modify those responses? What accounts for the phenomena shown in Figures 19 to 23? What are the details of the paralyzed limb with respect to its modes of growth and modeling, its responses to physical, biologic, and biochemical agents,

34 *Intermediary Organization of the Skeleton*

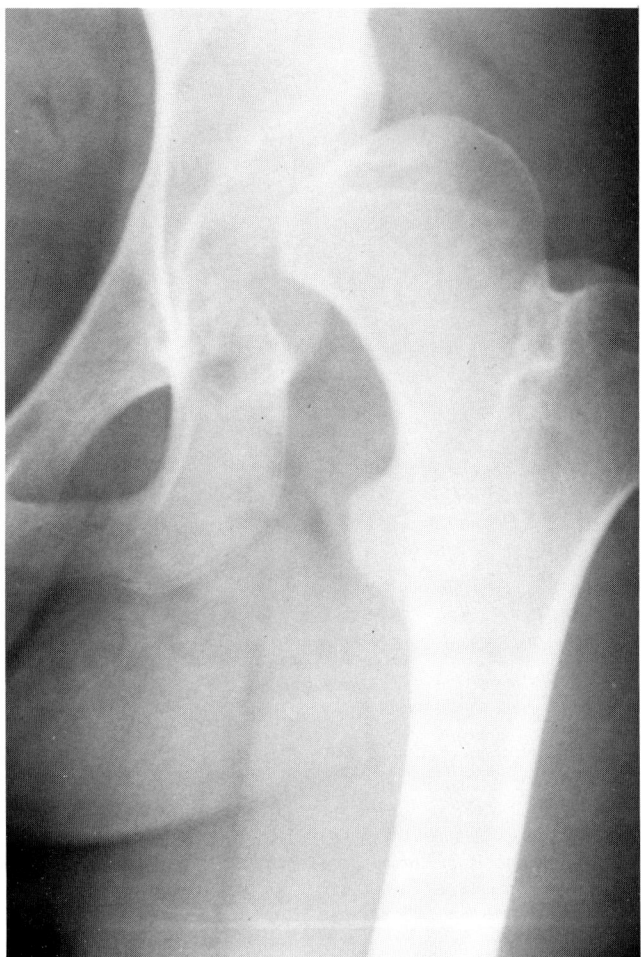

FIGURE 20. Positive feedback articular cartilage modeling. When she walks, this young spastic has insufficient medially pulling muscle strength (obturators, abductors, proximal adductors) and hip internal rotators (gluteus minimus and tensor fascia), but strongly acting vertically pulling muscle strength (hamstrings, psoas, long adductors) and hip external rotators (gluteus maximus). That imbalance caused chondral modeling effects at the growing hip and knee that led to an internal torsion of the distal relative to the proximal end of the femur. It also focused the hip loads on the outer margin of the acetabulum, which moved its unit loading down onto the growth-descending compression limb of the CGFR curve. While that also deloaded the more medial parts of the acetabulum onto the growth-ascending compression limb of that curve, the lesser slope of that limb causes the lateral growth of the medial socket to exceed the downwards growth of its roof. As a result, over time the facing of the socket becomes gradually more lateral, and under weight-bearing this femoral head slides or subluxes slightly laterally and upwards. It would progress to a full dislocation if not treated. One way to treat it would be to separate the bony socket from the pelvic bones above and medial to it and then rotate it as a unit, clockwise here, to cover the head (a Chiari osteotomy). Another way would be to rebalance the muscle loads carried by the limb by appropriately relocating their attachments and/or weakening the major offenders. Here then a minicourse in chondral modeling physiology, in which some basic science understanding can be carried productively into the clinic.

to sex and age, to race and climate? Do the circulation and innervation play passive or active roles in those phenomena?

Exactly how do altered neuromotor activity patterns affect chondral growth and modeling? What physical properties and space-time dimensions should be placed on

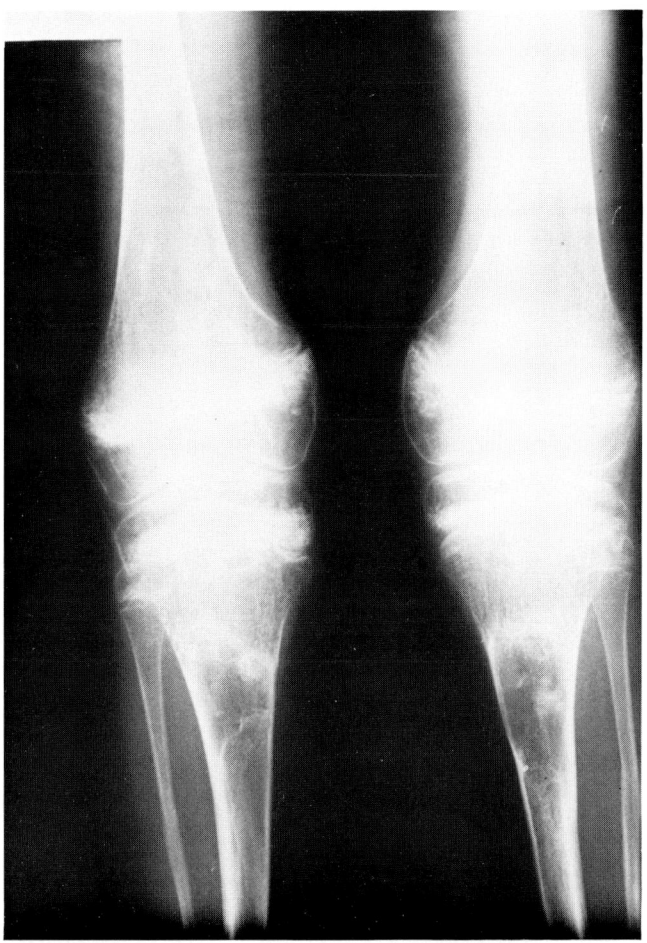

FIGURE 21. AP X-rays of knee and adjacent femurs and tibias. Currently being followed by the author, this 11-year-old girl has a rare congenital disorder named Stickler's disease. She has thin, weak muscles, abnormal dentition, reduced body weight, cartilage space narrowing of her major weight-bearing joints (compare the articular cartilage spaces here to those shown in normal knees in the figures in Chapter 7), a subluxing hip due to a positive feedback mode of articular cartilage modeling there, and she has developed knee deformities that led to earlier corrective tibial osteotomies. The regions of the epiphyseal plates are poorly defined, thickened, and mottled. Note the narrow outside diameters and thin cortexes of the femoral, fibular, and tibial diaphyses, a bone modeling effect that could be predicted on the basis of poor musculature and that probably is IO-extrinsic. Note the relatively great widths of the epiphyses relative to the diaphyses. Here L_2-mode growth, e.g., growth of the perichondral rings or envelopes, has exceeded the needs of longitudinal or L_1-mode growth, as though the vertical, load-carrying cartilage could not endure normal unit loading and so had to defocus that load over a larger area and more numerous L_2-domain chondral units. The above facts point to congenital or IO-intrinsic defects that involve cartilage and musculature, and that in the former tissue impair both its growth and modeling under large loads.

the two axes of the CGFR curve? In what units and other referents? What is the value of the chondral MES? Does it in fact have one? Does it differ in the different kinds of cartilage as suggested in Figure 24? How do the modeling game rules of hyaline cartilage and of fibrocartilage differ? Of elastic cartilage? How do their growths compare histologically, biochemically, and physically? What are the minimum hormone concentrations and/or changes thereof that are needed to enable chondral growth and modeling, and to regulate it, and in different parts of the body and different species?

36   *Intermediary Organization of the Skeleton*

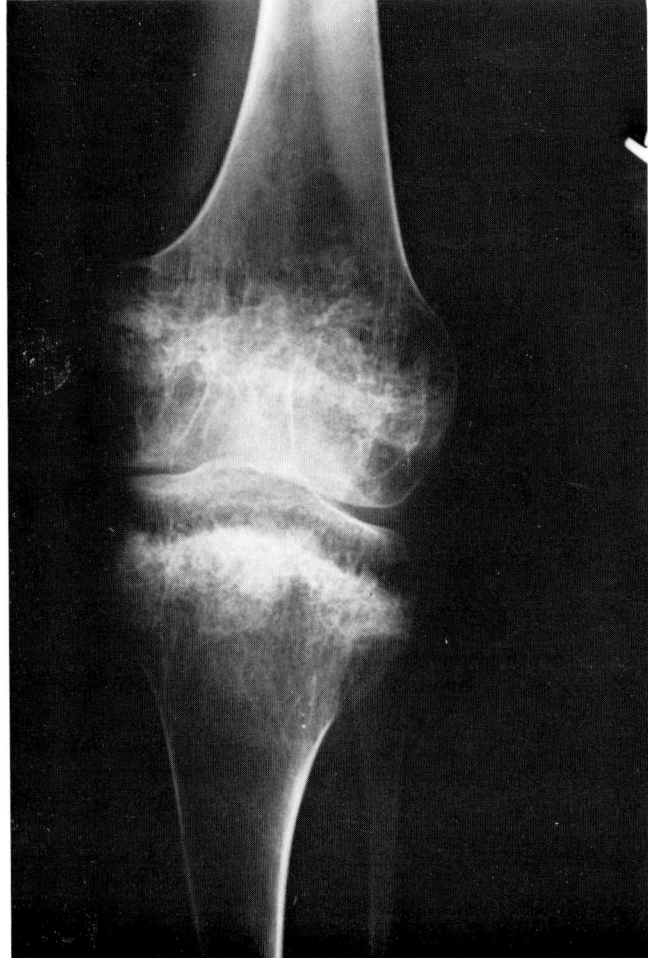

FIGURE 22. The left knee of the child in the previous illustration, this time exposed to show more clearly the epiphyseal plates and joint cartilage space.

What are the normal chondral micro- and macrorepair mechanisms in children and adults? What are their capacities? What enables and regulates them, and how? How does the RAP effect them, and what are the biological, biochemical, and physical details of a chondral RAP, and for each of the three kinds of cartilage?

What are the femoral, iliac, and tibial consequences, and in both bone and cartilage, of quadriceps paralysis, hamstring paralysis, abductor paralysis, and hip flexor paralysis early in life? In quadrupedal and bipedal beings? And similarly for the spine and the muscle groups that stabilize and actuate it? And likewise for adult-onset paralysis? What accounts for the kinds of phenomena shown in Figure 24?

What are the diseases due to abnormalities in the MES for the bone strain? What are the specifics of those abnormalities?

What else of importance that applies to the biological domain that concerns this chapter has been omitted, whether out of the author's ignorance or because it has not yet been perceived?

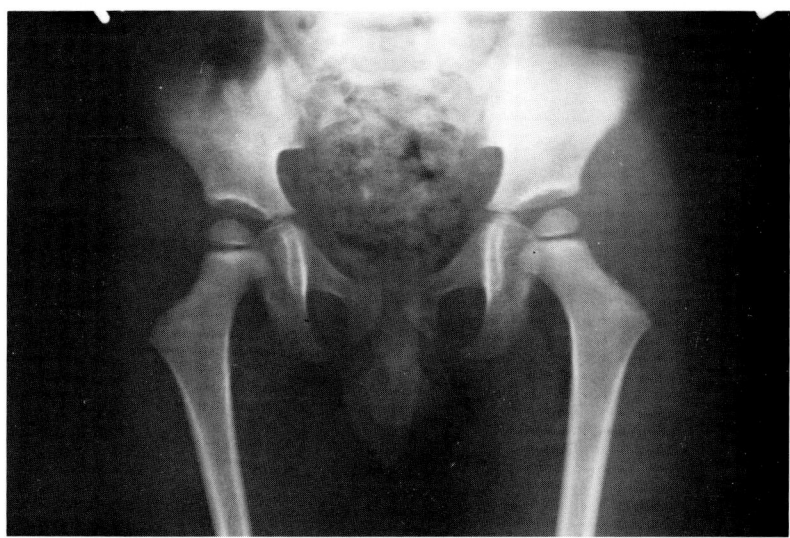

FIGURE 23. This child had muscular dystrophy of the Duchenne type with profound weakness of the hip and thigh muscles. The effects of this disease develop slowly over years without any accompanying RAP; due to the muscular weakness, the major loads on these hips and proximal femurs constitute body weight. The medially pulling muscles — obturators, abductors, and proximal adductors — have become ineffective. Given those facts, then the bone and chondral modeling laws would predict what is seen here. The capital epiphyseal plates have come to lie horizontally and the femoral diaphyses have subnormal outside diameters. Also, the neck-shaft angle is steeper than normal, i.e., a true coxa valga exists. What are the precise physical-chemical phenomena that translate the above loading factors into the observed effects on growing cartilage and bone? Streaming potentials and related effects look like a good bet at present, at least for bone, but if so, how do they affect cells, and which cells? Are other strain-related phenomena involved too?

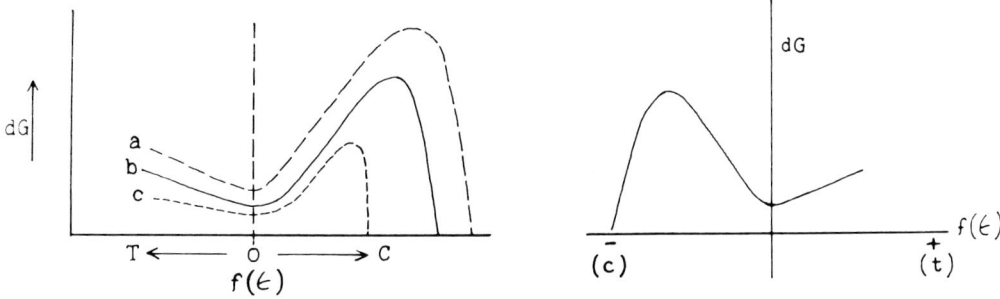

FIGURE 24. Different MES curves and graphing. *Left:* it seems likely to the author as well as to others that inherent differences may exist in the responsiveness to mechanical and endocrine agents of say, the distal femoral as compared to the phalangeal epiphyseal plates, articular cartilages, and tendon attachments to bone. The left drawing here shows three separate CGFR curves to illustrate the point that the magnitudes of the responses to a given load of three different articular cartilages (or epiphyseal or apophyseal plates) may differ, even though it is proposed that their curves should have the same shape. Thus, a loading magnitude that would arrest growth for the cartilage layer (c) would still lie on the growth-ascending limb of cartilage layer (a). These matters will require investigation in the future. It follows that some diseases should exist in which the responses of a given growing chondral plane are rendered abnormally insensitive to mechanical load effects, and others that are too sensitive as suggested, for example, by curves a, b, and c. Such abnormalities might raise or lower the MES value (the trough of the curve) and/or flatten or accentuate the peaks and/or shorten or lengthen the strain history range on the horizontal axis. This intriguing and new idea[267] could hardly be conceived without the theoretical and practical base of the material described in this chapter. *Right:* the basic CGFR curve shown as the solid line or (b) on the left could also be plotted in standard Cartesian coordinates as here, where positive values (tension strains) lie to the right of the vertical axis and negative ones (compression) to the left. Using Cartesian coordinates, or the author's, is only a matter of taste or convenience, for the information provided by a graph in either coordinates remains exactly the same.

Note 1

THE COMPOSITION AND CERTAIN OTHER PROPERTIES OF BONE, CARTILAGE, AND FIBROUS TISSUE[829]

A. Bone

A composite material, bone contains three major physical phases and a fourth, vital one. Forming about 39% of its bulk volume, its organic matrix contains about 95% Type I collagen and 5% proteoglycans and other organic, including noncollagenous proteinaceous, materials.[205,733,825] Forming about 49% of the volume of the bulk tissue, its inorganic mineral phase contains calcium hydroxyapatite crystals with minor fractions of carbonate, magnesium, sodium, and other substances. The organic phase provides most of the strength of a bone and the inorganic phase most of its stiffness, particularly in compression and shear. About 12% of the bulk bone volume represents a third phase, a fluid space. Part of that space represents osteocyte lacunae and canaliculae, plus vascular channels in compacta. A minor fraction of that space, the extracanalicular-lacunar or interstitial fluid volume, lies within the mineralized bony substance in innumerable ultramicroscopic clefts connected to the canaliculae that permeate all bone, so that somewhat like a wet sponge, squeezing the tissue can express some of that fluid and releasing it allows the fluid to return. As Figure 19 in Chapter 7 indicated, that little-studied interstitial fluid occupies about 3% of the volume of bulk mature bone, but it can occupy over 10% of the bulk of incompletely mineralized bone, which includes recently formed bone.

In the sense of a building material, two major types of bone occur, along with several minor types of interest to histologists, which need not be discussed here.

Woven bone (synonyms: fiber bone, reactive bone, primitive bone) lacks short- as well as long-range order (i.e., >0.1 mm), it can deposit *de novo* where no previous bone existed, and postnatally it appears in bone healing, in response to many irritants such as metastases, infection, and some bone tumors, and during the normal endochondral ossification process. In polarized light it shows a typical warp-woof pattern. Insignificant amounts occur in the normal adult human skeleton. Functionally speaking, woven bone can form relatively quickly in large amounts, and in locations where there is no preexisting bone. That is and to repeat, it can form *de novo*. All known osteosarcomas form primary woven bone. Woven bone does not seem capable of macromodeling in response to mechanical factors. It can mineralize more quickly than lamellar bone, and because of its anarchic long-range order its materials properties in bulk tend toward the isotropic state, and any anisotropy correlates more with the spatial organization of the capillaries that precede its deposition than with the local mechanical load and strain orientations.

Lamellar bone has both short- and long-range order and a characteristic lamellar pattern in polarized light. It comprises over 99% of the bone in a healthy adult human skeleton. For still unknown reasons it deposits only on preexisting bone (the LOBO property). As it forms, micromodeling factors align its preferred fiber grain parallel to the major local tension and compression bone strains. In life, lamellar bone always replaces living woven bone over periods of months by the BMU mechanism following local stimuli that seem in part inherent in the woven bone itself. When deformed or strained, dry lamellar bone can generate microvolt-range piezoelectric potentials on its surfaces, and fresh wet bone can develop millivolt-range streaming potentials as well as other effects that derive from the flow of its fluid phase, as described in the Discussion of Chapter 7. In that sense it can respond like a wet sponge that is squeezed. Mineral deposits begin to appear in newly formed bone matrix about 1 week after its formation and take 1 year or so to become complete. The present convention is to name unmineralized bone matrix osteoid, but after it has begun to mineralize it is called bone. In the mineralization process, the solid mineral deposits displace essentially equal volumes of water from the matrix without affecting the amount of matrix per unit volume of bone. Thus, per unit volume, young bone has less mineral and more water than old bone so it is less stiff, but both have similar amounts of organic matrix and similar lacunar-canalicular spaces. Lamellar bone also has an innervation in its vascular channels and on its endosteal and periosteal surfaces.

Functionally speaking, lamellar bone forms more slowly than woven bone and no known sarcoma forms primary lamellar bone, although some BMU-based replacement of woven by lamellar bone can occur in more slowly growing osteosarcomas. Lamellar bone apparently cannot form *de novo* (the LOBO property), whereas woven bone can. Micromodeling activity during its deposition causes its materials properties to become strongly anisotropic in a way that correlates with the local dynamic mechanical strain orientations during its deposition. As a result, the orientation of its maximum tension-compression stiffness as well as strength usually matches the orientation of the major mechanical loads it carries in life. Unlike woven bone, lamellar bone also has the macromodeling potential described in Chapter 7, and it can also somehow detect its own microdamage and enable a remodeling BMU that replaces the bone compromised thereby with new, undamaged bone. Its osteocytes form a kind of cytoplasmic syncytium that occupies the lacunar and canalicular spaces that riddle bone, and the original suggestion in 1963 to 1964[824] that that syncytium might detect microdamage and enable its repair begins to appear likely.[822] It may form one of the helper cell populations referred to in Note 1 in Chapter 7. The osteocyte syncytium in woven bone might play a similar role. Living lamellar bone is rigid, quite stiff, and strong. Normal lamellar bone possesses a major Hookean elastic property and a minor viscoelastic one, and it displays little evidence of gross creep, although osteomalacic bone can display significant amounts of creep in vivo. Both woven and lamellar bone show the "stiffness lag" phenomenon (the latter more than the former), meaning that while the formation of their organic matrixes and their initial mineralization typically occur within a couple of months (or half that for woven bone), full mineralization occurs more slowly and normally requires ~6 months. Until mineralization becomes complete the new bone remains unusually compliant under load so any older bone around it will carry most of the loads. This long-known phenomenon is of considerable importance in the reactions and tolerance of bone to load-bearing prostheses such as artificial hips, knees, and ankles, but those who design such devices have not yet begun to account for the stiffness lag for reasons given in Chapter 15.

Studies of the differences in chemical composition of lamellar and woven bone (and of another type named plexiform bone) are incomplete and the functional meanings of such differences are unknown at present.

B. Cartilage[830]

The composite three-phase material named cartilage has an unmineralized organic matrix, a volume of partly mobile extracellular interstitial fluid, and cells named chondrocytes that reside in lacunae. The body contains at least three kinds of cartilage: hyaline cartilage, fibrocartilage, and elastic cartilage. Hyaline cartilage forms articular cartilage, epiphyseal plates, and perichondral rings, and it lies at the bony attachments of tendons, ligaments, and fascia (where many histologists call it fibrocartilage). Fibrocartilage forms the menisci of joints such as the knee, pubic symphysis, and sternoclavicular joints, and also the intervertebral discs. Elastic cartilage occurs, for example, in the ear and nasal plates.

The organic matrix of hyaline cartilage forms about 20% of its volume (perhaps 30% for fibrocartilage), and extracellular water most of the remainder. The matrix contains about equal amounts of Type II collagen and hydrophilic proteoglycans that loosely bind the interstitial fluid phase (fibrocartilage has Type I collagen and a different proteoglycan spectrum). Small amounts of noncollagenous proteins and lipids are also present. The chondral matrix also acts somewhat like a molecular sieve, for gases and simple ions can diffuse through it more easily than large molecules such as enzymes and the growth-controlling hormones. Since the various molecules of the matrix contain many polar groups, electrical effects probably combine with mechanical ones to produce the observable sieve properties, and the polar or electrical charges on diffusing molecules probably modify their diffusion through chondral matrixes.

The interstitial fluid has relatively high lactate and bicarbonate concentrations and a low O_2 tension, partly due to the molecular sieve effect. The proteoglycan affinity for water causes mechanically unloaded cartilage to imbibe water and swell, while compression can express some of that water, shrink the tissue, compact its proteoglycans, and increase the apparent cell density, much like a sponge. That compaction could further impede the diffusion of regulatory, metabolite, and nutrient molecules through the matrix and may partly explain some of the responses of the tissue to growth-controlling chemical and endocrine agents under mechanical loads.

The chondrocytes in growing cartilage occupy 5 to 20% of its volume and lie in lacunae with fine processes that extend into the surrounding chondral matrix. Most cartilage lacks capillaries, lymphatics, other kinds of cells, and an innervation. The chondrocytes apparently oversee a life-long molecular-level turnover of the components of the organic matrix surrounding them. That turnover occurs much more rapidly during growth than in adult life and it annually turns over larger fractions of the proteoglycans than of the collagen. Some Type III collagen apparently arises around the time of chondral mineralization and the resorption of mineralized cartilage, and after trauma. The roles of such matters in micromodeling, macromodeling, molecular-level maintenance, and macrorepair processes are still unclear.

Functionally, all three kinds of cartilage can grow, and each possesses both the micromodeling and macromodeling potential. Each is quite compliant compared to bone and is also strongly viscoelastic. As far as ultimate strength is concerned, cartilage is much weaker than bone under shear, tension, and compression loads. Each is capable of much greater unit creep than bone, dentin, enamel, or cementum or than mature tendon and ligament.

Two other histological types of cartilage can occur in repair processes and in some neoplasms. They are a disorganized hyaline cartilage-like tissue and chondroosseoid. The differences in chemical composition of the various cartilage types are still incompletely studied and the functional roles of those differences are unknown at present.

C. Fibrous Tissue[831]

Fibrous tissues include the fibrillar materials that assemble to make tendon, ligament, fascia, periodontal ligament, and the sclerae. They are three-phase composite materials that contain an organic matrix, a volume of loosely bound extracellular fluid, various cells, an innervation, and vessels.

The matrix contains proteoglycans, some noncollagenous proteins and lipids, and mostly Type I collagen, all secreted by cells named fibroblasts. Those cells secrete the collagen first as long tropocollagen molecules that then assemble outside the cell in parallel, elongated arrays to form microfibrils that nest tropocollagen molecules side-to-side but with staggered ends. Thus, tropocollagen molecules have fixed lengths and molecular weights but their aggregates can have any greater length and weight. Chemical cross-links bind the sides of one tropocollagen molecule to the sides of its neighbors, which prevents them from sliding past each other and provides the bulk tensile strength and stiffness of the fibrils and their larger aggregates, including the collagen bundles that are visible in the light microscope. Collagen fibrils are enveloped in a hydrophilic proteoglycan gel made of elongated hyaluronic acid molecules bonded at one end via a link protein to a noncollagenous core protein. In the electron microscope the construct looks much like the bristles of a test tube brush bonded to a wire core. Collagen and proteoglycans form the chief components of the organic matrix of all fibrous tissues. In healing wounds, some Type III collagen is also formed, and as knowledge of collagen chemistry increases an increasing diversity of collagen types, structures, and biological associations thereof becomes evident. So far, however, the particular functional roles of the various collagen types and noncollagenous proteins remain unknown. Still, it seems safe to predict that they will be involved somehow in the enablement and regulation of the tissue-specific growth, modeling, remodeling, and repair activities that were defined in Chapter 3 as some of the basic functions of the skeletal IO, and possibly too in the molecular-level maintenance and putative creep-compensation mechanisms of the tissue.

The extracellular fluid of fibrous tissue binds loosely to the proteoglycan gel so some of it can flow in response to mechanical strains and the resulting hydrostatic pressure gradients. The gel also has some of the properties of a molecular sieve that can impede diffusion of large molecules more than of small ones, and its charged or polar groups could interact with the complementarily charged groups of diffusing molecules to modify their strictly mechanical diffuseability. This arrangement could create strain-induced streaming phenomena analogous to but not identical to those in bone.

Some, or perhaps many, of the fibroblasts that initially make a moiety of fibrous tissue then change to a less active and different kind of existence and become fibrocytes that reside in the mature tissue. They have tenuous cytoplasmic extensions that extend spider-like into the surrounding matrix. They oversee a slow molecular-level turnover of the components of the organic matrix enveloping them. They, like the osteocytes in lamellar bone, may also play the role of the signal man that senses any local microdamage and somehow sets in motion the events that repair it.

Like bone, fibrous tissues become stiffer or less compliant in tension as they age, probably due to increasing cross-linking of the fibers with time, so under equal unit loads young tissues strain more than old ones. For that reason, a "stiffness lag" similar to the one that occurs in bone probably occurs in newly formed fibrous tissue as well, for some time is needed after the initial deposition of a fibrous tissue mass for its stiffness to increase to the level of any surrounding, earlier-made fibrous tissue. Until it acquires normal tension stiffness, the older tissue around the new will continue to carry most of any applied local tension loads.

The chemical compositions of the various fibrous tissue structures of the body are still incompletely studied and known, and so are the functional roles of those differences.

Chapter 9

MECHANICAL DETERMINANTS OF FIBROUS TISSUE MODELING

"New opinions are always suspected, and usually opposed, without any other reason but because they are not already common."

(John Locke)

ABSTRACT

The vital mechanisms provided by the IO of living fibrous tissues have a micromodeling modality that aligns collagen fibers parallel to their tension loads and strains during their deposition. Another mechanism causes growth in the length of fibrous tissue structures (tendon, ligament, fascia) to occur at their ends, and a third causes mechanical tension loads equal to or larger than the MES limit of the tissue to increase the diameter and cross-section area of such structures, and thus their strength and stiffness, in response to a history of the peak dynamic tension loads. Special turnover mechanisms repair and eliminate mechanical microdamage of fibrous tissues to protect bulk structures from fatigue failures. The basic design appears to assign primary importance to fatigue failure and the tissue strains that most directly cause it, and lower importance to ultimate strength and mechanical stress.

I. INTRODUCTION

A. Preamble

When people first enter a field of science or clinical practice they tend to focus their attention on details, for much of their educational preparation will have involved memorizing facts, ideas, and algorithms provided by their predecessors, and in truth that exercise of memory is as essential to their work and future as the memorization of vocabulary, grammar, and idiom is to the students of a new language such as English, Danish, or Japanese. Accordingly, the clinical or scientific neophyte tends to look for the unique sign, symptom, blood test, chemical compound, pathological structure or stain, enzyme, or macromolecule that characterizes some condition and that alone causes it.

Experience instills an awareness that what is already known is but one grain of sand on a huge beach. An open mind and a desire to understand can slowly transform those simplistic objectives and views into larger and more complex subtler ones, and also more useful ones, through which run certain common threads. Those threads include looking for the operative basic mechanisms of the systems one works with, then trying to understand how they work and what purposes they achieve, and finally how they are and/or can be controlled. Those quests push at and move beyond the frontiers of knowledge, they tend to depend on sometimes subtle associations and circumstances for their advances rather than on the pathognomonic molecule, stain, X-ray finding or cell, and they tend to yield increasingly abstract but progressively more usable understanding. A microcrack in a bone is a readily demonstrable phenomenon, but the fact that it represents a result of a mechanical fatigue process is a concept. The latter cannot be photographed, but it is more useful than the former in understanding and dealing with some skeletal problems. A column of cells in an epiphyseal plate is another easily shown phenomenon, but that it provides the basis of a mechanism that controls longitudinal bone growth, some body proportions, and limb alignment comprise further concepts that are far more useful in the clinic and operating room than the basic phenomenon itself. While these concepts are real, they are abstract for they cannot be stained, analyzed chemically or mechanically or X-rayed. However, with the aid of such stains, analyses and X-rays obtained in the proper circumstances and in the proper sequence in time, at the proper age in the proper association with other things, and from the proper parts of the body, one can produce consistent and unequivocal evidence that the concepts signify genuine rather than invented biological phenomena.

The material provided in this book, including this chapter, belongs in the category just described, for it draws on the physical, temporal, and other circumstances in which associations of varied facts occur to reveal some of the basic mechanisms of the skeletal IO, the purposes those mechanisms seem to subserve, and to some extent what controls them in life. These facts were extracted from the laboratories of the clinic, the operating room, the histomorphometry and pathology workrooms, the chemistry and physiology departments, mathematical, physical, and biological knowledge and understanding, and mechanical engineering. As for what controls those mechanisms, we know relatively little in large part because the IO area is so new to most people in the field that their sights have not yet raised from the immediate particular details of these systems to their more distant and abstract, but more fundamental, purposes and natural controls. As the field begins to mature so should the aim of many of its investigators and clinicians. However, they must follow certain pioneers in that endeavor who undertook that quest when most other people could not see its value. Those pioneers must include Anderson, Courpron, Duncan, Epker, Harris, High, Jaworski, Heaney, Jee, Malluche, Melsen, Meunier, Mosekilde, Norrdin, Parfitt, Recker, Takahashi, and Teitelbaum, along with many of their collaborators.

B. General Remarks

Fibrous tissues have some clinically apparent properties that are provided directly by their IO rather than by the properties of their isolated cells or of the Type I collagen fibers that form their basic major structural component (see Note 1 in Chapter 8). Little study has focused on that IO so far, so accumulated knowledge of cellular- and molecular-level matters still cannot explain most clinical problems that relate to collagen-containing tissues.

Collagen comprises the basic structural component of all fibrous tissues and it serves in the body much as cotton or other fiber serves man.[825] A fibrillar material, much like cotton thread, it has great strength and stiffness under tension loading along its length, but in bulk it remains flexible when loaded in torque, flexure, compression, and shear. In the body it serves to make analogs of fabrics which provide tension strength and stiffness oriented in a plane in space (e.g., fascia, joint and organ capsules, adventitia, some basement membranes, periodontal ligament, dura and interosseous membrane), and of ropes which provide tension strength and stiffness oriented in a single line in space (e.g., tendons, ligaments). It also binds together the cells of muscles, loose connective tissue, and other soft tissue organs into anatomical units by providing strength in bulk in three dimensions. Figure 1 diagrams these three major L_3-domain constructs. The basic tensile strength and stiffness of collagen derives from cross-links that bind its elongated molecules together side-to-side but with ends overlapping or out of register. The same cross-links bind together bundles of those molecules into fibrils, and the latter into larger bundles that become visible in the light microscope. Three major types of collagen exist, named Types I, II, and III. Type I occurs in fibrous tissue and bone, and Type II in hyaline cartilage. Type III appears in many healing tissues, and also in skin. Figure 2 illustrates how cross-linking supplies tensile strength to collagen in bulk, as in a tendon. Table 1 lists some fibrous tissue structures according to their organizational level. Such structures and organs contain a variety of components be-

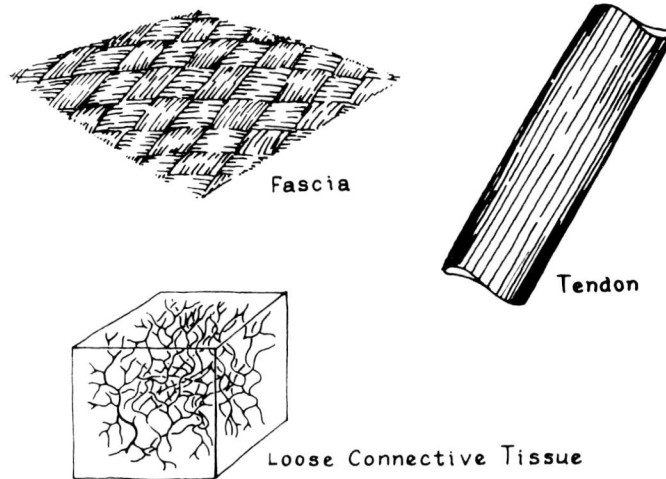

FIGURE 1. Modeling processes form collagen and the other components of living fibrous tissues into three different kinds of bulk structures. A sheet of fascia, like a length of cloth, provides tension rigidity and strength confined to a plane and it can serve to transfer tension loads as in the iliotibial band or various intermuscular septae, or to encapsulate structures and body compartments such as the liver and abdominal cavity. A ligament or tendon provides the same rigidity and strength, but confined to a single axis or line in space in order to transfer tension loads from one bone to another or from muscle to bone. A loose, three-dimensional network or mesh binds together the various cells and noncollagenous intercellular components of organs such as subcutaneous tissue, muscle, kidney, liver, lung, and gonads.

sides their tension-carrying collagen, including proteoglycans, loosely bound extracellular or interstitital fluid, and various kinds of cells, vessels, and nerve fibers, but those details are discussed elsewhere.[825] Collagen becomes stiffer in tension with aging, so under equal tension loads young fibrous tissues strain more than old ones. That probably relates to increasing numbers of cross-links per basic collagen molecule and fibril with age.

Note Bene: the basic function of collagen appears to lie in providing a strong and stiff material for carrying tension loads, a material that in bulk would remain stiff under tension parallel to its length but not under shear, torque, flexure, or compression. Like the other structural tissues of the body, fibrous tissues can grow, model, and turn over,[418] and special properties of those general functions become apparent within the different domains of biological organization. Special mechanisms provide those functions and attention turns to some of them next.

II. THE IO AND GROWTH, MODELING, AND TURNOVER

A. Growth

Scar formation in a healing wound provides a good example of pure, L_2-mode fibrous tissue growth, unmodified by any appreciable modeling.[825] It involves precursor cell proliferation nourished by newly formed capillaries and the subsequent production by some of the resulting daughter cells or fibroblasts of proteoglycans and new intercellular collagen that lacks long-range order, so the fiber grain is random over domains >0.1 mm. Both clone enlargement and clone multiplication occur, as described in

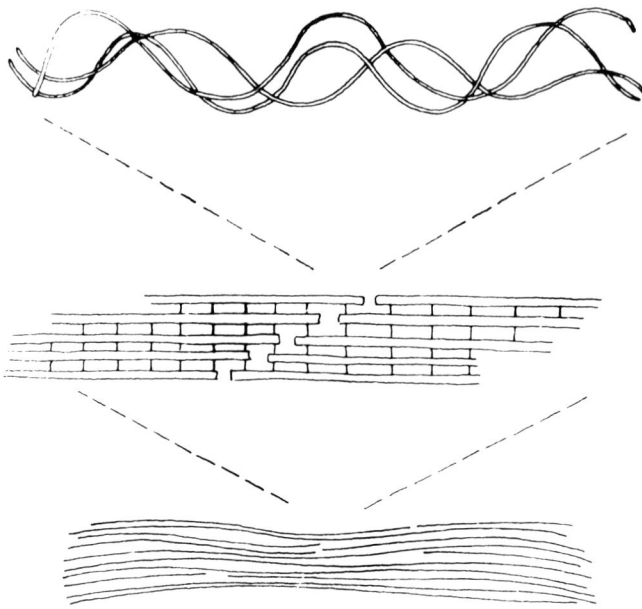

FIGURE 2. The basic collagen molecule synthesized inside fibroblasts assembles outside the cell, side-by-side with others but with staggered ends. Then chemical cross-links between the molecules bind them together, somewhat as shown in this diagram. When a tension load acts on such a construct, the strength and stiffness of the latter become functions of the relative numbers of intermolecular cross-links. In nature, many collagen molecules aligned side-to-side in the above manner form fibrils that can be seen in the electron microscope and that show a typical 640 Å banding. Groups of fibrils can then form larger bundles that become visible in the light microscope. It follows from the above that while the basic tropocollagen molecule contains a fixed number of amino acids and has a characteristic and constant molecular weight and length, its various assemblages have indeterminate numbers of collagen molecules and can be formed into virtually any length and width.

Chapters 2 and 3. Some of the new fibroblasts later become fibrocytes that reside in the newly constructed tissue for months to years, somewhat as some osteoblasts become osteocytes and some chondroblasts become chondrocytes.

Parenthetically, anarchic scar has several properties in common with woven bone. Both can form *de novo*, meaning where no similar tissue existed before; both the enabling stimuli are provided somehow by local injury or other threat; both can form in relatively large quantities and relatively quickly.[804] Both lack long-range order in or above L_2-space-time. Both seem to enable their own subsequent replacement by more highly organized and structurally superior lamellar bone and organized, mature, and uniformly grained fibrous tissue,[267] as in tendons and ligaments.

To return to growth now, in L_3-space-time and between birth and maturity all tendons, ligaments, capsules, and fascia grow in length in the direction parallel to their tension loads, noting that fascia may have two or sometimes three such preferred fiber orientations. While little discussed in clinical literature, that growth occurs preferentially at the ends of the structures rather than in their middles, somewhat like long bone growth in that respect. This is shown in Figure 3, and at skeletal maturity that

Table 1
SOME STRUCTURES AND FUNCTIONS OF FIBROUS TISSUES ACCORDING TO LEVEL OF BIOLOGIC ORGANIZATION

L_{sk} *Structures:* The intact, complete articulated skeleton including all tendons, ligaments, fascial sheets, joint capsules. *Functions:* locomotion, breathing, manipulation

L_3 (Upper IO) *Structures:* complete tendons, ligaments, fascial sheets, dura, joint capsules. *Functions:* transferring tension loads, containing muscles, and viscera; macromodeling; oriented growth in length; repair; the RAP

L_2 (Middle IO) *Structures:* the oriented tissues that form tendon, ligament, fascia, capsule, dura; all combine many L_1 elements. *Functions:* L_2-mode growth; BMU-based remodeling; the macromodeling mechanism; carrying tension loads

L_1 (Lower IO) *Structures:* the capillaries, fibroblast and fibrocyte populations, and supporting cells and innervation of individual biologic units of all fibrous tissues; collagen bundles; proteoglycans. *Functions:* the *de novo* formation or histogenesis of fibrous tissue; L_1-Mode growth; the creep-compensating mechanism and cell-mediated turnover of molecular components of the organic matrix; micromodeling

L_c (Cell level) *Structures:* individual precursor cells, fibroblasts, fibrocytes, endothelial cells, nerve endings. *Functions:* collagen and proteoglycan synthesis. *Note:* all organized associations of any kind of cells, their processes (e.g., dendrites), and their intercellular materials are peculiar to the IO

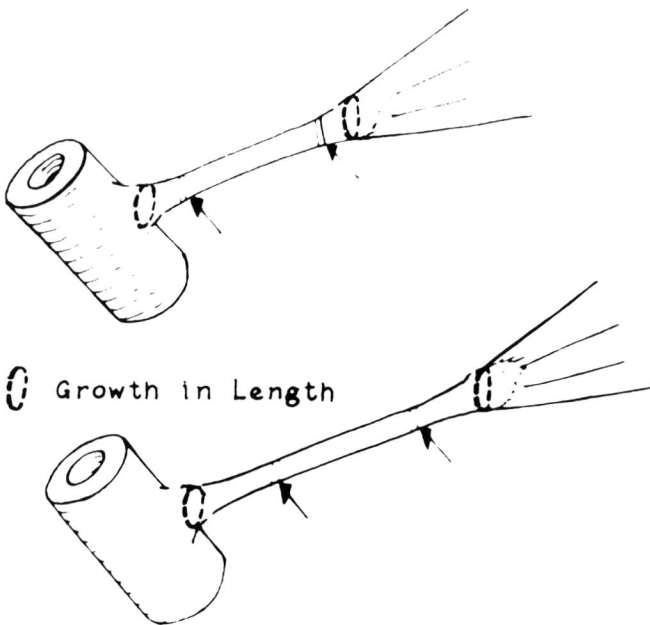

FIGURE 3. This tendon in a child connects a muscle to a bone in order to transfer the tension force created by muscle contraction to the bone. As the child grows in height, the tendon also grows in length but only at its ends. The two small index arrows show that after the tendon has grown in length between the left and the right drawing the separation between the index marks remains the same. As applies to bones also, the new length was not added in the middle of the tendon. Ligament and fascia grow in length by the same mechanism. A quite different mechanism accounts for increases in the diameter of tendons and ligaments, and the thickness and density of fascia.

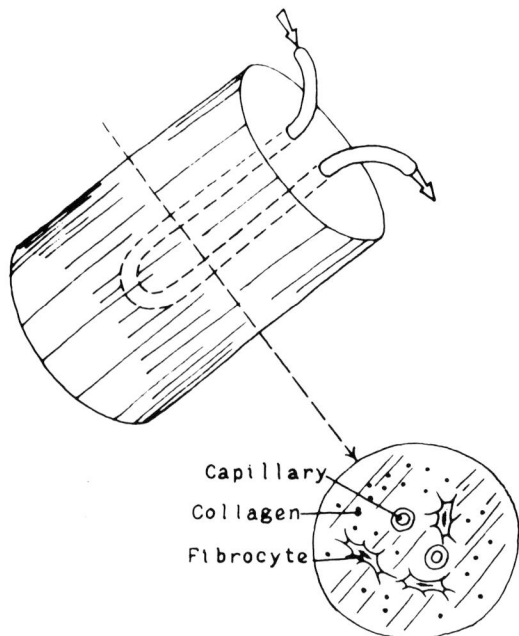

FIGURE 4. In a bulk fibrous tissue structure, here a tendon, the capillary and the cell populations of the tissue lie close to each other and both lie within the tissue itself, so that a local IO-extrinsic or "L" environment exists only for those cells close to the surface of the structure. Accordingly, circulating or "S" agents by-pass the "L" environment for most of the cells in these tissues, but they still have to traverse "M", the matrix itself to reach the resident cells, here fibrocytes. Since most fascia is quite thin, most of the cells in it are exposed to and potentially influenced by the "L" environment of the adjacent tissue, whether it be connective tissue, fat, muscle, synovia, or viscera.

growth ceases for practical purposes, just as longitudinal bone growth does, suggesting that the endocrine control of body growth generally also controls the growth in length of fibrous tissue structures, and that the growth-dependent class of cellular proliferation described in Chapter 4 supports it. The krσ property may also apply here, in that like each osteoblast or chondroblast, each new fibroblast may synthesize a given amount of new collagen before it disappears or becomes a resident fibrocyte.[267] In that case, that krσ property would have some of the same meanings concerning the mechanisms that control fibrous tissue growth, modeling, and turnover that apply to BMU-based bone remodeling.

As for its gating, the circulating agents that control L_3-mode longitudinal growth of fibrous tissue structures reach their cells via the resident capillaries. That by-passes the local "L" environment outside the tendon or ligament, because the capillaries that nourish growing regions of fibrous tissue structures, and the cells that proliferate to cause that growth, both lie inside the tissue. Figure 4 illustrates the situation. In contrast, the capillaries that nourish the cells that make bone grow and model lie on its "free" surfaces in contact with adjacent soft tissues. The state equations for pure, L_2-mode fibrous tissue growth could look like this:[263]

$$F(G) = (S) : \overrightarrow{(C:P:D:O:A:M:L)} \qquad (1)$$

and

$$F(G_f) = \begin{array}{c} S \\ \downarrow \end{array} \begin{array}{c} L \\ \downarrow \end{array} \longrightarrow \atop C:P:D:O:A:M \atop \uparrow\!\!\lrcorner \qquad (2)$$

While pure L_2-mode growth of fibrous tissues does not cause long-range order in its capillary, fibroblast, or collagen orientations, meaning over domains >0.1 mm, short-range intrinsic organization related to the associated capillaries occurs in actively forming scar as well as in most fibrous tissue tumors.

B. Modeling
Fibrous tissues reveal both micromodeling and macromodeling activities. They have different structural effects and may have somewhat different determinants.[267]

1. Micromodeling
An L_1-level phenomenon acts over a spatial domain of <0.1 mm, possibly in response to preferentially oriented cyclic tension strains to align the forming collagen parallel to those strains as the fibroblasts extrude it. The effect becomes visible in histologic sections viewed between crossed polars at ~400×. It can occur throughout life whenever new fibrous tissue is formed and it does not alter the amount of collagen in the tissue nor the external form or size of the affected fibrous tissue structure. It does increase the bulk tension strength and stiffness of the tissue parallel to its tension loads by orienting the fibers preferentially, as in Figure 5. Possibly smaller strains can more easily control this process than the macromodeling process described next.

2. Macromodeling
This L_3-level phenomenon adds new collagen to a structure in a way that increases only its total cross-section area and/or thickness, rather than its length, be it a tendon, ligament, or sheet of fascia.[236,267] That increase correspondingly increases the bulk strength and tension stiffness of the structure as will be described shortly. Nontrivial or supra-MES tension strains probably control macromodeling. It can occur throughout life, although much more slowly in adults than in children. The accompanying micromodeling orients the new collagen fibers parallel to the maximum local tension strains. The new collagenous tissue added by macromodeling can deposit along the whole length of the bulk structure, whereas its L_2-mode growth occurs only at its ends, as shown in Figure 3.

C. Turnover
Two quite different mechanisms can turn fibrous tissues over and both need more study than they have received so far.

1. Fibrous Tissue Remodeling BMU
While it remains arcane knowledge at present, turnover in normal and pathologic fibrous tissues can occur by an L_2-level, packet style ARF mechanism.[241,267] An initial microscopic focal destructive resorption of the tissue begins the process, as a batch of multinucleated giant cells appear and begin to tunnel through the tissue, usually parallel to its local tension strains. In the process they lyse the organic matrix ahead of

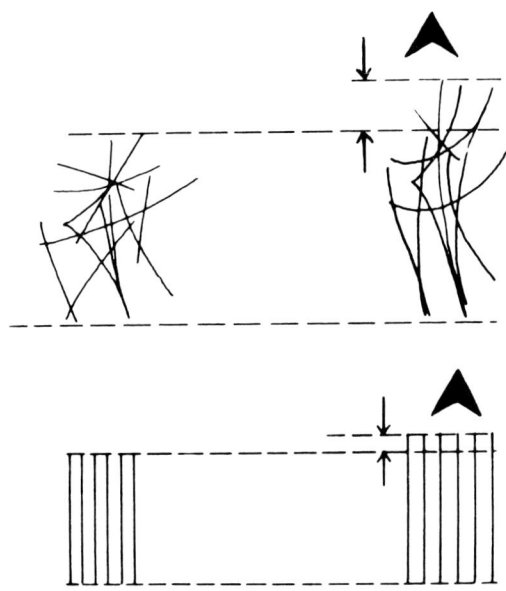

FIGURE 5. Micromodeling. When collagen deposits without accompanying micromodeling activity, the fibers align randomly in space, as in the scar diagrammed at the upper left. As a result, only a small fraction of all of the collagen in such a tissue aligns to carry a particular tension load, as on the right, and those fibers quickly overload and rupture. That makes the bulk tissue relatively weak and compliant under tension. When micromodeling aligns most of the collagen bundles parallel to a given axis, as at the lower left, then under a tension load parallel to that axis the tissue will have maximum strength and stiffness and minimum compliance, as on the right, and it will reveal a grain in the microscope. The fact that that grain normally always corresponds to the tension loads carried by the tissue implies that some particular effect of those loads controls the grain of the tissue, meaning its micromodeling in the terminology of this book. A good candidate for that controlling effect is the tension strain caused by tension loads.

and around them. A collection of new fibroblasts arises behind them and synthesizes new collagen and proteoglycans that replace the earlier removed matrix. Some of those blasts then transform into fibrocytes. A single growing capillary usually supplies both processes, and micromodeling orients the new collagen during its deposition so it parallels the local tension strains.[267] Due to the flexibility of bulk fibrous tissues, the resorption process creates a thin and poorly defined cleft in the tissue rather than the open tunnel created by secondary haversian resorption in bone. On cross-sections, the pursuing formation process deposits its new collagen in a fuzzily defined region rather than on a clearly defined surface. Those features probably concealed the integrated and ARF nature of this process from other observers, although its separate histological signatures have been known to clinical pathologists for decades. Such events imply some enabling stimulus that activates each new fibrous tissue turnover packet (see Note 2 in Chapter 2). Hence, an activation → resorption → formation (ARF) sequence occurs here, similar in that abstract respect to the BMU-based bone remodeling processes described in Chapter 4.[267]

Trivial amounts of this quantized turnover appear in the fibrous tissues of short-lived and small animals such as the rat, chick, and mouse.[267] It becomes more apparent in the adults of long-lived species such as man and dog, and most apparent during the internal remodeling of recently healed tendon and fascial lacerations, ligament ruptures, and scars. It escaped earlier recognition for some of the same reasons the quantized turnover of lamellar bone did: investigators like to study rapidly growing, cheap and short-lived animals that usually die before the activity in question becomes prominent.[267] Also, fibrous tissue repair processes have been studied histologically for only relatively brief periods after injury.

The above remodeling mechanism can routinely replace scar with functionally normal tendon, ligament, and fascia in adults as well as in children, although it takes 1 year or more to complete that replacement; biomechanical competence of the scar must develop first, as described later.[267] Little or nothing is known about the factors that control this BMU-based remodeling of fibrous tissue structures, but the author suspects that mechanical microdamage plays an important role in it, as it clearly does for bone.

2. Molecular-Level Turnover

In L_1-space-time fibrocytes in already existing fibrous tissues oversee a slow life-long molecular-level turnover of their collagen, proteoglycans, and other organic components. Various radiobiological studies, including those of Klein and colleagues,[405] indicate that in rats normally about 1 to 5% of the local collagen and 50% of the local proteoglycans can turn over annually, and that the resident fibrocyte population turns over slowly as well.

With respect to their function in fibrous tissues and in essence:

- Growth determines their length.
- Macromodeling determines their cross-section size.
- The two turnover processes maintain subsequent functional competence.

The above represent some basic vital mechanisms in fibrous tissues, so attention turns now to how some mechanical phenomena can guide the behavior of those mechanisms in the L_3-domain that becomes apparent in clinical medicine and orthopedics, and then to several other matters that also have direct clinical relevance.

III. FIBROUS TISSUE MODELING PHENOMENA

A. General Facts
1. Control of Tendon Length

Considering an intact tendon as representative for the moment of all bulk or L_3-domain fibrous tissue structures, its length is independent of the volume, contractile range, diameter, and frequency and strength of contraction of its muscle.[804] It grows in length at its ends, like a bone, and only during the period of general body growth, so factors other than its mechanical usage normally determine its length as well as the lengths of all other fibrous tissue structures, including ligament, fascia, and interosseous membrane. As a relation, and remembering that tendon serves here only to illustrate the general case for all bulk fibrous tissue structures or L_3-domain cases, one could write the following relations:

$$\text{tendon length} = F(\text{growth at the ends})$$

$$\text{tendon length} \neq F(\text{mechanical loads}) \quad (3)$$

2. Control of Bulk Strength

Three facts apply here: (1) the strength of a normal tendon, ligament, or fascial sheet relates directly to its cross-section area or thickness, and normally it always fits the peak typical loads it carries;[241,267] (2) in a limb that has been paralyzed since birth or infancy the tendons remain tenuous in diameter (they do show some diametric growth, however) but those in the normal limb develop normal diameters and cross-section areas; (3) however, the blood supply brings the same growth-controlling circulating agents to both limbs.

Such facts imply that the mechanical loads it carries can somehow control the total cross-section area of a tendon, ligament, or fascial sheet, and in such a way that strong muscles have thicker tendons than weak ones.[236,267] As a relation then one could write:

$$\text{tendon strength} = F(\text{diameter} + \text{grain orientation}) \qquad (4)$$

It should be noted here that while the earlier mentioned cross-linking can also affect tendon strength, the focus of this chapter lies above the cell level so it can consider cross-linking to be a constant.

Such elementary facts have suggested the following fibrous tissue macromodeling axioms. The reader may accept them as reasonable inferences for they fit truly numerous facts that appear in the laboratories of clinical medicine, orthopedic surgery, and pathology, but they still await some form of direct experimental proof.

B. The Dynamic Peak Strain Axiom

Probably dynamic tension strains rather than stress or constant strains most directly govern fibrous tissue macromodeling and micromodeling.[267] Furthermore, larger loads and strains, and possibly higher strain rates, appear to have disproportionately greater effects than small strains and slow strain rates on the structural adaptations of fibrous tissues to their mechanical usage.[267]

C. The Time-Averaging Axiom

Like bone, this modeling system somehow averages its peak strains and strain rates over some period of time, and it then produces the changes in its bulk strength that are needed to fit that history.[236,267] Such adaptations probably take 1 month to begin or get in motion, and may take 1 year or longer to become complete.

D. Tension-Creep Compensation

A constant load on fibrous tissue structures in vitro causes a gradual and permanent elongation in creep, as the collagen fibrils and bundles slowly slide past each other, probably due to disruption of some of their cross-links.[804] If not corrected, then over the lifespan of a man that creep should ultimately stretch out all ligaments, tendons, and fascia to the point that their mechanical usefulness was destroyed. Since that does not happen in life (something like it can occur in lathyrism, Ehlers-Danlos syndrome, rheumatoid arthritis, and inguinal hernia), some vital mechanism must correct it[267] and one may infer reasonably that the resident fibrocytes and the turnover of the molecular components of the tissue that they cause probably oversee the creep compensation mechanism too.

E. The Minimum Effective Strain (MES)

A minimum or threshold amount of tension strain probably is needed to activate this macromodeling process.[267] That MES probably is a range with upper and lower limits, as it is in other tissues and biological systems, rather than a sharp step function. Below the lower limit, tension strains would virtually never evoke macromodeling; above the upper limit they virtually always would and they might saturate and enable an SOS type of response too; they would evoke macromodeling with increasing regularity as the signal strength rose between the lower and upper limits of the MES range.

As applies to bone and cartilage then, a normal tendon or ligament should be biomechanically adapted, and according to the load partition principle described in Chapter 15 and in Note 1 in Chapter 7. Accordingly, a complete fit between the architecture of, say, the anterior cruciate ligament of the knee and its mechanical loads should be observed only when all of its separate major load collections are accounted for, including the fact that those separate collections do not act simultaneously in time but rather can and often do act at separate moments.[267] In a tendon such as the flexor pollicis longus, the major loading collections may number only one, but in the aforesaid cruciate ligament, many separate kinds of motions of the femur on the tibia are involved, including the obvious hinge motion plus relative rotation and several relative translation loads. The above discussion explains why extant studies have shown that the distribution of strength along the length of that ligament does not match one for one the loads caused by any single kind of displacement of the femur on the tibia.[267]

No vital study so far suggests the magnitude of the MES for fibrous tissues. Normal mature tendon can rupture at ~0.07 unit strain in tension, which is more than twice the tension strain needed to fracture mature bone, so if the bone properties provide a useful clue, then the MES for fibrous tissues could also lie in the range of 3 to 10% of its rupture strain, or approximately 0.002 to 0.007 unit tension strain, the same as 0.2 to 0.7% elongations.

As a relation one could now write:

$$\text{fibrous tissue macromodeling} = F(\text{repeated strains} \geqslant \text{MES}) \tag{5}$$

To repeat and whatever the details, smaller strains than the above may control the micromodeling process. If so, the latter could occur without macromodeling, but when the latter occurs then the former should always occur too,[658] as in fact it does. As for the cases of bone and cartilage, diseases may well exist in which the set-point of the tension MES for fibrous tissue is set too high, and others in which it is set too low. The former would have the clinically observable effect of producing tendons, ligaments, and fascia that are unusually thin or tenuous, and the latter the opposite. Those structures would be too thick in relation to their mechanical usage. As an example, an MES set too high does provide a reasonable explanation of the thin fascia, tendons, and dermis in many patients with osteogenesis imperfecta, and it was noted in Chapter 7 that such a phenomenon could also explain many of the bone features in that group of diseases.

F. The Stretch Hypertrophy Axiom

Nontrivial tension loads that repeatedly overstrain a fibrous tissue cause its cells to begin to make more collagen and proteoglycan. That added collagen increases the diameter and cross-section area of the intact structure all along its length, which of course also augments its bulk tension stiffness and strength in proportion.[236,267] That process stops when the added stiffness lowers the unit tension strains under the same loads below the lower limit of the MES, noting that here as in bone, a time lag, the "stiffness lag", separates the deposition of new tissue from its development of a structural stiffness and strength equivalent to the preexisting tissue around it. Thereby, the nontrivial dynamic tension loads it carries over some period of time, and at each point along its length, determine the cross-section area, the bulk tensile strength and stiffness, and the strength and stiffness of a unit length of an intact fibrous tissue structure.[236,237] Figure 6 illustrates this mechanism. In that negative feedback system the error of the system constitutes excessive tension strain, and it responds to that error by adding material to increase the bulk stiffness and reduce subsequent tension strain. As a relation one could write:

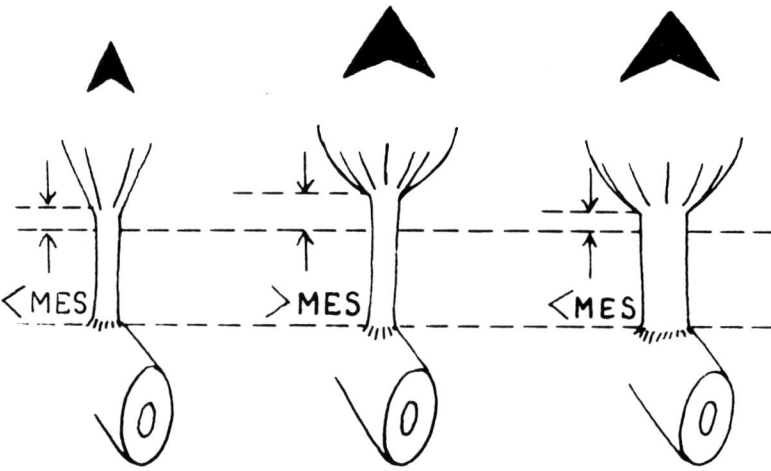

FIGURE 6. Macromodeling. In the normal, biomechanically adapted tendon in the left drawing, the peak contractile forces of the muscle above are never or are rarely large enough to stretch the tendon in tension enough to equal the lower limit of the MES for the issue. The graphs of that MES or tension strain should have essentially the same form as that for the bone MES described in Chapter 7 and illustrated in Figure 4 of that chapter. If the strength of the muscle increases significantly, as in the middle drawing, whether due to growth or to exercise against maximal resistance, then it can cause the tension strains in the tendon to equal or exceed the MES for macromodeling of this tissue. That evokes or enables macromodeling activity within the tissue that leads to the addition of new collagen throughout its length. That in turn increases its diameter and cross-section area, and after its stiffness lag has passed, its bulk strength and stiffness increase in direct proportion to the increase in cross-section area. That process stops when the same muscle loads no longer elongate a unit length of the tendon equal to or above the lower limit of the MES for the tissue. An astute reader will appreciate that since the transition from trivial to nontrivial strains follows a curve as in Figure 4 in Chapter 7, rather than representing a true step function, there probably is no absolute cut-off point for macromodeling of bone and fibrous tissue, so some "sublimal" macromodeling activity could occur in both tissues well into adult life. As examples, that phenomenon might partly explain the gradual increases in outside diameter of most human bones in adult life and certain cases of carpal tunnel compression of the median nerve.

$$\text{repeated strain} > \text{MES} \rightarrow \text{increase in diameter} \rightarrow \text{increased strength/stiffness} \qquad (6)$$

This basic stretch-hypertrophy mechanism arises in the middle IO of fibrous tissues and it represents a kind of operational brick, a basic intraskeletal mechanism. When "S"-mode mechanical factors pattern the signals that enable it throughout an intact tendon, ligament, or fascial sheet, that patterning creates the gross anatomical structures and configurations one associates with particular tendons, ligaments, and fascial sheets in the dissecting room. That L_3-domain macromodeling therefore represents the construction of higher-order structures, such as walls or posts, out of the basic L_2-domain bricks or basic mechanism. It should also be subject to modification by "L"-gated agents such as the nature of adjacent tissues, biochemical milieus, and the like. This stretch hypertrophy macromodeling rule has an effect similar to one observed in the bone macromodeling system. That is that increasing strains can enable the fibrous tissue system, but acute disuse does not appear to, probably because that disuse does not generate the necessary enabling signals which should be $>$ MES. However, one or more different vital systems do exist in fibrous tissues that can gradually remove chronically disused fibrous tissues. Their nature remains unknown at present but may depend more upon direct effects of fibrocytes than upon the BMU-based mechanism described earlier. Like bone, the evidence implies that fibrous tissues have one system that can increase their size in response to increased mechanical usage and another or others that are inhibited by normal usage but which are released under conditions of chronic disuse.

G. State Equations for Fibrous Tissue Modeling

To summarize, when a series of nontrivial tension loads or "S" causes strains of a fibrous tissues structure or "M", to equal or exceed the lower limit of the MES, that generates a cellular signal within the structure (perhaps streaming effects of its interstitial fluid, analogous to those known to occur in lamellar bone and mentioned in Chapter 7) that activates or enables the functions of its precursor cells. Some of the new daughter cells made by those precursor cells then become fibroblasts that add new collagen (and the accompanying proteoglycans), which increases the diameter and bulk strength and stiffness of the structure. Or, load → structure → strain ≥MES → activated precursor cells → new fibroblasts → new collagen → greater stiffness. At least some of those fibroblasts then become resident fibrocytes in the new tissue they helped to make. In symbols,

$$F(M_f) = (S):(M:P:D:C:A:C:L) \quad (7)$$

and

$$F(M_f) = P:C:D:O:A:M \quad (8)$$

As indicated, the intercellular material or "M" of this system becomes the primary gate for its mechanically controlled macromodeling, as it does also for bone and cartilage. The local environment or "L" at the surface of the tendon or ligament may have little influence on fibrous tissue macromodeling because the latter probably occurs inside the tissue rather than on its external surface. Note, however, that the surface layers of tendon and ligament can obtain a significant part of their nutrition by diffusion from the surrounding soft tissues, or in the case of a tendon within a sheath, by diffusion through the synovial fluid itself.[472,473] The above matters require more study; some have already begun. Again, and as for the cases of bone and chondral growth and modeling, one problem with previous research has been a lack of a reasonably detailed and successful body of theory that could serve as a basis for making particular predictions to test experimentally, or for setting up experiments to study the implications of the theory.

H. Rules of Thumb

Table 2 lists some biomechanical rules of thumb for fibrous tissues. Their L_1- to L_3-domain growth, turnover, and modeling activities govern their postnatal behavior and some of their functions that can cause clinical problems. It may not be a coincidence that with respect to the evolution of life, fibrous tissues are the oldest of structural materials of the vertebrate and also seem to have the smallest variety of IO-intrinisic and IO-extrinsic malfunctions. In a manner of speaking, nature has had more evolutionary time to debug fibrous tissue than she has to debug bone, cartilage, or teeth.

Table 2
BIOMECHANICAL FUNCTIONS FOR FIBROUS TISSUE MODELING

1. Strain rather than stress most directly controls fibrous tissue modeling.
2. Macromodeling determines the cross-section size, unit-length stiffness, and bulk strength of intact tendons, ligaments and fascia.
3. Local, nontrivial, dynamic, and repeated tension strains control macromodeling; constant or trivial strains do not.
4. The cross-section size of fibrous tissue structures accurately reflects the typical peak dynamic, nontrivial tension loads they carried during the recent past.
5. Micromodeling aligns the preferred fiber orientation of fibrous tissues parallel to their typical tension strains during their formation.
6. Both kinds of modeling can act throughout life.
7. Remodeling repairs fatigue-induced microdamage.
8. The scar that heals a complete rupture or incision needs 6 to 8 weeks to attain biomechanical competence.
9. A RAP can accelerate the above processes.

IV. DISCUSSION

A. Fatigue and Microdamage

Clinical and laboratory evidence as well as theoretical considerations suggest that microscopic mechanical damage should occur in repeatedly loaded and strained fibrous tissues. No known structural *material* whatsoever, whether of living or artificial origin, fails to develop microdamage due to mechanical fatigue processes.[804] Living intact biological *structures* fail in fatigue so rarely that it becomes necessary to infer the existence of a vital mechanism(s) that specifically repairs such microdamage before it can accumulate to the point of threatening bulk mechanical integrity.[236,267]

The L_2-domain ARF-type remodeling process in fibrous tissue probably provides that function, at least in part; like lamellar bone, both modeling and remodeling in fibrous tissues probably collaborate to maintain the mechanical integrity of the structures made from those tissues.[267] Macromodeling would reduce the incidence of new microdamage to some acceptable level by setting an upper limit to the peak strains of the structure. Micromodeling would orient any new collagen fibers parallel to the tension loads so the tissue could carry that load most efficiently. BMU-based remodeling would repair any microdamage that did develop by replacing damaged with new tissue before the damage could accumulate to form a burden large enough to threaten bulk mechanical integrity.[236,267] The molecular-level turnover could also participate in correcting molecular-level damage as well as creep (and those two might be the same phenomenon) before it grew to light microscopic-level size. However, the exact roles of strain rate, frequency, range, and magnitude in such matters remain unknown.

That mode of collaboration by the modeling and remodeling processes suggests that in fibrous tissues, as well as in bone, fatigue failures became a major threat during the evolution of the tissue, so that protection from failure in fatigue became the primary objective, and protection from failure under single catastrophic overloads became a secondary objective.[267] It also suggests that the trivial amounts of BMU-based remodeling found in the bony and fibrous tissue structures of short-lived animals as well as human infants means only that it takes 2 or more years for normal new bone and new fibrous tissue to begin to develop enough microdamage to evoke its BMU-based repair. Therefore, mice, and similarly, short-lived animals simply may not live long enough to accumulate threatening burdens of microdamage in their fibrous tissues or bones, and for that reason few observers have perceived the phenomenon.[267]

As described in Note 4 in Chapter 2, if some activity merits the status of a function (as defined in this book), then its malfunction must cause a disease. Indeed, under some circumstances the cellular mechanisms that detect and repair fibrous tissue microdamage can apparently become "deaf" to it in life, whereupon it begins to accumulate until a perfectly normal final load ruptures the remaining intact tissue.[267] That phenomenon probably causes many of the spontaneous tendon ruptures that can occur in otherwise healthy people, and also in patients with rheumatoid arthritis and in some patients treated with adrenalcorticalsteroid analogs. In the past 35 years the writer has repaired many such ruptures within 48 hr of the rupture, and without exception, at operation the ends of such ruptured tendons were edematous, soft, and somewhat attenuated. Histological sections of that material revealed that structural deterioration had been going on for many weeks prior to the final rupture, and also that an inadequate local repair response to that structural deterioration had arisen. The tendons always healed well after the surgical repair. The writer believes that reflects a saturation effect of the kind described in Note 2 in Chapter 2, one in which the final rupture plus the surgical repair provided an intense enough stimulus to enable the previously dormant macrohealing process, whereupon its continuum phase functioned perfectly normally.[267] That obtunded repair of microdamage probably also explains the attenuation of many surgical ligament grafts made of transplanted fascia or tendon.[267]

Microdamage probably underlies some of the painful tendons and the shin splints that arise in sports enthusiasts, especially those who participate vigorously but irregularly, or who train too quickly and intensely.[236,267] As applies to bone modeling and remodeling, the analogous fibrous tissue reparative mechanisms or individual BMU need time, typically some months, to achieve their intended results. The stiffness lag applies here, for it takes additional weeks or months for newly formed collagen lying within older collagen to attain sufficient rigidity to begin to carry its share of the load on an injured tendon or ligament. For such reasons, a highly motivated and aggressive athlete can challenge those mechanisms with a vigor that saturates their capacity to react and adapt to such challenge. The time needed for such adaptations to occur would be a function of the σ values for the biological and physical processes that are involved.[267]

The above matters relate to the biomechanical competence of fibrous tissues so that matter will be described next.

B. Biomechanical Competence of Fibrous Tissues

After a gross rupture of a tendon or ligament occurs, the thereby enabled *primary healing process* fills the defect with scar that bonds to the edges of the defect, much as fracture callus causes the primary healing of a fracture. Like the trabeculae in a fracture callus, the collagen in freshly formed scar lacks any preferred orientation over domains >0.1 mm. That random orientation makes the bulk scar weak and compliant under tension loads when compared to the strength of a similar amount of tendon loaded parallel to its grain, as shown earlier in Figure 5. The newly formed scar then undergoes some kind of internal maturation that becomes complete in some 6 to 9 weeks. If the scar carries significant tension loads before that maturation has occurred, then it stretches out progressively over time, and Figure 7 illustrates a fairly common clinical example of this. If loading resumes after that maturation the scar does not stretch and the earlier described ARF mechanism begins to remodel it internally to replace it with new tissue.[267] Due to the associated micromodeling, the fibers of that new tissue align parallel to the local tension loads and strains. After 1 to 2 years, a highly oriented tissue very like the original one replaces the anarchic scar and restores essentially normal bulk strength and stiffness. That represents the *secondary healing process.* Of course the ubiquitous RAP described in Chapter 11 normally accelerates those processes to reduce the healing time, which otherwise would take much longer.

As a relation, one could write the above processes and sequences thus, where $F(h)$ means the healing of a fibrous tissue structure:

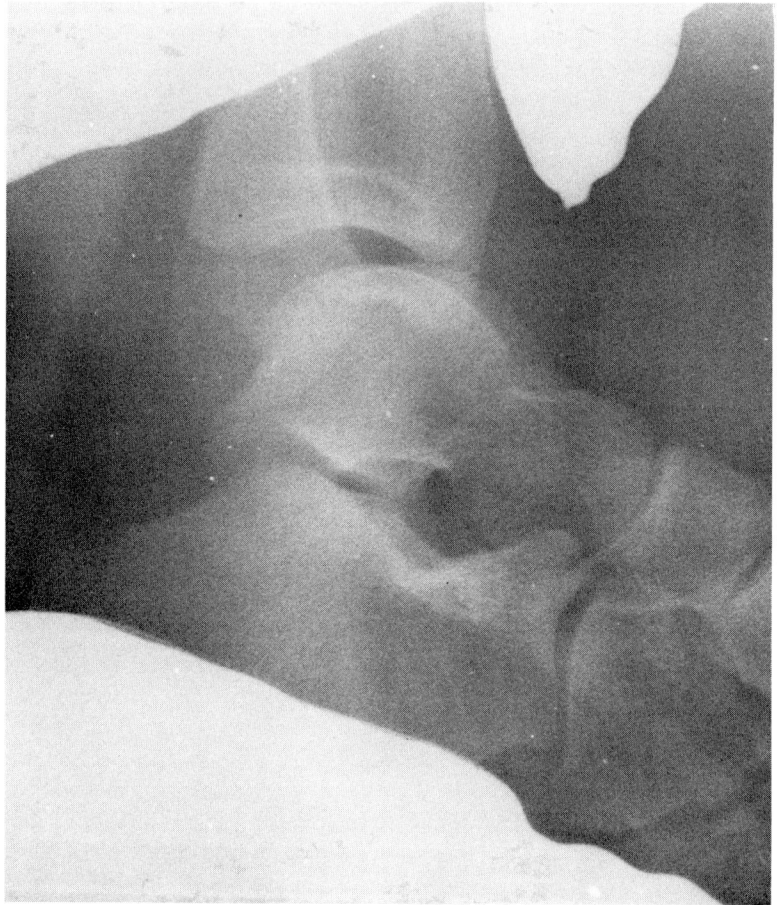

FIGURE 7. Biomechanical incompetence. Lateral X-ray view of an athletic young adult male's ankle. He had received a severe bimalleolar sprain some months before which significantly stretched the medial and lateral collateral ligaments. He was treated elsewhere with a short leg walking cast for 4 weeks and then released to unprotected activity. By 4 weeks the healing ligaments had not yet attained the biomechanical competence described in the text. Furthermore, in a short leg cast, considerable internal and external rotation of the talus in the ankle mortise can occur during the day. That combination of inadequate early immobilization and premature return to function led to slack collateral ligaments and a so-called anterior-unstable ankle. In running sports, that caused enough problems to lead to this X-ray and a subsequent surgical reconstruction by the author's technique (which was successful).

$$F(H) = injury \longrightarrow scar \begin{matrix} \nearrow RAP \dashrightarrow \searrow \\ \longrightarrow remodeling \\ \searrow modeling \end{matrix} \qquad (9)$$

$$\underbrace{}_{1° \text{ healing}} \underbrace{}_{2° \text{ healing}}$$

This complex process resembles in many operational ways the one that heals bone,

Table 3
SUGGESTED FUNCTIONS OF FIBROUS TISSUES

(L_o) Organ level		Transfer tension loads from muscle to bone and conversely. Stabilize joints. Contain joints, muscles, or other organs
(L_3) Upper IO		To provide tension-carrying structures of rod-like or planar form. Provide a loose three-dimensional mesh for holding cells of an organ together. L_3-Domain macromodling keyed to mechanical usage. The RAP Macrorepair, biomechanical competence. Oriented L_3-mode growth in length, keyed to general body growth. L_3-Mode MCN, LBO, and transient steady-state properties. ? kro property?
(L_2) Middle IO		Provides a tension-carrying tissue. Provides bulk rigidity in tension, compliance in flexure, torque, shear, compression. Provides L_2-mode MCN, LBO, and transient steady-state phenomena. ? kro property?
(L_1) Lower IO		Histogenesis of fibrous tissue. Micromodeling to align fibers parallel to tension loads. L_1-Mode growth. L_1-Mode turnover of matrix macromolecules. Provision of L_1-mode MCN, LBO, and transient steady-state properties
(L_c) Cell level		To provide synthesis of matrix macromolecules by fibroblasts, turnover of same by fibrocytes. Endothelial and related cells to form capillaries, lymphatics, nerve fibers, and other supporting cells

as discussed in Chapter 12. As applies to that latter process, this one probably also involves a sequence of enabling events and agents that set the separate processes in motion, and then separate agents that regulate the behavior of the resulting continua during their functionally active spans of time.

Accordingly, and again as applies to lamellar bone, biomechanical competence means the modeling and remodeling mechanisms of a skeletal tissue can respond appropriately to overstrain and to microdamage.[267] Those responses protect an intact structure from rupture or fracture under normal usage.

Table 3 lists what appear to be some of the functions of fibrous tissues that arise at each level in the organizational ladder, using the word function again in the upwards-tending, one-way sense defined in Chapter 2.

C. Conclusion

The above material weaves a logical — and probably basically correct — synthesis or fabric from many otherwise seemingly disjointed facts and observations. The various threads of that fabric come from histology, mechanics, biochemistry, control system theory, clinical experience, gross anatomy, and the recently perceived skeletal IO. However, some of the above relationships are too new to have received detailed study, so while they are logical and probably correct in essence, they await further experimental study.

One philosophical point has begun to emerge from the growing understanding of this system. When investigators began to study its mechanical properties and the relationship of structure to mechanical function, along with the vocabulary of mechanical engineering they also borrowed some of its default judgments. Those judgments included the assumptions that the fibrous tissue structures of the body were designed primarily to provide sufficient ultimate strength under single large loads according to the principal stresses generated in those tissues by those loads.[267] Furthermore, it was assumed that each tendon, ligament, and fascial sheet represented a construction that blindly executed a predrawn master blueprint independently of subsequent mechanical usage in life, exactly as men design and build buildings and bridges.

A different perception of the goals embodied in fibrous tissue anatomy — and in bony and chondral anatomy too — has begun to gel that makes sense. As Carter independently surmised for the case of bone, the primary purpose in such designs seems

Table 4
SOME APPARENT PURPOSES OF THE MACROMODELING MECHANISMS

Domain	Functions
L_2	To allow local growth to respond to local mechanical loads
	To allow local growth to respond to other juxtaposed tissues
	To orient L_2-mode turnover mechanisms according to mechanical strains
	To ignore strains below the MES
L_3	To orient structures according to their nontrivial mechanical usage
	To orient structures according to their innervation and blood supply
	To limit production of new fatigue damage from mechanical usage
	To restore mechanically efficient structure after primary healing
	To limit typical peak mechanical strains

to consist of making bulk fibrous tissue structures immune to fatigue failure even though they are made of fatigue-prone materials. The writer's observation in 1964 that strain rather than stress was the major factor in that regard is proving correct. Strain controls macromodeling by a feedback mechanism that can enlarge the cross-section size and stiffness of fibrous tissue structures, to limit the amount of new microdamage that will develop during usage over 1 month or 1 year. In that regard, the MES determines the difference between the trivial that the system can ignore, and the nontrivial that could threaten future structural integrity and that it must adapt to.

When for whatever reason — increased rate of production of new microdamage and/or impaired ability to repair it — that repair mechanism becomes saturated, then gross fatigue failures or ruptures can occur, and they do. Also, the local proliferation of new and archaic fibrous tissue — scar — can also become enabled, and often does.

Consequently, the new biomechanics that begins to gel out of informational chaos emphasizes strain, the MES, and fatigue, and it assigns subsidiary roles to the 1963 focus on ultimate strength and tension stress. Even though much still remains to be learned about these tissues and their IO-level functions and mechanisms, the above material represents a great change from the views and knowledge that characterized these matters in 1963.

Table 4 lists some apparent general purposes of the macromodeling mechanisms described in this and the two previous chapters.

V. ILLUSTRATIVE QUESTIONS

As in other chapters of this book, this one focuses on the properties of the IO and their relation to the organ, so the cellular, biochemical, and immunologic matters that concern the bulk of current fibrous tissue research receive little consideration here. That omission is deliberate but it does not reflect any disregard for the quality or value of such work. As the writer sees it, the problem is rather that we will not know what to do with most of that information until it is known how the cell relates to the IO. Similarly, even a complete knowledge of the properties of renal cells would remain of little use until the properties of nephrons, and how cells fit into them, were also known. While in renal matters that knowledge of the IO, the nephron, already exists, in skeletal matters its analogs have just been discovered and are still not well understood.

Consequently, this book focuses on that gap, that missing link of knowledge that concerns the skeletal IO. For that reason, the sampling of questions that follows may not really intrigue many established cellular or molecular biologists in this field, but they do pertain to the IO of fibrous tissues, they have direct clinical relevance, they must be answered, they will not go away until we have answered them, and likely those answers will have considerable bearing on some of the diseases of fibrous tissues that concern clinical medicine and surgery.

What locates fibrous tissue growth at its ends? What are the details of the process? What mechanisms and what physical, biological, and biochemical agents affect it and how? Does the process itself, and the effects of the above agents, differ in different parts of the body, in different ages, species, and sexes?

What signals enable and regulate micromodeling, macromodeling, the molecular and BMU-based turnover processes, and the repair process in these tissues? How do they work? What modifies their effects? What specific diseases arise from their particular malfunctions? Do piezoelectric and/or streaming effects have roles in the physiology of these tissues?

In different fibrous tissue structures, what are the differences in the kinds of collagen, proteoglycans, and noncollagenous proteins they contain? Or are they all the same in that respect? What is the role of the noncollagenous proteins in fibrous tissues? And the innervation and lymphatics? What kinds of interactions occur between fibrocytes and precursor cells on the one hand, and the components of the organic matrix on the other?

What are the details of the growth of the bony attachments of fibrous tissue structures? And of their muscle attachments, as shown in Figure 8? What couples the growth in length of ligaments so precisely to that of growth in length of the bone and in height of the epiphysis?

What is the value of the MES for mechanically controlled modeling in these tissues? And for micromodeling as well as macromodeling? Is it the same for tendon, ligament, and dura? Is it special in any of the above respects? How do strain rate, frequency, range, and magnitude relate to each other and to both forms of modeling of fibrous tissues? What are the fatigue limits of living fibrous tissues? What is the nature of the creep compensating mechanism? What potentiates it and what impairs it? How does it work?

How do the types of collagen in various fibrous tissue structures relate to their growth, micromodeling, macromodeling, repair, and creep compensation mechanisms?

What are the diseases of MES itself, and what are the details of the abnormal MES ranges? What separate mechanisms become prominent under chronic disuse of previously normally used fibrous tissues?

What else of importance to the domains that concern this chapter do we need to know? What does it omit out of the ignorance of the writer or because we have not yet perceived it or stumbled across it?

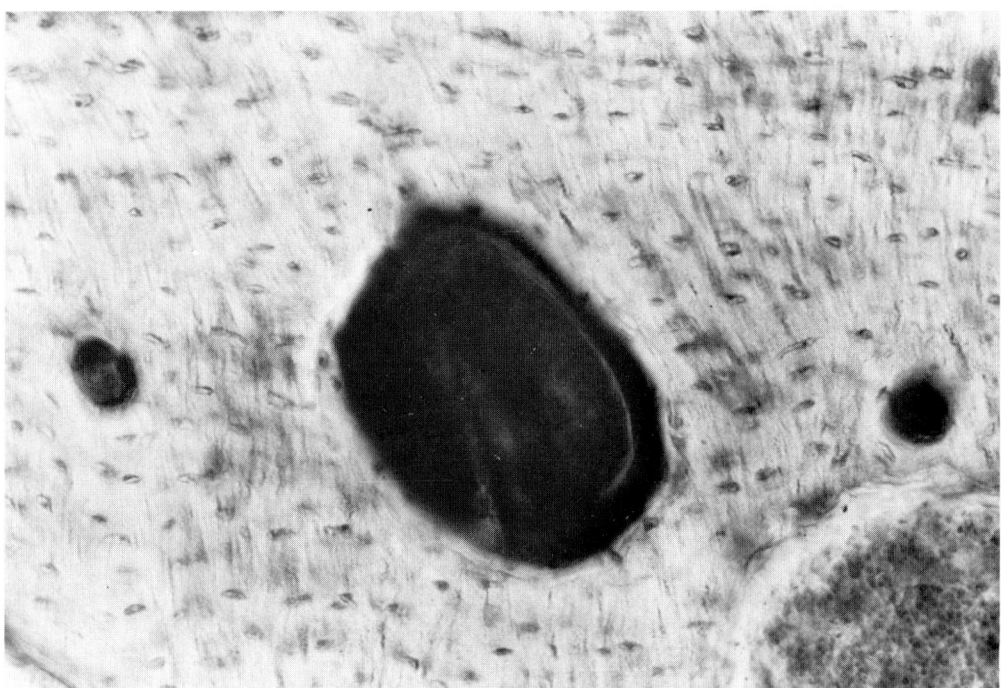

FIGURE 8. A fibrous tissue micromodeling microcosm. Undecalcified cross-section of human long bone compacta taken from a region beneath a fleshy muscle origin. About 200×, periosteum above. One newly forming secondary osteon lies in the center of the field, its haversian canal lined with a densely stained osteoid seam. Circumferential lamellae occupy the remainder of the field, plus the small canals of two primary osteons. The many dark, nearly vertical lines represent Sharpey's fibers, which are collagen bundles that run upwards to the periosteal surface and continue into the overlying muscle itself.[379,676] The fibers have the function of anchoring the muscle to the bone, and thus of transferring the muscular tension loads to the bone. Similar fibers anchor tendons, ligaments, and fascia to bone. The tension loads applied to them do not pull these fibers out because they are encrusted and embedded with mineral crystals that provide a miniature shear locking mechanism by engaging the mineralized bone around them, in essence the same mechanism that holds a threaded nut on a bolt.[241] This is a result of what this book calls micromodeling. The microanatomy and organization shown here was established during the deposition of the tissue, but precisely how, and in response to what L_1-domain IO extrinsic factors? How does a similar mechanism become established when a tendon or muscle is transferred surgically to a new bony attachment (for it does)? Without this simple but poorly studied and understood mechanism, no vertebrate could locomote, breathe, talk, write, play, or reproduce, many orthopedic surgical procedures could not succeed, and many kinds of musculoskeletal injuries could never heal in a way that restored normal function.

Note 1

THE ATTUNED PERCEPTIONS

Once his mind, his cognitive faculty, has become attuned to the significance of some previously not understood phenomenon, an observer can then quickly perceive and recognize its tracks, its tell-tales, even though the mind or eye may have run across but ignored those tracks innumerable times before and for many years. Seven examples will illustrate some aspects of this phenomenon which affect much frontline research and the general diffusion and acceptance of really new ideas. It has cropped up many times with respect to the evidence of the functions and importance of the skeletal IO, and it will probably continue to do so for some years to come. It can be frustrating, and at times hard for someone new to the research game to handle. As elsewhere in these Notes, it may help young people to handle it better if they know its name and nature, and if they realize that not all people have equal abilities to *understand* a novel way of looking at problems, even though they can accept the idea and the evidence that that way seems to work well when it is put to use. To begin, we will examine a histological example.

A. The ARF Sequence

As described in Note 2 in Chapter 4, by 1960 the author realized from clinical evidence that something connected bone resorption to formation in man, because neither in disease nor in response to prolonged treatment with drugs did human skeletons ever completely vanish due to enhanced "resorption", or keep accumulating bone without limit due to "formation". In a couple of years another idea grew out of that observation. In secondary haversian bone remodeling inside the compacta, R has to occur before F to make a hole to put the F in, and osteoclasts make that hole. What if that same sequence occurred elsewhere? Or perhaps the reverse? A check of cement lines on the various skeletal envelopes began which revealed within less than 1 week that a bone resorption process usually precedes a bone formation process. The rest has become history and Note 2 in Chapter 4 outlined some of it.

It came about partly because a nagging curiosity concerning the clinical observation had attuned cognition to some possibly relevant phenomenon, and partly because by chance the author had both the special materials and the special methods needed to reveal the explanation of the phenomenon. The author must have looked at literally millions of cementing lines before that attunement and totally failed to perceive the message they contained.

Once attuned, and then given access to the evidence, the eye and mind can then perceive both facts and meaning they previously overlooked.

B. Communicating the ARF Sequence

About 1960, a fellow from another institution came to spend 4 months in the writer's laboratory to learn the new techniques of sectioning, staining, and measurement that were involved in the then embryonic bone histomorphometry. In passing one day he was told that in human adults bone remodeling nearly always occurred in packets that coupled a following F process to an initial R process. He immediately and strongly objected to that idea, and also refused to accept the nature of the histological evidence for that conclusion as relayed to him verbally. Keep in mind that by then he had been looking at that very evidence through the microscope for most of the working day for a full 2 months.

Accordingly, he was assigned the task of classifying 500 cement lines as smooth or scalloped, and on his own selection of the slides of the hundreds of adult human bones contained in the library of the laboratory at the time.

He admitted 1 week later that the evidence was exactly as described, *but he still could not see why it meant that* A → R → F. When the matter reached the shouting stage the author gave up trying to instruct this man, who spent the next 8 or 10 years of his career telling others in all sincerity that the ARF idea was fallacious. He became known as a dedicated worker and many influential people believed him for a while, but then something happened; suddenly he began to investigate and write about the mechanism that tethered the R to the F in human bone remodeling. As an aside, some of those publications distinctly suggest that he or others discovered the phenomenon. None of them refer to the author's original published descriptions of the matter.

At least on the first exposure to the idea, not all minds can perceive in its clear telltales, the existence and meaning of a new phenomenon, and the more persistently one tries to open such a mind to that existence and meaning, the more stubbornly it may reject it.

C. Sigma, Transients, and Steady States

By 1964 enough had been learned about the properties of BMU-based lamellar bone remodeling to permit some productive speculation about some of their organ-level effects. Some of them were collected and described in 1966,[230] and again in 1973 at the symposium on Clinical Manifestations of Metabolic Bone Disease at Henry Ford Hospital[G] in Detroit. Most of the clinically apparent phenomena that had been predicted from the nascent IO theory had in fact been seen by many clinicians in the past on many occasions but their true meaning and significance had never been recognized, or even suspected. However, since that 1973 meeting they have become relatively common knowledge, and the terms "transient" and "steady state" (borrowed from Norbert Wiener) have become a part of the jargon of most people who deal today with metabolic bone disease, its research, and with some maxillofacial problems. For example, informed people now recognize that when a human subject's bone balance turns positive after receiving an estrogen, that signifies a depression of the activation of new remodeling BMU, as originally described in 1964.[230] That in turn leads to an initial deficit of new resorption centers, while previously activated BMU switch over to bone formation to keep the level of that activity at its previous rate. In a predictable period of time, the σ_r period for the resorption phase, then bone formation begins to fall also, as the previously activated BMU finish up their bone-forming stages but no new ones replace them. That represents one of a group of so-called activation effects, and Figure 9 uses a ladder graph to indicate the changes in mineralized bone mass (ΔB) and in the bone balance rate (dB) that should be observed as the result of such treatment, and that have been observed in fact in man. The final, steady-state effect of continuous administration of estrogen becomes an equal depression of both resorption and formation, and that state will not develop until one σ period after the treatment began, the σ_r value being that appropriate to the treated subject, and possibly also having an abnormally prolonged value. The drug does not affect "bone resorption" directly and unilaterally as some people still believe, but as Caputo et al., among others, have shown, does not.[109]

As one result of the above, studies that are designed and interpreted without accounting for the temporal properties of the remodeling system draw prompt criticism from referees and audiences today.

Once attuned to the basic idea, then if its evidence is widespread, and if it has a form that many can perceive and understand, many will promptly recognize and incorporate the basic idea into their thinking, even though they remained totally unaware of that meaning or its evidence before.

D. A True Clinical Episode

An elderly man in previously good health except for a prostatectomy developed sud-

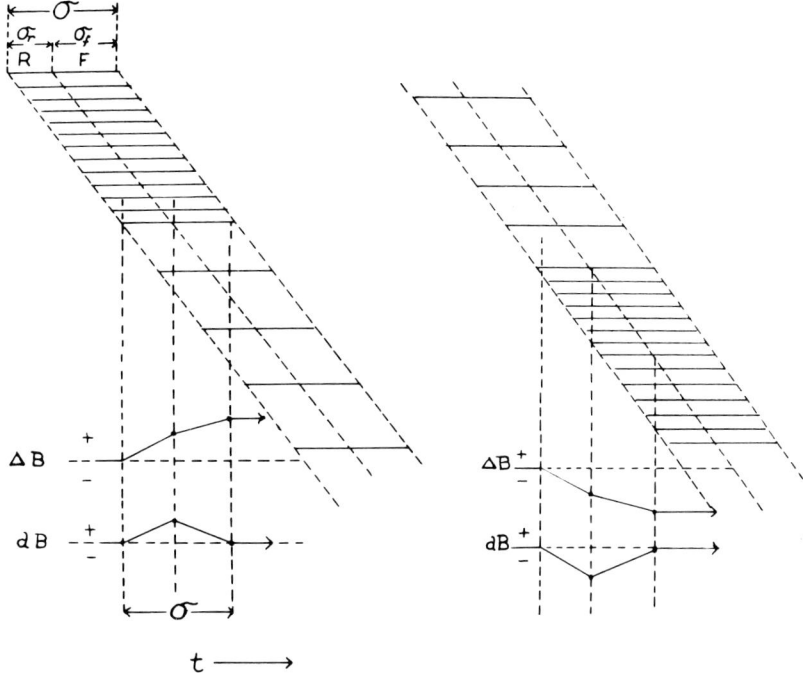

FIGURE 9. BMU activation effects in intact humans. The left-hand ladder graph diagrams the depressive effects of estrogen and certain other agents on activation of bone remodeling BMU. It leads to a net but noncontinuable small gain in bone, and to the transient changes in the rate of gain/loss shown in the dB graph below. Figures 28 and 29 in Note 2 in Chapter 6 also diagrammed these effects. The right-hand graph shows the effects of increasing BMU activation. They are the mirror image of the effects of decreasing it. Agents that can exert such effects in the intact subject probably include parathyroid and thyroid hormones, growth hormone, and some prostaglandins. However, and as indicated in a recent superb review by Parfitt,[572] the actual cell-level biochemical and biological events involved in BMU activation are probably quite complicated, and as the author suggests in Note 1 in Chapter 7, one or more kinds of "helper cells" may be involved in transmitting the primary signals for a new BMU to the responding L_2-domain remodeling unit. Note that probably few known agents can selectively affect only the activation parameter of remodeling BMU. Most known ones can also affect the σ and ΔB · BMU parameters and they also demonstrate "L"-gated effects and dose-response and duration of treatment-response effects, all of which implies MCN-mode effects, analogous to those described in Chapter 3.

den severe diarrhea, then low back pain, abdominal pain and distension, fever, and leukocytosis. He was hospitalized for further studies, and abdominal films showed much gas in the bowel and gallstones, while lumbar spine films were reported as normal except for a degenerated L4-5 disc. His urine contained leukocytes and both blood and urine cultured out *Escherichia coli*. He then developed Gram-negative septicemia and shock which responded satisfactorily to appropriate treatment. His fever decreased on antibiotics and his backache improved during the bed rest enforced by his illness, but some fever persisted, so his gallbladder was removed. It did not contain pus. He improved subjectively and briefly, but then the fever and backache recurred and blood cultures again revealed *E. coli*. By this time, 3 weeks had passed and he had been evaluated and treated by a family practitioner, two competent internists, two competent general surgeons, and a competent urologist, while his various X-rays had been reviewed several times by three competent radiologists.

At this point an orthopedist was asked to evaluate the above-mentioned disc space degeneration as a possible cause of the backache, and that evening he relayed the above history to the author, who suggested at once that the patient should have a disc space infection around the thoracolumbar junction. To end this story, the appropriate new studies proved that he did; it was treated appropriately and he recovered.

How did the writer make the diagnosis without even seeing the patient, when 10 other physicians failed? Simple. Such conditions are uncommon; in a city of $\sim$100,000 population where this happened, perhaps 1 will occur every 10 years. Because of its rarity, none of the other physicians had ever seen such a case before. For 21 years the writer practiced and taught in large medical and training centers where the referral mechanism filters out the mundane and concentrates the unusual. As a result, he had seen $\sim$30 such cases; they tend to follow a pattern, and the clinical facts abstracted above duplicated that pattern and suggested the nature of the problem. Thus, attuned perception rather than exceptional intelligence suggested the correct diagnosis.

It may take years and hard work to learn the tell-tales of a phenomenon, but once learned they can be recognized effortlessly in many new and confusing situations.

E. The On-Off Phenomenon in Lamellar Bone Remodeling

In 1960 several human adults studied in the author's laboratory had received tetracyclines 1 to 3 weeks prior to bone sampling. The resulting bone labels lay at the exact periphery of some of their osteoid seams, while in other seams in the same samples the expected amount of mineralized new bone intervened between the seams and the labels. The former seams were not unusually thick. The meaning seemed novel at the time but clear: both the lamellar bone matrix-forming and -mineralizing activities of osteoblasts could cease for a while in some bone-forming centers,[220] and also the osteoblasts had to play a direct role at least in initiating the mineralization of new osteoid. One significance of that latter conclusion lies in the widespread belief at the time — 1958 to 1968 — that that mineralization depended directly on the serum $Ca \times PO_4$ ion product, so it reflected a physical-chemical rather than a cellularly driven process. An illustrative discussion on that matter between the writer and Howard Rasmussen occurred at the osteomalacia symposium held at the Chateau d'Artigny in France in 1964.

With perceptions attuned to that On-Off idea, further human and animal material over the next 6 years or so revealed that that phenomenon, also known for a time as "Up time-Down time", arose mostly in aged humans and in certain diseases. It occurred seldom in children and in small laboratory animals, but it could also occur in the formation drifts associated with human bone macromodeling in the latter half of the growth period. As examples, in a typical symptomatic postmenopausal osteoporotic woman, $\sim$80% of her osteoid surface can be in the "Off" state at a typical moment,[385] while in chronic renal failure and renal dialysis patients $\sim$85% of the osteoid surface has been found in that "Off" state.[250]

By 1967 the accumulated evidence appeared conclusive to the writer,[251] although keep in mind that it came from a kind, volume, and variety of tetracycline-labeled bone in the slide library of his laboratory that did not exist anywhere else in the world at the time. While the basic information was clear and unequivocal, it was also arcane and inaccessible to most other authorities.

The next problem lay in informing other histomorphometrists and internists interested in such matters about the existence of that feature and some of its possible implications. Efforts to do so proved unsuccessful. One might even say they bombed. People simply would not take seriously a strange phenomenon that had groups of "A" cells switching on and off in unison in response to completely unknown factors and independently of other groups of "A" cells in the same bone section (the astronomer, Hubble, once said in essence that things we cannot even imagine, let alone accept as speculative possibilities, actually occur in natural systems). Most people would not accept the fact itself until a logical and acceptable explanation was provided for it, and no such explanation existed then, or even now. Consequently, while the On-Off phenomenon occurs regularly in many severe human diseases, it remained unstudied by anyone else until 1978. In that year, the writer devised and circulated to the σ bone group a theoretical correction for something that histomorphometrists call the "labeling escape error" that is described elsewhere,[262] and to which Smith of Salt Lake City contributed. That stimulated Melsen in Denmark, Recker in Omaha, Meunier in Lyon, and Parfitt in Detroit to evaluate that correction in their own by then sizable libraries of tissue time marked human material, for the phenomenon was important in improving the absolute accuracy of formation and resorption rates measured histomorphometrically in bone. They found that the correction procedure did work, but also that considerably more single labeled surface occurred in some of their material than the labeling escape process could account for, while an On-Off process could explain it. The definitive proof of the matter finally come from an elegant study done by Takahashi and colleagues at Niigata.[341,719] They used serial but different-colored tissue time markers deposited in vivo in adult dogs, and their histomorphometric analysis of that material, partly in press and partly published at this writing,[341,719] proved directly that "On-Off" states do exist, so the matter now begins to receive wider attention. Parenthetically, the On-Off phenomenon in lamellar bone formation also proves that the osteoblast directly initiates the mineralization of osteoid. Ironically, that fact has now become accepted but the origin of its proof has been forgotten or ignored.

The attuned perceptions do allow one to recognize a previously unrecognized phenomenon, provided he has access to the kind of material that contains its evidence and the motivation to check it out. If he lacks the access or the motivation, then his problem reduces to believing someone else's message rather than personally verifying it. Since even today only five laboratories in the world known to the author have the kind of material and expertise at hand that can demonstrate the truth of this matter, one can understand why it took about 25 years for the "On-Off" phenomenon to become respectable. Also, attuned perceptions are essential, but not alone sufficient, in spreading the knowledge of a new phenomenon. As Chapter 5 indicated, an On-Off mechanism probably appears in many other IO activities and skeletal tissues, but anyone not specifically looking for it will probably not recognize its tracks. A similar feature occurs in the skin, and also in the nephron, the efferent capillary of which can shut down or reduce blood flow to minimal levels for varying time periods, then open up again, then shut down again, and so on, with consequent effects on the function of that particular nephron. The capillary of a remodeling BMU might control the On-Off behavior of its bone formation phase (and resorption phase?) in the same manner.

F. The Closed Mind

Recently a department head asked the writer to advise a young orthopedic surgeon who planned to do some part-time research aimed at the adaptations of living bones to the mechanical and physical effects of artificial joints and fracture fixation devices. He trained at a famous institution that has a reputation in this field for energetically and promptly disseminating the original contributions of its own staff, but for slowly and seldom acknowledging those made by their competitors. This enthusiastic lad planned to do some surgical bone procedures on growing rodents, hoping that the information gained about "stress shielding" effects would help to redesign joint endoprostheses and fracture fixation devices intended for human use. Of course, the surgical procedures would have evoked a RAP, as well as changes in the bone growth and modeling of any rapidly growing animal. However, in his training this lad had never heard of the RAP or of bone modeling, so he refused to accept either their existence or the thesis that his experiments should include controls for their effects before facts that might apply to humans could be extracted from the experiments. He also refused to accept the idea that strain rather than stress might turn out to be the important factor. As far as the writer knows, he has gone ahead with his original plans, and if so, when the final report crosses the desk of some informed reviewer it will be rejected precisely because it lacks those controls.

This lad was victimized by two things that appear not infrequently in this field of research, that many more young people in it will encounter, and that already victimizes some of them. First, his prestigious teachers had successfully imbued in him the idea that since they were the best, there was little point in listening to others, and one of their own major functions was to correct the mistakes of inept outsiders. Second, this young man, like many others, including the author some 4 decades ago, lacked the experience that forcibly brings home the facts that he does not know it all yet, nor even any significant fraction of it, so that when formal training ends more still remains to learn than the previous 25 years of schooling have already crammed into the brain.

Some minds and perceptions will remain deliberately and stubbornly closed to some kinds of new facts and ideas, particularly when they come from unfamiliar or somehow threatening or otherwise unacceptable sources. Such minds cannot evaluate ideas on their own merits. Rather, they attach to an idea their evaluation of its sources. They lack the special eyes and ears that experience can lend to older people, and they also deny out of hand the existence of those eyes and ears simply because they have no personal knowledge of them. Aristotle summarized it thus for such young men: "The young think they know everything and assert it confidently." Boorstin (see Preface) put it even more aptly.

G. The krσ Property

One of the earlier students of the IO of bone, now a respected authority in his own right, started on the course that took him there in the author's old laboratory at Henry Ford Hospital. By way of background, he is of professorial stature and aristocratic descent, he speaks five languages and reads two more, and besides the intelligence and dedication revealed by those facts he is honest, warm, and generous. His work in the author's laboratory and with its material and dynamic histomorphometric techniques quickly revealed to him the reality and importance of what was later named the IO of lamellar bone, yet his prior education had planted deeply in his mind the concept that where the skeleton was concerned, cell → organ.

As one result he could not understand how tissue and higher-level bone turnover in steady states depended directly on BMU activation but not on the cell-level vigor of osteoclasts and osteoblasts. He knew about the data shown in Figure 11 in Chapter 4; in fact, he actually measured some of it, so he knew from personal observation that those facts were correct, that cell- and tissue-level turnover were independent matters. Although he tried, he could not understand why tissue-level bone turnover should directly follow BMU activation rather than the cell-level appositional rate which directly indicates cell-level bone formation. Remember that he approached that problem as a friend, not as a hostile critic of the methodology, the insights it provides, or its creator. If a friend could have such problems, imagine the reactions of a critic or enemy.

The matter is important, and many others less generously endowed than this man in the Betz cell department have had similar problems in understanding it. To explain it requires taking Samuel Johnson's advice and finding an instructive analogy in what most people already know of other things, so one such analogy follows. Table 5 lists the parameters of the analogy on the left, and on the right, what they stand for in terms of annual lamellar bone turnover in a typical human adult skeleton. However, the analogy can be adapted to many other physiological systems.

Table 5

This analog	Stands for
The nation	The typical adult bony skeleton
A person	The typical BMU or bone moiety
Growth	Histological activity of the BMU
Maturation	The end of histologic activity in the BMU
Mature body size/mass	Amount of bone in a completed BSU
A birth	Activation of a new BMU
A death	Removal of an old BSU by a new BMU
Adult life	The time between the end of the activities of the BMU and the removal of the BSU it formed
Mean lifespan	The period between the formation of a new bone packet and its removal by later remodeling

An analogy — Consider a nation with neither immigration nor emigration, so it forms a closed system. It has a reasonably stable human population of 1 million or 10^6 souls, their average body weight equals 50 kg, they grow to maturity in 20 years, and their mean total lifespan equals 60 years. Since the population stays relatively stable, birth rates equal death rates, and here "birth" shall actually mean conception, for in this wonderful land there are no miscarriages or abortions, so each conception leads to the guaranteed birth of a normal and lustily squalling infant. Since that is a true population problem, the author's 1964 population equation applies in the steady state,[229] or:

$$A = \mu\sigma \qquad (10)$$

meaning the population at any moment (A), which we already know, equals the mean normal birth rate (μ), which we do not know yet, times the mean lifespan (σ) which we do. Accordingly, $10^6 = \mu \cdot 60$.

Solving for μ by dividing both sides by 60 shows that the birth rate is 16,667, or rounded off, 1.7×10^4 annual births. To repeat, an identical number of people also die annually in this nation. Accordingly, the total annual turnover of human protoplasm due to births and deaths in that nation equals 16,667 × 50 kg, or 833,350 kg, or rounding off, 8.3×10^5 kg/year. That figure would be analogous to the annual bone turnover in a typical healthy adult human skeleton, which as a matter of interest and in gravimetric units would approximate $\sim$ 0.6 kg/year, or in volumetric units, $\sim$ 0.25 ℓ/year. Now do some simple thought experiments on this system to see how its various parameters affect it.

First, the effect of growth rate — keep everything constant in the above except the growth rate. Each person began as a fertilized ovum with negligible mass when compared to the mass of an adult, and during subsequent embryogenesis and postnatal growth each ovum eventually grew to a mass of 50 kg, which stabilized at that value at the end of growth at age 20 years. Now if each person finished growth in, say, 5 years instead of the normal 20, but still weighed 50 kg at maturity, then the individual growth rate, analogous to cell-level bone formation or the appositional rate, would be four times greater than normal. Since the body weight leveled off at 50 kg, growth ceased in one fourth of the usual 20 years. Here, the steady-state annual national turnover of human protoplasm due to births and deaths would remain at the original 8.3×10^5 kg figure. Remember, the birth rate is unchanged here at 1.7×10^4 persons per year. If, on the other hand, each person took a full 60 years to grow to adult body size, then the individual growth rate would fall to one third of normal, but it would continue three times longer than normal to attain the 50-kg value so 8.3×10^5 kg of human protoplasm would still turn over annually from births and deaths.

Thus, the growth rate alone of individuals in a steady-state population does not affect birth rates, death rates, or the turnover of the population entities themselves, given the steady state and a constant size or mass of the mature entity.

With respect to bone remodeling the above growth rate is the analog of the bone appositional rate (literally and rigorously, the radial closure rate) as measured between two tissue-time markers, while the maturation period is the analog of σ for bone remodeling packets, and the mature body weight is the analog of the amount of bone in the typical completed BMU.[229]

Now instead of changing the growth rate (which does not appear in the population equation), experiment with changes in the birth rate and lifespan.

Second, the effect of mean lifespan — hold everything else constant but vary that lifespan. If the birth rate stays constant at 1.7×10^4 births per year, but the lifespan doubles to 120 years, then the national population or A will begin to increase as new births continue, but deaths are held in abeyance until the oldest citizens attain the honorable age of 120 years. After the new σ period of that system has elapsed (120 years), its population will stabilize at a new value of 2×10^6 persons, twice the original. However, the annual turnover of people and body mass will still equal the birth or the death rate times the mass of the typical mature person, which will still be 1.7×10^4 persons supplying a total of 8.3×10^5 kg. If the lifespan halves to 30 years, then the population will immediately begin to decrease, but after the new σ value of the altered system has elapsed (30 years here) it will again stabilize, but now at the new figure of 0.5×10^6 souls, or half the original. Both birth and death rates will again become equal at the original 1.7×10^4 figure, so the turnover of human body mass will still remain at 8.3×10^5 kg/year.

Thus, changing the mean individual lifespan alone does not alter the steady-state annual turnover of the individuals in a population. The bone remodeling analog of that lifespan is the σ value of the remodeling BMU, or in different words, the time taken to finish making the typical new remodeling packet, i.e., the growth period. That packet of course represents the individual in our imaginary nation. However, changing that lifespan can change the turnover transiently during the time the system takes to attain a new steady state after one of its basic determinants has been changed.

Third, the effect of birth rate — hold everything else constant and change only the birth rate or μ. If μ is halved to 0.85×10^4, then after a transient period of 60 years (the σ value of this system) the population or A will also have reduced to half its original value or to 0.5×10^6 souls, at which time both birth and death rates will have become equal again at the new value of 0.85×10^4. The turnover of body mass will also have reduced to half its original value, or to 4.25×10^5 kg/year, because 0.85×10^4 deaths times 50 kg equals 4.25×10^5 kg. That happens because the birth rate (which equals the number of fertilized ova in this system) defines the number of new *individuals* created each year, but their initial *mass* is zero for practical purposes (compared to an adult's 50-kg mass, what is the mass of one fertilized ovum?), while each death constitutes a loss of 50 kg of living body from the total population. That 50 kg accumulated of course between conception and maturity, meaning during growth. For similar reasons, doubling the original birth rate will double the new, steady-state turnover of tissue via births and deaths to 16.6×10^5 kg/year, since that is the value of 3.4×10^4 times 50 kg.

Thus, in steady-state situations the μ function determines the turnover of individuals in any population. The μ function is the birth rate or number of new entities that enter the system in unit time. If the amount of material in the typical individual of a population is constant (as it is essentially in lamellar bone BMU) then the total *amount* of material (which is not the same thing as the number of individuals in a population) turned over annually in the system also depends directly and proportionally on the μ function.

Fourth, the effect of individual size — now make one final experiment with this system. Keep everything else constant at the original values ($\mu = 1.7 \times 10^4$, $\sigma = 60$ years, growth period = 20 years), but halve the mass of a typical adult in this straw nation to 25 kg. Then, after the new steady state has developed (which will take 60 years, for as noted in Chapter 3, the lifespan or σ value of a system equals that period of time that must elapse after challenging or changing it before the steady-state effect of that challenge can arise) the birth and death rates or μ will still remain at 1.7×10^4 persons per year, the population remains fixed at 1×10^6 persons, their mean lifespan remains steady at 60 years, but the annual turnover of human *mass* due to births and deaths has now reduced to half its original value, or to 4.15×10^5 kg/year. Doubling the typical person's original mass of 50 to 100 kg would then double that annual turnover mass, again without changing the birth rate or death rate, or the lifespan, or the total number in the nation or other pool.

Above then lies a major meaning of the "k" term in the krσ property. When the amount of material in the completed individual units of any system is constant or essentially so, then the steady-state turnover of those units directly depends on the μ function and is unaffected by the σ function. In different words, because the amount of "work" or other quantity per *completed* unit in the system is a constant or "k", the mean rate at which it is produced or "r", times the duration of the action of that rate or σ, is also a constant. If the rate goes up, then it acts over a shorter time. If the rate declines then it simply acts longer. Odd? Certainly (the fundamental paper in this respect was published by Landeros and Frost in 1964[439]). Thus, "krσ". In skeletal physiological dynamics the σ function is one uncommon but perfectly valid way to express cell-level vigor. It expresses it as the complete period of time taken to finish a "natural" unit of "work", rather than as a rate per unit of absolute time in centimeter-gram-second units.

The friend in question earlier pondered and puzzled over the above matters for over 10 years. They represent an unfamiliar way to look at turnover dynamics and system balances (of which more is said later in Note 1 in Chapter 10). Quite suddenly one day, he finally understood how that way of analyzing such problems applies to BMU-based bone remodeling. The above relationships also apply to the Frostian balances described in Note 1 in Chapter 10.

However, in most real physiologic systems, analogs of immigration and emigration do occur, for example during BMU-based turnover of bone, mineralized cartilage, or fibrous tissues, or during growth where added tissue mass is the analog of immigration. Also, other kinds of molecular-level or L_m-domain turnover can occur in many tissues. Consequently, one should try to account for such events, and in all applicable domains, both in experimental designs and in interpreting clinical and experimental data.

Even the willing and eager mind may need time to understand a novel way of analyzing common problems in spite of the fact that it can personally verify and accept the evidence that that novel way yields consistently correct predictions of the behavior of some system. As noted elsewhere in this book, things do happen in the living skeletal IO that nobody even imagined, let alone suspected, before they were perceived.

Chapter 10

THE IO AND HOMEOSTASIS

"Set down before fact as a little child, be prepared to give up every preconceived notion, follow humbly wherever and whatever abysses nature leads, or you will learn nothing."

(Thomas H. Huxley)

ABSTRACT

The skeleton contains at least three separate systems that participate in blood-bone exchange to help maintain homeostasis in the blood. They include a histological system made up of osteoclast and osteoblast populations which are formed into integrated modeling and remodeling IO units, plus a bone surface-lining cell system and a percolation system. They have different IO origins, react at different speeds, and provide widely differing total buffering capacities. They also act concurrently and probably continually.

I. INTRODUCTION

A. Preamble

The Ptolemeian concept of cosmology with its planetary, lunar, and solar cycles and epicycles became and remained the orthodox view and description of such matters until truer descriptions were offered by Copernicus, Galileo, and Kepler, and then explained by Newton. Those newer ideas contradicted long-established dogma to which even the Catholic church lent impressive support by labeling its critics as heretics and backing the label up by the Inquisition and death by fire at the stake.

That slice of history only provides another example of the natural tendency of human beings to identify with certain of their ideas, and then to perceive and react to threats to the correctness of those ideas as though they threatened or impugned instead the values, morality, and worth of the persons who held them or of the deities they worshipped. That tendency has unloosed innumerable horrors on innocent people, horrors committed in the names of justice and truth, and simply naming them can turn to their pages in history. Examples are the Crusades, Masada, the Inquisition, the Huegenots, the Gestapo, the Cheka and KGB, the "final solution", the dungeons of Iran, *los desaparecidos*.

The modern scientific community is far less corporal in such reactions, for when it does chastise its heretics it usually substitutes for the rack, the stake, and boiling oil, more subtle instruments such as ridicule, rejection, denial of credit and publication, and the yawn. Yet while that community is saner and wiser than its forbears, it can still reveal the tendency men have to assume that a truth must be the whole truth if it is genuine. So, if some particular truth can be shown to have flaws or to fail to account for some apparently pertinent fact, then there is a tendency to reject it entirely. In contemporary argot, baby is thrown out with the bath water.

This chapter deals with a class of bone-blood interactions around which a fairly rigid dogma developed after the mid 1920s. Like Ptolemy's view of cosmology, the facts that expose that dogma as inadequate surfaced long after the dogma itself solidified, so any threat to the catholic scope of the dogma now also threatens some complicated conceptual edifices that physiologists have built on it in efforts to explain diverse biochemical and clinical phenomena.[826] Those edifices occupy prominent places in most contemporary standard texts that describe the role of the bony skeleton in the homeostasis of the blood. However, they rest on a foundation that is flawed, although more by errors of omission than of commission, by assuming that a truth was therefore the whole truth. That foundation is described briefly next.

B. Pre-IO Concepts

The discovery of the relationships between parathyroid adenomas, the parathyroid hormone, serum calcium, and the calcium "reservoir" role of the skeleton in and before the 1920s[827] had much to do with the realization by physiologists that the skeleton played an important role in acid-base physiology and in regulating the concentration of calcium and certain other ions in the blood. In those years, and because hyperparathyroidism caused by an autonomously functioning parathyroid adenoma represented a newly recognized disease and a relatively rare one too, the typical patient found to have that disease had strikingly abnormal bone changes, including an osteopenia, and often vertebral crush fractures as well as other kinds of fractures, plus histopathologic evidence of increased numbers of osteoclasts in bone associated with fibrosis of juxtaosseous marrow, increased numbers of osteoblasts, and often evidence of an osteomalacia.[361,362,700]

Combine the above features with the elevated serum calcium, the elevated urinary excretion of calcium, and the obviously large reservoir of calcium in the mineralized skeleton of such patients, and combine them too with the 19th century atomistic view of the relationship of bone cells to intact bones, and the concept of the pathophysiologic mechanism of the hypercalcemia that evolved seems inevitable. It was a reasonable start, given the evidence and the methods for studying the problem available at the time.

That concept is described next.[490] The parathyroid adenoma secreted increased amounts of the parathyroid hormone, which the blood then distributed to the bones and the osteoclasts lying therein. The hormone directly stimulated the resorption of mineralized bone by those osteoclasts, and since those cells solubilized all of the various components of bone, the calcium released by that process entered the blood in increased amounts to raise the serum calcium ion concentration. The blood then took that calcium to the kidney which excreted it, and that combination of mechanisms led to the net loss of bone that occurred in most victims of the disease who were recognized as such before about 1940. The reader will recognize here the fundamental assumption described in Chapter 1 of cell → organ.

Up to about 1964, many investigations of that basic concept or of matters that related directly to it produced much diverse information that supported it. They produced little information that could have suggested that the basic concept oversimplified the problem, that while the concept contained some truth, it lacked even more.[A,L,449]

Oversimplify it, it did. Figure 1 illustrates a recent and illustrative experiment concerning parathyroid effects on the intact skeleton which was done by Takahashi and associates in Niigata, Japan for other reasons, but the experiment bears directly on the present matter by revealing how much the above dogma omitted from its considerations. This chapter will outline how the problem looks now, with the aid of newer information provided by newer methods and formed on the anvil of newer and better concepts. In reality, much more happens in the skeleton in this respect than was generally suspected even as late as 1975,[N] and it only began to become evident when the above pre-1964 views could be accepted as only part of the truth rather than all of it. Only then could we begin to sit down before further facts as little children and learn anew, as Huxley advised.

As with other chapters in this book, this one emphasizes the IO mechanisms that are

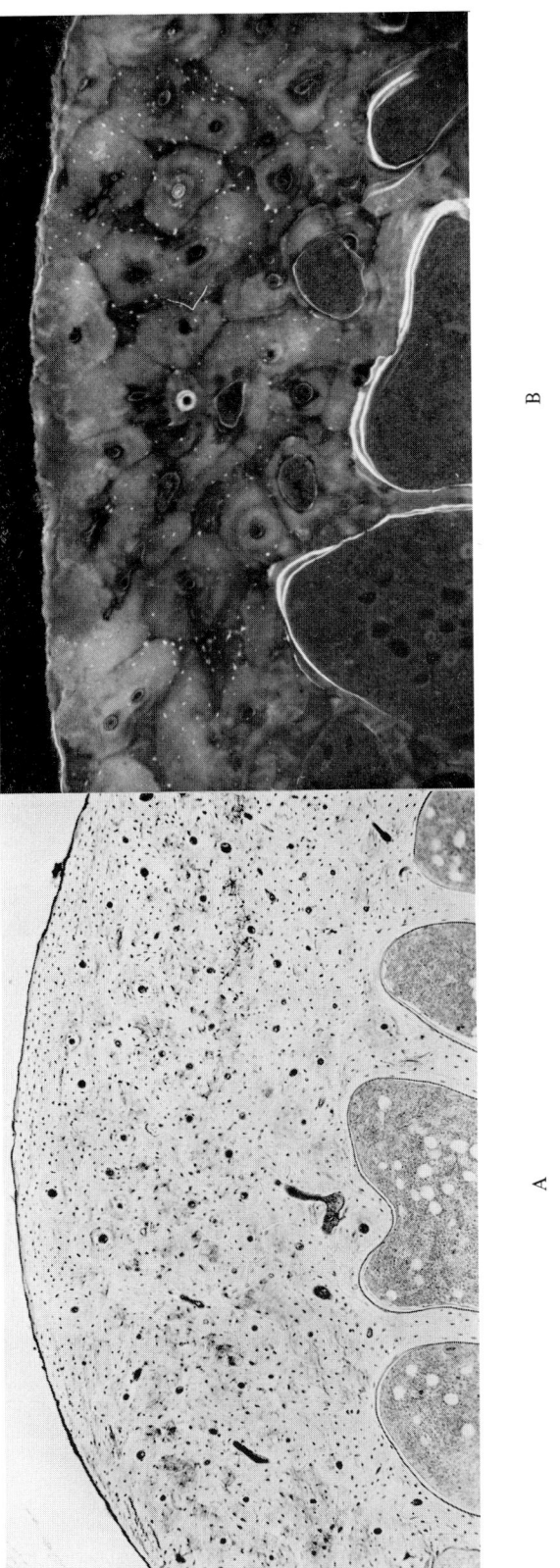

A   B

FIGURE 1. Photomicrographs of undecalcified bone biopsies of young adult dogs, both control and experimental. The transmitted light photomicrographs appear on the left and the blue-light fluorescence ones on the right. Each dog received a double tetracycline bone label at the end of the experiment, followed by the biopsy. The experimental animals received a synthetic human parathyroid hormone preparation (the 1-34 fragment) parenterally daily for 4 months. The control animals received injections of the vehicle only. The treated animals did not develop hypercalcemia or lose weight or appetite during the experiment. A and B: control dog, rib compacta, cross-section, periosteum above and marrow below, about 40X, revealing a normal intracortical porosity, one labeled secondary osteon, and three sets of labels on the cortical-endosteal envelope.

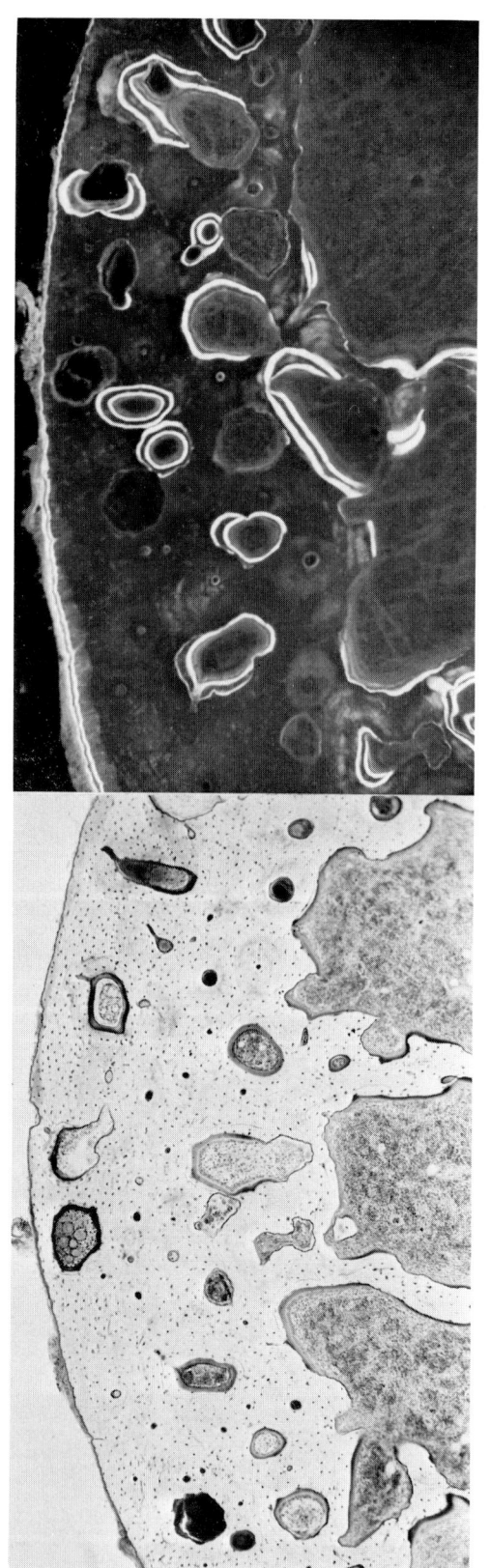

FIGURE 1, C and D. Same biopsy site of an experimental animal. Greatly increased number of actively forming new bone packets are apparent and the intracortical porosity or remodeling space has enlarged in proportion. The labels are also slightly farther apart due to an increased appositional rate. Note that the labels are a direct demonstration of bone *formation*, that in these adult dogs BMU-based remodeling is the game so each locus of formation was preceded by one of resorption, and that a slight amount of periosteal modeling rather than remodeling was enabled during the experiment (which illustrates an envelope-specific effect).

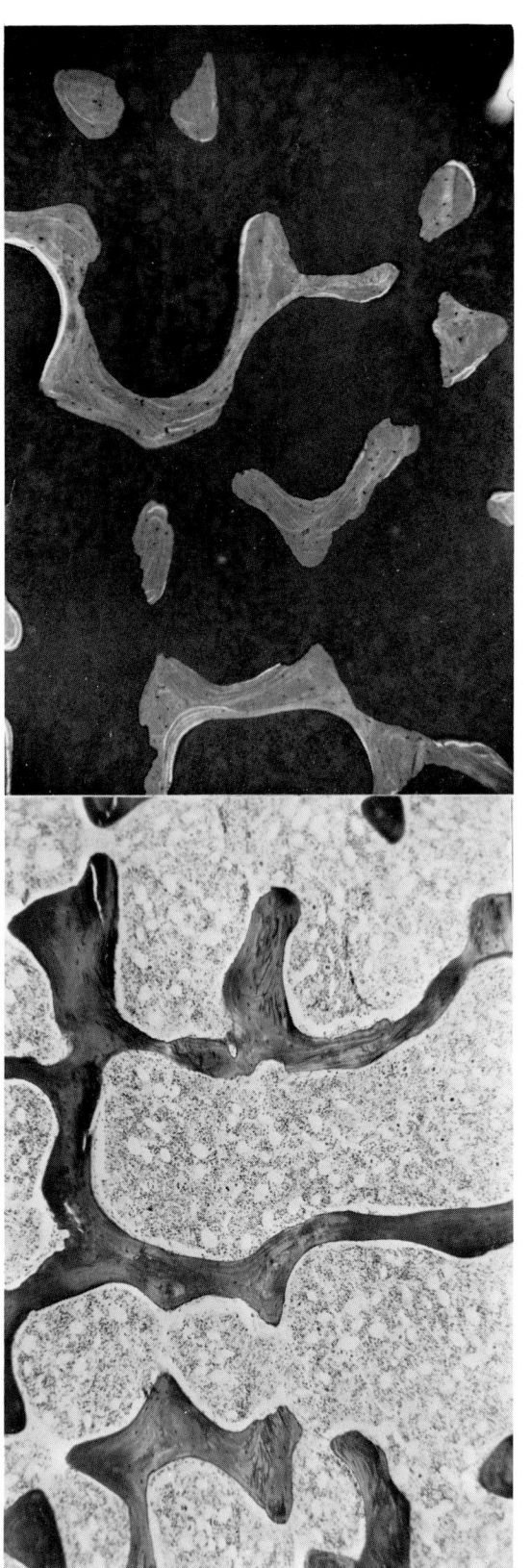

FIGURE 1, E and F. Iliac trabecular bone from a control animal.

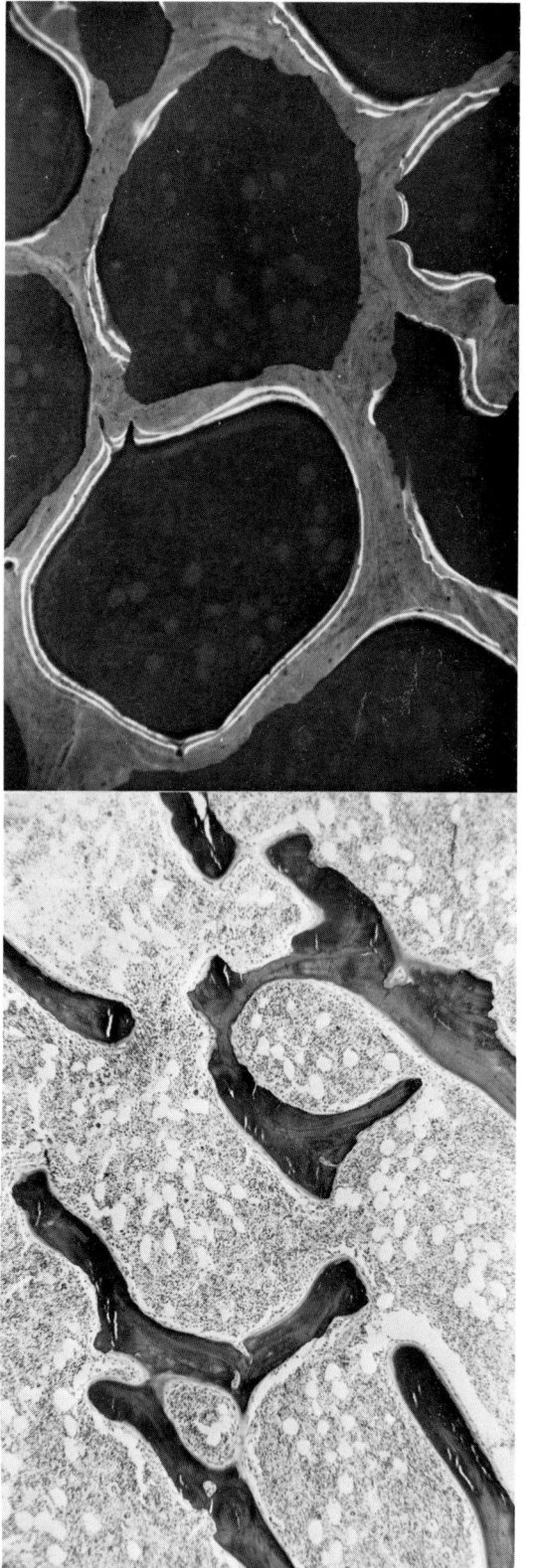

G   H

FIGURE 1, G and H. Bone from the same site of a treated animal. The same kinds of changes have occurred here as in the compacta. Note that the increased number of remodeling sites reflects an increase in BMU activation, which is an L_2-domain matter. The increased appositional rate is an L_c-domain matter, while the parallel increase in bone resorption (because we look here at an ARF-mode bone turnover process) represents an MCN effect. The effects produced by the parathyroid hormone here, whether directly or/and indirectly, include essentially equal increases in both resorption and formation at the tissue level, an increase in formation at the cell level, a reversible and small net bone loss due to an increased remodeling space, which in turn arose from an increase in BMU activation. Finally, no marrow fibrosis occurred in these animals. This elegant experiment proves directly that any interpretation of the blood-bone effects of parathyroid hormone as consequences only of an effect on existing osteoclasts grossly understates the real effects of the drug and the real responses of the living intact skeleton to it. (Photomicrographs courtesy of Drs. J. Inoue, T. Haba and H. Takahashi, Niigata).

involved and it will say little about the cellular and molecular details of the specific endocrine and biochemical mechanisms. The reason is the same as elsewhere in this book, and a brief analogy can explain it.

Blazing a trail through previously unknown territory consists of finding and marking the major features of the terrain: its larger streams, mountain ranges, plains, forests, and deserts, the most obvious features of its climate, fauna, and flora. That initial crude charting includes the relationships of such features to each other, to the compass, and the altimeter, and their approximate horizontal extents. Others who follow that blaze will then add the details, the more accurate measurements of topography, soil, climate, fauna, and flora. In doing so they will have the benefit of the record of their predecessors, by knowing that if they want major mountains they should go there, if great plains then here, and if large rivers then somewhere else.

Since this book blazes a scientific trail through the IO of the skeleton (something that has not yet been attempted by others), those who follow must add the details and corrections that better knowledge may provide.

II. THE SKELETAL ROLE IN HOMEOSTASIS

At least three separate intraskeletal systems seem to participate in controlling the homeostatic interchanges between bone and blood.[267] Those systems represent basic mechanisms of the IO that chemical, physical-chemical, and cellular processes control in order to achieve the results apparently intended by nature. Those systems differ from each other in kind, in the rapidity of their responses to challenge, in the length of time each needs to achieve a steady-state response to a challenge, and in the total amount of buffering capacity each offers to the blood, or in different words, in the sizes of their respective sink-reservoir capacities. In the process of outlining them below, a large number and variety of facts that seemed to lack any obvious connection, even in 1970, will assemble into an organized whole.

Those three systems could be named the *histological system* composed of osteoclasts, osteoblasts, and their backup activities in L_1- and L_2-space-time, the *bone surface-lining cell system* that contains a layer of flattened "lining cells" on free bone surfaces (as elsewhere in this book, "free bone surfaces" mean those not covered by cartilage, active bone resorption, or osteoid), and the *percolation system* that contains the bulk of the mineralized bone and which is riddled by osteocytes and their canaliculae.

The text will describe them in that order and then synthesize a whole from them. In reading that material, keep in mind that initial efforts to understand these matters repeated an error that has often occurred in all biological research. That is, people often tried to explain too much with too little, and seldom admitted that the existing knowledge of the system could not explain fully what was known about it. We, all of us, have repeatedly manifested a reluctance to admit the existence of "X", the still unknown but also important, and that with only rare exceptions *a* truth is not automatically the *whole* truth. That is true even though few people would contest the observation as a general remark.

A. The Histological System
This comprises the two different kinds of long-known effector or "A" cells of bone, osteoclasts and osteoblasts, and also the backup activities described in Chapters 3 and 4 that produce and continually renew them.

1. Osteoclasts
These multinucleated cells lie on a bone surface and somehow solubilize first the superficial layer of bone mineral, and then the densely collagenized organic matrix

previously shielded from enzymatic digestion by its solid-phase mineral deposits.[379] Solubilizing bone mineral requires an input of protons, e.g., of hydronium ions, from the "L" environment at the bone surface, an input that the osteoclast somehow supplies via its brush border region that lies directly against any bone surface undergoing active resorption by an osteoclast (see Parfitt in Reference 822). Those protons must ultimately come from the blood flowing through the local efferent capillary. In qualitative terms sufficient for present needs, solid bone mineral contains Mg, Ca, PO_4, CO_3, and OH ions. When solubilized it looks like this, by the addition of hydrogen ions from the blood: Mg^{++}, Ca^{++}, $HPO_4=$, HCO_3^-, and H_2O. That solubilization of bone mineral then releases its content of calcium, magnesium, phosphate, and carbonate into the adjacent extracellular fluid. Some of those ions probably pass through the overlying osteoclast — the writer suspected for years they may actually provide something the clast needs in order to do its chemical tasks — and all of them ultimately enter the local afferent capillary blood. The following relation encodes the kinds of chemical changes involved in bone resorption:

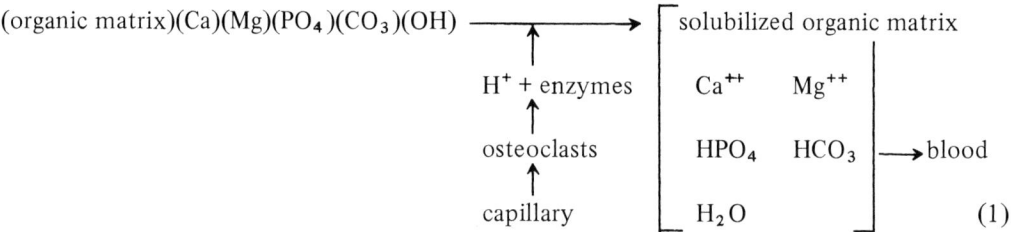

As Bonucci and Jaworski have described, an osteoclast represents a fusion of mononuclear cells that probably come directly from the circulating blood, and indirectly and possibly from the bone marrow generally, so it contains several nuclei of different ages, and while the intact clast can function as an entity for several weeks, its population of cell nuclei turns over during its lifespan as some older nuclei die or leave it and are replaced by new ones.[80,374,376] Figure 2 diagrams this situation as seen in a remodeling BMU on trabecular surface. Of course, regulatory action on an existing clast can begin to change the speed of its activities within a few minutes and clasts might attain steady state responses in L_e-space-time within 30 min or less.

In sum — An actively functioning osteoclast removes protons from the local capillary while it donates calcium, magnesium, phosphate, and carbonate to that blood.

The total number of clasts in the whole skeleton at any moment sets a limit on the total amount of bone mineral they can solubilize per hour by the above mechanism. Also, the total number of clasts in a unit amount of bone (in ABV referent) determines the total amount of bone mineral that can be solubilized per hour per unit amount of bone by that same mechanism. The bottom inset of Figure 2 diagrams those effects. Consequently, infants and young animals that have larger numbers of clasts per unit ABV of bone than adults (and in a ratio of ~100:1) can respond more vigorously than adults to agents that act on already existing clasts, e.g., calcitonin. In part for such reasons their serum calcium values can change more dramatically than those of adults to doses of a drug or hormone that appear equivalent otherwise in milligrams per kilogram body weight referent (see Note 1 in Chapter 6).

It also follows that the whole osteopenic skeleton will respond less vigorously than a normal one to such challenges, other things being equal, because in ABV referent an osteopenic skeleton by definition has fewer unit amounts of bone in it than normal.

The above description presents only one part of this story. Osteoclasts do not, repeat not, arise randomly as unassociated cells in response to systemic or local regulation, nor does a given osteoclast function indefinitely in the skeleton. With only rare excep-

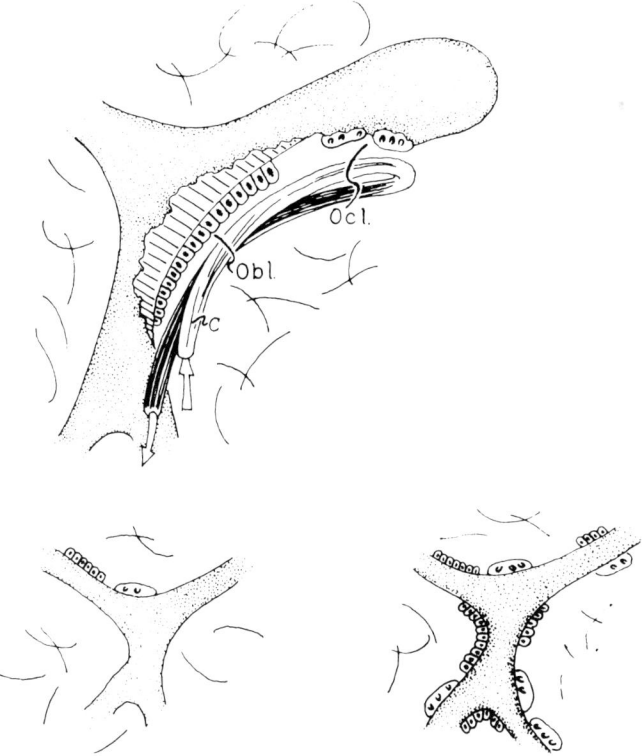

FIGURE 2. This figure diagrams the basic relationships that apply to the bone remodeling BMU as they pertain to homeostasis. We look at a trabeculum as it might appear in a bone biopsy section at ~150×. *Top:* blood enters via the capillary and flows over the BMU from right to left. It first passes over the osteoclasts (ocl), where it donates the protons needed to solubilize bone mineral and accepts the resulting dissolved mineral ions and molecules. It continues to flow over the downstream formation site on the left, where osteoblasts (obl) remove those same kinds of ions and molecules to deposit them in the new bone at the same time that the precipitation of mineral returns protons to the capillary. Thus, some of the resorbed material upstream is redeposited or reutilized locally by the downstream formation process and it does not return to the general circulation. That local form of reutilization does not occur in the bone modeling processes, for the resorption and formation drifts that produce modeling are exclusively AR or exclusively AF rather than ARF mechanisms. *Bottom:* left, in the presence of lethargic BMU activation, relatively few clasts and blasts exist in the skeleton to respond to hormones, drugs, or other agents, so such a skeleton will appear "refractory" to many agents that affect clasts and/or blasts. When BMU activation is high as on the right, (for whatever reason) the skeleton contains large populations of clasts and blasts and will prove very responsive to many such agents.

tions, clasts arise as parts of modeling IO entities in children, or as parts of some kind of remodeling BMU as described in Chapter 4 and as shown in Figure 1. If a particular challenge — say an increase in parathyroid hormone concentration in the blood — should increase the "P" or precursor cell activity that makes new osteoclasts, the resulting increased numbers of clasts will increase total bone resorption proportionally, although only after the time delay or σ period for that particular process has elapsed. That σ represents the lead time for that system, typically 1 to several days, during which the "P" response occurs, followed by any subsequent "C:D:O" effects that are needed to increase the population of "A" or clasts, to organize those cells in packets on bone surfaces, and to supply each packet with its own functioning capillary. Thus, this "cell population" effect and property of regulation of this system that was first described in 1964[229] should take 1 week or so to attain a steady-state response to a competent challenge in a child, and it might take ~2 weeks in adults.

In principle of course, such clast-active agents could also act on the other terms in the basic-state equations of the IO, e.g., on the "C:D:O" terms themselves, in ways that reduced or increased the total value of σ for the intact system and that also changed the total number of functional clasts in the skeleton. It is known that both parathyroid hormone and calcitonin do act on capillaries.[118]

In life many, if not most, agents that can affect bone resorption probably affect each of the above subsystems within L_1-domain space-time, so after a pulse dose of such an agent the observed changes in the serum calcium concentration could show variations with time that at present are unpredictable from direct actions on existing osteoclasts alone so they must be found by observation. However, as John Paul Jones said, the story has only just begun.

2. Osteoblasts

As the properties of the "On-Off" phenomenon first revealed, these cells somehow initiate the mineralization of recently formed organic bone matrix. To make the soluble ionic components of bone mineral contained in the interstitial fluid bathing all bone surfaces deposit in solid form requires removing protons, e.g., hydronium ions, from those solubilized ions and donating them to the extracellular fluid, while at the same time the mineral ions themselves are withdrawn from the blood and sequestered in the solid-phase mineral deposits, probably as a defective calcium polyphosphate that then converts to a defective hydroxyapatite that contains calcium, magnesium, phosphate, carbonate, and hydroxyl groups. As the reader will have noted, those deposition processes reverse, and probably almost exactly, the mineral ion processes involved in resorbing mineralized bone. Accordingly, the complement to Relation 1 might be written thus:

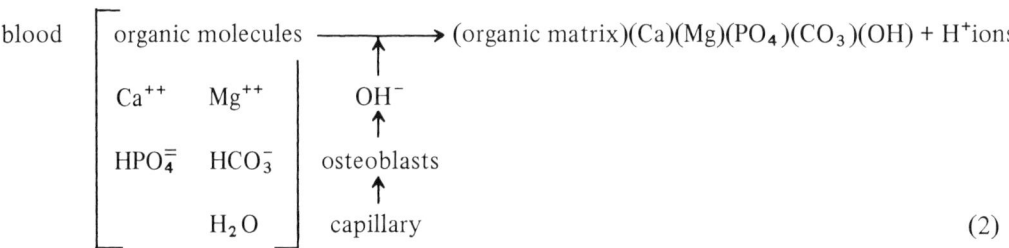

(2)

Given the above, then the same possible control responses that apply to the clast system apply also to the blast system, i.e., regulation can act directly on the existing blasts[395] and/or on their backup terms.[267] The former effect could appear within minutes of an effective challenge while the latter might take ~2 months (the normal value of σ_f in man) to achieve a steady-state response to a challenge.[267] The features illustrated in Figure 1 reveal that the parathyroid hormone treatment did change the activity of existing osteoblasts in the intact subject; it invigorated them, thereby potentiating the homeostatic effects encoded in Relation 2. That provides direct evidence, and in essence proof, that trying to explain a homeostatic response as a result of osteoclast activity and regulation alone is a fallacy. There is also now good evidence that the same kind of complex response to this hormone occurs in man.[324] A major apparent difference between the clast and blast systems constitutes the facts that increasing blast vigor increases the yield to the blood of protons at the same time that it removes the mineral ions named above from the blood; decreasing blast vigor has the opposite effects. To repeat, the steady-state response of the existing blast system may require ~2 months to develop, or some fourfold or more longer than for the clast system.

3. The Remodeling BMU

When the clast and blast activities couple in the form of remodeling BMU, as in Figure 2, an additional property of the system becomes apparent that was already known to Rowland, among others, in 1964.[652] It relates to the fact that the same capillary usually supplies both the R and F stages of a remodeling BMU. The efferent limb of the capillary supplies the resorption activity first, and afterwards its downstream afferent limb then supplies the formation activity.

Accordingly, the capillary supplying the resorption front loses protons but gains the mineral ions, and as it flows more slowly over the usually longer and larger formation surface it then regains those protons and returns those mineral ions and cations to the actively mineralizing new bone. Figure 2 diagrammed that phenomenon. Since the local perfusion is probably relatively rapid,[344,345,563] the differences in pH and in mineral ion concentrations at the efferent and afferent ends of the capillary are probably small. In different words, of any 100 protons brought to a resorption process by the capillary, normally perhaps 5 or fewer are removed by the resorption activity and the other 95 or more pass right on through. The same effect probably occurs in the formation process as well. As one result of the above relationships, the blood actually returned to the general circulation by the local capillary should contain proton and mineral ion excesses or deficits that relate directly to the value of the local $\Delta B \cdot$ BMU parameter.[267,658] If that parameter has a value of zero, then the only difference between the composition of the blood at the beginning of the efferent limb and at the end of the afferent limb should be those caused by the metabolic activities of the various cells supplied in between those points. When the $\Delta B \cdot$ BMU has a negative value, as on the normal cortical-endosteal and trabecular surfaces, meaning when somewhat more bone is resorbed than formed per completed BMU, then the end of the afferent loop will contain a corresponding deficit of protons, the same thing as an excess of base, and a corresponding surfeit of the mineral ions. When the $\Delta B \cdot$ BMU is positive, as on the normal periosteal envelope, the opposites occur.

Clearly, that $\Delta B \cdot$ BMU-related phenomenon could serve to buffer chronic metabolic acidoses and alkaloses, and it may be one of the mechanisms responsible for the observed tendency of the skeleton to become osteopenic in states of chronic metabolic acidosis.[267] That also suggests that both generalized and local metabolic acidosis or alkalosis may represent L_2-domain IO-extrinsic factors that can influence the value of the $\Delta B \cdot$ BMU, meaning that they might control the bone balance of the BMU-based bone remodeling process.[658] It also suggests special histomorphometric studies to evaluate the ideas of investigators such as Barzel, who proposed that systemic acid-base phenomena might participate both in the pathogenesis of some osteoporoses and in their prevention and treatment.[59,449] The coherence treatment stratagem described in Chapter 5 might transform his ideas into a practical therapeutic reality.

The contributions of blasts and of the intact remodeling BMU to the homeostasis of the blood have received little attention so far, but as Figure 1 reveals, they must be considered in the future. Most of the study of the histological homeostatic system has focused on the atomistically conceived roles of already existing and unassociated osteoclasts, and to the virtual exclusion of the backup activities encoded in the basic state equations of the IO, and of their complementary osteoblast activities.[828]

In addition to the above little-discussed but quite real homeostatic properties, it should be clear that at any one moment in a typical incoherent adult human skeleton, one population of BMU exists that engages in resorption, and another in formation, as shown in the ladder graphs in Chapters 5 and 6, because each member of those populations began at a different time from all of the others.[267] Accordingly, in the incoherent skeleton a regulatory agent will affect all the different BMU terms encoded by Relation 19 in Chapter 4 at the same time, although in different individual BMU.

If that agent is present continuously rather than as a transient pulse, then each term of every BMU activated after treatment with that agent began would also come under its influence, as described in Chapter 5.

It follows from the above that the possible true effects of a given regulatory challenge on the homeostatic functions of this system are too numerous to predict accurately from the fragmentary information on hand at present.[658] The experiment by Takahashi's group shown in Figure 1 tends to prove the point, for if the classical atomistic view of parathyroid hormone action on bone were correct (Relations 10 and 11 in Chapter 3 encoded that view), then bone would steadily disappear on *all* bone envelopes, which would become totally covered with Howship's lacunae without any change in bone formation at the cell or tissue level; however, no such things happened in the Takahashi study. Accordingly, the true responses of the intact system must be observed in the living subject, and since few people have attempted to do that so far (they include High in the U.S.,[327-331] Takahashi in Japan[614-621] and colleagues), all extant hypotheses of how this system contributes to the steady-state homeostatic needs of the living organism represent unproven speculations. One might encode the above facts in this manner:

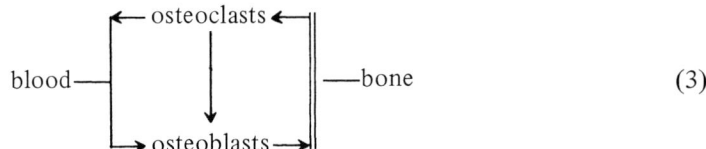 (3)

Note: In terms of its total capacity, the above system could draw in principle on the entire mineral burden in the skeleton, and as many analysts have noted, in relation to any moment-to-moment needs of the blood that burden represents an immense sink or reservoir. However, to draw on the whole skeletal mineral burden would take a number of years at minimum, for the cellular activities that are involved are relatively slow and so is the $\Delta B \cdot BMU$ mechanism itself, for it normally turns over approximately 29 parts of bone for every 1 part that it removes permanently.[263,267] While the sink/reservoir capacity of this system is immense when compared to the size of the sink/reservoir systems found in the circulating blood itself at any moment, it can usually be drawn on or added to rather slowly.

It will be interesting to see how this system really works when appropriate information on the matter becomes available from properly designed experiments. Attention now turns to a quite different but concurrently acting system.

B. The Bone Surface-Lining Cell System

Both compacta and spongiosa expose their free surfaces to juxtaposed soft tissues, which are "L" agents according to the notation used in this book. Excepting surfaces covered by cartilage, and possibly some kinds of soft tissue coverings of periosteal surfaces, and also excepting surfaces covered by actively evolving bone resorption or formation processes, a layer of flattened cells covers the remaining bone surfaces. It is called the *surface-lining cell layer* and it covers the free adult bone surfaces incompletely, but those of infants and animals of equivalent age reasonably completely. Those cells somehow oversee a turnover of mineral in the underlying bone to a depth of only a few microns.[238,571,572] In the process, those cells also oversee the exchange of the mineral in that shallow bone layer wth the adjacent extracellular fluid and blood.[544,545] Figure 3 diagrams the situation.

As for the behavior of that surface, numerous autoradiographic studies of β-emitting

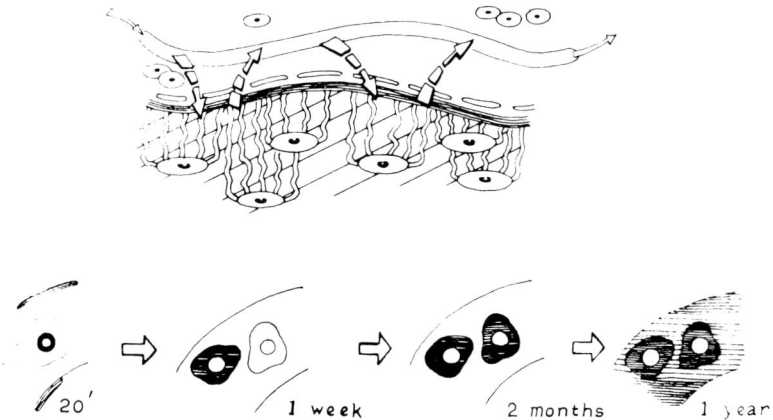

FIGURE 3. *Top:* a diagram of a typical bone surface lined with flattened "lining" cells. Bone with osteocytes below, capillary above. A bone-seeking tracer, e.g., radiocalcium, enters via the capillary, leaves it, and deposits in the thin surface layer of bone, darkened here. Then, 20 min or so later, that same tracer disappears, presumably by reentering the capillary and thereby the general circulation. *Bottom:* these diagrams show the time-related changes in distribution of a bolus of ^{45}Ca injected into the general circulation as they would appear in a typical cross-section of compacta, demonstrated by autoradiography (^{45}Ca is a β-emitter). It appears first on thin surface laminae, but on only a fraction of the total free bone surfaces, and then disappears. A few osteons take it up 1 week or so later; in a couple of months more of them have; in 1 year most of them have, and the interstitial lamellae have become labeled.

bone-seeking isotopes, including ^{45}Ca, have revealed that within minutes of the injection of a bolus of the isotope into the general circulation, a significant fraction of that free surface has taken up the isotope intensely, and to a depth of a few microns as mentioned.[238] Then, 1 or more hours later, that isotope has disappeared nearly completely from those locations, although in the same bone and in the same time period the loci of active new bone formation have accumulated progressively increasing amounts of the isotope, and continue to do so. Those phenomena do not occur in dead bone.[622]

This rapidly turning over, thin bone surface layer contains ∼1% of the total skeletal calcium, and the active turnover in it that was just described does not seem to occur in the absence of the flattened surface lining cells. It provides a rapidly exchangeable pool of bone mineral and buffer that can apparently react to a homeostatic challenge within a few minutes. However, it has a relatively small sink/reservoir capacity that can fill or deplete within perhaps 1 hr or so. A number of studies, particularly by Talmage, Matthews and colleagues,[482,571] have shown that its overlying, flattened cells can respond quickly and sensitively to parathyroid hormone, calcitonin, and other blood-borne agents.

C. The Percolation System

Most authorities have tended to view the mineralized fraction of bone as a highly impermeable material that contains and is riddled only by the voids that contain osteocytes and their canalicular processes: the osteocyte lacunae and the canaliculae.[297,319,643,761] Some authorities have speculated that the walls of those voids provide another surface at which significant bone-blood exchange occurs, and there is probably some truth to that view, which has been proposed at various times by Baud,[63] Belanger and Wassermann,[761] Robinson,[643] and Heller-Steinberg,[319] as well as Frost.[222] Some basic facts suggest an even larger picture, first proposed and described in 1971 by Arnold et al.[40] The following paragraphs will record some of the basic facts first, and then their synthesis, for that minimodel of the analytical process illustrates how an understanding can gel from once apparently unrelated facts.

As noted in Chapter 7, excluding its vascular spaces — haversian and Volkmann's canals — fully mineralized healthy lamellar bone contains by volume about 45% mineral, 40% organic matrix, and 15% water. Of that 15% water, about 12% fills the histological voids comprising the osteocyte lacunae and their connecting canaliculae.[344] The remaining 3% lies within the mineralized tissue but outside of the osteocyte lacunar and canalicular lumens. The author calls it the *interstitial bone fluid* and Arnold has shown that most of it can be removed by simple centrifugation,[36,37] which means it is not chemically bound to the mineralized bone.[42] Studies and a recent analysis of those matters by Johnson[822] indicate that some theoretical reasons as well as hard evidence do exist to believe this interstitial fluid is mobile as described here and that the mechanical bone strains caused by normal mechanical usage could be another factor (besides the one postulated later in this chapter) in promoting flow of that fluid. However, that latter idea was first proposed by the author[238] and Arnold et al.[40] in 1971 and 1973 on the basis of the evidence reported below. Figure 19 in Chapter 7 graphed the above volume distributions. In incompletely mineralized bone, the amount of organic matrix and the osteocyte-lacunar-canalicular volumes do not change. Rather, additional interstitial water fills any space in the organic matrix that is unoccupied by solid-phase mineral deposits. As a result, the interstitial bone fluid volume can be as large as ∼45% of the total volume of unmineralized osteoid, and it steadily shrinks during mineralization to the above-described ∼3% limit as mineralization progresses, observing that that mineralization can take 1 year or so to become complete (the stiffness lag described in Chapter 7). Should the osteocytes in that bone die, then much of that remaining fluid will also become displaced by mineral and a region of micropetrosis will develop.[226] Figure 4, lower right, provides an example.

Isotope distribution and exchange studies have shown that the water in living bone can exchange completely with the blood in ∼20 min.[238] Similar studies show that even after 20 years some of the bone calcium and phosphate still have not exchanged with the blood.[297]

After parenteral injection of a bolus of ^{45}Ca, and as noted earlier, a progressive and little-reversible uptake occurs in new bone-forming centers, and a separate uptake followed by a rapid clearing occurs at some of the free and histologically inactive resting surfaces.[34] Initially no significant accumulation of isotope is apparent in the bone beneath those surfaces. Later — in days rather than hours — autoradiographs of cross-sections of diaphyses show that certain secondary osteons begin to show a diffuse label of the calcium, but without any noticeable gradient that would relate to the location of the haversian canal, the cement line, or the resident osteocytes. As the time between injection of the label and skeletal sampling increases, more and more such osteons, and then more of the interstitial lamellae between them, also begin to show diffuse labels. In time — after several years — most of the bone will finally show a faint, diffuse label.[238] Such phenomena have been described and discussed by Groer and Marshall,[297] and by Rowland[652] and Figure 3, bottom, diagrams them.

Between 1956 and 1964 the writer became interested in and studied the diffusion of varied ions in fresh, wet human bone, using a variety of ex vivo and in vivo techniques. Most of those experiments were never reported because the in vivo tetracycline labeling excitement had developed and learning to use it and to understand what it revealed had overwhelmed and gradually supplanted most other interests. The results of the diffusion experiments were informative and consistent, and although they remain arcane, they apply directly to the present matter.[225,267]

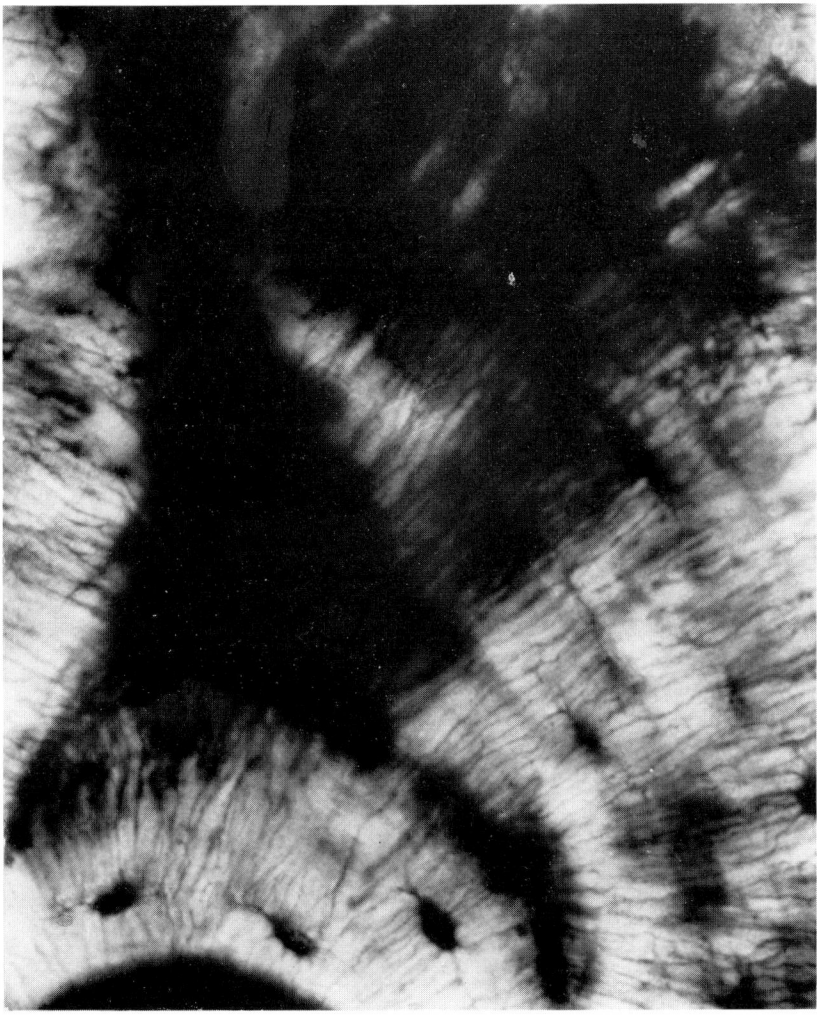

A

FIGURE 4. (A) Undecalcified cross-section of a human secondary osteon sectioned and stained by the writer's techniques,[217,218] about 400×. The patient had secondary hyperparathyroidism. The dark oval blurs are osteocytes and the fine lines radiating up and to the right are canaliculae. The dark blur extending upwards is a region of feathered bone: incompletely mineralized and thus permeable to the dye molecules. (B) A longitudinal undecalcified section of fresh human compacta treated briefly with a hot aqueous, dilute permanganate solution, about 900×. Each oval blur is a canaliculus. The longitudinal axis runs from the upper left to the lower right, and the elongation of the stained canalicular halos in cross-section along that axis reveals greater diffusion along it than perpendicularly to that axis. (C) Undecalcified cross-section of human compacta, stained with basic fuchsin, about 100×. The patient had a chronic metabolic acidosis. The dark regions represent feathered bone, some involving secondary osteons and some in the interstitial lamellae between the osteons. These low-density, feathered regions extend for 1 mm or so along the longitudinal axis of the bone. (D) In contrast to the left photomicrograph, this one, at about 200× is of normal compacta from a healthy but elderly man. No low-density bone is present; however, the clear areas in parts of the interstitial lamellae are dead micropetrotic bone as described in 1960. All of the osteocytes in it have died and their lacunae and canaliculae have become completely filled with solid mineral.

In brief, both simple anions and simple cations will diffuse passively into fresh, wet human bone in vitro.[222] As expected, they can diffuse through the canalicular and lacunar network, although quite slowly in vitro,[218] probably because the cytoplasm of

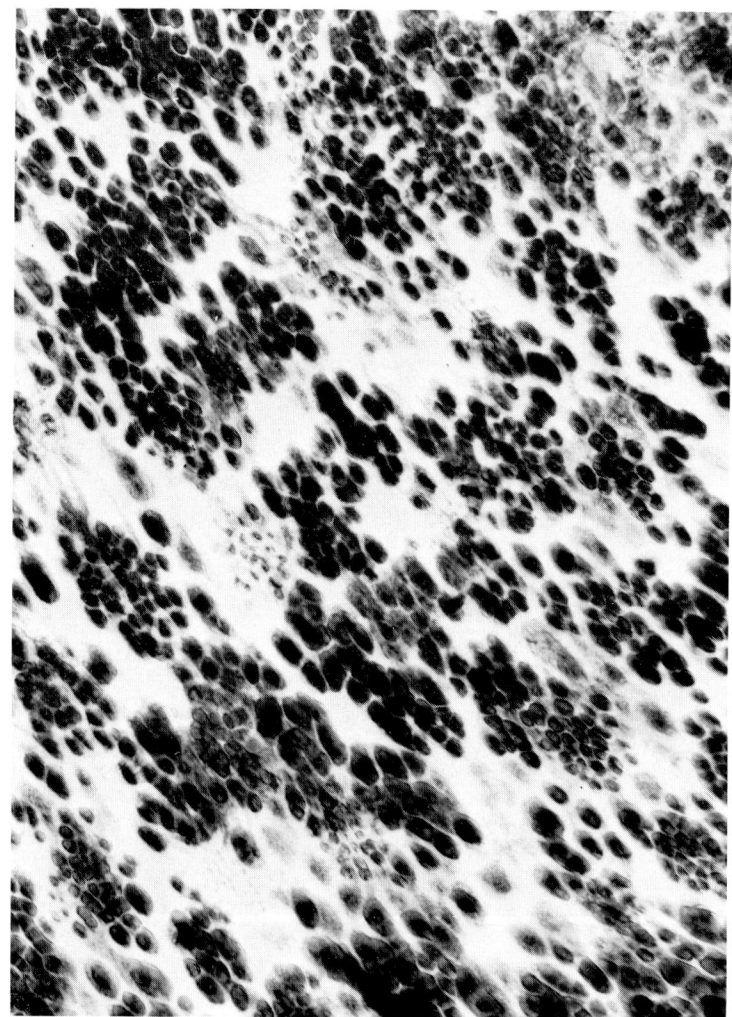

FIGURE 4B

the recently expired osteocytes forms a barrier to free diffusion in the spaces they occupy. The voids inside the canaliculae and osteocyte lacunae are quite small, for a typical canalicular diameter in human lamellar bone would approximate 350 to 400 nm or 0.35 to 0.4 μm. Figures 5 and 6 illustrate those spaces. In 24-hr periods, those ions could also diffuse several microns into the pericanalicular and perilacunar bone itself, and more readily parallel to the grain of the bone than across it.[267] In vitro, that diffusion occurred in bone that was alive in the patient, and it did not occur in bone that had been dead in vivo, as in micropetrotic bone and in sequestrae.[267] Figure 4, top right, provides one illustration of the effect.

A molecular sieve effect was also clearly present, in that if given time the larger molecules of numerous dyes used in histological work could enter the lacunar and canalicular network itself, but not the "fully" mineralized bone around them, even though the much smaller mineral ions could do so.[267] Figures 5, 6, and 7 illustrate that effect.

George Vose, among others, showed by electron microscopy in the 1960s that numerous fine clefts of ultramicroscopic dimensions can run longitudinally through la-

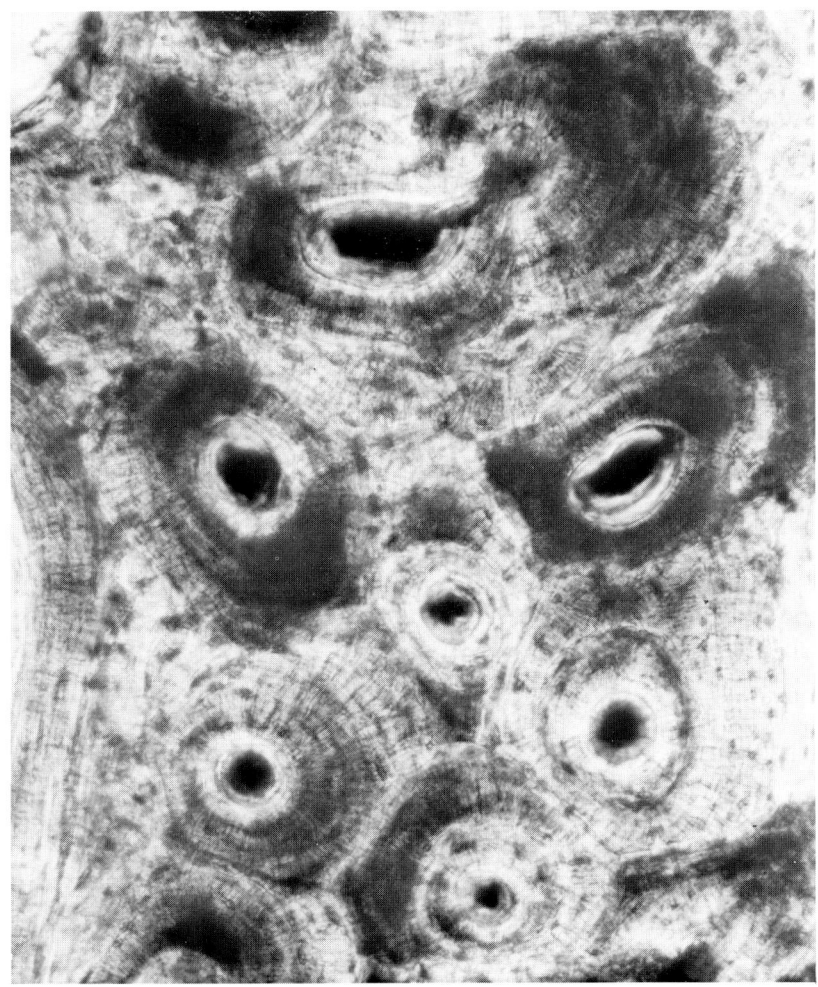

FIGURE 4C

mellar bone and tend to radiate outwards from individual canaliculae. The ionic diffusion the writer observed probably occurred at least partly along those clefts, which according to their diffusion seen in the light microscope had lengths on the same order as the typical diameter of a canaliculus, or in some cases the longer distance along the longitudinal bone axis between one canaliculum and its nearest neighbor.

Both personal studies and a number of microradiographic studies by others have shown that in so-called mature bone, different degrees of mineralization can occur, and some of those differences can relate in the geographic sense to the location of the local osteocytes while others do not.[224,397] Both tend to associate with certain chronic disease states. Those mineralization abnormalities comprise the *halo volume*[222] and *feathering*[274] phenomena described by Frost around 1960, and observed by a number of others since including Laval-Jeantet in France, Van der SluysVeer in Holland, and Glorieux in Canada. However, they still await systematic study and correlation with disease and known chronic homeostatic disorders. The important fact here is that they do not associate in any obvious way with histological new bone formation or resorption. Figure 4, left top and bottom, illustrates two examples.

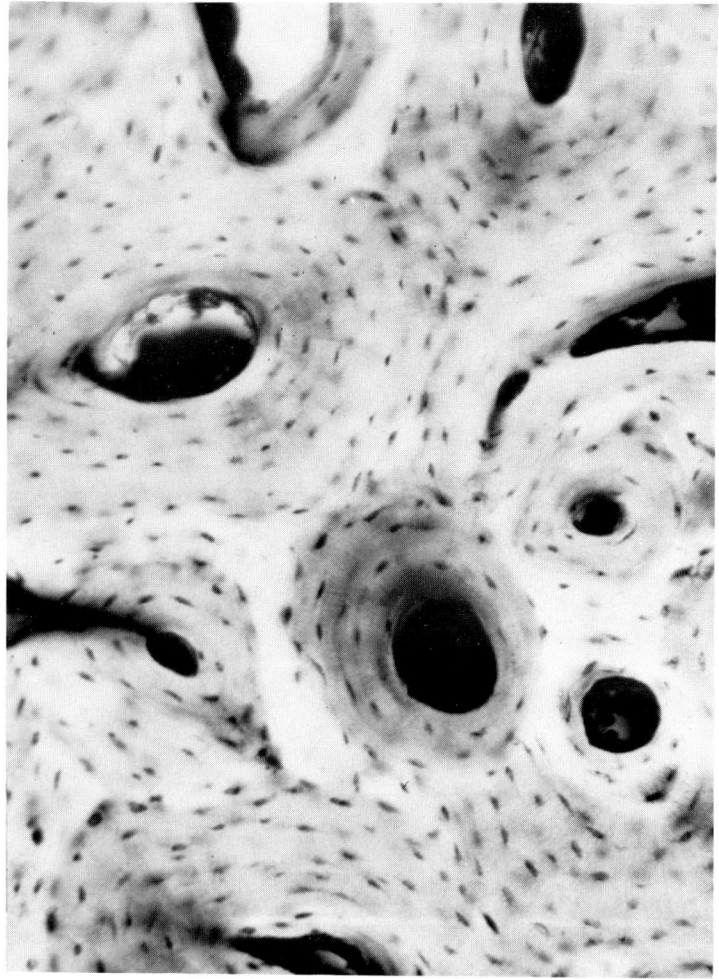

FIGURE 4D

When bone biopsies are taken from a person who still has tetracycline antibiotic circulating in the blood, whole groups of osteocyte lacunae and their canaliculae can show a tetracycline label on their bony walls, while adjacent groups of osteocytes and canaliculae can show little or no such label. When the bone biopsy is deferred until 2 or more days after the last administration of the antibiotic to the patient, then such lacunar and canalicular wall labeling rarely persists.[232] That means it is brief and quickly reversible and it implies an active circulation of extracellular fluid into and out of those regions. Also, Harris (in a personal communication) as well as the author noted before 1965 that in subjects given a 1- or 2-day tetracycline bone label, low-density feathered bone could take up a diffuse tetracycline label in vivo, as well as the basic fuchsin molecule in vitro, and that in vivo label showed no evidence of concentration around osteocyte lacunae. It was truly diffuse in the bone lying between lacunae and canaliculae. Other investigators subsequently observed the same phenomenon independently. That proves that an extracanalicular-extralacunar circulation of fluid occurred in that bone in life.[225,267]

The hagfish provides an example of a vertebrate that lacks living osteocytes within

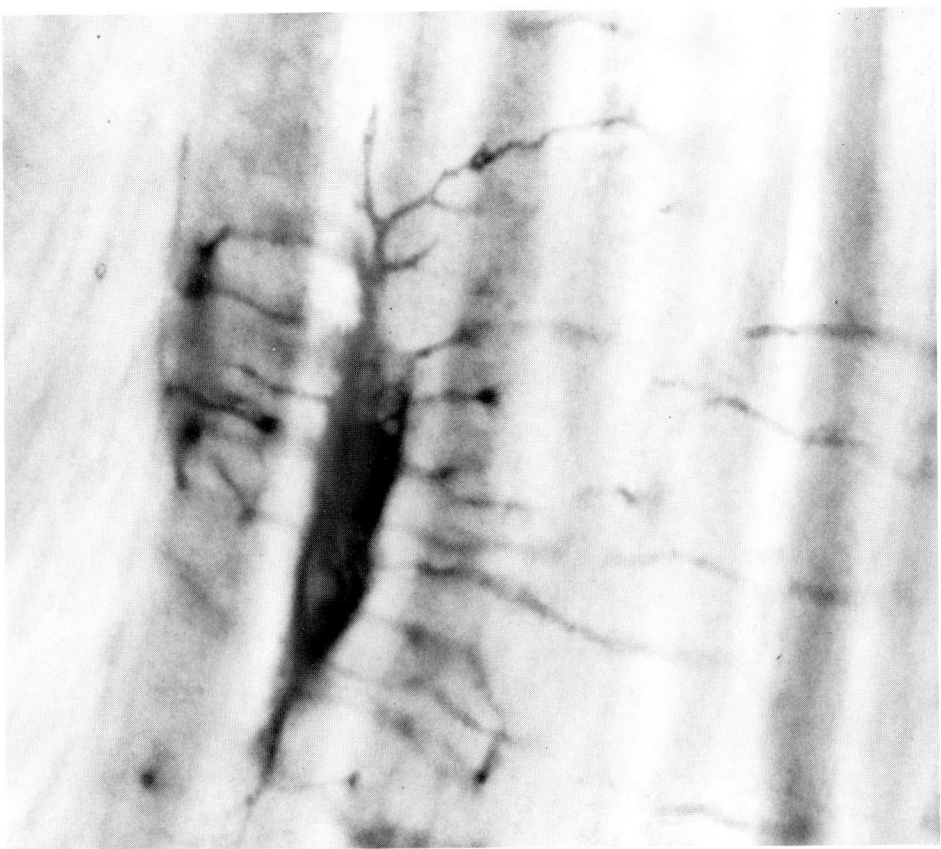

FIGURE 5. Undecalcified human bone, cross-section, Villanueva stain, about 1200×. A single osteocyte lacuna lies in the field, its spider-like but hollow canaliculae radiating to the right and left. Given the approximate lacunar dimensions of 12 × 35 μm, the small diameters of the canaliculae can be appreciated; they are less than the wavelength of most visible light.

its bone, and tracer studies in vivo have shown that its bones also lack the capacity to form the diffuse deposits of radiocalcium that can occur in mammals.

1. Interpretation

This melding of the above evidence into a whole comes in parts,[267,658] which are discussed in the following paragraphs.

At any given moment, a particular bone domain exists, probably supplied by one capillary and including the osteocytes and lining cells supplied by that capillary. That domain is "On" at the moment, as a result of which its osteocytes perform some function that actively moves or draws interstitial bone fluid into the lacunae and canaliculae from the bone that surrounds them. The fluid then returns via the canaliculae and the free bone surface to the local capillary. The interstitial fluid came from and travelled through the extracanalicular bone, and it came ultimately from other bone domains that were adjacent to the "On" domain along the longitudinal grain of the bone. It entered those latter domains from their nearest supply of extracellular, but also extraosseous, fluid, and ultimately from their own local capillary.[40]

In other words, the interstitial bone fluid *percolates* through the bone, much as ground water percolates through the earth, probably in part at least via the aforesaid

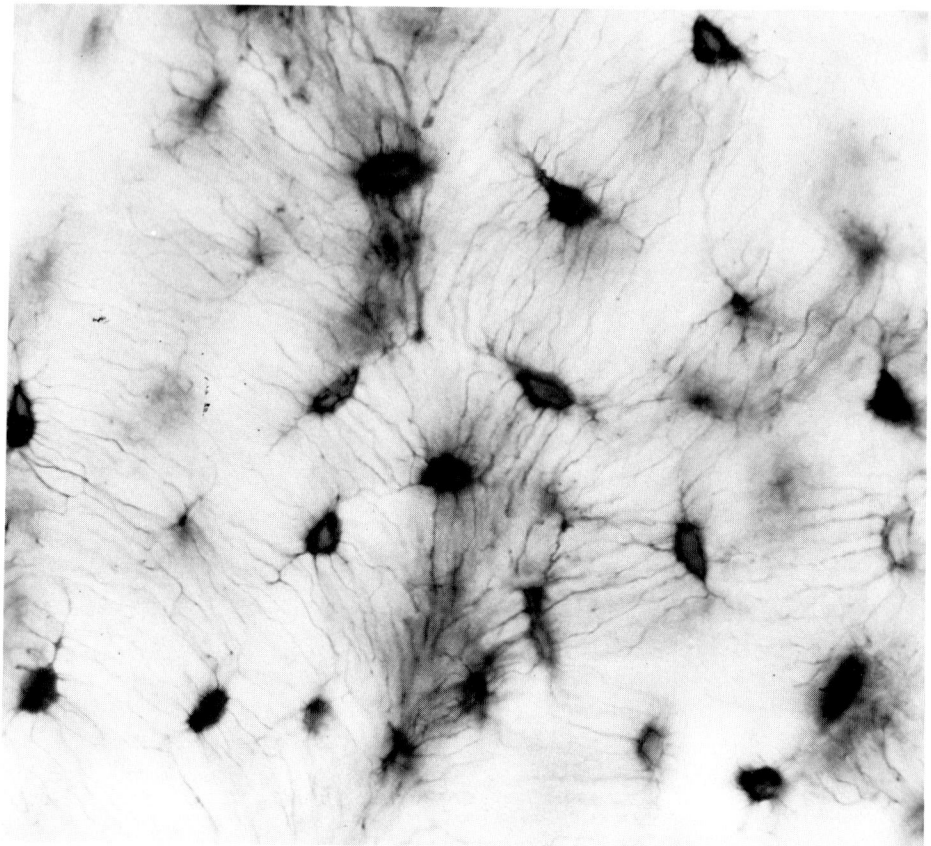

FIGURE 6. Undecalcified cross-section of human lamellar bone, basic fuchsin, transmitted light, about 600×, taken with a red-minus filter to enhance contrast. The canal of this upper quadrant of a secondary osteon lies below the field at 6 o'clock. The fuchsin has permeated and stained the walls of the spaces of the osteocytes and their lacunae, but the dye molecules could not enter the mineralized matrix due to a molecular sieve effect. Consequently, that matrix appears clear. For reference, the major dimension of the osteocyte lacunae here is ~15 to 20 μm.

ultramicroscopic clefts.[267] In that process it becomes exposed to a vast surface of bone mineral, and therefore to a significant fraction of the total mineral burden in the skeleton. During percolation the ions dissolved in the percolating fluid can exchange with the solid-phase mineral, an exchange process that may be mostly physical-chemical rather than vital.

At a later day or week, the original "On" domain then changes to the "Off" state, while its adjacent neighbors switch to the "On" state so the direction of the above percolation process reverses.[267] Over 1 year, most of those domains in healthy intact bone could spend approximately equal time periods in each state. Figure 8 diagrams how this mechanism could work.

As the interstitial fluid percolates in that manner, its ions and water can exchange with ions and water on the surfaces of the mineral deposits that are readily accessible to the fluid. However, in any one "pass" through the percolation bed only a small fraction of, e.g., the calcium in the interstitial fluid, will exchange with the mineral deposits it flows between. For every 100 radiocalcium atoms that might enter the percolation bed, perhaps 1 to 5 of them exchange with stable calcium in the solid mineral deposits in a single pass through the bed, and the remaining 95 to 99 atoms pass on

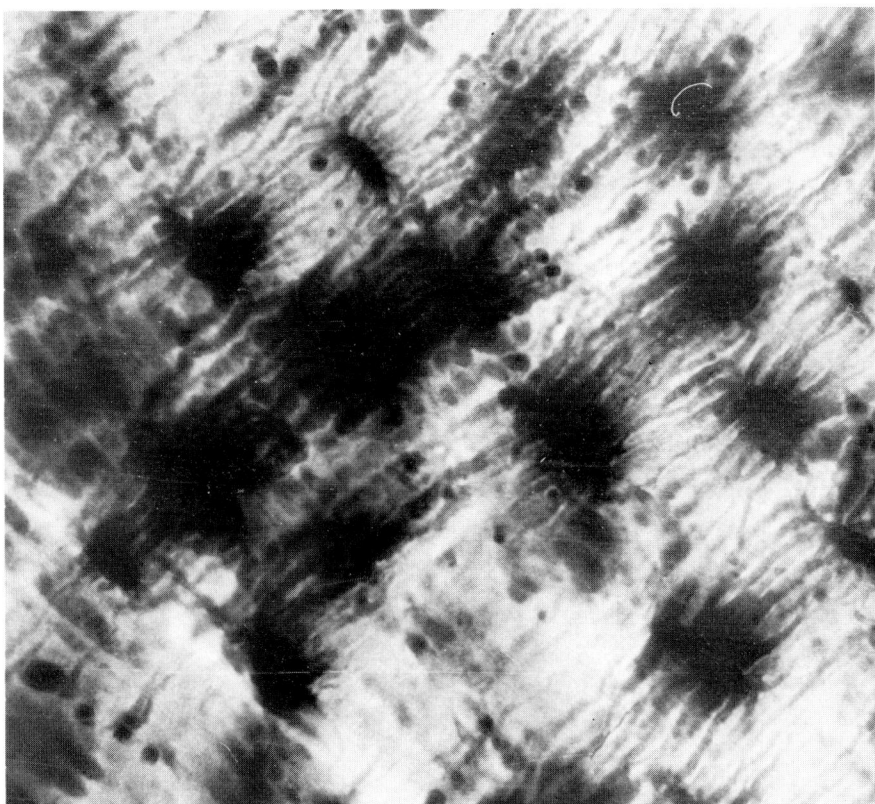

FIGURE 7. A fresh undecalcified human bone cross-section of compacta exposed briefly (~30 sec) to dilute hot aqueous permanganate. The magnification is nearly the same as in Figure 6. The spider-like extensions represent diffusion of permanganate ions into the bony walls of the canaliculae, the penetration ranging from 1 to 3 μm. Diffusion into the halo volume around the osteocyte lacunae penetrated more deeply, from 2 to 10 μm. This bone nevertheless would be quite impermeable to hot basic fuchsin, the molecules of which are much larger than the permanganate ion. Numerous experiments of this kind in the author's former laboratories at Yale University and at Henry Ford Hospital revealed the still little-discussed molecular sieve effect mentioned in the text.

through the bed to reenter the blood at the downstream end of the percolation pathway.[267] As one result of that mechanism, the solid-phase mineral at the entrance and exit of the percolation pathway would "see" about the same amount of radiocalcium in the fluid flowing past them per hour and therefore would take up equal amounts so that an autoradiograph of bone that had been labeled in that manner would show only a small concentration gradient between the sites of entrance and exit of the percolating fluid. Figure 9 diagrams that effect.

What could drive such a mechanism? It should consume energy, and it seems reasonable to assign that role at least in part to the osteocyte, particularly since dead bone — meaning that in which the osteocytes have died — does not accumulate diffuse burdens of bone-seeking isotopes.[238] Rather, it develops a shallow layer of uptake on its free surfaces in contact with living juxtaposed soft tissues. The author has speculated elsewhere that an osmotic pump might drive this system by making use of the molecular sieve property of the bone enveloping it.[238] However, the matter remains enigmatic.

It was also speculated in 1973 that the mechanical strains caused in bone by normal

94 *Intermediary Organization of the Skeleton*

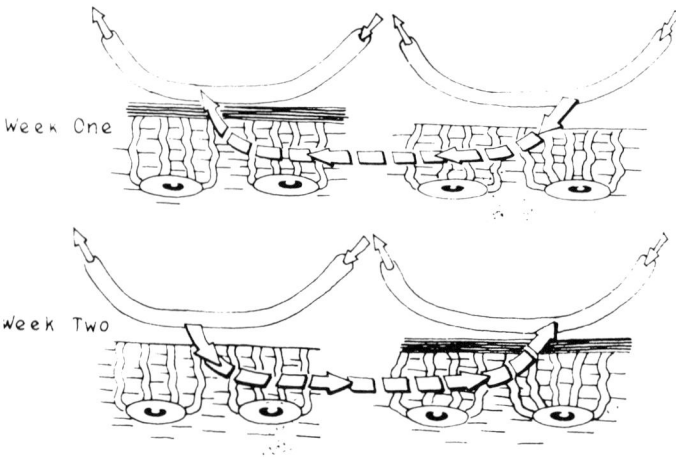

FIGURE 8. "On-Off" states. *Top row:* here extracellular fluid arising from the capillary on the right enters the bone, moves through it some variable distance along its longitudinal axis towards the left, and emerges to reenter another capillary. *Bottom row:* the author suspects that 1 week or so later the above directions could reverse, so that adjacent "On" and "Off" domains normally exist with respect to the percolation process.

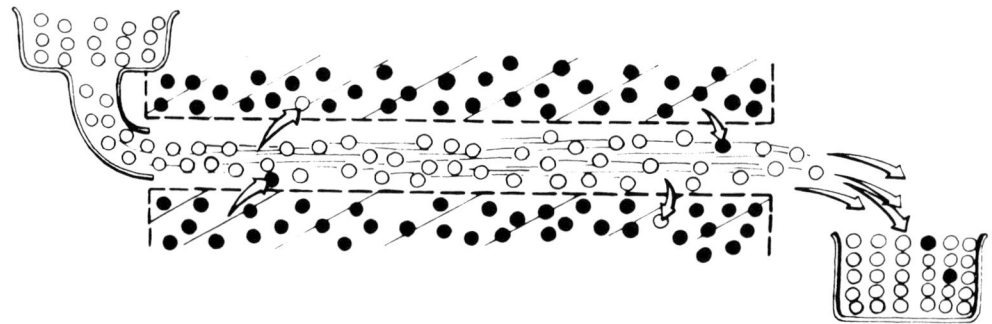

FIGURE 9. Here fluid flows through an ultramicroscopic cleft in bone during the percolation process discussed in the text. It contains radiocalcium ions that enter from the general circulation on the left. In its pass or transit through this cleft an occasional radiocalcium atom enters the surrounding bone mineral to exchange with preexisting cold calcium. The flow is so fast compared to that exchange that the bone on the right, where the fluid exits the cleft or pathway, "sees" about the same amount of radiocalcium per hour as the bone at the entering cleft or pathway on the left, so it accepts about the same amount of isotope. Hence, an autoradiograph would show no significant or very little difference in isotope uptake at the entrance and exit. Furthermore, the major gradients associated with this process should become most clearly visible on *longitudinal* rather than cross-sections, and so far other investigators have not studied the process in longitudinal sections.

daily activities could aid in "pumping" the interstitial bone fluid through the percolation bed[238] due to the strain-amplifying nature of the microanatomy of the interstitial bone fluid which was described in Chapter 7. Recently, Pollack et al.[822] and Johnson et al.[822] have had similar thoughts. Such a process has some implications about the effects of mechanical usage on the materials properties of bone that would be essentially independent of its chemical and anatomical composition, and some experimental evidence suggests that such effects do occur. However, they lie apart from the present concerns.

Table 1
PROPERTIES OF THE SKELETAL
HOMEOSTATIC SYSTEMS

System	Capacity	Response time	Duration
Osteoclasts (existing)	Small	Minutes	2 weeks
Osteoblasts (existing)	Small	Minutes	2 months
Remodeling BMU	Immense	Weeks	Years
Surface-lining system	Small	Minutes	Hours
Percolation system	Immense	Minutes	Years

While the above mechanism is novel in the sense that nothing like it has been proposed by others so far, a considerable body of diverse evidence suggests that it is probably correct in its essentials. If so, it will require more perceptive experiments and the use of newer techniques to study its details and to make any corrections of the scheme that are necessary; at present, however, their nature is not obvious.

As for its L_o-domain properties, the percolation mechanism offers both a vast surface and a vast quantity of the bone mineral to the extracellular fluid fraction that percolates through it, and thence to the blood. Within about 20 min it could achieve a steady-state response to a homeostatic challenge — one that could react very efficiently to changes in the pH of the extracellular fluid. To repeat, the total buffering capacity offered by that mechanism to the blood is huge, comprising a sizeable fraction of the whole mineral burden of the lamellar bone fraction of the skeleton. It should function more efficiently in the low-density bone of infants and children than in the more highly mineralized bone of the elderly, and it would be quite independent of the histological R and F activities.

III. SYNTHESIS

As noted in 1971[40] and 1973,[238] at least three separate intraskeletal mechanisms allow the skeleton to participate in maintaining the homeostasis of the blood. The evidence for each of those systems is firm, even though some of it is arcane in that it is unknown and/or not available to many authorities who have become interested in such matters. Reproducing the evidence in any one laboratory would require the use of tracer distribution studies, autoradiography of fresh, undecalcified sections, special staining and chemical treatments of fresh, wet human bone samples, and quantitative histomorphometry.

The three systems in question form the aforesaid *histological, surface,* and *percolation* systems. Each appears to act concurrently in life, and probably continually too. They appear to respond to sudden pulse challenges at different speeds. The magnitudes of their buffering capacities probably differ; the mechanisms of each also differ, yet each depends on IO entities that show quantum-like properties with regard to structure and function. Those properties suggest that each system may respond to different controlling mechanisms and a variety of agents could affect the performance of each.

Table 1 summarizes the above material and the following relation encodes it:

$$\text{blood} \leftrightarrow \begin{cases} \rightarrow \text{remodeling} \leftrightarrow \\ \rightarrow \text{surface-lining cell} \leftrightarrow \\ \rightarrow \text{percolation} \leftrightarrow \end{cases} \| \text{---bone} \quad (4)$$

The writer knows of no studies to date that have tried to measure and compare the relative speeds and capacities of those three systems, so the crude estimates offered above are speculative and await hard data supplied by perceptively designed and interpreted future experiments. Also, no studies have been published that reveal how each responds as an intact system to commonly known skeletally active factors such as pH, osmolality, electrolyte disturbances, hormones, vitamins, nutrients, and the waste products of cellular metabolism.

Until such systems-oriented evidence becomes available, it remains naive to try to account for the skeletal role in homeostasis as a function of only one of those systems, and even more so, as a function of only one part of any of them, such as osteoclasts.

IV. ILLUSTRATIVE QUESTIONS

Little change has occurred in the knowledge of the homeostatic role of the skeletal IO since the writer last discussed the problem in 1973,[237,238] so most of the questions that seemed pertinent then remain unanswered today, although some new ones have arisen as well. Those questions merit some attention from the clinical and research communities, particularly with regard to acid-base and fluid and electrolyte homeostasis. A sampling follows.

What relative magnitudes of sink/reservoir and buffering capacity do the three systems actually supply? What biological, physical, and chemical factors control them as continua? And how? What are their true reaction times? And durations? What drives the surface and percolation systems? What effects do age, sex, kind of bone, dietary and physical activity histories, commonly used drugs, and climate have on them? How does each respond in disease and to different kinds of homeostatic challenge? How might one study them, collectively and separately? What animal models would be appropriate for extrapolation to man? And which ones inappropriate?

What is the precise nature of the differing kinds of feathering and halo volume phenomena? What physical, biochemical, and cellular phenomena underlie them? What systemic factors in the body at large do they reflect, if any? How ubiquitous are "On-Off" phenomena in this system? What other kinds of mineralization abnormalities that are not the result of osteoblastic and osteoclastic activities will turn up when people begin to search for them with appropriate methods? Figure 10 shows one such abnormality.

Exactly how and when does each system respond to a change in pH, osmolality, mineral ion concentrations, and gas tensions? And to all hormones, nutrients, and varied drugs? What kinds of acute and chronic problems can malfunctions of those three systems cause, singly and in combinations?

What else that pertains to the concerns of this chapter do we need to know about this bone-blood interface system, whether because we have not yet seen its significance, or we have not yet stumbled across it? And what errors does the above synthesis contain?

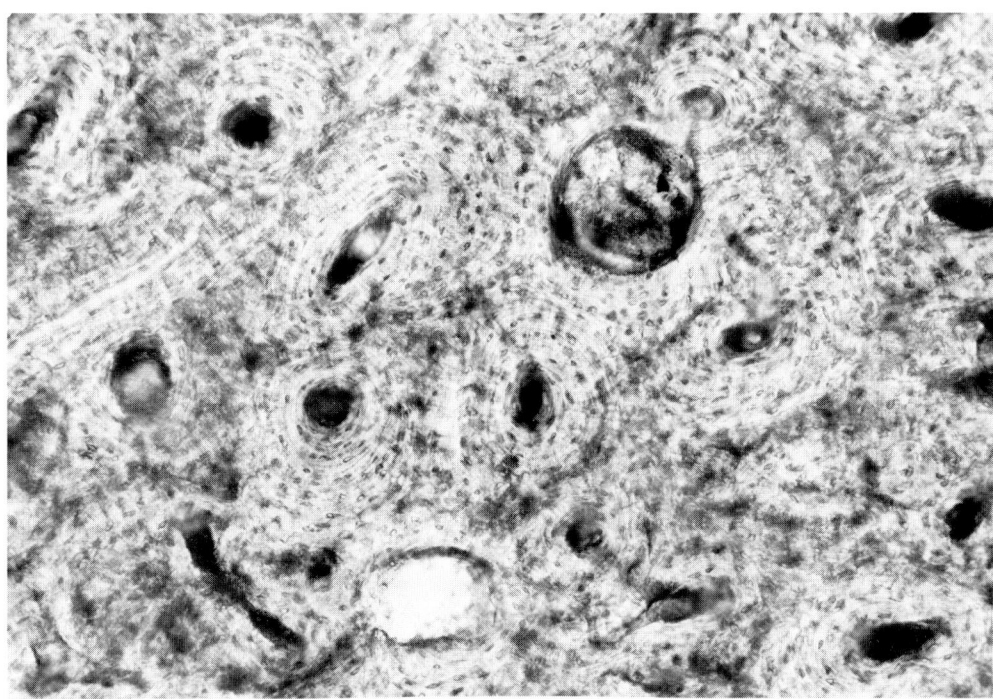

FIGURE 10. Hand-ground, fresh, wet, undecalcified cross-section of human compacta, about 125×. Transmitted light with substage numerical aperture reduced to 20% of that of the microscope objective. That causes a phase contrast effect that brings out a pronounced difference in the refractive index of perilacunar bone around many of the osteocyte lacunae. The blurs that pepper the field reflect that difference rather than any stains, and lack of penetration of the stain molecules into these regions proves they are fully mineralized. Hence, another kind of mineralization abnormality not apparent in conventionally prepared sections (the decalcifying solution removes the above evidence completely) but present in this adult with a chronic metabolic alkalosis and not associated with any known form of histological osteomalacia.

Note 1

SOME PROPERTIES OF PHYSIOLOGICAL BALANCES

Such balances occur widely in animals and plants from the molecular level up to the intact organism. Since balances apply to much of the material in this book, a word about them is appropriate. Some of this material was presented in 1964.[229] There is nothing original or controversial in the material below but people whose formal educations lay in branches of medicine or biology have not had to learn that material with the same rigor that chemists or physicists have, while those with backgrounds in the physical sciences often have skimpy ideas of the properties of real physiological systems. Between the two, some of the hard facts of skeletal physiological balances often become overlooked.

This Note will consider separately the nature of a balance and its regulation. A balance can be analyzed conventionally as a sum of opposed rates, or from a less well known population dynamic viewpoint, the so-called Frostian balance. The former is discussed first.

A. The Nature of a Balance

This always involves a system, compartment, or *pool* which some material can enter and also leave, and in which both of those processes usually, but not always, occur at the same time. The material could be cell nuclei, phosphate ions, bone tissue, proteoglycan, collagen, or sodium. Let R signify the rate of bone resorption or of any other process that can remove a material from a system or pool, and let F signify the rate of bone formation or of any other process that can fill up a system or a pool with the same material (so R signifies a minus quantity and F a positive one). Let B signify the balance between the two, the difference if any. Then one can write:

$$B = F + R \tag{5}$$

noting again that in terms of the pertinent arithmetic the +R actually means +(−R). Accordingly, all of the calcium, phosphate, hormone, vitamin, bone, collagen, proteoglycan, blood, tooth, or any other substance, structure, or entity in some pool at some moment, B, will equal all that was added to it over past time (F), less all that was removed from it over past time (R) regardless of the specific processes and the number thereof that contribute to each, and also regardless of the numerical and dimensional referents used to express their quantities (e.g., integrals, derivatives, deltas, mass, volume, equivalents, cells, people, molecules, counts). Of course, consistent units and referents must be used throughout.

A basic property of all balances and the first axiom of balances is that *an observed net loss taken alone signifies only the fact of that loss*. A loss by itself does not reveal its mechanism, and the same applies to a gain.[229,324] In symbols, to illustrate the point and to repeat:

$$B = F + (-R)$$

If some study now provides a number for B, regardless of whether it is dB (meaning a rate) or ΔB (meaning a net change over some arbitrary time interval) what is the value of R responsible for that change to a new value for B? To find R, solve the above expression for it by subtracting F from both sides, thus:

$$B - F = F + (-R) - F \tag{6}$$

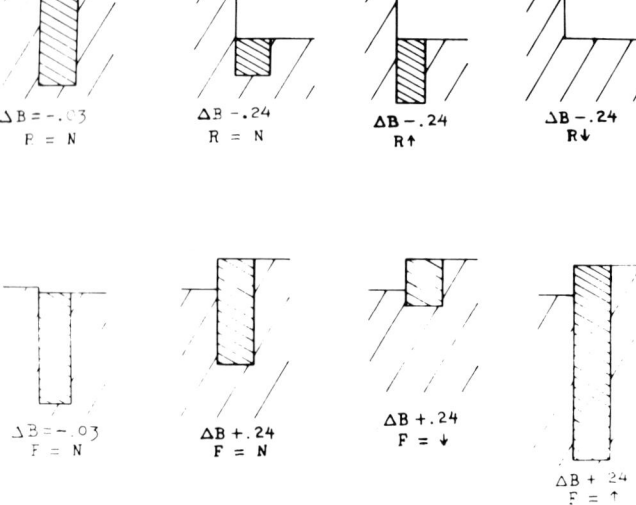

FIGURE 11. More balance facts. *Top row:* a normal net rate of loss from some pool of −0.03 can increase 13× in the presence of a normal, increased, or decreased absolute rate of loss. The −0.03 value approximates the size of the decrement of bone removed permanently by the typical completed adult bone remodeling BMU when averaged over all four skeletal envelopes. The actual values, of course, will vary on the different bone envelopes, with age, and in differing circumstances. One salient but previously unappreciated intelligence contained in that observation is the fact that in human adults approximately 30× more bone tissue turns over annually than is irreversibly lost. As one result, large changes in net loss can occur in the presence of small changes in turnover and in the absolute rates of the underlying R and F activities. *Bottom row:* a net loss from the pool can switch to a large gain in the presence of normal, decreased, or increased absolute rate of filling or influx.

Since the +F and −F on the right cancel each other to equal zero, eliminate them to obtain B − F = +(−R), and let +(−R) be signified simply by R, and one can write B − F = R. Simply change sides (if 3 − 2 = 1, then also 1 = 3 − 2), and write:

$$R = B - F \qquad (7)$$

To repeat, here B is known but what is the number that belongs to R? To get it one must also measure F. Until that is done one cannot tell from the information about B alone what happened to R.

As Figure 11 illustrates, in balance situations a *net* loss can increase in the presence of normal, increased, or even decreased R, depending on how F responds. The same reasoning and conclusions apply to a net gain. Since most known balance phenomena and systems in human physiology connect the F to the R mechanism by some form of coupling, it follows that if one of them changes in a particular direction the other often does too, although not necessarily equally, allowing for transient phenomena to disappear first so steady states can become established. That observation applies not only to bone turnover; it applies also and only as examples to the body's burdens of calcium, phosphate, magnesium, carbonate, hydroxyproline, proteoglycans, synovial fluid, collagen, Gla protein, alkaline phosphatase, acid phosphatase and other enzymes, fluoride, lead, cartilage, fibrous tissue, cementum, and trace element burdens.

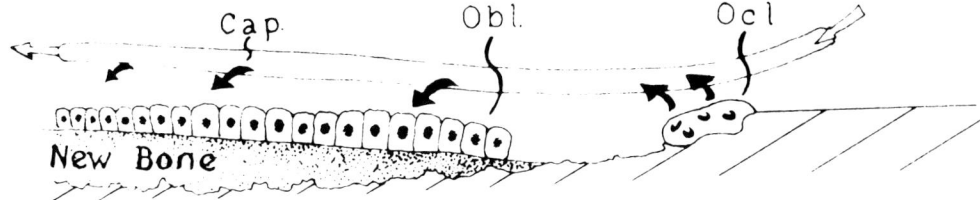

FIGURE 12. This diagram of a histologically active bone remodeling BMU shows the capillary of the system, with the blood entering from the right and supplying the resorption front first. The blood first picks up the solubilized components of the underlying bone that have been freed by the clasts, and it then flows down over the following bone formation phase where some of the previously solubilized material is redeposited in the newly forming bone and does not return to the general circulation. On the other hand, observe that in the bone resorption drifts that occur during bone modeling, all of the resorbed material returns to the general circulation.

The coupling between the R and F processes can be *direct* (as in the case of the various remodeling BMU) or *indirect* (as in the case of the steroid-induced increased net bone loss). There the drug causes an enhanced renal calcium leak and a decreased intestinal calcium absorption that both offset the enhanced entrance of skeletal calcium into the blood, so the serum calcium concentration is little affected.

That coupling can act *simultaneously* on the R and F processes of a system as in the steroid effect above, or in the exchange of gases and mineral ions between blood on the one hand and bone, synovial fluid, cartilage, or tooth on the other. The coupling can act *sequentially* in time as it does in the remodeling BMU, and in some situations it can act in both ways. In the remodeling case the summation in the whole skeleton of the momentary drug effects on the separate resorption and formation packets of a temporally incoherent BMU population can suggest to an observer of the intact human that the two effects are simultaneous and independent in L_2-space-time and at lower levels, when in fact they may not be.

The second axiom of balances, which follows from the above, is *a change in any balance never alone reveals the changes in the absolute rates of the R and F processes that caused it.*

B. Reutilization Phenomena

These can affect studies of balance phenomena and at least two of them occur, one local and the other systemic.

1. Local Reutilization

In the remodeling BMU that turns over the adult's lamellar bone and the growing child's spongiosa, the arrangement of the capillary blood flow relative to the coupled resorption and formation fronts supplied by that capillary causes the material solubilized at the resorption front to enter the capillary there. It then moves downstream to the formation front where some of it deposits in the newly forming tissue, as shown in this chapter and in Figure 12. As a result, the blood returned to the general circulation by that capillary contains somewhat less of the organic and inorganic products of local bone resorption (including any tracer burden) than it contained when it left the resorption front to flow over the formation front. For example, radium redistribution studies reported by Rowland in 1963[652] revealed that fact clearly. Exactly how much less of the resorption products remains in the blood leaving the downstream formation front remains uncertain. However, the slowness of the normal resorbing and forming processes relative to the speed of blood flow through the capillary suggest that normally,

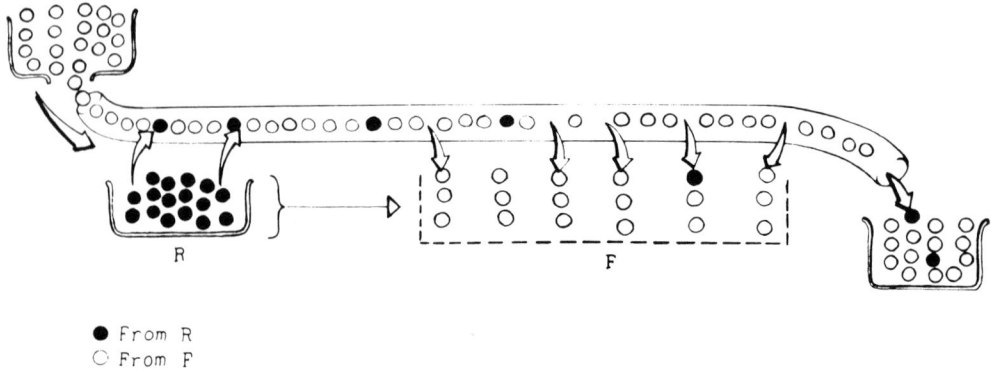

FIGURE 13. The mechanism of local reutilization in bone remodeling. In this diagram, the blood flowing from left to right over the resorption front and the formation front of a bone remodeling BMU carries the soluble components of bone, indicated by the open circles. The components in the bone itself are black dots. When capillary flow is rapid, the bone-derived components supplied to the blood by the resorption front become diluted in the inflowing blood, and only a fraction of those bone-derived components will redeposit in the downstream coupled formation process.

and due to mixing with the contents of the incoming capillary, much of the resorbed material passes on through into the general circulation, and the formation process extracts only a minor fraction of it. Figure 13 illustrates the mechanism.

Various physiological challenges could change that situation, both greatly and promptly, including those that relate to the adrenalcorticosteroid effects that often appear in experimental animals subjected to psychologic and physiological stresses in the course of experiments that involve repeated injections, surgical procedures, or other painful and/or frightening manipulations. Such stress-related effects on balances are seldom accounted for but they have appeared rather clearly in many in vivo studies, including some of physical immobilization effects during orbit-simulating situations on monkeys and rats.[342,786]

2. Systemic Reutilization

Here material that enters the blood from one source then distributes in the general circulation to other more remote locations, where it is withdrawn. Thus, calcium entering the blood from the radius, and due to bone resorption there, may then go to the kidney, to the sweat, to the femur, or into the digestive juices.

Klein[417,418] has done a number of ingenious experiments with such systems in recent years and in various laboratory animals. Considering the blood as the pool under study for the moment, the first or F effect tends to elevate the concentration of the material in the general circulation, and the second or R effect tends to lower it. Because this forms another balance situation, a rise or fall in the blood level reveals nothing by itself about the changes in the two absolute rates that underlie it, as shown in Figure 11. The typical habit of interpreting a sudden rise in the serum calcium as directly reflecting only an increase in bone resorption is illogical and often erroneous. A decrease in bone mineralization at ongoing formation fronts would have exactly the same effect, yet it is frequently ignored or discounted as an explanation, and typically without justification based on direct evidence. Figure 1 provides direct and firm evidence that those possibilities are real and should be taken seriously in future work and analyses.

C. Push-Pull Phenomena

Many push-pull mechanisms exist in the systems that control the body's balances, and from the cell level to the man. For example, putting a man on adrenalcorticalsteroids can significantly increase net bone tissue and mineral loss from the body, and considerably elevate the 24-hr urinary calcium and phosphorus excretion, yet the serum calcium and phosphate concentrations seldom rise. Whether directly or indirectly, the steroid also increases the urinary calcium and phosphate excretions, probably by reducing their tubular reabsorption, and it decreases calcium absorption by the gut at the same time that its effect on the $\Delta B \cdot BMU$ mechanism increases net bone loss. Those opposing effects usually match well enough that the serum calcium and phosphate change little or not at all,[300,302,507,581,692,764] yet the total amount of skeletal calcium and phosphate transported per hour by the blood from bone to kidney increases in proportion to the net renal loss.

Push-pull control of balance mechanisms occurs in many physiological systems. It often involves the coordination by systemic and/or local regulation of functionally and anatomically separate mechanisms, as in the bone, gut, and renal situations just described. Sometimes, as in the bone modeling and remodeling systems, certain transient effects can appear when one part of those systems reacts more promptly than others that "compensate" for it. Such transient effects can appear clearly when some kind of pulse treatment or other challenge establishes a temporary state of temporal coherence in a system, whereupon some of its quantum-like properties can become unmasked in the form of a series of transients that can be seen at the organ level and above. The transients described in Chapter 5, or shown by Hangartner et al.[304] and Kimmel[416] belong in that class, as do the serum calcium transient following parathyroid hormone injection that led Copp to what turned out to be calcitonin effects on calcium homeostasis,[131] and the RAP-induced osteopenia that can follow certain injuries to bone and soft tissues, conventionally termed "disuse osteoporosis".[230,237]

In general terms, the above facts mean that in some pools, some of the balance processes can vary independently of the pool size and of the amount of material in the pool, so they could be termed independently variable and bidirectional. For example, the movement of calcium between the blood and the gut, or total bone turnover, can each increase or decrease regardless of the level of calcium in the blood and whether that level is constant or changing. However, other processes are irreversibly unidirectional, as in the loss of calcium via the urine, feces, or sweat, in the accretion of mineral in growing enamel and dentin, or in the addition of length to a growing bone. The matter of space-time domain also becomes important here. For example, in L_2-space-time the bone deposited by a formation drift is an irreversibly unidirectional process, so in that domain and with respect to either the underlying bone or the blood that supplies the process, it is unidirectional. In L_o-space-time, however, in the intact bone, different resorption and formation drifts occur simultaneously on different parts of the surfaces of the bone and the organ sums up those effects. The resulting net gains and losses will then be bidirectionally and independently variable in that domain. Accordingly, the third and fourth axioms of balances can be stated as: *define the space-time and organizational domains of the balance situation, and any change in the net amount or concentration of some material in a pool must be characterized as the result of a unidirectional process, or of a balance process.*

Unidirectional processes would include the formation of enamel in the tooth, the addition of new length to a growing bone or tendon, and loss of calcium or related substances in the urine, feces, and sweat.

Table 2 lists some general balance situations that pertain to the material in this book. To repeat, the prevalent practice of interpreting an increased net loss of any substance, including bone, as due only to an increase in the absolute rate of the R factor, including bone resorption, is naive. This applies to interpretations of changes in the serum concentrations of skeletally important ions and molecules, of changes in the ash weights

Table 2
SOME SKELETAL BALANCE PHENOMENA

All changes in whole body and whole bone, joint, ligament or tooth calcium, phosphate, related ions, plus the molecular components of the organic matrixes, including proteoglycans, collagen, hydroxyproline, proline, and noncollagenous proteins.

All changes in bone, chondral, and tendon size, alignment and thickness, in the quantity of synovial fluid, the amount of spongiosa, the marrow cavity diameter, or intracortical porosity.

Skeletal uptake and/or release of all tracer elements and labeled organic compounds, including isotopes of calcium, strontium, fluoride, technecium, gallium, phosphate, sulfate, plus labeled amino acids and other organic compounds, and tissue time markers such as tetracyclines and DCAF.

Serum, joint fluid, and bone fluid concentrations of calcium, magnesium, phosphate, alkaline and acid phosphatases, GLA, hydroxyproline, fluoride, and any other tracer element or molecule or other substance, mineral or organic.

The number of clasts and blasts in bone, and of nuclei per osteoclast, the number of cells in a cartilage column of a growth plate, or in the whole plate, the thickness of a plate or bone, the number of leukocytes in a unit volume of tissue.

The amount of microdamage in a bone, tendon, or articular cartilage.

of whole bones or poorly chosen fractions thereof, of changes in the nitrogen, phosphorus, sulfate, proline, or related contents of whole bones, muscles, or joints (or poorly chosen fractions thereof), of changes in the whole body burden of calcium or other elements observed by neutron activation or other means, of changes in the skeletal burdens of any and all radioactive tracer materials, and of any kind of change shown by any kind of metabolic balance study.

To reiterate, the above matters are not in dispute, although they are often overlooked.

D. Population Logistics and "Frostian" Balances

While it is not profound, this different way to think about the balance situations in the body has considerable analytical power, although some may find it novel and hard to understand and use at first.

As an example, an interesting study by Nakamura and Konda[536] of calcitonin-treated rats found that the number of nuclei per osteoclast in the metaphyseal spongiosa tended to increase (note that the organizational domain is specified). That might suggest that the drug caused more nuclei to enter the typical osteoclast in unit time. However, that is another balance problem, for osteoclasts continually add new nuclei (F), and shed or otherwise lose old ones (R),[376] so the number in a given cell at any moment (B), reflects the balance between those added and those lost over past time, or B = F + R. As shown in Figure 11, the number of nuclei per osteoclast can increase even in the presence of an actual decrease in the number that enter it in unit time.

This is also a population problem, for in a steady state the mean number of nuclei per cell (A), will equal the number that enter it in unit time (μ) times the mean functional lifespan or residence time in the cell of the typical nucleus (σ). Thus:

$$A = \mu\sigma \qquad (8)$$

As applies to the $\Delta B \cdot BMU$ parameter described in Chapters 4 and 5, increased numbers of nuclei per osteoclast can occur in the presence of increased rates of addition and removal, or of decreased rates, or of combinations of those two rates, as indicated in Figure 11.

In fact, the increased nuclei per clast observed by Jee and others after chronic diphosphonate treatment (EHDP) of rodents probably reflected a decreased addition of new nuclei in unit time (μ decreased), but an even more prolonged residence time of the typical nucleus in the cell (σ greatly prolonged) before it left or otherwise disappeared.[77,292,378] In terms of the population dynamics described in Note 1 in Chapter 9, the birth rate (entrance of new nuclei into the cell) could go down, but if the lifespan (σ) prolongs proportionally even more, then the population (A) will actually increase in the face of a lowered birth rate. Similar relationships probably apply to the increases noted by Jaworski et al. in osteoclasts in experimental canine renal osteodystrophy.[369] An analogy presented in 1966 can explain such population situations, which are important but which many clinicians and physical science acquaintances of the author found difficult to understand. They apply with particular aptness wherever the content of a universe of study can be considered as a population (A), the typical individual of which has a discrete finite size regardless of whether or not that size is a "natural" unit determined in nature, as it is for the size of secondary osteons, teeth, and chondral columns in a growth plate, or if it is determined by man as a milligram, millimole, or other arbitrary and constant unit of measure.

Thus, assume a room that a given number of people enter per hour (signified by μ). Assume further that after an average length of stay (σ) each person then leaves the room. How many people will the room contain (A) in a steady state in which 10 enter it per hour (μ), and each remains for 1.5 hr (σ) before leaving? The answer is seen in Relation 8,[229] or A = 10 × 1.5, or A = 15, and in the steady state in this situation, for each new person that enters the room another one already there leaves it.

If σ increases to 3 hr, meaning each person stays in the room 3 hr before leaving it, but μ remains unchanged, then when the new steady state has developed (which will not happen until one σ period or 3 hr has elapsed after changing the parameters of the system, which are μ and σ) the room will contain 30 people, since now A = 10 × 3, = 30. If the number that enter it is then reduced to a third of its original value so μ now equals 10/3 or $3^1/_3$, but σ is increased to 6 hr, then the population in the room will finally stabilize at A = 10/3 × 6, = 20 people, but again not until one σ period for the new system (now 6 hr) has elapsed.

Note Bene: The above relationships can be applied to all physiologic and biochemical balances. The universe of study or pool, instead of a room, may be a deciliter or any other unit of blood, urine, bone, or any other body compartment, component, compound, tissue, or organ from the L_m- to the L_{is}-domains. Instead of people, the birth rate (μ) may apply to population units such as nuclei, organelles, cells, BMU, infants, millimoles, photon density units, grams, or milliliters that enter the universe in unit time to form its population, A. The mean lifetime (σ) of a typical unit in the universe of study may be expressed in any convenient units of time; σ and μ are the basic determinants of all population balances.

One value of this way of looking at balances lies in the fact that the σ value of the system under consideration reveals how long to wait after challenging or otherwise changing the system before the steady-state effects on A can be seen. That time period or lead time equals exactly the value of σ. That σ value also allows one to distinguish transients from steady-state effects (because steady states cannot develop until one σ period has elapsed after changing or challenging the system). This is no minor matter because only steady-state effects can cause or/and cure most skeletal disease. The relation B = F + R provides no clue to those values, or even that they exist. Finally, σ provides a mean age for the population units of the system, something that cannot be obtained from the relation B = F + R. That mean age equals, quite simply, $\sigma/2$.

The F or filling term in earlier paragraphs is analogous to the birth rate term (μ) just described, and both are true independent variables. From the population viewpoint, the earlier R term has become a dependent variable, a kind of death rate, for it depends on the value of both A and σ, and in the sense algebraically of A/σ. That is, loss of material from the system, the rate of R, will increase if its content increases, and also if the mean lifespan (σ) of a typical unit of A decreases (note that the mean lifespan (σ) is not the same thing as mean age). From this point of view, a unilateral increase in bone resorption would reflect the effect of decreasing the mean lifespan (σ) of the typical bone tissue moiety. This way of looking at balances is especially appropriate for systems in which the inherent properties of the unit can affect its lifespan in the system. That applies, for example, to the functional lifespan of cells, which do have some sort of predetermined lifespan, but the cell environment can modify it. That property does not apply to a milligram of stable calcium, the lifespan of which in the blood will be determined solely by the independent R and F processes that put it there and remove it.

Accordingly, while it is conventional to think of physiologic and biochemical balances as a function of the relation:

$$B = F + R$$

and that way of looking at them has its undisputed uses, it is also informative, and often more so, to learn to think of them in terms of the population equation described for histomorphometric work with bone remodeling BMU in 1964,[229] or $A = \mu \sigma$.

The fifth axiom of balances, which follows from the above, is *when the contents of a pool are expressed as a population of units, the factors that control their steady-state pool balances are μ and σ.*

E. The Size of the Unit

One more property of population balances requires discussion. There is no problem when the unit of the population in question is defined as a constant unit of measure, such as a milliequivalent, a cell, or an organ. Such units are constants — some natural and some arbitrary. The amount of bone in the typical individual BSU, the amount of cartilage in one typical chondral column in a growth plate, the amount of RNA in a typical cell, or the size of the $\Delta B \cdot$ BMU, can each vary, and in the latter case considerably and from positive to negative values. In each case, the value of A may provide the number of population units in the system but to obtain the absolute quantities in grams, milliliters, or other such units provided by a given number of population units requires multiplying the total number of units by the quantity of material contained in the typical individual unit. That quantity might consist of the DNA in a nucleus, the amount of bone in an osteon or a femur, the amount of proteoglycan in a joint cartilage, the number of radioactive atoms in a milligram of bulk material, etc.

Accordingly, the annual amount of bone lost from the trabecular envelope can be expressed as the population or total number of $\Delta B \cdot$ BMU decrements or bites taken out of that envelope annually (which in the steady state will be exactly μ) times the amount of bone represented by the typical individual decrement and expressed in cubic millimeters or milligrams. In symbols, where ΔV signifies the volume of bone gained or lost in μ discrete increments or decrements and v the volume of bone gained or lost per typical individual increment or decrement (which is one way of expressing $\Delta B \cdot$ BMU), positive values signifying an addition and negative values signifying losses, then:

$$\Delta V = v\mu \qquad (9)$$

Parenthetically, in the referent of the whole skeleton the balance term, B, in earlier paragraphs of this Note is identically equal to ΔV, so it follows that over unit time $v\mu$ = F + R, while μ itself = A/σ. Since already available methods permit measuring B, F,

A, and σ directly in the intact system, both diseased and healthy, real values for μ, R, and v can be found in principle, and in fact one could estimate them from already published data. By such means it can be computed that an approximate value for the typical ΔB · BMU in healthy human adults is ∼-0.0015 mm^3 of bone when all four skeletal envelopes are lumped together.

To return to the ΔB · BMU problem, apply Relation 9 to the situation in which a postmenopausal woman begins to lose bone some three times faster than before menopause. That means that ΔV is three times more negative than before, so the vμ product is as well. That product can become more negative by enlarging the value of v, the decrement per typical completed BMU, or by enlarging μ, the annual number of such decrements, or by various combinations of each. Since v is itself a function of the relationship of R and F per completed remodeling BMU in the sense that at the BMU level or referent ΔB · BMU = Rσ$_r$ + Fσ$_f$, recalling that R actually equals (-R), Relation 9 contains all of the independent variables that from a population viewpoint determine annual gains or losses of bone. That relation also tends to hide the absolute resorption rate, R, thereby making it difficult to perpetuate some of the more naive analytical invocations of that rate described in Note 2 in Chapter 7 and in Note 1 in Chapter 10. That could be useful, for most physiologists and clinicians need to learn that in human adults, steady-state global losses/gains of lamellar bone, the skeletal bone balance, usually do depend directly on the independent variable, ΔB · BMU, but usually do not depend on the absolute resorption rate alone. In transient situations, of course, phenomena such as those shown in Chapters 3, 5, and 6 come into play and can complicate understanding a given situation.

From the above comes the sixth axiom of balances, which states that *in a steady state the amount of material (v) in the typical unit of a population, times the number of those units in the population (A), controls the amount of material in a pool, in the sense of V = Av.*

In the above net bone loss example, v would equal exactly the value of ΔB · BMU, assuming both are expressed in the same referents and dimensions.

1. Amount and Concentration

The amount of some material in a pool, whether expressed as a population of units, A, or in measure such as grams or liters, is not its concentration in the pool. To obtain that latter value, the value of ΔV must be divided by the size of the pool expressed in suitable units. Thus, the concentration, C, of some material in a population pool of size P equals the number of units in the pool, A, times the amount of material in the typical population unit, v, divided by the size of the pool, or:

$$C = Av/P$$

From the above comes a seventh axiom of balances: *in steady states the concentration of a material in a pool depends on the number of units of the material in the pool, the size of the pool, and the amount of material in each unit.*

It follows from the above that a steady-state change in the concentration of some agent in a pool need not reflect a corresponding change in its R or F parameters. For example, the serum concentration can rise in the presence of normal R and F activities if the residence time of the typical unit in the pool prolongs and/or if the size of the pool independently decreases. The converses also apply; Parfitt discussed some such phenomena and showed that at least some of them do occur in life.[571] They underlie some examples of hyponatremia, hypernatremia, and hypercalcemia encountered in clinical medicine.

Chapter 11

THE REGIONAL ACCELERATORY PHENOMENON (RAP)

ABSTRACT

This recently recognized but common complex reaction of the mammalian body to diverse noxious stimuli distributes regionally in the anatomical sense, involving both soft and hard tissues. It consists of an acceleration of most ongoing normal vital tissue processes that potentiates and accelerates tissue healing and defensive processes. An obtunded RAP can retard healing and cause a lower resistance to infection, mechanical abuse, and other threats to functional competence.

If it is ignored in experimental design, the phenomenon can also cause major errors of comission, omission, and interpretation in studies of metabolic bone disease and of the effects of mechanical, endocrinologic, biochemical, and other factors on virtually all aspects of skeletal physiology.

I. INTRODUCTION

A. Preamble

Most of the published orthodox articles of science are as dry as ancient papyrus in a Sahara noon. They bristle with cool and emotionless numbers and methods, with subjects and propositi instead of troubled/joyous men and women, with means, standard errors, and *t*-tests and with desiccated citations instead of friends with ailing parents, children with marginal grades or adjustment problems, and with enemies.

Real science, however, is more, far more than that, and different. Its sterile individual reports are like some of the individual sentences in Homer's *Iliad,* Dostoyevski's *Anna Karenina,* or Rand's *The Fountainhead,* in that taken alone they provide no hint of the joys and agonies, the tears and laughter, the failures and triumphs, the crimes, cowardice, and courage, the plain grandeur that the whole fabric of the thing contains and rests upon and that becomes revealed bit by bit as its pages turn and its chapters fall behind.

From that perspective science has certain similarities with great literature. People do it, not automatons, and they have families and trouble paying their taxes and plumbers. They strive, work, and worry. They sweat in fear in the solitude of some nights. They succeed for a while and then they fail. Sometimes they make honest errors, for which sometimes they are inexplicably rewarded, at others illogically and viciously punished, and at other times they achieve triumphs, small or large, for which again they are sometimes rewarded, at others punished. Some deserve and receive the Nobel or various equivalents. Others, like Semmelweiss, die with the bitter knowledge that they were right all the while that their colleagues showered them with calumny and ridicule. Some with great scientific talent are allowed or actually helped to use it, while others are forced to set it aside simply to survive in our complex society.

Out of that troubled and unpredictable soil grows a structure, a product of science and technology, that is great and grand, and perhaps in part because of its very blemishes. A straight line or circle may be a perfect thing, but the geometrically random jagged coasts of New England and Japan are both fascinating and beautiful. Because of that structure, the open fracture that in 100 B.C. or 1600 A.D. would have led to an agonizing and protracted death for a young man and destitution and starvation for his young wife and child, has become a matter of a few months inconvenience today. The diphtheria, smallpox, pneumonia, septicemia, and plague that decimated whole populations in Roman times and the Middle Ages have become nuisances prevented or cured with a needle or a few scribbles on a prescription pad. In short, because of that structure, people today have been freed from many of the random and mindless cruelties of Dame Fortune's playing with the dice of disease, weather, public health, economics, and tyranny, so they have far greater opportunities today to create their own joys and sorrows, their own triumphs and failures.

It takes time and scars on the soul, and a ken that reaches over centuries and into other men's moccasins to see such things hidden behind the desiccated verbiage of any particular contemporary scientific article. That includes these chapters, which endeavor to add other notes to that structure. This brings up the RAP, a behavioral entity that has been around longer than man, but which remained unperceived as such until quite recently.

B. General Remarks

The ubiquitous regional acceleratory phenomen affects both hard and soft tissues and it appears to explain some features of certain clinical affections and problems.[261] It has also perturbed many experimental studies of metabolic bone disease and of mechanical, endocrine, and other effects on skeletal physiology, for when unrecognized its features are usually assigned to other causes. Many of its manifestations have been known for centuries, but the author first recognized it as an entity.[267] Thus, of necessity this chapter will reflect his experience and biases. The text will describe the phenomenon first, and then some of its clinically apparent effects and then its bearing on certain kinds of research, the design of experiments, and the interpretation of their results.

II. THE RAP

A. Characteristics of the RAP

The following paragraphs will describe in sequence the causes, nature, anatomical distribution, and probable roles of the RAP.

1. The Causes

Clinical and pathologic experience has identified many of the factors that can cause a RAP, and some of them will be listed shortly. However, the L_c- through L_3-domain details of the physical, biochemical, and biological mechanisms that transduce its gross inciting or enabling stimuli to its effects on varied aspects of tissue dynamics remain enigmatic, although recently some promising information has been reported in that regard. In interest here, Kolář et al.[423] were clearly aware of the RAP as an ubiquitous postinjury phenomenon by 1965, but they termed it the "general metabolic shift in mineralized tissues following local injury" and they did a variety of animal experiments to study the effects of the phenomenon on soft as well as hard tissues.

As for its causes, in a normal body any *regional noxious stimulus* of sufficient magnitude can evoke a RAP.[261,267] The size of the affected region and the intensity of its response vary directly with the magnitude and nature of that stimulus to different degrees in different individuals.[267,510]

The effective noxious stimuli include, but are not limited to, crushing injury, contusions[760] and fractures of all kinds, bone operations of any kind,[345,678] arthrotomy, arteriotomy, burns, acute peripheral denervation and acute paralysis of central origin,[1,286] infarcts, infections of soft tissues,[292] bones, and joints, certain tumors and metastases, and most noninfectious inflammatory processes, including rheumatoid arthritis,[93,336] lupus, dermatomyositis, rheumatic fever, pseudogout, and Reiter's disease. Acute disuse can also cause it.[327,403,467,483,738]

2. The Nature of the RAP[261]

Once evoked, most ongoing regional soft and hard tissue vital processes then accel-

Table 1
SOME TISSUES AND ACTIVITIES AFFECTED BY A TYPICAL RAP

Tissues: Woven bone, lamellar bone, spongiosa, compacta, articular cartilage, epiphyseal and apophyseal cartilage, teeth, tooth socket, gingiva, skin, subcutaneous fat, fascia, tendon, ligament, synovia, joint capsule, muscle, vascular tree, innervation

Activities: Metabolism and activities of differentiated cells, precursor cell activities, differentiation of cells, perfusion, growth of teeth, skin, hair, and nails, longitudinal and transverse growth of bone and cartilage, BMU-based remodeling of lamellar bone, primary spongiosa, calcified cartilage, and all fibrous tissues, microrepair and macrorepair of bone and fibrous tissues

erate above normal values.[267,451,804] Those collective accelerations and their effects comprise the RAP, and they include and span the L_c- to the L_{os}-levels of biological organization. They include perfusion,[93,563] growth of skin, bone, cartilage,[91] and hair, BMU-based turnover of woven bone,[550] lamellar bone,[550] and fibrous tissues plus turnover of the components of cartilage matrix, synovial fluid, and the molecular components of soft and hard tissues, chondral and bone macromodeling including corrections of bony malunions in children, skin epithelialization, cicatrization, all soft tissue and bone healing,[17,71] and the metabolism of all of the regional cell populations. Thus, the metabolic activities of chondrocytes, osteocytes, fibrocytes, and synovial cells tend to increase in the neighborhood of such noxious stimuli. Table 1 lists some of the tissues affected by a RAP and Figure 1 shows an example. Defensive tissue reactions to infection, abrasion, and contusion also accelerate when a RAP occurs, and other processes not named here may also accelerate. As a relation the above may be written thus:

$$\text{RAP} = (\text{ongoing activities} \xrightarrow[\text{noxious stimulus}]{} \text{accelerated activities}) \qquad (1)$$

Because of those increases, an affected region typically develops erythema and edema, it feels warm to the touch, and it proves warm on thermography. The accelerated local bone turnover and perfusion increase the local uptake of radioactive bone-seeking agents such as those of technetium, phosphate, gallium, calcium, strontium, and fluoride, which causes the "hot" regions often found in bone scans of cases of acute and chronic osteomyelitis, in actively healing and recently healed fractures, in some bony nonunions, in the presence of joint inflammation of any cause, in Sudek's atrophy, and in the presence of some bone metastases. Photon absorption studies and routine X-rays can reveal a significantly decreased regional bone density due to the locally increased remodeling space that accompanies any increase in BMU-based bone remodeling, as in Figure 2. A joint contracture too rigid to respond to wedging casts or to traction can respond more readily during the 3 months or so after a major local operation, a phenomenon that suggests that the RAP made the capsular and related tissues somehow more plastic for a time. Kolář et al. summarized some of the above features in a monograph in 1965,[423] and the work they quote makes it clear that the phenomena and other associations are well known. What was missing was the recognition that they all had a common denominator.

Clinically, as well as histologically, the particular cause of a RAP can imprint or superimpose its own features on the concomitant general features of the RAP, so their combination can often appear characteristic and even diagnostic. Such characteristic imprints can include necrosis, hypertrophy, pus, monocytic or eosinophilic infiltrates, granulation tissue, woven bone, Langhan's giant cells, amyloid, lymphocytic infiltra-

112 *Intermediary Organization of the Skeleton*

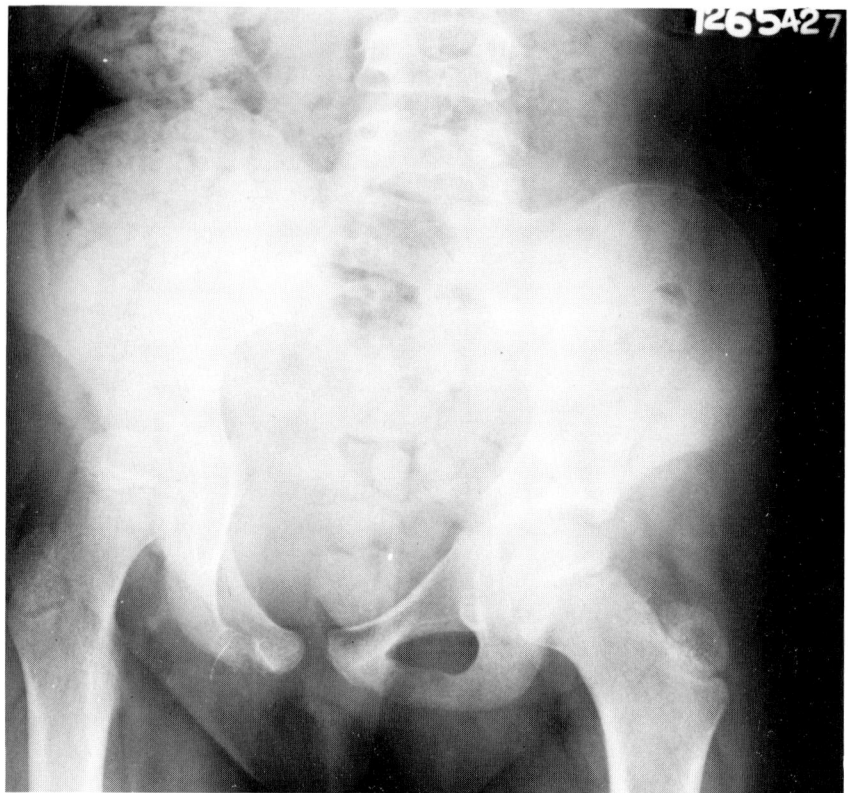

FIGURE 1. RAP effect on longitudinal bone growth. This adolescent girl has had Still's disease or juvenile rheumatoid arthritis for several years and it has involved her right hip (on the reader's left), but not the other one. The disease produced a RAP in the region of the involved hip and that accelerated regional chondral growth in the epiphyseal plate, making the femoral neck longer than in the normal hip. Due to pain, muscle loads on the involved hip reduced substantially, which left the unchanged vertical load of body weight unmodified by the medially directed component of the hip muscles. Thus, vertical loads were larger relative to the medial loads on this hip. As a result, the capital epiphyseal plate has come to lie more horizontally than in the normal hip, an effect that is predictable from the chondral modeling laws described in Chapter 8. Of course bone modeling contoured the femoral neck to fit the dictates of chondral growth and modeling. (Case courtesy of Drs. C. L. Mitchell, La Jolla, and H. Duncan, Detroit.)

tion, endothelial thickening, hemosiderin, and scar. Such imprints usually allow the pathologist to distinguish between trauma, acute and chronic infection or fungal infection, immune reactions, necrosis, and metastases, as examples. Table 2 lists some of the signs of a RAP in four skeletal tissues and Figure 3 provides an example of a characteristic histological imprint. The above manifestations include many of the classical signs of inflammation that physicians have known under that name for more than 2000 years. However, the writer proposed that those signs represent only the first recognized case of stereotyped and more general phenomena that occur in all mammals, and probably in all vertebrates. Analogs may also occur in higher forms of plant life.

3. Its Anatomical Distribution[261,267]

It involves the region where its stimulus arose, such as a knee, wrist, arm, foot, or hip. The involvement includes both the hard and soft tissue regional components, as

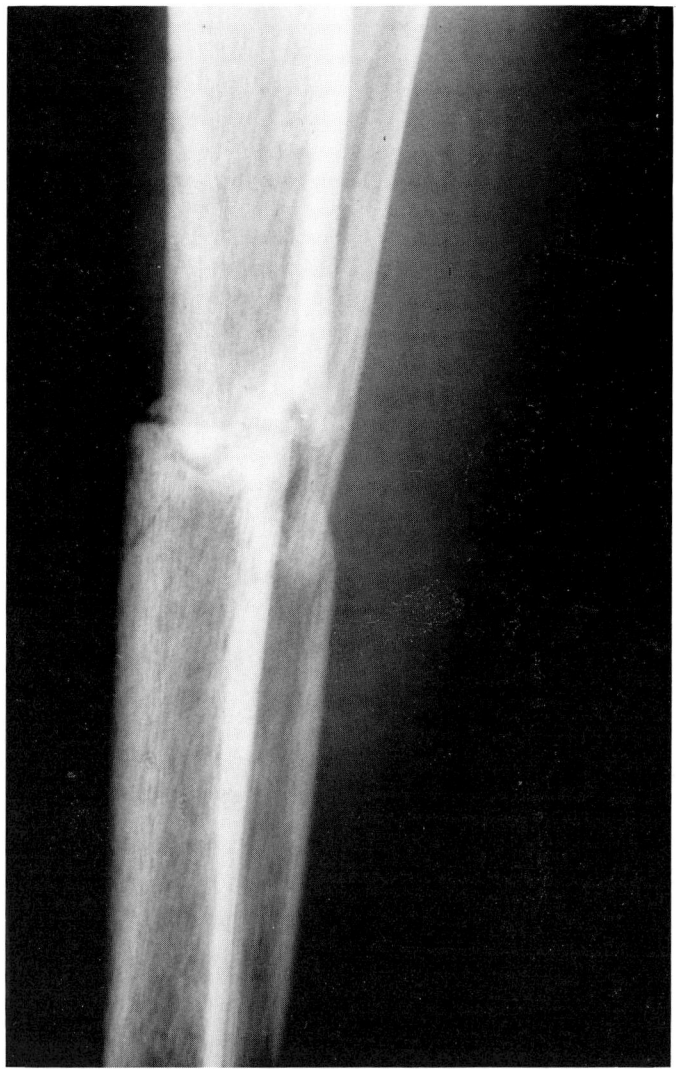

FIGURE 2. RAP effect on compact bone. Lateral X-ray of the tibia and fibula of an adult with a fracture of both bones that has gone on to a nonunion. This X-ray was taken about 5 months after the initial injury. No significant amounts of fracture callus have appeared. Note the extensive longitudinal tunneling of the compacta. Each radiolucent longitudinal line represents an evolving secondary osteon inside the compacta and their numbers are greatly increased over normal, by ~50×. The resulting increase in the remodeling space makes the bone appear less dense on the X-ray and a photon absorption study would reveal significantly decreased amounts of mineral. A bone scan with any gamma-emitting bone-seeking isotope (technetium, gallium, fluorine, calcium, strontium) would show the entire tibial diaphysis as "hot" compared to the tibia in the normal leg. In this patient the injury stimulated an adequate RAP but it did not also enable the formation of adequate amounts of local healing callus.

already noted. After an acute paraplegia, hemiplegia, or monoplegia, whether due to trauma, poliomyelitis, or other acute disease, and whether due to lower motor neuron or to central lesions, the RAP can affect the whole paralyzed part of the body. The

Table 2
RAP: SOME SIGNS IN SOME SKELETAL TISSUES

Bone (lamellar)	Increased cortical porosity, trabecular thinning, marrow cavity expansion; increased numbers of osteoclasts, and osteoblasts after σ_r period; increased perfusion, bone resorption, and formation; increased bulk mechanical compliance, decreased ultimate strength, and breaking energy; increased macromodeling capability
Hyaline cartilage	Increased cell division, matrix synthesis, and turnover; increased mechanical compliance, growth rate, macromodeling capability; increased alcian blue staining and water content; increased macromodeling capability
Fibrous tissue	Increased cell division, matrix synthesis, and turnover; increased mechanical compliance, growth rate, macromodeling capability; increased water content
Synovia	Increased cell turnover, perfusion, and synthesis and absorption of synovial fluid macromolecular components as well as increased turnover of its water and electrolytes

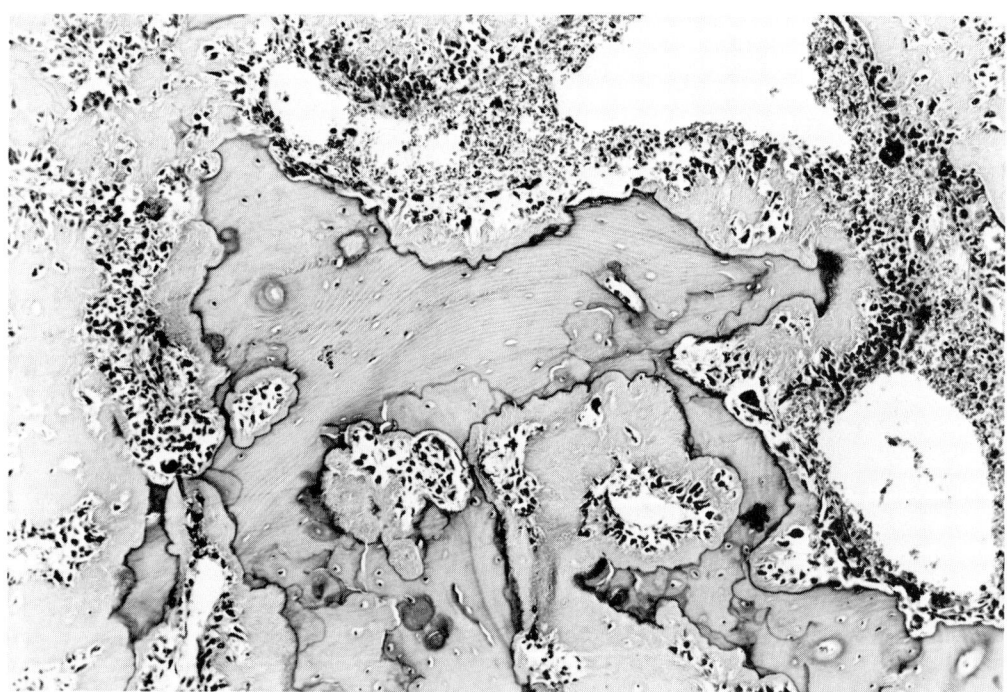

FIGURE 3. A hematoxylin and eosin stained decalcified section of an osteosarcoma that developed in Pagetoid bone. The middle of the field contains a region of the preexisting nonmalignant Pagetoid bone and the scalloped thin dark lines are cementing lines of the reversal type that took the hematoxylin stain. The pattern is the typical florid and anarchic one usually associated with Paget's disease, but sometimes also with congenital lues and with intense local infection, and it is called the "mosaic pattern". This tissue is very vascular and its perfusion is increased. An example here then of a RAP with superimposed features which suggest to the pathologist the nature of the underlying disease. (Photomicrograph courtesy of Dr. K. Wu, Detroit.)

size of the involved region depends at least in large part on the magnitude of the initiating noxious stimulus, for larger stimuli tend to cause RAPs that involve larger body regions. The transition from involved to uninvolved regions is gradual rather than abrupt, and the anatomical distribution of a RAP tends to follow certain features of the regional vascular anatomy and innervation.[267] Erratic abscopic involvement can occur, i.e., changes in tissue dynamics in other parts of the body in response to severe stimuli.

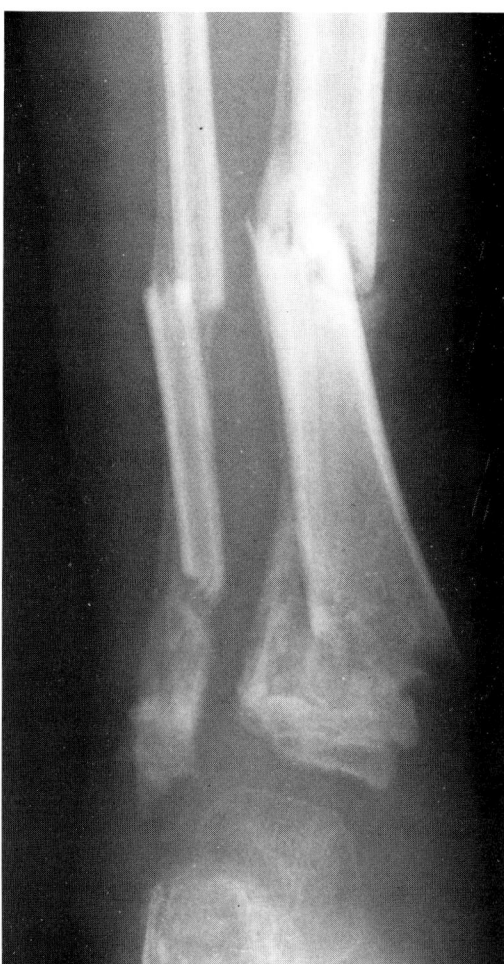 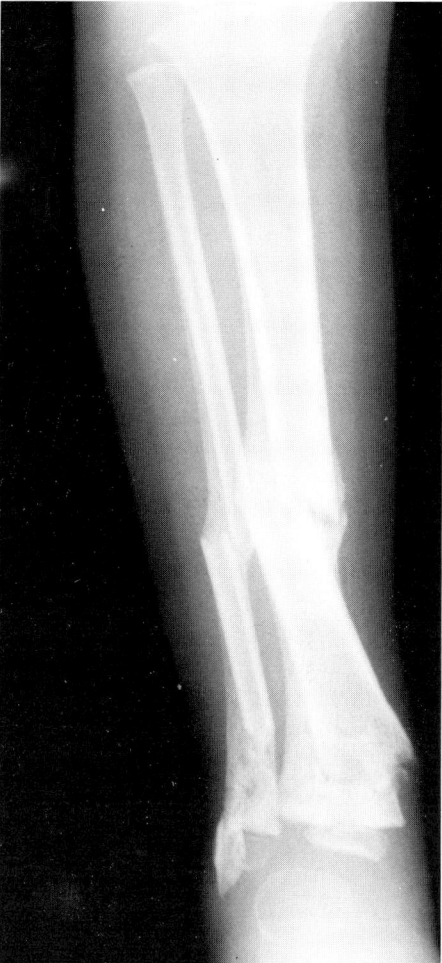

FIGURE 4. The left X-ray of this young child's complex fracture of the tibia and fibula was taken about 1 month after injury and the right one between 3 and 4 months. The left one shows severe osteopenias of the spongiosa of the talus and metaphyses of the tibia and fibula. The right one shows considerable improvement because as the RAP subsided, the increased BMU activation that caused the increased remodeling space that caused the osteopenia on the left also subsided, whereupon the greatly increased remodeling space on the left began to fill back up again. Hence an injury-induced but inherently reversible osteopenia. While some authors persist in attributing such severe and rapidly arising osteopenias to mechanical disuse, no orthopedist of any experience would expect to see it in a perfectly normal leg placed in the same cast for the same period of time. As Seneca said, it is hard to unlearn what one has spent long in learning.

4. Its Duration

A single stimulus such as a femoral fracture, an acute attack of pseudogout, a meniscectomy of the knee, or a gunshot wound typically evokes a RAP that lasts about 4 to 8 months in bone, somewhat less in soft tissues, and as already noted, longer for severe than for minor stimuli and often longer for adults than children.[267] Figure 4 illustrates a quickly resolving bone RAP in a child. However, after an acute paralysis, such as from a brachial plexus injury or after a severe burn, it can last from 6 months to 2 years or more, and for commensurate periods it can accelerate any bone loss caused separately by mechanical deloading and other factors.[261,267] Thereby, it can predispose to hypercalciuria and GU tract lithiasis in such patients. When the causative stimulus persists for prolonged periods, as it can in active rheumatoid arthritis, as the result of a series of bone operations or in the presence of an osteoid osteoma, the RAP can persist similarly without any evidence known to the writer that its duration has a natural limit.

Thus, a RAP has a characteristic σ value, a lead time that separates the time of the enabling stimulus from the time when the characteristic dynamic responses of the regional tissues have developed.

The SOS role of the RAP — The author believes the RAP probably evolved to accelerate the healing of injuries and the elaboration of tissue defense reactions to local infection, infarction, mechanical abuse, and other noxious processes and agents.[214,261,267] Much evidence, some of it given shortly, supports that idea. Such a role suggests an SOS-type phenomenon that could increase the chances of survival of a species during the course of evolution in physically harsh and competitive environments.

Two matters deserve comment before considering the clinical manifestations of a RAP. First, once begun, a RAP tends to strongly dominate other ongoing regional processes as well as those endocrine, mechanical, drug, and other effects that would tend to affect those activities in healthy subjects.[261,267] For example, in the arthritic hip shown in Figure 1, the stimulating RAP effect on growth exceeds the depression in growth that results from reduced loading and mechanical usage. That dominance also fits an SOS role. Second, the RAP as an integrated process exists apart and separately from those other agents and processes that determine *if* the various growth, physiologic, healing, defensive, or other subprocesses will occur, and also apart from the agencies that determine *when* and *where,* and in part (but probably not wholly) how *abundantly* they will occur. The primary apparent effect of the RAP then is to accelerate already ongoing processes. As a relation then:

$$F(\text{RAP}) = \begin{bmatrix} \text{IO potentials} \longrightarrow \text{ongoing processes} \\ \uparrow \qquad\qquad\qquad \uparrow \\ \text{enabling} \qquad\qquad \text{noxious} \\ \text{agents} \qquad\qquad\quad \text{agents} \end{bmatrix} \longrightarrow \text{accelerated processes or RAP} \qquad (2)$$

Two collections of clinical situations illustrate that probable SOS role. One collection reflects the positive effects of an existing RAP, the other the effects of its impairment or absence.

B. Clinical Examples of RAP Effects
1. Potentiated Bone Healing

In the type of fracture nonunion, called a biological failure by the writer[247] and an atrophic or oligotrophic nonunion by others (and illustrated in Figure 2),[127,138] fracture callus arises too slowly and scantily to provide satisfactory union, but not as the result of any known error in treatment. (The next chapter goes into this matter in more detail.) Schenk[666] and many physicians in the Swiss AO group have taught the orthopedic community that such nonunions can heal, given intimate apposition over a large surface that is immobilized or fixed rigidly enough to allow less than 50 μm or so of interfragment motion. Various types of devices implanted surgically can provide that fixation. Union of the bone then occurs, not by callus production, but by BMU-based remodeling that "knits" the fragments together with numerous secondary osteons that cross the fracture interface.[613]

As the writer found in the early 1960s,[230] normally, remodeling turns over less than

Table 3
APPARENT CLINICAL EXAMPLES OF ABNORMAL RAPs

Affection	RAP increased or decreased	Domains of defect
Delayed fracture healing	D	Macro
Delayed healing of arthrodeses	D	Macro
Migratory osteoporosis	I	Macro
Post-traumatic osteodystrophy	I	Macro
Delayed wound healing	D	Macro
Charcot joint	D	Macro
Leprosy	D	Macro
Denervated tissue (sensory)	D	Macro
Looser's zones	D	Macro
Spontaneous fractures (but before the final, complete fracture)	D	Micro
Spontaneous tendon ruptures (but before the final, complete rupture)	D	Micro
Restricted perfusion	D	Macro
Idiopathic	D	Macro

5% of the adult human tibial compacta annually. Thus, if no other factor acted, then after such fixation less than 5% of the tibial fracture interface would become bridged by secondary osteons in the first year and it would take about 20 years to complete bridging it. Good union typically occurs within about 6 months in such cases.[804] The reason lies in the fact that the operation itself (and whether an intramedullary nailing, a compression plating, or a securely fixed sliding bone graft) caused its own RAP that accelerated the local bone turnover 10 to 50 times or more above normal, and usually for longer than 1 year. That acceleration proportionally and directly accelerates the above BMU-based healing process. In that sense, therefore, the RAP fits the requirements of an SOS phenomenon.

Such observations suggest this intriguing and novel — but probably correct — idea: normal fracture healing routinely benefits from and depends on an accompanying RAP. If so, then some delayed unions not due to any apparent inadequacy of treatment could reflect the consequence of an obtunded or absent RAP. It follows that the often beneficial effects of bone grafting, certain other surgical procedures,[140] or electrical stimulation on a delayed union could result from a new RAP evoked thereby, in part at least.[88,127,795] Studies by Takahashi and colleagues[714,721] have shown exactly such effects of the electrical stimulation of living bone. Those authors have published the first report of an experimental design that included a definition of the effects of a RAP and a control for it that allowed the effects of the experimental variable to be corrected for the concurrent RAP effects.[721] The next chapter will summarize some observations on those and related matters.

Table 3 lists some putative examples of abnormal RAPs encountered in clinical practice. The following paragraphs discuss some of them.

2. The Pathological RAP

Two clinical entities seem to illustrate pathological or "runaway" RAPs. In *Sudek's atrophy,* sometimes termed post-traumatic osteodystrophy,[237] an acute injury originally evokes a typical RAP but the local accelerations characteristic of it persist long after the original injury has healed.[61,713] In the writer's experience, sympathetic nerve blocks usually cure this condition, but corticosteroids and related agents do not.[261] The other and relatively rare entity originally described by Frost, Frame, and Duncan, *regional migratory osteoporosis,* and termed "algodystrophy" by European authors,[29,43,440,681] also usually follows a local injury, but it can also arise spontaneously.

When due to an injury it too persists long after that injury has healed. Intriguingly, sympathetic blocks do not cure it while corticosteroids and some other prostaglandin inhibitors do,[261] and in one study about 20% of its victims had overt or latent diabetes.[43]

3. Arthrofibrosis

Joint stiffening due to diffuse periarticular fibrosis of the joint capsule and ligaments (but seldom of synovial gliding mechanisms) can follow such noxious stimuli as regional surgery, local trauma or infection, noninfectious inflammation as in rheumatoid arthritis, and the influence of a regional osteoid osteoma.[804] All such factors can evoke a RAP which, among other things, increases collagen turnover, including the production of new collagen in the regional connective, fascial, capsular, and ligamentous tissues. Lacking frequent range of motion exercises, that fibrosis could then limit the normal flexibility of the capsule and ligaments to produce variable losses of motion. The next few examples will illustrate proposed clinical examples of obtunded or absent RAPs.

4. Neuropathic Soft Tissue Problems

In diabetics with significant peripheral neuropathy but with good perfusion, and in nondiabetic patients with denervated limbs or with severe peripheral neuropathy of other origin (biochemical, mechanical, postfrostbite, nutritional, pernicious anemia), the affected tissues react inadequately to wounds, mechanical abuse, and/or infections.[503] In healthy tissues such lesions promptly evoke a RAP, which includes the classical signs of inflammation plus accelerated perfusion, turnover, and metabolism of local skin and underlying soft tissues. That RAP accelerates the local healing. In the neuropathic affections discussed here, those accelerations prove obtunded or even absent, so relatively little erythema, increased perfusion, or edema occur; the increases arise more slowly than normally, while healing time prolongs, local resistance to infection and mechanical abrasion and pressure declines, and an infection can extend faster through a tissue than its local defensive processes can wall it off.[261,267] Proof of the regional rather than the systemic origin of those matters lies in the fact that no such problems arise in the upper or uninvolved body parts of the same person.

Most clinicians and physiologists tend to dismiss all such problems as the direct and sole result of impaired local blood supply caused by several well-known forms of vascular disease that certainly can occur in diabetic subjects. However, the author has reason to believe that impaired perfusion does not explain all such healing problems, and the kind just described accounts for most of that latter group. Parenthetically, note that poor blood supply, by limiting perfusion of the region, will of itself prevent a normal RAP response to a local injury because most, if not all, of the accelerated activities of a RAP require the support of increased regional perfusion and cannot occur or perhaps persist without it. Thus, in these cases and as a relation this situation obtains:

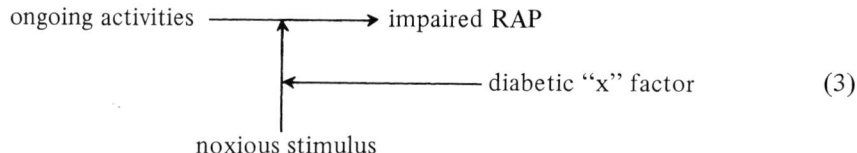

(3)

5. The Charcot Joint

Whether due to diabetes, lues, syrinx,[635] other CNS disease,[171] or traumatic lesions, typical Charcot joints show considerable local bone and joint destruction and gross instability, but in relation to the amount of anatomical damage that can be shown by X-ray and clinical examination they also show relatively mild edema, heat, swelling, and little discomfort.[119,261,776] Static histological examination of tissues from Charcot joints reveals the typical kinds of tissue responses to those stimuli that one would expect to see in a healthy person with equivalent structural damage. Those responses include fibrosis, dilation of capillaries and an increase in their number, production of new woven bone, a low-grade inflammatory response, resorption of hard and soft tissue debris, and edema.

The *static* histology misses an important dynamic feature, and in the writer's experience a consistent one. That is, given equal destruction, daily usage, and instability, then the periarticular and articular as well as the more distant regional tissues of a Charcot joint develop far less response per week than otherwise normal joints would, for in healthy tissues equivalent articular abuse would promptly evoke massive swelling, edema, erythema, and greatly increased regional perfusion, plus abundant fibrosis and reactive bone formation.[261,267] Thus, the rate of response to micro- and macrodamage in a Charcot joint reduces greatly below normal.

That observation suggests that an obtunded RAP plays a role in Charcot joint pathophysiology. That, plus the associated impairment of deep pain sensation, allows daily activities to create new tissue damage faster than the lethargic repair processes can react to and heal it, which in turn leads to the progressive destruction of tissue that characterizes the affection. Similar phenomena can occur in the rare syndrome of congenital absence of pain.

The above thesis can be tested, and in effect the author has done that with some 20 such joints since 1965. To understand the treatment rationale, paraphrase the above material thus: in a Charcot joint the tissues retain the ability to heal, but they can only do it very slowly. As a result, normal usage can create new damage faster than the lethargic healing mechanisms can cope with it, in which case it will accumulate. As a simple relation the matter would look like this:

$$\text{Charcot joint} = \text{microdamage production} > \text{microdamage repair} \qquad (4)$$

If true, then to arrest further destruction of a Charcot joint one would only need to deload it mechanically, and long enough to permit the lethargic healing processes to catch up to demand. Or as a relation one needs to achieve this situation:

$$\text{healing and stabilization} = \text{microdamage production} < \text{repair} \qquad (5)$$

To repeat, the author has done that for some 20 such joints since the above concepts gelled around 1964, and so far at least all of them healed and returned to normal activity. Several of them several years later developed a new episode of destruction of the original joint that required another course of deloading, to which they responded in the same fashion as originally. Figure 5 illustrates an example in one of the author's recent cases.

Parenthetically, a clinically obtunded RAP of neurogenic origin usually accompanies local sensory (and especially pain) rather than motor denervation.[267] It was rare or nonexistent in victims of anterior poliomyelitis,[804] where the motor innervation was lost but the sensory innervation remained. That observation also applies to denervation due to central or peripheral nervous system trauma, CNS disease, and peripheral neuropathies. Those facts strongly implicate the sensory, and particularly the pain, innervation in evoking and controlling the RAP; varied other clinical observations as well

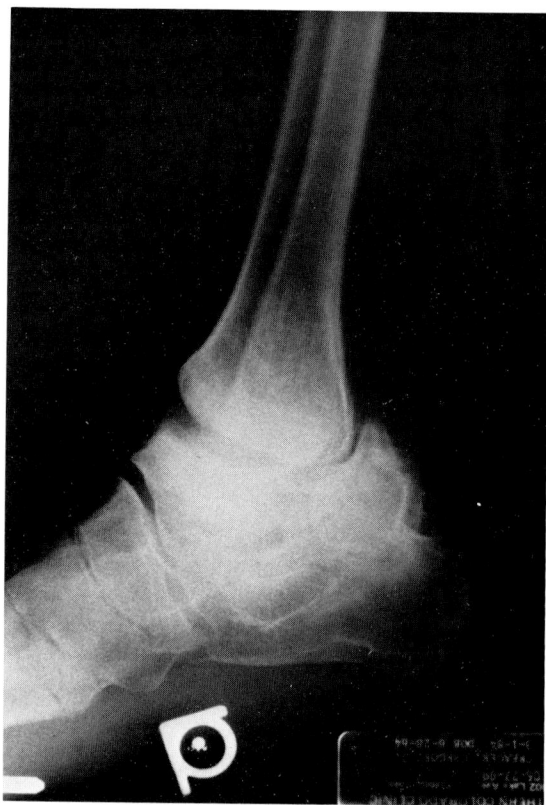

FIGURE 5. A lateral X-ray of an ankle of a 16-year-old lad who has a congenitally absent pain sensation from mid-leg down on both sides. At several times during his walking years the boy developed episodes of typical Charcot joint destruction. They were deloaded by other physicians since the parents refused to consider amputation. Each time, the damage stabilized and repaired sufficiently during the period of deloading to allow resumption of weight bearing. As a result, by age 16 most of the astragalus is gone and a nearthrosis (a new joint) has formed between the tibia above and the talus and calcaneus below. Note that the curvature of this new ankle joint is the opposite of normal, for its concavity faces proximally rather than distally. Chondral modeling of growing articular cartilage sitting on collapsing spongiosa produced that shape during his growth in response to subsidence of the talar dome. The author was brought into this scene by an episode of active destruction in the other ankle. It required a major and unorthodox surgical procedure to replace the calcaneus under the tibia. It was handled according to the guide rules outlined in the text and it did well.

as some experimental data obtained from laboratory animals fit that idea.[1,90] For example, in paralysis due to spinal cord injury at the cervical or thoracic levels, in which lower limb innervation remains intact, lower limb wounds and fractures typically heal well (but not always), and wound infections usually respond satisfactorily to treatment. However, after regional peripheral denervation, reduced periosteal and longitudinal overgrowth of the amputation stump tend to occur, while impaired healing and resistance to infections in such denervated regions has become well known to clinicians. As suggested in Note 1 in Chapter 7, such phenomena could be interpreted as a result of some increase in the local set-point of the MES mechanism(s) that evokes and regulates a RAP.

6. Rheumatoid Phenomena

The tendency for excessive local collagen production to cause joint stiffness, postulated earlier as a positive RAP effect in the active stage of rheumatoid arthritis (RA), usually occurs early in the disease and during active and severe inflammation. In late RA cases and therefore usually in treated ones too, as well as in some rheumatologic variants, the opposite can occur: bone can erode,[651] ligaments and capsules can stretch so joints become lax and sublux, and microscopic structural damage in tendons accumulates unrepaired or so slowly repaired that spontaneous complete ruptures finally occur. The classical external signs of inflammation and other RAP manifestations remain unimpressive. Indeed, the clinical picture here repeatedly is one of considerable internal damage with minimal external evidence of any local tissue reaction to it. Those facts suggest that an impaired RAP may play a role in their genesis, and that it could reflect either a local increase in the set-point of the MES mechanism that normally activates a RAP, a blunted perception of the local MES signals, a blunted ability to respond to it, or some combination. What separate roles the disease process itself and its treatment may play in such impairments remain obscure, as does an explanation for the additional facts that the same patient may heal skin and bone properly. A spontaneously ruptured rheumatoid tendon usually heals well following open repair, which might reflect the fact that the micro- and macrorepair process are fundamentally different, have different enabling and MES mechanisms, and respond to different regulating agents as suggested in Note 2 in Chapter 2. Figure 6 shows a rheumatoid foot X-ray that illustrates several of the points discussed above.

7. Obtunded RAP of Unknown Cause

As an orthopedic surgeon, the writer has had the opportunity to study in the clinic the behavior of two particular osseous RAP features since about 1962, when the RAP was recognized as a distinct pathophysiologic entity. Those features include delayed but ultimately successful bone unions and the development of an increased remodeling space in compact bone (as shown in Figure 2) following a fracture, other injury, or infection. That space can be seen on regular X-ray films of good resolution as a longitudinal tunneling or cavitation within the cortex. Those clinical observations produced the following evidence that still awaits a satisfactory explanation.

First, most cases of an obtunded RAP seen in a typical U.S. general orthopedic practice have no presently known or discernable cause. The known causes described earlier collectively represent a minor fraction of all the examples the author has perceived in his patients. The larger group of "idiopathic obtunded RAPs" do not appear to associate with any currently known form of diabetes or other endocrinopathy, neuropathy, nutritional deficiency, or drug toxicity.[267]

Second, in the same person, say a victim of a motor vehicle accident who sustained 20 fractures, 19 of them can heal promptly and normally but the 20th does not due to an obtunded RAP, and possibly as well to other factors that are discussed in the next chapter.[267] This phenomenon usually affects adults and seldom occurs in children. To the author it suggests that some sort of cycling of the local tissues can occur, so they alternate between a state of competence to respond to a major injury and a temporary state of incompetence, the latter occupying only a small part of the year.[658] The reader may recognize here another example of an "On-Off" system, analogous in an abstract sense to those referred to in Chapter 4 and elsewhere in this book but probably different in its details. As relations that matter might be written:[658]

$$[\text{Challenge} \longrightarrow \text{Tissues in "On" state}] = \text{competent healing response}$$

$$[\text{Challenge} \longrightarrow \text{Tissues in "Off" state}] = \text{defective healing response} \qquad (6)$$

122 *Intermediary Organization of the Skeleton*

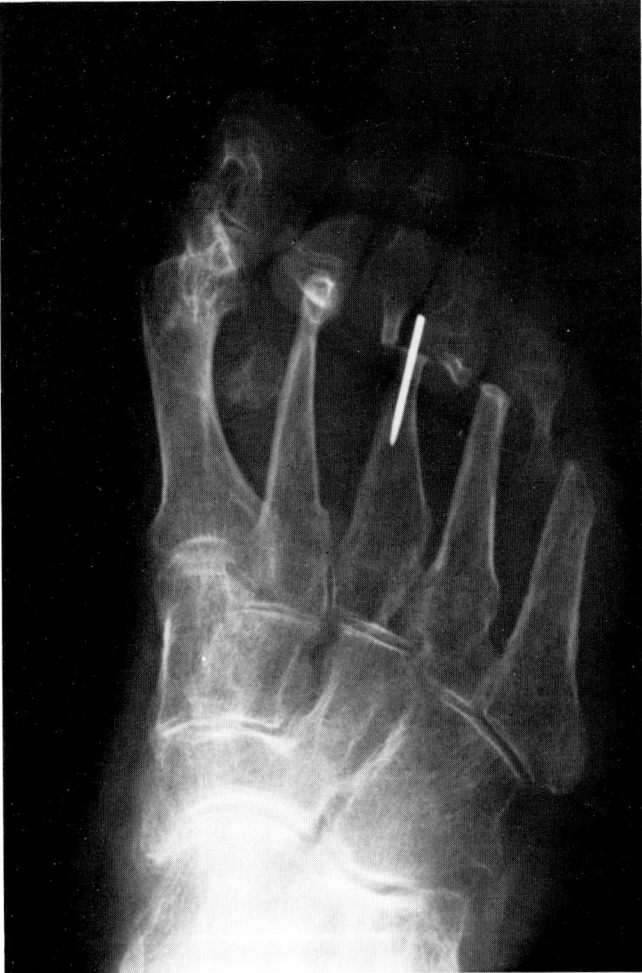

FIGURE 6. X-ray of a rheumatoid foot. It has had extensive disease in the past in the metatarsophalangeal joints which led to a bunionectomy and to a surgical resection of the metatarsal heads and proximal parts of the proximal phalanges. Note the following. First, a diffuse osteopenia, no doubt due to a combination of disuse due to pain, and the acceleratory effects of the disease-induced and postoperative RAP on the net bone loss usually induced by deloading. Second, the outside and inside diameters of the distal metatarsals have decreased, especially in the second and third X-rays. This represents a modeling effect that was enabled somehow in adult life by the RAP and mechanical deloading.

Third, not infrequently, when a nonunion of the kind just described subsequently undergoes a corrective operation, that operation will evoke a perfectly adequate RAP so healing occurs quickly (and the patient is impressed by the skill of his orthopedist). Here the temporary "incompetence" of the local tissues was resolved by the time the final operation was done, and the local tissues switched in the interim from the "Off" to the "On" state.

Fourth, certain sites in the skeleton appear more disposed than others to develop poor RAPs after an injury. They include the long known ones at the junction of the middle and lower thirds of the tibia, femur, humerus and radius, and the pars interarticularis of the vertebrae.[804] Considerable evidence that cannot be reviewed here has led the author to conclude that the conventional explanation for such problems, a poor local blood supply, is usually incorrect although not always so, so some more subtle biological/biochemical factors must be involved. For such reasons the bony tissues in such regions spend a larger part of the year in the "Off" state than other regions.[658]

C. The RAP in Research

Except for demographic matters, probably no area of skeletal research has escaped the perturbing effects of the RAP on experimental design, execution, and interpretation. Like the transient steady-state distinction, therefore, good future research cannot be done in ignorance of the existence and nature of the process. It was discovered initially during the work that led to the development of the dynamic histomorphometry methodology described in Note 1 in Chapter 4 and a brief review of that discovery should be instructive.

1. The RAP in Histomorphometry[261]

About 1960, the author removed the 6th and 7th ribs at autopsy from a patient who had undergone a thoracotomy 7 years before by Dr. Conrad Lam of Henry Ford Hospital. The originally resected 6th rib had regenerated completely as far as the naked eye could tell, but histomorphometric measurements showed evidence of a much higher bone turnover in it than in the uninjured adjacent rib. That suggested that something, its nature obscure but somehow related to the original resection, still acted and increased the turnover of bone in that rib above the normal for the adjacent rib and of the skeleton generally. With perceptions attuned thereby, numerous subsequent personal observations proved that significant injury of any type to a bone will usually accelerate its turnover for 1 year or more.

For a budding histomorphometrist using the light microscope to study the space-time domains of cells and osteons, that phenomenon meant that biopsying the same bone before and after some treatment to determine the effect of the treatment on bone physiology and its dynamics would yield real but perturbed data. The same conclusion would apply to any other cause of a local RAP. The writer developed the technique of serial rib biopsy, and others the technique of biopsying first one iliac crest and then the other, to avoid such errors. Also the writer originally, and then others in the o bone group, circulated that information verbally among other histomorphometrists, beginning at the first Sun Valley Bone Workshop organized by Jee in 1964. As one result, today most histomorphometrists around the world realize that serial biopsies of the same bone preclude useful comparison of pre- to post-treatment effects on bone turnover parameters.

Note Bene: The potentially perturbing effects of the RAP on experimental data have remained unknown to most investigators interested in other kinds of skeletal problems, including biomechanicians, biochemists, anatomists, endocrinologists, and pathologists, which has created some confusion and misunderstandings; some examples follow. They became apparent gradually as the author's early perceptions of RAP effects in the L_3-domain expanded to include larger domains: the organ and the patient.

2. Longitudinal Bone Growth

In children, bone growth often accelerates for some time after a major fracture or surgical procedure on a bone.[698] With erratic success surgeons have tried to use the phenomenon to stimulate the growth of a child's short bone,[423] e.g., by periosteal stripping[38,410] or by implanting beef bone or ivory pegs in it. The ensuring growth accelerations represent typical positive RAP effects.

Growth acceleration can also follow acute paralysis, whether it is due to trauma, poliomyelitis, or other effects,[639,640] as well as local irritants such as chronic osteomyelitis, a foreign body, or, again, an osteoid osteoma,[6] and also during juvenile RA as in Figure 1.[336] Some observers attributed those growth effects to mechanical deloading. However, and regardless of the details, the ultimate steady-state effects on limb growth of mechanical deloading without any complicating RAP appear in children paralyzed early in life, including by myelomeningocele.[662] The invariable end result is a short limb. Unquestionably, an acute paralysis can produce the earlier described features of a RAP in both laboratory animals and in man. In those situations, a RAP could transiently stimulate growth (i.e., for about 1 year, often less, sometimes more) and, while it lasted, it would dominate the separate but concurrent, oppositely acting, and weaker effect, a growth retardation due to deloading. When the dominant RAP subsided, then the still persisting depressive effect of deloading could become apparent.

The RAP also explains at least some of the otherwise confusing results of applying experimental forces across growing epiphyseal plates to determine the effects of compression or tension loads on their growth rate.[30,601,712] The surgical implantation of the hardware chosen to apply those forces also evoked a RAP that tended to accelerate growth, as did also the often present pin tract motion, inflammation, and pain, while the mechanical forces exerted their own effects, which can vary according to their magnitude and nature as described in Chapters 7 and 8. The investigators saw their variable net result, mixed in unknown proportions.

3. Bone Mechanical Loading Experiments

The interest of the orthopedic community in mechanical effects on bone architecture has led to many experiments that involve surgically implanting hardware at each end of a bone, and then connecting the hardware by springs, rubber bands, or other restraints to compress the bone uniaxially.[822] Some investigators attributed the resulting increased outside diameter of the bone and its increased vascular porosity solely to the increased mechanical compression. However, they usually also committed two basic experimental errors. First, they omitted the necessary control for the surgical procedure itself: identical devices implanted in the opposite limb but with the mechanical restraints disconnected.[822] That control would allow one to distinguish the effects of the RAP from the added effects of the mechanical force, and such surgical procedures do cause RAPs in their own right. Second, they mistook the formation of new, plexiform, or woven bone on the periosteal envelope that such experiments have usually evoked as representative of the structural adaptation of a normal skeleton to a physiological increase in compression loading or a change in its pattern (see Lanyon's article in Reference W). The plexiform bone, however, represents a pathologic or SOS response to a pathologic stimulus.[804] It is the protective response of the system to something that overloaded or saturated its physiologic mechanisms, and it is not representative of the circumferential lamellae that form during an osteoblastic drift in response to some change in the loading of a bone. Relation 11 in Note 1 in Chapter 7 diagrams those relationships.

Consequently, little has come from such experiments that will withstand critical examination, and in fact they have not revealed how normal bone architecture in a healthy child responds to the mechanical factors that affect it during normal growth and development.

Some of the same problems afflict studies of mechanically deloading long bones by plate fixation to observe the effects of "stress" on bone healing and other physiological processes.[151,534,730] There, one could distinguish RAP effects from stress-strain ones simply by implanting identical devices on each femur but backing the screws off a full turn on one of them so that it carries no mechanical load. That elementary control has usually been omitted. Also, comparisons of devices made of different materials and in different shapes and sizes require separate controls for the effects of each, but so far those controls have been omitted by most investigators doing such work.[7] It follows that conclusions based on such experiments are suspect and such data cannot be interpreted until better experiments are done. The conclusions of various authors reporting such work could ultimately prove correct of course, but if so the experiments published so far in support of those conclusions do not prove it.

III. COMMENT

A. Some System Properties of the RAP

A typical RAP following a Colle's fracture of the wrist somehow involves and coordinates the regional control of numerous separate factors that exist and function in healthy, uninjured tissues.[261] For example, it increases the metabolic activity of most regional cells in both soft and hard tissues, an L_c-level effect. It increases BMU-based remodeling at all ages and regional growth and modeling in children, and it can enable a limited amount of modeling of bone and cartilage in adults (examples appear in osteophytes, degenerative joint disease, and fractures; see Figures 6 to 8), all of which span the L_2- to L_o-space-time domains. It also increases regional perfusion.

In other words, a RAP does not create any new *kind* of activity in individual tissues. It simply depresses the accelerator and/or releases the brake on many or all of the activities normally present in tissues as potential responses or active processes, and it also coordinates those activities of the various regional tissues to create a new, integrated L_o-domain process that has as its purpose the goal of potentiating survival. As a highly integrated process, it associates separate activities that span the L_c to L_o and higher levels of biological organization. It follows that to study it effectively, one must measure its various constituent rates and compare them before and after as well as during the RAP; this should be done in the intact individual. Here is another situation where the concepts inherent in coherence research should help to understand and learn how to control the process. Table 1 summarized some of the above material.

B. Experimental Design and the RAP

The foregoing material should make it clear that when any invasive procedure becomes a part of an experiment addressed to the physiology of the skeleton, the design of that experiment should include means to control and/or correct for the effects of the RAP caused by that procedure. Otherwise, its effects will mix in unknown proportions with those of the intended dependent variable and will probably be interpreted as due to the latter.

When they involve anything else that could cause a RAP, then that control should be provided in all other studies of skeletal growth, modeling, remodeling, repair, and homeostasis. An effort should probably be made to provide them for organ and tissue culture work too, and also in any experiments with intact animals that subject them to considerable psychological stress. To repeat, Takahashi et al.[721] published the first example of such experimental controls. All of which means that many things it was thought were already understood about the responses of skeletal tissues and structures to diverse challenges will have to be looked at again.

C. Experimental Studies of the RAP

High and colleagues[331] have provided published experimental data specifically applicable to RAP effects. Evoked in canine rib by periosteal stripping, dynamic histomorphometric analyses of the stripped plus the adjacent uninjured ribs revealed significantly accelerated turnover of the stripped ones 6 weeks later, apparently unmodified by concurrent treatment with several other agents. The design of her experiments elim-

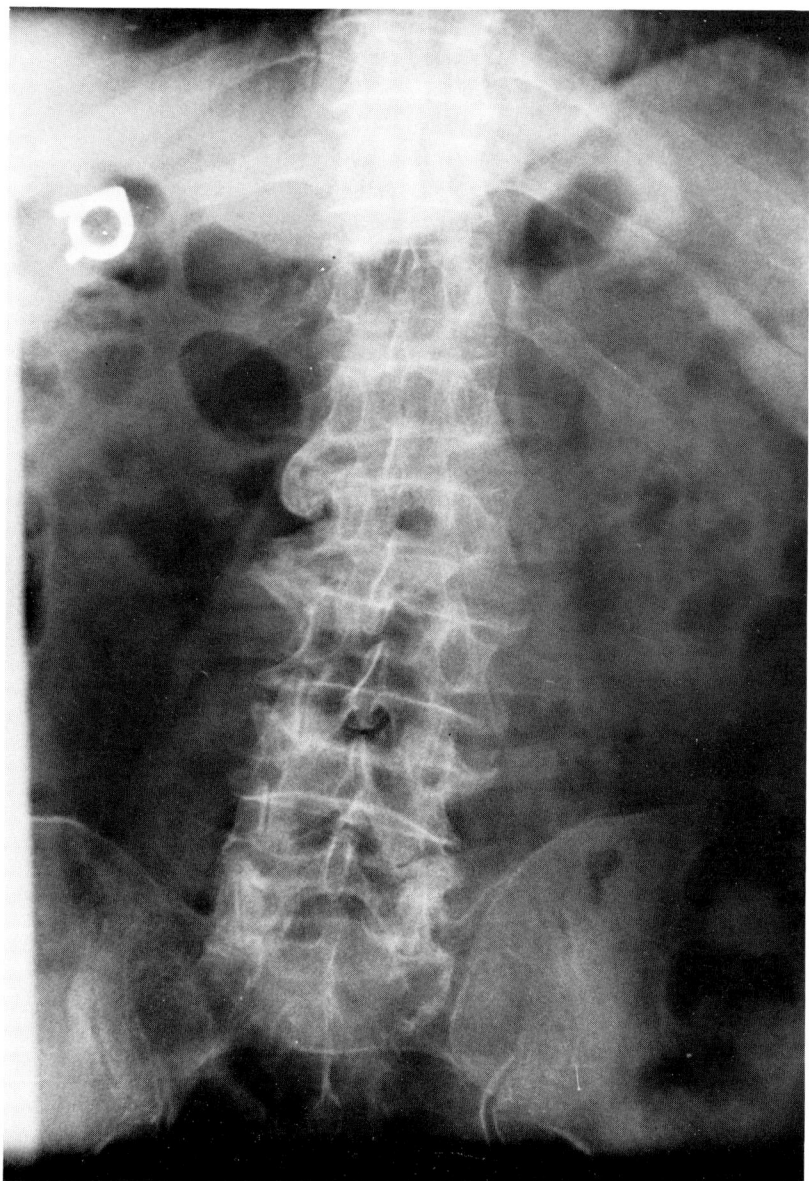

FIGURE 7. RAP-induced chondral growth and modeling in the adult. This AP X-ray of an older adult lumbosacral spine shows large osteophytes projecting off the margins of the end plates of the vertebral bodies. Here, something presently unknown set off a local RAP which, among other things, increased local tissue turnover. This also involves increasing local cell turnover, which of course involves increased "P" cell activities. That reactivated chondral growth at the margins of the vertebral end plates, which are normally covered with cartilage. The new cartilage acted like an epiphyseal plate in that it mineralized at its base, which then enabled the endochondral ossification process. This process proceeded to replace the mineralized new cartilage with bone, although much more slowly than it occurs in childhood. The bone then becomes apparent on an X-ray as a spur or osteophyte. Hence, in an older adult a local RAP has somehow enabled purely local chondral growth, chondral modeling, and endochondral ossification.

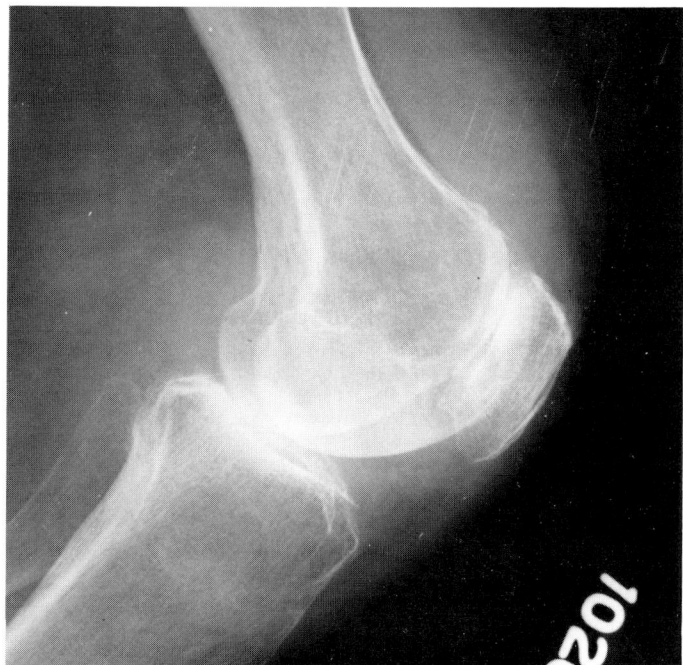

FIGURE 8. RAP-induced bone modeling in the adult. This patient had a chronic (e.g., long-standing) synovitis of the knee. The synovia was also voluminous or bulky. It caused the oblong soft tissue density above the patella and to the right of the lower femur. Over the years some modeling occurred in the underlying anterior femoral cortex, and a periosteal resorption drift and a cortical-endosteal formation drift moved it posteriorly (to the left here) to make room for the bulky synovia. The inflammation of the synovia probably caused a local RAP that enabled those bone drifts, at least in part. Such adaptations require many years to occur in adults, whereas they can occur in months in children. (From Frost, H. M., *Bone Modeling and Skeletal Modeling Errors*, Charles C Thomas, Springfield, Ill., 1973. With permission.)

inated mechanical loading effects as a significant factor in the results, so bone RAP effects need not necessarily follow mechanical deloading alone. On the other hand, other uniquely informative experiments by Jaworski and Uhthoff[372,738] revealed that a RAP can in fact also result from deloading and furthermore, that it can differ significantly in kind, severity, and anatomical location according to the subject's age, which suggests that multiple enabling factors and parallel gating can be involved in evoking a RAP.

D. Causative Mechanisms of the RAP

These remain conjectural at present, although the above material suggests a multifactorial causation that includes the anatomy and competence of the regional blood supply, its sympathetic innervation, and the regional sensory (particularly pain) innervation, plus mechanical loading, plus direct injury to bony and soft tissues, plus the gamut of local biochemical and biological factors already known to associate with injury, repair, metastasis, and inflammation.[261] A partial list of such potentially involved factors would include prostaglandins, osteonectin, leukotrienes, histamine, lysozymes, axon flow products, OAF, and several growth and angiogenic factors. It seems likely to the author that many different kinds of agents are involved in causing the RAP, and it could turn out that it has distinctive features at different levels of biological organization, so that, for example, one collection of factors might enable and regulate L_2-domain and lower-level responses, and different ones the L_o- and higher-level responses. Hans Selye's generalized stress reaction, which involves responses of the adrenal cortex,[49] becomes a regular feature of the RAP when it is evoked by extensive injuries to soft and hard tissues, yet the systemic stress reaction does not occur in response to minor injuries — such as a splinter in the hand or an ankle sprain — so in that respect evidence already suggests that different magnitudes of response can associate with different enabling and regulating factors.

Takahashi and colleagues[721] have shown that typical bone RAPs are produced by electrical stimulation of bone with commercial equipment used to treat fracture healing problems, and they provided good controls for the RAP. Those controls revealed that the surgical implantation of the electrodes caused its own RAP, to which the electrical effects became added.

E. A Final Common Pathway?

Somewhat like the lower motor neuron or BMU-based bone remodeling, the RAP may represent a kind of final common pathway for an appropriate physiologic expression of a protective reaction to diverse challenges. Its particular clinical, biochemical, and histologic features can also accompany considerable overlay by the special features of a particular cause, such as trauma, tumor, or infection. Clearly, however, it reflects the direct effects of the activities of the IO more than any properties of unassociated cells. That suggests that effective study of it will require the use of intact subjects rather than ex vivo systems.

F. The RAP: It is Ubiquitous

Finally, the RAP has been described here with an emphasis on its skeletal manifestations and effects, but it occurs also in the viscera of the abdominal, thoracic and intracranial cavities, in the soft tissues of the oral and nasopharyngeal regions, the eye, and the middle ear, so it represents an ubiquitous and anatomically and biologically general phenomenon rather than one peculiar to the skeleton. Since it was first formally described in an indexed journal quite recently,[261] it may take some time before it is appreciated by those who are currently unaware of it but should know and understand its effects in order to improve the quality of their experimental designs.

The next chapter will describe certain aspects of fracture healing, a process which is so regularly accompanied by a RAP that only recently were the two processes perceived as separate.[261,267]

IV. ILLUSTRATIVE QUESTIONS

Exactly what agents — physical, biological, and biochemical — enable a RAP and then regulate its continuum phase? And for each of the regional tissues that it involves? What are the roles of the regional sympathetic innervation, the sensory innervation, the motor innervation, and the vascular tree in enabling and regulating a RAP? Are they causative, permissive, both, or what? What are the precise histological, metabolic, perfusion, tissue turnover, modeling, and growth effects of a RAP? And in different ages, sexes, species, and parts of the body? What are the homeostatic effects of a major RAP, if any? What accounts for the phenomena shown in Figure 9?

What causes the not uncommon idiopathic obtunded RAP? And the pathologically protracted ones?

How should one study the phenomenon? How to measure it and with what tech-

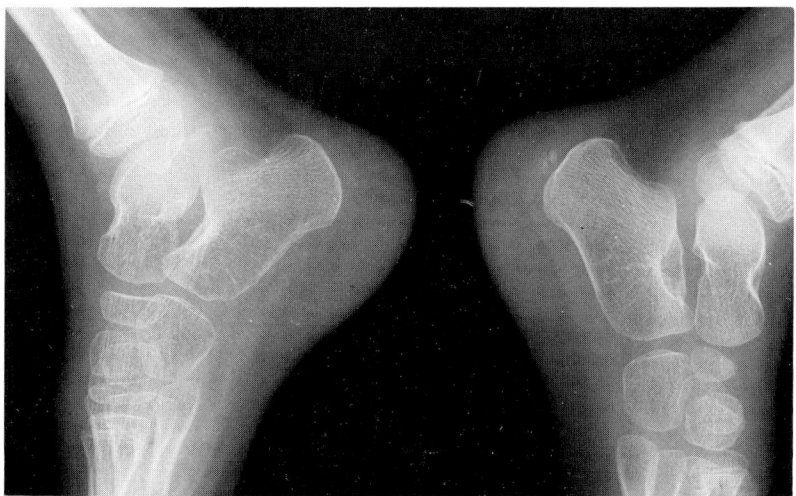

FIGURE 9. Lateral X-rays of both feet and ankles of a youngster who had Still's disease (juvenile rheumatoid arthritis) involving the joints of the foot on the reader's left. The resulting accompanying RAP has led to accelerated maturation of the ossification centers in the mid-tarsal bones, an increase in the size of the talus, and an increase in the heights of the distal tibial epiphysis. Therefore, it also increased local growth. What mechanisms — biological, biochemical, and physical-chemical — cause these effects, and precisely what mechanisms within the tissues respond to them? (From Frost, H. M., *Bone Remodeling and Its Relation to Metabolic Bone Disease,* Charles C Thomas, Springfield, Ill., 1973. With permission.) Case courtesy Dr. H. Duncan, Henry Ford Hospital.

niques? Could it also serve to aid growth, healing, modeling, remodeling, and diagnosis and treatment, in clinical situations? And if so, how? How could one evoke one deliberately, and, contrarily, suppress one that already acts? What controls its duration and extent? And how? How did something so ubiquitous escape recognition as a specific behavioral entity for so long? What effects do the endocrines and commonly used drugs have on it? And in different tissues, species, sexes, and ages? Where do prostaglandins, OAF, growth factors, axon flow phenomena, and related and similarly currently enigmatic substances and processes fit into the picture?

What are the basic differences, if any, between RAPs that involve tissue domains of L_2 and below, and L_o, and L_{sk}?

What else of importance to this subject and the domain of this chapter do we need to know that this chapter omits, whether out of the writer's ignorance, or because we have not yet perceived its significance, or have not yet discovered it?

Note 1

ON DRUG AND ENDOCRINE EFFECTS ON THE SKELETAL IO

In effect, the perception of the operational importance of the skeletal IO, the properties of temporal coherence and of the MCN and MES properties, opens a new area of endocrinologic and pharmacologic action and research that has considerable potential for the health sciences. A great deal is known about the effects of many agents when given continuously to the intact subject as well as to cell and tissue culture systems, but today virtually nothing is known about their selective effects on the living mechanisms of the intact IO, nor about those effects when the agents are given for brief and intermittent periods rather than prolonged and continuous ones or/and in different doses. As one example, prolonged treatment with the diphosphonate etidronate does not cause a progressive and usefully positive bone balance,[208,209] yet brief periods of treatment with that agent timed to act only on the resorption stage of coherent bone remodeling BMU appear to have had large positive trabecular bone balance effects in the London, Ontario human osteoporosis experiment.[24] Also, Hesp et al.,[324] as well as others, find quite different effects in man of small doses of parathyroid hormone compared to large ones. The precise nature of those IO effects cannot be deduced solely and accurately from the organ-level effects of continuous treatment, nor from the L_{is}-level properties of steady-state disease, nor from the responses of cells in tissue and organ culture systems. The preliminary success of that Ontario ADFR treatment provides a demonstration of the contention that perfectly ordinary drugs with which the medical profession has accumulated vast and seemingly exhaustive experience nevertheless can have different and useful effects when the manner and timing of their administration and withdrawal fit the temporal properties of the bone remodeling BMU.[267,658] Since those effects were predicted in advance from theory rather than being observed in action as a result of serendipity, of chance, they lend support to that theory and suggest that it is not as esoteric as many clinicians and investigators who lack personal knowledge of it assumed it to be.

Accordingly, the same kind of strategy, basically a coherence study or plan of research, could and should be made of the effects of all known skeletally active hormones, drugs, and other agents on the many physiological mechanisms that are provided by the IO entities of all skeletal tissues, which are described in Chapters 3 to 5, and that have never before been studied as specific problems in temporally coherent situations.[267,658] That advice acquires significant weight from the fact that the malfunctions of those IO entities directly cause many troublesome clinical affections that cost the peoples of the developed nations of this planet scores of billions of dollars annually in terms of disability, lost earnings, taxes, and direct medical expenses.

Table 4 lists some of the IO mechanisms that would seem at present to deserve such specific studies in the future, and Table 5 lists a few skeletal disorders that are prevalent in people in the developed nations and which produce large drains on national economies and on the efforts and income of working citizens.

In contemplating and designing such studies, certain general features should be kept in mind. In all such studies it will be assumed here that temporally coherent populations of IO units can be created by some stratagem so that the sequential activities of those units will then follow in a predictable order, at predictable times, and in usefully synchronous states.

Drugs given for brief periods, say 1 to 10 days, can have different effects on IO units in kind as well as in magnitude than those caused by prolonged treatment at the same daily dosage.

Table 4
SOME SKELETAL IO MECHANISMS, MALFUNCTIONS OF WHICH CAN DIRECTLY CAUSE HUMAN DISEASE

Lamellar bone remodeling BMU	Lamellar bone macromodeling
Formation of primary spongiosa	Lamellar bone micromodeling
Formation of secondary spongiosa	Fibrous tissue macromodeling
Histogenesis of basic skeletal tissues	Fibrous tissue micromodeling
The MES for growth, modeling, remodeling, repair, homeostasis	Chondral macromodeling
	Chondral micromodeling
Growth, L_1-mode of all skeletal tissues	Growth, L_2-mode, of all skeletal tissues
Microdamage of all skeletal tissues	Repair of microdamage in all tissues
Turnover of fibrous tissues, L_c- and L_2-modes	Mechanical creep in all structural tissues
Turnover of cartilage	Creep compensation in fibrous tissues
Skeletal role in homeostasis	Creep compensation in cartilage
Biomechanical competence of bone	Biomechanical competence of fibrous tissues
The CGFR response characteristic	The stretch-hypertrophy axiom
Strain-averaging mechanism	Strain-induced signals
The activation of an IO mechanism	Regulation of IO continua
Saturation mechanisms	SOS mechanisms
Load focusing mechanisms	Load defocusing mechanisms
MCN, LBO, LOBO effects	$kr\sigma$, On-Off effects

Table 5
SOME PREVALENT DISEASES THAT DERIVE FROM SKELETAL IO MALFUNCTIONS

All kinds of osteoporosis	Degenerative joint disease
All kinds of osteomalacia	Rheumatoid arthritis
Delayed and nonunions of fractures	Children's limb malalignments
Delayed and nonunions of arthrodeses	Chronic back pain
Fractures due to mechanical incompetence	Disorders of skeletal growth
Failed endoprostheses, skeletal	Failed endoprostheses, vascular
Dental caries	Periodontal disease

Drugs given briefly and intermittently, say for a few days once every month or two, can also have different effects in kind as well as magnitude than they do when given continuously at the same daily dosage.

Drugs given both briefly and intermittently may not create lasting resistance to their effects on skeletal tissues and cells, whereas they often do when given continuously for prolonged periods.

A drug which has harmful effects when given continuously and for prolonged periods may prove safe when given for brief intermittent periods and may have different kinds of effects on IO entities. That is particularly true for coherent IO populations, which offer the possibility of circumventing the ubiquitous MCN effects that associate with the treatment of temporally incoherent states and/or with any treatment equal to or longer in duration than the σ value of the IO units in question.

The sequential use of two or more drugs can have different kinds of effects on the IO units of all skeletal tissues than their simultaneous administration, providing the timings of their delivery and withdrawal fit the temporal and sequential properties of the coherent populations of the IO units that the treatment manipulates, and provided they act on different nodes or activities in the internal sequences of those units.[584] Improperly timed administration of an agent should usually evoke an MCN effect that will vitiate or nullify a desired therapeutic response.

When given for brief intervals, different daily doses of the same drug could exert selective effects on some of the different activities encoded in the terms of the basic state equations of the IO (Relations 3 and 4 in Chapter 3). That may apply to incoherent populations of IO units as well as to coherent ones. Thus, as an example of such possibilities (not a specific proposal), magnesium given briefly at one dose level might act primarily and directly on the capillary but at a higher dose and/or for longer periods it might affect the "P" and/or "A" cells too. Some published evidence fits that idea, although it does not prove it. For example, work by Parsons[576] and High[327-331] among others suggests that a given amount of parathyroid hormone in a 24-hr period can and probably does have different effects on both bone remodeling and on the skeletal contribution to homeostasis when it is delivered as a single bolus in contrast to when it is given at a constant rate by infusion over the whole 24-hr period.

To confine a drug effect to a brief period, say over 1 to perhaps 12 days, the half-life of the drug in the blood or/and in the tissues under treatment should be sufficiently shorter than the intended duration of action of the drug to avoid prolonging the drug effect beyond its intended cut off time.[267] For example, vitamin D_3 would seem to be a poor choice to activate a "pulse" of bone remodeling because the liver stores and subsequently releases it over a period of 4 months or so which, from the standpoint of the σ value of the bone remodeling BMU, is of the same order of time; and from the standpoint of the σ_r period is several times longer than that. The metabolite, 1,25-dihydroxyvitamin D_3, has a half-life in the tissues of much less than 1 day; in that respect at least it could form a good activator for a pulse of bone remodeling, for the day after its delivery stopped no significant amounts of it would remain in the tissues or body fluids. Fluoride tends to accumulate in the mineral phase of the skeleton so that after a period of treatment with it, a significant level of fluoride will persist in the general circulation.[85] Lead and some other materials have the same property. Also, during the bone remodeling cycle, when a resorption process erodes a previously fluoridated moiety of bone, the relatively higher fluoride concentrations in the downstream capillary of that BMU will expose the corresponding bone formation center to larger doses of fluoride than exist in the general circulation at the same time. Other drugs can have analogous properties, including etidronate, thyroxine, lead, and some long-acting adrenalcorticosteroid and estrogen analogs.

Somewhat analogous situations concerning endocrine and drug effects should be considered for the other four of the skeletal quintet of basic tissues and their higher-order constructs, of course with appropriate changes in σ values, sequences, and timing.[267]

Very likely in the coming decades, what are named coherence treatment and coherence research here will expand and become particular and versatile branches of pharmacology, therapeutics, and research, whether or not the terminology that signifies them here survives intact. It is already clear that as general procedures those techniques can be formalized and expressed in relatively simple mathematical formats and terms. However, doing so is a task for another time and place.

In doing such work it will prove helpful to keep several ideas in mind. First, enablement and continuum regulation are usually somewhat different activities which may involve different biochemical and physical messengers. Second, many biological responses divide naturally into trivial, physiologic, and pathologic ranges. Third, separate kinds of MES mechanisms may be involved in some of those transitions. In one way the MES concept is like the concept of σ and its associated ideas of transient and steady-state behavior: a "sleeper". It is deceptively simple, yet it will probably be found to permeate all areas of skeletal physiology and disease, and again, whether the present terminology survives or not.

Chapter 12

THE BONE REPAIR PROCESS

"When all else fails do a history and physical."

(H. M. F. about 1965)

ABSTRACT

The bone macrorepair process associates a series of special IO mechanisms in fixed sequences following an enabling or activating stimulus. The intact process heals fractures, osteotomies, and arthrodeses; parts of it underlie the reactions of bone to many diseases, including infection, some metastases, Paget's disease of bone, and myositis ossificans traumatica and progressiva. While the details remain obscure, special local tissue signals evoke each process and control the transitions of one to the next, while some systemic factors can potentiate and/or depress those processes and transitions. Failures of these mechanisms produce a variety of clinical problems and situations.

Effective research into the bone macrorepair process will probably require approaching it as a complex and integrated system in the intact subject.

I. INTRODUCTION

A. A True Ministory

A young plumber enjoying an outing on his motorcycle made unexpected contact with another vehicle whereupon that vehicle became the hammer, the engine of the cycle became an anvil on wheels, and the leg of the young man astride the work. Severe, comminuted, open fractures of both leg bones plus skin loss and other soft tissue injuries ensued.

He had good primary care. The skin grafts and other wounds healed and so did all but one of his fractures. Over the next 8 years he underwent eight tibial bone grafting operations by four orthopedic surgeons of the region. All operations failed to achieve bony union; gross deformity developed and he appeared in the author's office one afternoon walking on his lateral malleolus requesting an amputation and an artificial leg so he could return to work. Instead, he was advised to submit to another operation; some 4 months later he did and 6 months after that he returned to work and continues to work 9 years later on a sound limb.

This chapter concerns some basic bone biologic matters that intruded into clinical practice and made the first eight operations fail, but made the ninth succeed.

B. Background

For over 50 years, recently graduated medical students have had a tendency when faced with a clinical problem to begin ordering many tests in order to find the diagnosis (the writer was no exception). However, that strategy leads to confusion far more often than to accurate diagnosis. And when "test confusion" reigns, an adequate history and physical examination will often reveal or lead directly to a correct diagnosis. As a teacher, the author confronted the "test and more test" strategy anew each year with new orthopedic residents, which led to the above aphorism. It has a point that goes beyond its wry humor, for in effect it observes that when one is confused by the leaves, a better perception of the forest can be had by backing off and taking a look at the

Table 1
REPAIR MECHANISMS IN MAMMALIAN STRUCTURAL MATERIALS

Separate mechanisms counteract molecular, microscopic, and macroscopic deterioration and injury.
Molecular-level mechanisms turn over the chemical components of the tissue under the control of the cells in residence, e.g., osteocytes, chondrocytes, fibrocytes, without necessarily producing new cells or adding new tissue. One role: minimizes the development of fatigue-induced molecular-level mechanical damage in bone, cartilage, teeth, and fibrous tissue.
Microdamage repair mechanisms remove and then replace with new tissue of same kind, fatigue-induced microdamaged regions of tissue. Requires producing new cells and new organic matrix. One role: prevents accumulation of microdamage in bone and fibrous tissue.
Macrorepair mechanisms bridge gross ruptures or fractures, first with scar or callus, respectively. Remodeling then replaces the scar or callus with the original kind of tissue. One role: repairs gross failures of bulk tissues.
Each mechanism has its own properties, enabling agents, regulation, and disease spectrum.
The RAP potentiates each of the above processes.

whole thing, whereupon the way to go to find the proper leaves to study or test often becomes apparent.

This chapter will approach the bone repair process in somewhat the same way: a grand view first, then down to some of the details. Both fibrous tissue and bone have evolved mechanisms for healing injury that show many similarities but which bear little resemblance to the different mechanisms that repair chondral tissues. Both bone and fibrous tissue demonstrate at least two levels of repair in the operational sense and in intact animals. Microdamage usually results from mechanical fatigue, and specific L_2-level IO entities repair it, as mentioned in Chapters 7 and 9. Grosser injury activates or enables the macrorepair process, a different and larger-scale phenomenon that combines several L_2- and L_3-domain IO entities to form specific sequences.

The following text will discuss the bone macrorepair mechanism as an illustrative case that has generated considerable contemporary interest and study. The discussion will use much of the material presented in earlier chapters to help explain how normal healing occurs and how it can malfunction to cause clinical problems. Table 1 summarizes some of the facts that relate to micro- and macrorepair in man. They are important because their malfunctions probably cause over 1 million human bone repair problems annually in the U.S. alone, which makes them major medical and socioeconomic problems throughout the world as well as in the U.S.

The sequential biological processes that heal ordinary fractures form a basic "SOS"-type of response to a variety of threats besides trauma, and they contribute significantly to the survival of mammals as species as well as individuals. They play central roles in delayed and nonunions of fractures and of arthrodeses of appendicular and spinal joints, as well as in various forms of heterotopic ossification, in Paget's disease, in certain bony metastases, and in Looser's zones. Although those processes are well described, their determinants remain poorly understood, so extant descriptions emphasize their anatomy and chemistry and underestimate and/or misunderstand many of their systems properties. However, recent developments promise to change that situation. Table 2 lists some clinical examples of bone repair problems. The factors behind them will only be outlined in this chapter in four parts: (1) the biologic processes of bone repair; (2) some clinical problems that relate to malfunctions of those processes; (3) how the two could fit together; (4) some treatment- and research-oriented comments.

Table 2
CLINICAL EXAMPLES OF MALFUNCTIONS OF THE BONE MACROREPAIR MECHANISM

Delayed union or established nonunion of a fracture
Delayed healing of an arthrodesis of an appendicular joint
Established failure of an arthrodesis of an appendicular joint
Failures of spine fusions in children and adults
Looser's zones, e.g., pseudofractures
Failed bone grafting operation for fracture or to fill a bony defect
Failed bone transplantation, autogenous or other
Delayed or established failure of union of corrective osteotomy
Spontaneous fractures
Myositis ossificans, traumatica, and progressiva
Heterotopic ossification in nonskeletal tissues
Myelofibrosis; osteoblastic metastases

II. BIOLOGIC PROCESSES

A. The Processes

From the broad biological viewpoint, six major subprocesses, each a system unto itself, associate to form the complete typical macrorepair process.

1. The healing process begins with an injury that damages both a bone and its adjoining soft tissues, thereby creating a special collection of enabling signals that have the function of activating or switching on, and also in part of determining the nature of, subsequent cellular events.[510]
2. Those signals then enable a regional acceleratory phenomenon (RAP) that speeds up all subsequent events as described in Chapter 11, and it usually lasts well past the primary healing stage. In effect, once initiated, the RAP becomes one of the continuum mechanisms described in Note 2 in Chapter 2.
3. Those signals also evoke the histogenesis or production by the injured but viable local tissues of a soft granulation tissue that fills the injured region.[613]
4. That tissue then becomes transformed into a hard callus composed of mineralized woven bone and hyaline cartilage. At the end of this primary healing phase (or 1° phase), a mechanical union has developed, some function usually resumes, the biomechanical competence described in Chapter 7 has developed, and some 6 to 15 weeks have elapsed since the fracture.[765]
5. Then packaged, BMU-based remodeling replaces that callus piecemeal and internally with normal lamellar bone that orients according to the local mechanical forces.[666] Observe here that the remodeling process now serves yet a fourth purpose in nature. Earlier chapters described its role in development (replacing growing cartilage with bone by the endochondral ossification mechanism), in lamellar bone maintenance (repair of microdamage), and in homeostasis. Here it is involved in macrorepair, so in a way it also resembles one of the letters of our alphabet. The same letter *a*, for example, is used in the words for artist, war, Saturn, analogy, nation, and grass, all quite different things.
6. In children, an accompanying but separate macromodeling process recontours the periosteal and cortical-endosteal surfaces of the bone in response to its mechanical usage, and according to biomechanical principles that we now partly understand and that were described in Chapter 7. Here this process too has been bent by nature to a new purpose, assuming the reader will grant that shaping a bone during growth to fit it to its normal mechanical usage fulfills a different

purpose than restoring an efficient architecture after some injury has created a harmful anatomical distortion. At the end of this secondary healing phase (or 2° phase), a self-maintaining and biomechanically competent "weld" composed of normal lamellar bone has restored full bony continuity, a continuous marrow cavity has been restored, and some 1 to 4 years have elapsed since the fracture occurred.[608]

As a relation, the above sequences may be written:[267]

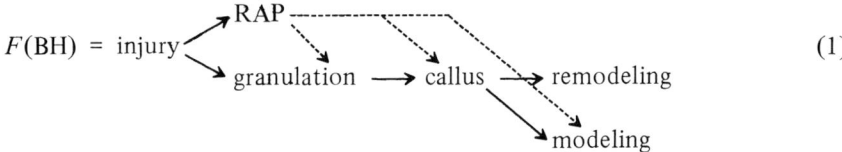
(1)

Here $F(BH)$ signifies some function of bone healing. To an orthopedist who treats a fracture and to a pathologist who sometimes examines and evaluates biopsy and postmortem material from fractures or other bone repair processes, the above events appear a little less complicated. For example, to the orthopedist, X-rays reveal the fracture and the subsequent callus as it mineralizes, while on clinical examination the fracture gradually becomes mechanically rigid and nontender. Further X-rays over the next 1 to 4 years reveal evidence of the final remodeling and modeling activities. Since most clinicians do not yet recognize a bone RAP by its roentgenographic tracks, for them the apparent bone repair sequences are encoded in this relation:

$$F(BH) = [\text{injury} \longrightarrow 1° \text{ healing} \longrightarrow 2° \text{ healing}] \qquad (2)$$

The above six subprocesses will now be considered in more detail.

B. The Biology

First, to repeat, the macrorepair process normally begins with the injury,[760] but other mechanisms can also initiate it, including infarction, infection, thermal injury, chemical injury, and the poorly understood biological stimuli involved in some metastases, in heterotopic bone formation, in Paget's disease of bone, and in some rare bone diseases such as myositis ossificans progressiva.

Table 3 lists a sampling of some of the nontraumatic clinical affections that involve and/or enable the basic bone repair process as a part of their pathology and course. Whichever the case, the initial stimulus enables two different processes by means of locally generated biological, physical, and biochemical signals that affect the local capillaries, the local supporting cells including several kinds of precursor cells, interstitial leukocytes, fibrocytes, and platelets, and often the local innervation.[138] The nature of those enabling signals and the mechanisms of action remain largely enigmatic at present, but they have the MES and switch-like properties discussed in Note 2 in Chapter 2 and Note 1 in Chapter 7.

Second, the initial stimulus also sets off a RAP that speeds up all subsequent local healing processes and probably also augments their amplitude. It normally lasts well past the primary healing stage into the second or remodeling and modeling stages before it subsides.[267] However, it can last much longer in chronic infection and Paget's disease. Since it involves all local tissues, both hard and soft, and from the L_c- to the L_o-domain, and in an integrated and reproducible fashion, it should be considered an L_o- or organ-level phenomenon.[267]

Third, the injury or other initial stimulus also generates specific local enabling sig-

Table 3
NONTRAUMATIC CLINICAL SITUATIONS IN WHICH THE BONE REPAIR SUBPROCESSES OCCUR

Affection	Bone repair stage initiated	Subsequent natural stages follow
Osteoblastic metastases (breast, prostate, thyroid, stomach)	Histogenesis of woven bone	Yes
Myositis ossificans, progressiva congenita	Histogenesis of woven bone	Yes
Osteosarcoma, parosteal	Histogenesis of woven bone	Depressed
Osteochondroma	Histogenesis of woven bone	Yes
Paget's disease of bone	Local resorption	Yes
Giant cell tumor of bone	Local resorption	Inhibited
Acute osteomyelitis	Histogenesis of granulation tissue	Yes
Osteoid osteoma of compacta	Modeling-remodeling	NA
Rheumatoid arthritis	Remodeling	NA
Fibrous dysplasia of bone	Histogenesis of granulation and bone	Yes
Lamellar periosteal osteoma	Modeling	NA

Note: NA means not applicable.

nals that cause the different kinds of local precursor cells to proliferate, and some of their daughter cells to differentiate into cells that make the new capillaries and supporting cells, innervation, lymphatics, and intercellular substances that form a soft granulation tissue. That tissue also contains clastic cells that remove debris, clot, and devitalized soft tissue and that resorb exposed bone surfaces. One could call that spatially anarchic aspect of the granulation tissue phase, the stage of local resorption. The tissue that results from that histogenesis occupies the injured tissue volume and replaces any hematoma.[138] Typically in humans, this granulation tissue histogenetic process takes about 3 weeks and it probably has L_2-level roots. The proliferation of new capillaries represents one major aspect of it, and another represents the production of new local precursor and differentiated cell populations. The details of the origins of those cells are still obscure. In sum, the typical granulation tissue contains the following constituents:

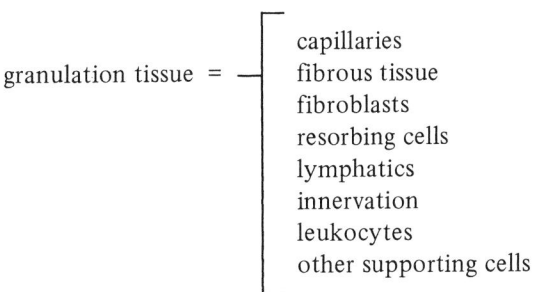

Fourth, then more and perhaps partly different local signals make some of the aforesaid new precursor cells in that granulation tissue proliferate and cause some of their daughter cells to differentiate into osteoblasts, and often chondroblasts too, which produce their corresponding organic matrixes of woven bone and disorganized hyaline cartilage, and then cause those matrixes to mineralize. The calcium in that mineral makes this tissue apparent on routine X-rays. The nature of the signals involved here

140 *Intermediary Organization of the Skeleton*

also remains largely enigmatic at present but they are probably complex. The complex of mineralized woven bone and variable amounts of hyaline cartilage, plus the supporting capillaries and other cells, constitutes the callus, which in effect solders the preexisting fragments of bone together at cement lines that arise where (and after) local osteoclasts (which also differentiate from the above precursor cell activity) partly resorb the exposed surfaces of the bony fragments. In sum, the typical callus contains the following constituents (see also Figures 1 and 2):

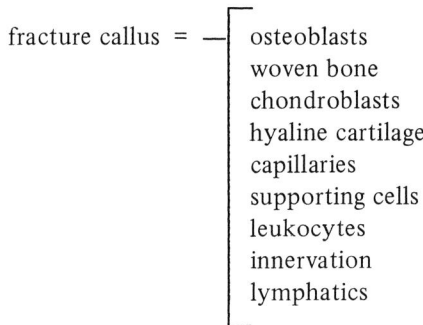

The interface at cement lines where new bone deposits on old bone represents a mechanically efficient biological weld that joins the old to the new, both soundly and in an aqueous medium. The chemical and physical nature of that weld is not yet understood but it is crucial to the mechanical strength provided by the bone micro- and macrorepair processes, as well as to that of normal bone itself. See Figure 3. To repeat, at the end of this biological stage, function usually resumes, some 6 to 15 weeks have passed since the initial injury, and the 1° healing phase is complete. The production of that callus can be considered an L_3-level phenomenon.

Fifth, then, probably in response to further specific local enabling signals, the BMU-based bone remodeling process described in Chapter 4 begins to replace the mineralized and disorganized callus piecemeal with spatially organized lamellar bone. In that process, micromodeling causes the grain of the new tissue to conform spatially to the orientation of the local mechanical strains induced by mechanical usage of the part. That replacement requires the initial production by appropriate precursor cells of groups of osteoclasts that resorb packets of preexisting hard tissue, following which other precursor cells produce groups of new osteoblasts that fill the resulting voids with new lamellar bone. To repeat, the new is welded at cement lines to the older bone. Figure 4 provides polarized light photomicrographs of woven and lamellar bone. Due to the ongoing RAP, this remodeling process proceeds rapidly in the first months after injury (a RAP can accelerate normal local bone turnover by more than 50 times), but as the RAP subsides the remodeling gradually slows, so the local bone turnover rate usually returns to normal some 6 months to 4 years after the initial injury. This replacement of the spatially anarchic callus by spatially oriented lamellar bone uses the L_2-level remodeling process described in Chapter 4 as a basic mechanism or brick to create the more highly integrated L_3-level process.[666]

It should also be recorded that further processes reestablish a marrow cavity in the fractured bone and replace the callus that originally filled it with marrow tissue that is more normal for the region.[613] This process occurs as well in a number of steps that need not be described here, but Figure 16, middle, in Chapter 7 shows an example in the leg of the child illustrated in Figure 2 of this chapter.

Sixth, in children, and while the internal replacement of the callus by lamellar bone remodeling proceeds, the external contours of the injured region, including the callus,

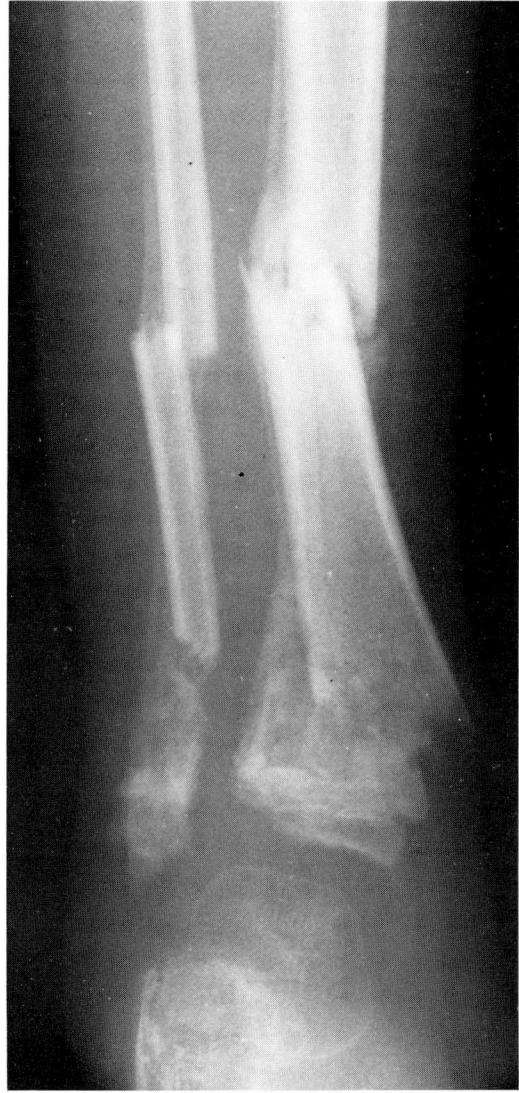

FIGURE 1. AP X-ray of a complex fracture of both bones of the leg in a young child, about 2 months after injury. The original fragments have become embedded in callus, the organic matrix of which had mineralized sufficiently that the calcium in it can cast an X-ray shadow. That shadow proves that the preceding biological processes were enabled and proceeded properly. Note also the obvious osteopenia of the metaphyses and talus. It represents a reversible increase in the remodeling space due in turn to an increase in regional BMU-based bone turnover, probably by a factor greater than 20, due in turn to the severe RAP caused by the injury.

gradually become reshaped by the separate macromodeling process, i.e., by resorption and formation drifts that arise and act on biochemically appropriate parts of the bone surfaces. Where mechanical usage causes convex-tending flexure, resorption drifts arise and shave bone away. Where that usage causes concave-tending flexure, formation drifts arise and build the local surface up. The macromodeling process re-

142 *Intermediary Organization of the Skeleton*

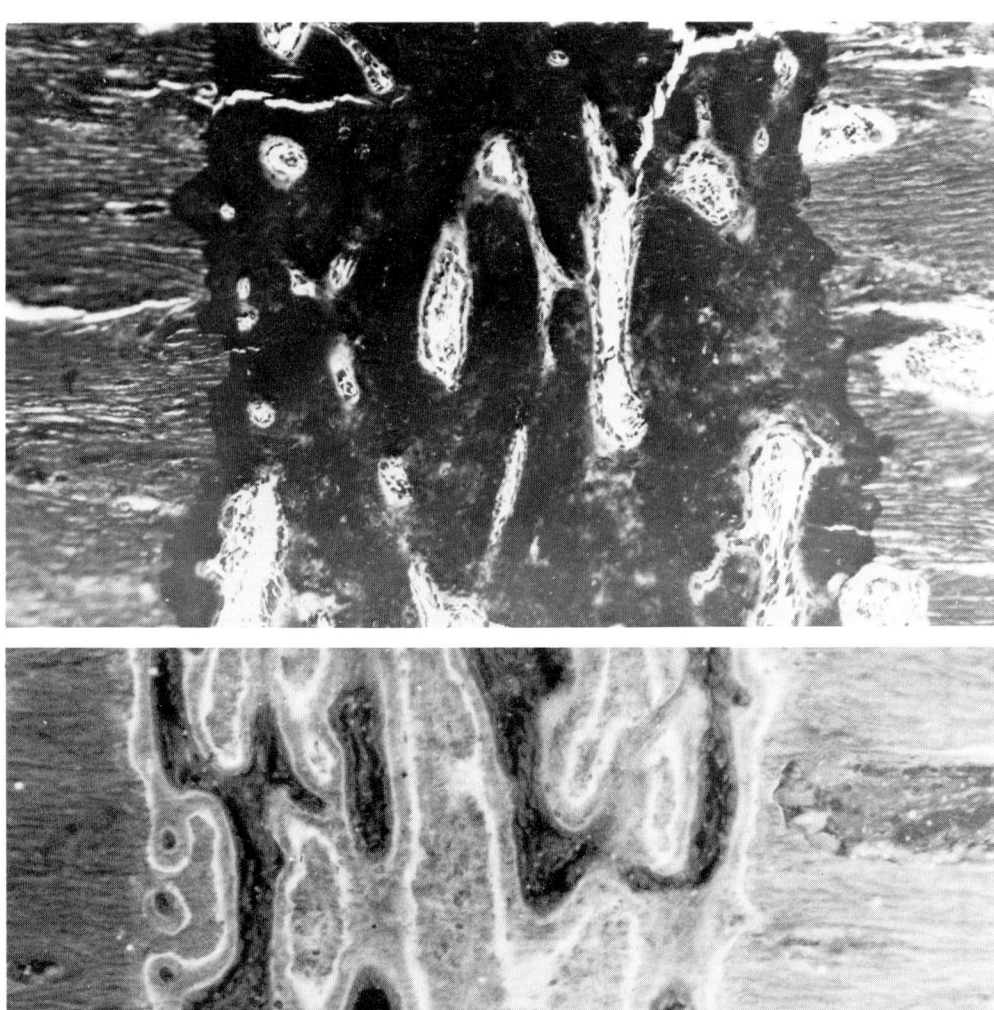

FIGURE 2. Photomicrographs of undecalcified longitudinal sections through an experimental osteotomy of the canine femur. The longitudinal axis runs from left to right. The tissue was embedded in plastic and then cut on a Jung microtome. The animals were given tetracycline bone tissue time markers prior to sacrifice; sacrifice occurred about 6 weeks after the osteotomy. *Top:* transmitted light, about 130×. A gap lies between the ends of the fragments and it has become filled with unmineralized callus so it is overstained; layers of osteoblasts can be seen lining the clear vascular channels. *Bottom:* the same specimen but in blue light fluorescence. The bright bands are tetracycline labels and they show the location of the mineralization front in the woven bone in the callus about 1 week prior to sacrifice. Active woven bone deposition is in process and near an end, whereupon replacement by lamellar bone and the BMU process will begin (photomicrographs courtesy Dr. Colin Anderson, London, Ontario).

stores a mechanically efficient gross bony configuration or architecture, and in young children it can restore a completely normal gross architecture to even an originally severely deformed bone, as illustrated in Figures 2 and 16 in Chapter 7. While remodeling can occur at all ages in man, modeling is most effective in children and becomes

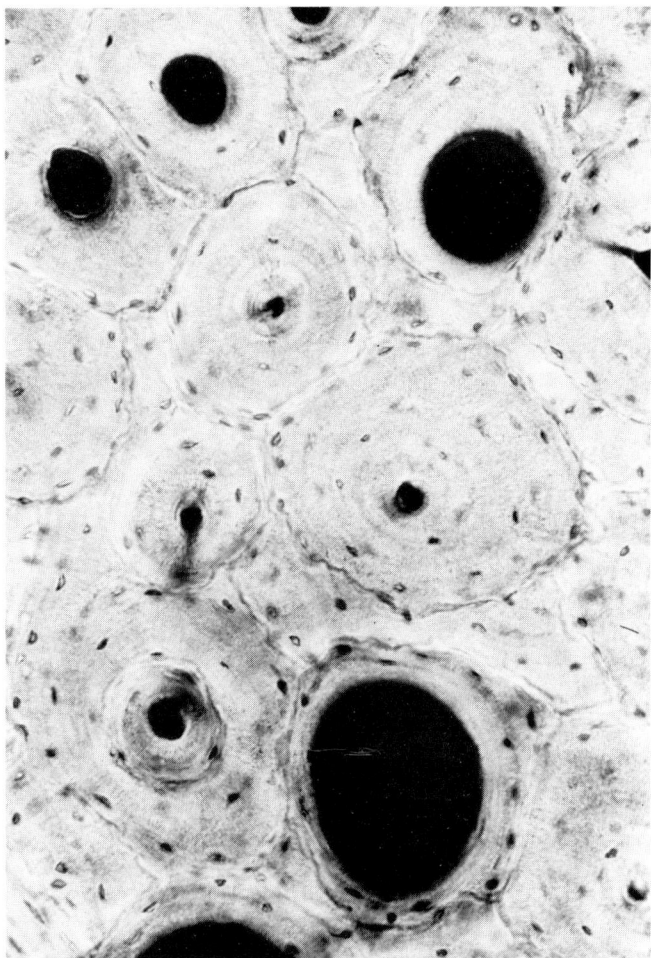

FIGURE 3. Adult human compacta, undecalcified cross-section, transmitted light, about 190×. The field shows a number of secondary haversion systems and the crenelated or scalloped cement lines at their periphery where they are mechanically "welded" to the preexisting bone within which they formed. That weld is one of the unsolved mysteries of bone. It occurs and persists in an aqeous environment with impressive durability because when microdamage occurs it usually does not occur and remain within the confines of the plane of the cement line, which it would if that line were deficient in fatigue endurance or strength. That cement line weld joins fracture callus to the preexisting bone, and the lamellar bone that replaces the callus or circumferential lamellae to its host, and it appears to endure under subsequent usage better than the bone itself.

relatively ineffective in normal adults, although a local RAP can apparently enable some macromodeling in adults.

This macromodeling process also uses the basic bone macromodeling mechanisms described in Chapters 4 and 7 as a brick to create an integrated L_3-level process that can reshape a part — although a major part — of an intact bone. Local mechanical strains probably control it, and the physical loads applied to the bone cause the strains. The transducer that converts that strain into biological signals that control modeling

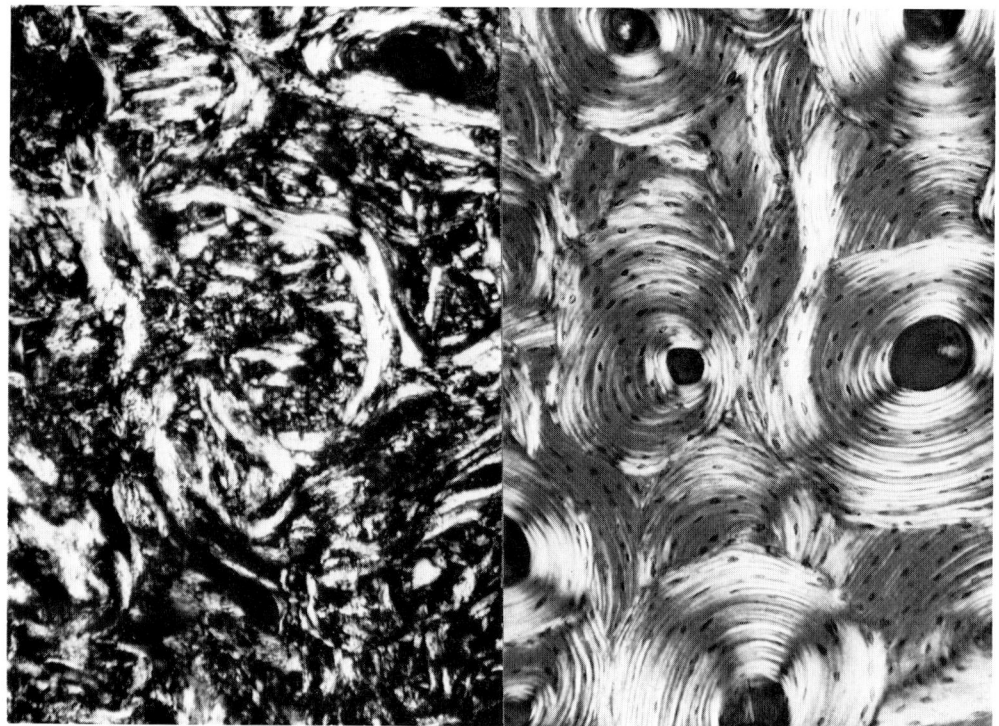

FIGURE 4. Photomicrographs of human bone, undecalcified sections, polarized light, about 200×. *Left:* woven bone showing the warp-and-woof pattern that characterizes its lack of L_2-domain or higher organization. A section cut through this tissue in any direction would look much the same. Its architecture is independent of the local mechanical loads and their strains. *Right:* compacta, cross-section, showing nearly complete replacement of the original circumferential lamellae by secondary osteons. The architecture of those structures faithfully reflects the spatial orientation of the local principal strains and the mechanical loads that caused them.

probably represents streaming potentials and related effects. The recently formed callus as well as new bone deposited by the modeling process is much less stiff than mature bone because it still lacks its full mineral deposits (the stiffness lag) and it also models more than mature bone. In bone healing this property of locally potentiated modeling by both the RAP and the presence of hypercompliant bone has obvious survival value. In sum, the completely healed fracture will contain the following constituents:

fully healed fracture = ⎡ lamellar bone compacta
⎢ lamellar bone spongiosa
⎢ normal marrow tissue
⎢ normal periosteum
⎢ capillaries
⎢ innervation
⎣ supporting cells

1. Temporal Overlap and Smearing

Considerable temporal overlap of the above processes often appears in bulk tissues. Thus, different parts of a single histological section of a given healing fracture or arthrodesis can show several different stages in progress. However, in each microscopic domain and under normal circumstances the same invariant sequences occur, although the changes tend to arise and subside gradually rather than as sharp step functions.

2. Biomechanical Competence

During the time taken for a rigid callus to form, unprotected function of the limb in which the fracture exists can lead to the complications of progressive displacement, angulation, and overriding of a fracture. However, once the callus has become rigid, then the gradual resumption of function does not lead to such complications. Rather, it leads to the modeling and remodeling activities described above and one may infer in those activities the purpose of seeing that normal function of the bone can continue for the remainder of life. Hence, one may assume that the transition from a plastic to an enduring rigid bone signals the achievement of biomechanical competence. Achieving it can take anywhere from a couple of weeks for a greenstick fracture in an infant to 1 year or more for a femoral or tibial shaft fracture in an adult.

In sum, then, the four major groups of subprocesses that are involved in normal bone macrorepair include the injury, the 1° and 2° healing stages, and the overlapping and concurrent RAP. The 1° stage includes the subprocesses that form granulation tissue and then transform it into callus. The 2° stage includes the modeling and remodeling processes, and the RAP oversees and speeds up all of the other processes and possibly augments their amplitude as well. Figure 5 summarizes the above facts. As an aside, Swiss workers interested in the treatment of fractures by means of internal fixation have named *primary healing* the BMU-based remodeling described above, which is termed the 2° healing stage here, and they named *secondary healing* the initial production of the callus that this text terms the 1° stage.[613] This problem of nomenclature is mentioned to avoid confusion and to make clear the meaning of terminology used in this chapter. A comparison of the AO and present terminology then would look like this:

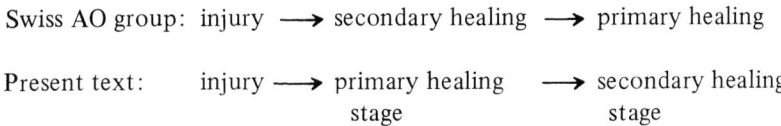

As a relation, the processes, terminology, and timing shown in Figure 5 could be written more concisely with F(BH) standing for any function of bone healing:

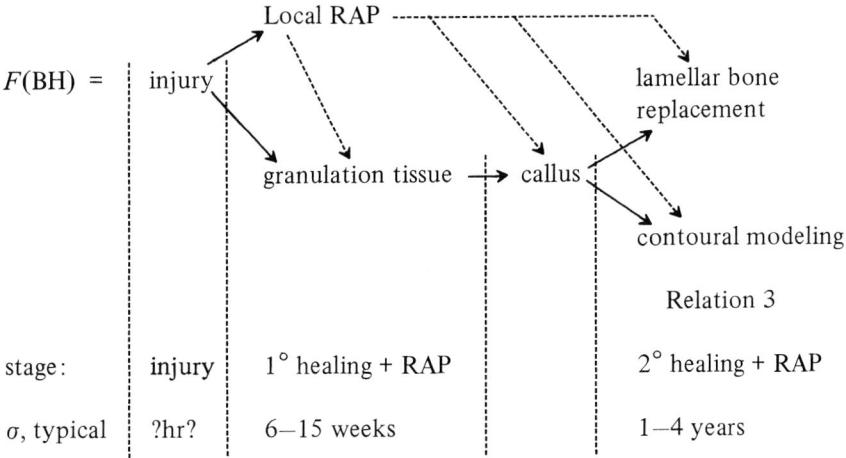

Relation 3

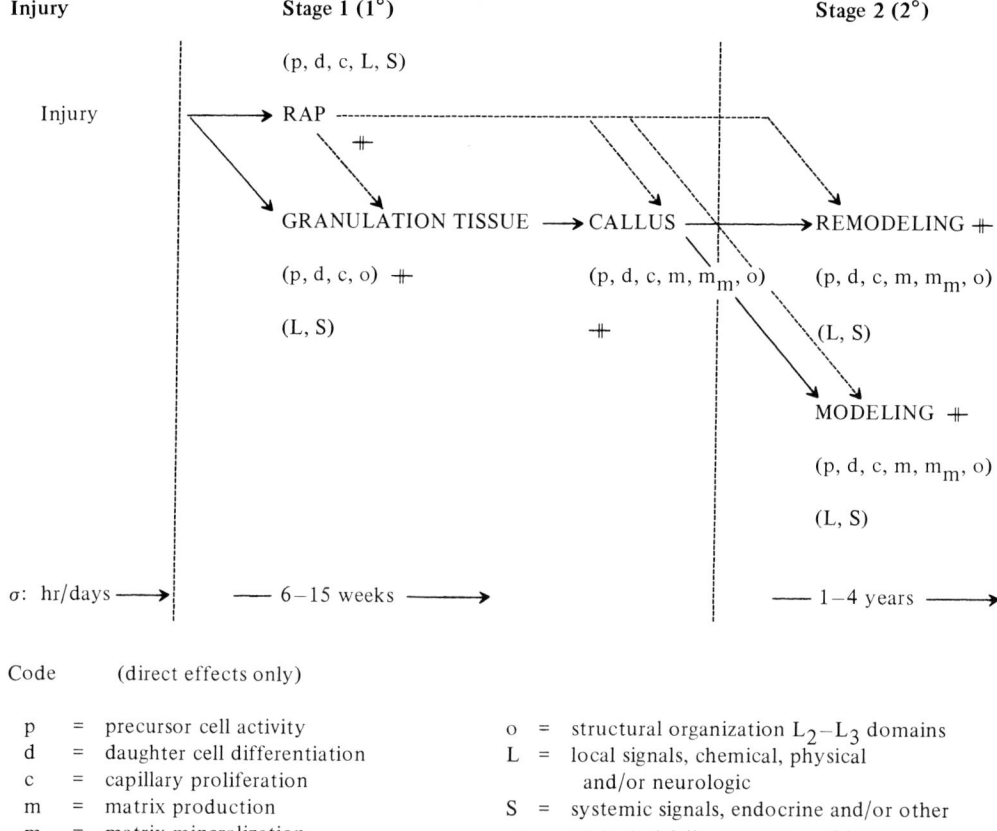

FIGURE 5. This figure summarizes the general features of the normal human bone repair process. The coded letters indicate the nature of the enabling and regulating agents thought to be involved in the various subsystems of the intact process. The fact that these subsystems occur in sequence, and in a nearly invariant order, leaves little doubt that coupling phenomena occur, possibly quite complex in their determinants, and in the abstract sense analogous to the coupling observed in the ARF sequence of the remodeling BMU itself. The above phenomena are real rather than hypothetical, but so far we know very little about the enabling and regulatory agents that control them (revised from a manuscript circulated to the σ bone group in 1983).

Note Bene: If the above sequences should arrest at any stage, then all subsequent stages will usually fail to appear, which implies that each stage somehow supplies the local signals that are needed to evoke or enable its successor to couple the two together in sequence.

Several further properties of the above macrorepair process bear on its problems as found in clinical practice. First, the immediate stimuli that evoke or enable and then regulate any given subprocess arise in the inherent properties of the preceding process and any tissue it creates, at least in part.[430] Second, some systemic factors, including nutrition, innervation, some hormones and drugs, and some diseases, can condition the responses evoked by such stimuli by potentiating responses in some cases and depressing them in others. For example, a second injury a few weeks after the original one can potentiate the speed and volume of subsequent healing.[140,156] Third, mechanical loads and the strains they cause in a healing bone probably also affect those processes, although the details are unclear and their effects on each subprocess seem to differ, one from the other. Fourth, so far, few laboratory and clinical investigations of the bone repair process have taken the above sequences and related matters into account in their design, reporting, or interpretation, so the voluminous literature on the subject can seem rather confusing to a student.

3. Normal Variations of Traumatically Induced Bone Repair

Children can develop a type of greenstick fracture called an infraction or torus, in which the bone deforms and buckles plastically but does not completely disrupt. These fractures heal well, promptly, and nearly invariably, but without the initial production of grossly apparent granulation tissue. The phase of callus production appears to be directly stimulated by some property of such fractures. A similar phenomenon occurs in the healing of so-called march fractures (see Figures 20 and 21 in Chapter 7), which represent fatigue failures of whole bones, often arising in vigorously training people (children as well as adults), but which can also arise during normal activities in patients on corticosteroids and other so-called immunosuppressive drugs, many of which interfere with cell proliferative activity required by all major stages of fracture healing.

4. The General Biologic Functions Involved

The above facts describe a highly organized and sequential arrangement of some of the fundamentally distinct general skeletal processes that were described in earlier chapters. For example, the function of a fracture — or tendon — callus involves enabling the histogenesis of specific kinds of tissue and then regulating their continuum growth phases, which involve growth-independent precursor cell activities (because they can occur in adults as well as children). Then an enablement of a form of BMU-based remodeling replaces the anarchic callus with structurally oriented lamellar bone, which implies micromodeling, the continuum phase of which is then also regulated somehow. Finally, in children, and to a minor degree in adults, local macromodeling recontours the fracture in accordance with the dictates of the local MES and the bone modeling laws.

Thus, histogenesis, growth, micro- and macromodeling all occur in an organized sequential and stereotyped fashion in the bone healing process, whether it is evoked by trauma or by diseases such as those listed in Table 3. Each of these general functions arises in IO entities that follow the common plan of the basic state equations presented in Chapter 3.

III. CLINICAL PROBLEMS RELATED TO THE BONE REPAIR MECHANISM

Two broad categories of these problems occur. One includes problems related to the healing of traumatic injuries and the other includes problems in which some disease process rather than a physical,chemical, or thermal injury activates what is termed herein the repair mechanism. We will look first at the former.

A. Clinical Problems in the Repair of Bone Trauma

Using fracture healing as the illustrative case, these problems can also divide into two broad subclasses: *technical failures*[241] due primarily to inadequacies in treatment, as illustrated in Figure 6 and *biological failures*[241] due primarily to malfunction(s) of one or more of the biological processes of healing that are not due to the fracture treatment as in Figure 7. Other authors sometimes refer to the above as hypertrophic and atrophic nonunions.[127] The discussion ignores technical failures henceforth since they lie in the realm of practical clinical orthopedics rather than in the more basic

FIGURE 6. A technical failure of bone union. An united clavicle fracture in an adult male. Some 3 months postinjury, abundant callus has formed on the ends of both fragments, but a pseudarthrosis or plane of false motion separates them. The presence of the callus proves that the biology here was intrinsically normal in its potential and responses. This patient's many injuries, however, included cerebral ones, such that he refused or removed all appropriate treatment of this fracture for the first 3 months postinjury. The resulting combination of excessive motion and displacement of the fragments caused the nonunion, with a pseudarthrosis or false new joint separating the ends of the bone and the two masses of callus. Here the repair biology tried to do its job, but it was prevented from succeeding by nonbiologic problems. This and other X-ray illustrations in this and other chapters of the book illustrate an important and useful but seldom used matter, i.e., the important events that caused the findings in this X-ray occurred in the first 2 weeks or so after the fracture. They no longer exist, so this situation was already determined at the end of that initial time. Further time had to pass in order to make the nonunion apparent on the X-ray. We should learn in looking at such an X-ray not only to recognize the existence of an established reunion, but also to perceive its original and no longer present cause(s). Like a histological section or a man's face, the X-ray is also a kind of historical record of the structures it reveals, and if one takes the trouble to learn the language in which it is written, then one can perceive the intelligence, the information it provides.

biological realm that concerns this chapter. However, it must be noted here that the same general process that heals a fracture also heals surgical osteotomies and attempts to fuse or arthrodese any joint of the appendicular and axial skeletons. The clinical diagnosis and distinction between technical and biological failures is direct and simple. X-rays of a technical failure show normal amounts of callus at times after injury when one would expect to see it. The presence of callus proves that the biology was competent and that it tried as in Figure 6. X-rays of biological failures reveal at the same postinjury time that little or no callus has formed. The absence of callus in such cases proves that the biology of the healing process was incompetent, whatever the reason, as in Figures 7 and 8. The biological failures that concern this chapter can themselves be dissected or resolved into two major subgroups: one related to continuum processes and the other to enablement.[267]

1. Rate Problems

These represent fractures, arthrodeses, and/or related problems that heal too slowly

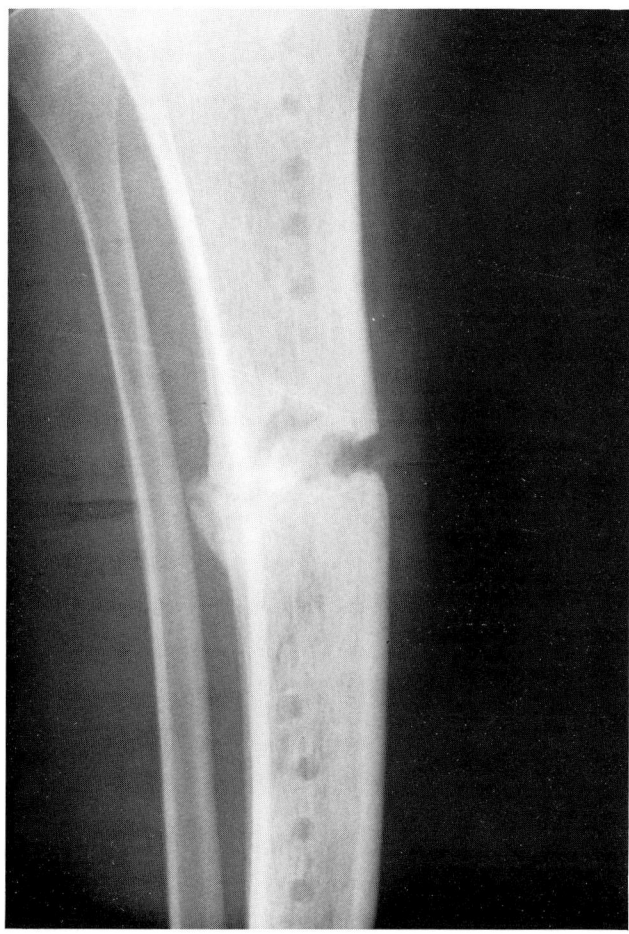

FIGURE 7. This X-ray illustrates a delayed union of a tibial fracture in an adult some 5 months postinjury and an unsuccessful treatment with an external fixation device now removed (the pins of which caused the round holes in each fragment). Only a trivial amount of callus has formed. Note the longitudinal tunneling of the two major fragments (compare to Figure 8 top and bottom right, which show no tunneling and thus no RAP). It reveals that a perfectly adequate RAP developed. This is one kind of biological failure of bone union due to inadequate callus production in the presence of a normal RAP.

but that, given time and proper management, ultimately do heal satisfactorily with adequate amounts of callus; they follow the sequences described above. Thus, this group presents primarily a rate abnormality, and it is often caused by an inadequate local RAP or by other problems involved in regulating the continuum phases of the healing subprocesses.[267,503] Probably over 50% of the biologically delayed unions seen in clinical practice arise all or in part from this cause (see Figure 8, top right). As noted elsewhere, the determinants of the ubiquitous RAP remain obscure, but bone grafting and related operations and the currently popular electrical stimulation of slowly healing and ununited fractures[62,88] can evoke a vigorous new RAP and/or potentiate the original one to accelerate local healing without materially altering its nature or sequences.

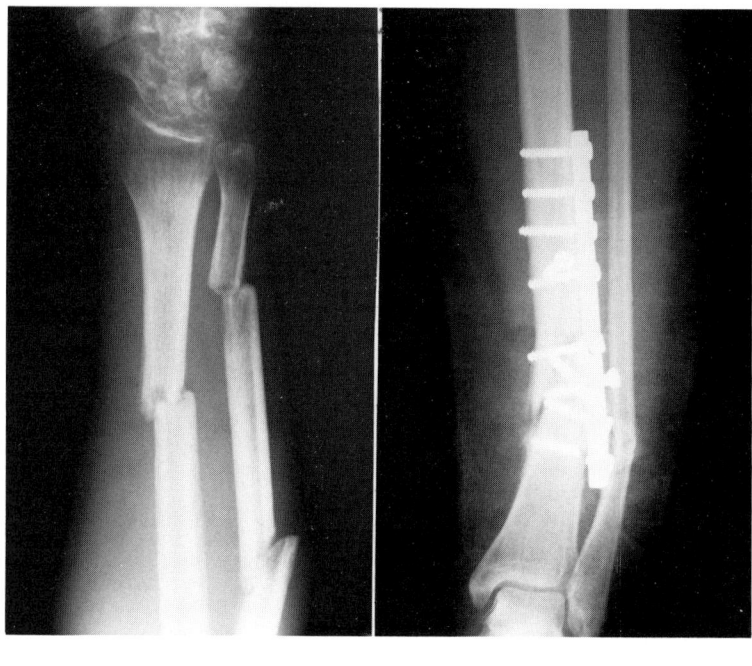

A B

FIGURE 8. (A) AP X-ray of the forearm of a 26-year-old male who sustained multiple fractures 5 months before this X-ray. The X-ray shows evidence of a vigorous RAP, with longitudinal tunneling of the cortexes of the bones and pronounced osteopenia of the metaphyses, due to an increased remodeling space, yet no callus whatsoever is visible. This represents a biological failure of the primary stage of bone healing. The adequate RAP suggested that healing by the BMU-based remodeling mechanism would prove satisfactory, so the three fractures were subjected to compression plating and they healed soundly within 3 months. (B) This 28-year-old woman sustained this fracture skiing some 5 months before this X-ray was taken. In the interim she underwent the open reduction and internal fixation shown plus a wound infection and three local skin grafts, yet no evidence of a local RAP had appeared, nor was there other than trivial amounts of callus. The hardware was removed to eliminate dead space and to allow soft tissue healing, which occurred. Then, an internally fixed sliding bone graft was done and 9 months was allowed for subsequent but slow healing by the retarded BMU-based remodeling mechanism, which in turn was due to the inadequate RAP. The strategy succeeded. (C) The young man's femoral fracture at 2 months postinjury and postintramedullary nailing shows both abundant callus and evidence of a vigorous local RAP, as shown by the longitudinal tunneling of the compacta. He resumed weight bearing 2 months later and work 2 months after that. (D) At 6 months postinjury neither the fibular nor tibial fracture show any evidence of callus whatsoever; an inadequate RAP has developed as well. Here too a grafting procedure that requires the ability to form a callus to succeed would probably fail, while a procedure that depends on a RAP-enhanced BMU remodeling mechanism would need longer than normal to succeed. Recognition and judicious evaluation of the meaning of such simple factors make the difference between poor and high rates of success in managing such problems. Success is not the result of the manual skill, dexterity, and speed of the surgeon. Tissues are usually kind, even to ham-handed operators. The success depends much more on what was done to whom, when, and for what reason.

2. Problems of Quantity and Kind

This group includes three problems that relate primarily to the ultimate quantity and kind of healing activity rather than to rates. They are described next.

a. Insufficient Callus

Callus arises in insufficient amounts or not at all, as shown in Figure 8. This can follow insufficient enablement and subsequent production of granulation tissue, in

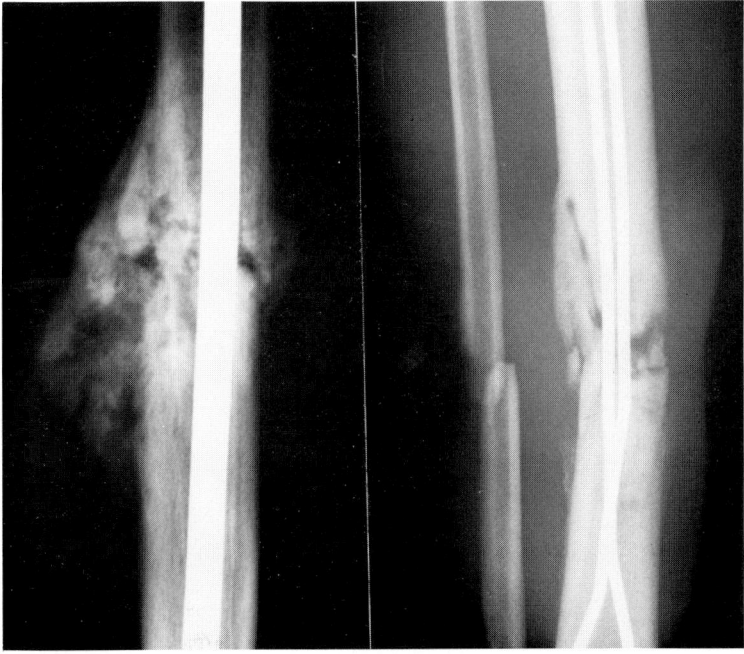

FIGURE 8C   FIGURE 8D

which case a commensurate reduction in the subsequent amount of callus may follow. It can also follow a maldifferentiation of the cell populations in the granulation tissue that transform that tissue into a fibrous tissue scar rather than into woven bone and hyaline cartilage matrixes. That scar (see Figure 9) prevents any effective subsequent local bone modeling or remodeling, so in the terms of the IO notation used in this text it represents a closed gate for both those activities and it is an "L" term with respect to both of them. It too is a problem of enablement rather than the regulation of a continuum process. Thus, when "$\not{L}$" represents scar and $F(BR)$ a function of bone remodeling:

$$F(BR) = (S : \not{L})(C\ P\ D\ O\ A\ M) \tag{3}$$

In the author's experience, the above two problems, those related to rate and those related to ultimate quantity, do not always accompany each other, so while others have not dissected them in that way they are listed here as separate biological matters. In somewhat different words, the author is convinced that enabling healing is one game, the game of histogenesis, and guiding it once it has been set in motion is another game, the game of the continuum processes described in Note 2 in Chapter 2. Each has its own rules in addition to whatever else the two may have in common. As a relation one could write the above propositions thus:

$$F(BH) = \begin{bmatrix} \text{histogenesis} \longrightarrow \text{continuum phase} \\ \uparrow \qquad\qquad\qquad \uparrow \\ \text{enabling} \qquad\qquad \text{regulating} \\ \text{agents} \qquad\qquad\quad \text{agents} \end{bmatrix} \tag{4}$$

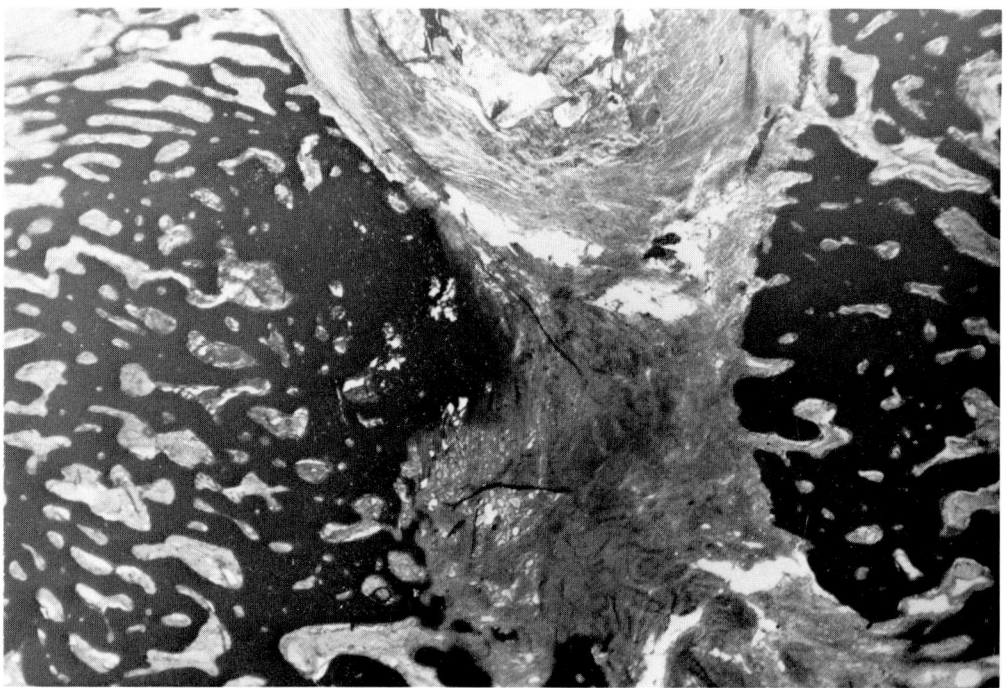

FIGURE 9. Photomicrograph of a longitudinal undecalcified section of healing long bone fracture in an experimental healthy dog. The longitudinal axis of the bone is horizontal here; the fracture plane lies in the middle and it is vertical. The heavy dark bars on either side are woven bone trabeculae which have filled much of the original fracture gap, but which left a zone of fibrous tissue or scar in the middle. This represents the 1° healing stage described in the text, and it also illustrates a threatened nonunion of the technical type, due here to distraction, meaning too large a gap between the fractured ends of the original bone. That distraction caused cell differentiation in the local granulation tissue to move toward the production of fibroblasts rather than osteoblasts and chondroblasts. Galen Hoover (Tacoma, Wash.) spoke of this as the OJD, or osteoblast jumping distance, and the term stuck (photomicrograph courtesy Dr. Colin Anderson, London).

Accordingly, the enabling features determine *if* a repair process or subprocess will occur, and *where, when,* and to some degree *how much*. The regulation of the subsequent continuum process(es) then determines *how quickly* the enabled processes will appear, and to some degree also *how much*. To repeat, it is not conventional to analyze the healing process in this way (the same features apply to fibrous tissue, chondral, synovial, and skin healing), but since 1965 the author has become convinced that the analysis is correct in its essentials.

Concerning the young plumber in the opening ministry, he never showed adequate amounts of callus at the tibial nonunion site after any of his first eight operations. That revealed three things. First, for whatever reason, that site could not form callus in adequate amounts. Second, therefore, any operation that required the production of callus to succed would probably fail (as all eight previous operations did). Third, therefore, a procedure was indicated that did not require callus to succeed. The writer did a securely fixed sliding bone graft, and relied on the competent BMU-based remodeling mechanism to heal the fracture, as described earlier in this chapter and in Chapter 11.

b. Impaired Mineralization

In another healing problem, inadequate mineralization may occur in the newly formed woven bone and cartilage organic matrixes. This can happen in many kinds of osteomalacia and in vitamin D-deficiency rickets. It is a major cause of Looser's zones in osteomalacia, which represent complete fatigue failures that cannot heal because of the metabolic disorder.[237] It can also occur on occasion without any currently known cause, and when it does it causes a great prolongation in the stiffness lag, meaning in the time a moiety of newly formed bone needs to become well enough mineralized, and thus stiff enough, to begin accepting its share of the regional mechanical loads.[267] When that mineralization does not occur or is markedly retarded in speed, then the subsequent remodeling and modeling processes do not occur either because, and whatever the details, they require a previously elaborated mineralized bone matrix to enable them.[267] Typically, when the underlying metabolic disorder is corrected, such nonunions then proceed to heal, beginning by mineralizing the already accumulated local bone and chondral matrix.

c. A Remodeling Disorder

In yet another problem, callus can form and mineralize in adequate amounts and at the appropriate rate, but a defect in the ensuing lamellar bone replacement by the BMU-based remodeling process retards the deposition of new lamellar bone but not the preceding resorption. That accumulated resorption space temporarily weakens the callus so it begins to deform plastically or even refractures. That happens, therefore, during the 2° stage of healing and in response to normal mechanical usage. The writer has seen this rare phenomenon only seven times in 40 years of practice. Because of its rarity it remains unknown to most clinicians — and attorneys — so it has led to misconceived and, unfortunately, sometimes successful malpractice suits.

B. The Bone Repair Mechanism in Disease

The same histological and biological processes and sequences evoked by an injury can also be evoked by nontraumatic factors associated with affections such as Paget's disease of bone, osteomyelitis, myositis ossificans, osteoblastic mestastases, certain osteomas of bone, osteosarcomas, pulmonary hypertrophic osteoarthropathy, and myelofibrosis, as examples.[608] Some factor(s) associated with those affections can initiate or enable the repair response at its beginning at the 1° stage in some instances (e.g., myositis ossificians traumatica),[25] in other cases at the beginning of the stage or woven bone production (e.g., Paget's disease),[55,274] and in yet others at the beginning of the modeling state (e.g., osteoma).[6] In some cases, an accompanying RAP is also evoked (Paget's disease, osteomyelitis, osteoid osteoma, myositis ossificans post-traumatica) and in others it is not (periosteal lamellar bone-forming osteomas, hyperostosis frontalis interna). Whatever the combination and beginning stage, the processes and sequences that normally follow that stage as shown in Relation 3 do then follow it and in proper order. Reactivation of the initiating event by the disease in a region already partly evolved on subsequent steps can create an initially confusing mosaic pattern of overlapping histologic, biochemical, and radiologic features in the region (e.g., Paget's disease, chronic osteomelitis, congenital lues). Also, some specific diseases (infection, metastases, osteomalacia) can imprint their special features on the common basic histological and biochemical events to create patterns that are diagnostic, that appear unique, and that have tended to hide from most observers the common denominators considered here.

IV. CLINICAL-BIOLOGICAL SYNTHESIS

A. The Skeletal IO

The above problems arise directly from malfunctions of the skeletal intermediary organization (IO), and only indirectly — if often ultimately — from cell-level malfunctions too.[263]

As a complete process, fracture healing represents an L_o- or organ-level phenomenon that inherently associates soft with hard tissue phenomena.[804] It has direct roots in the L_2- and L_3-levels of the IO, and less direct ones at L_1 and L_c, so one should not extrapolate cell-level effects directly to the organ in this system any more than he would with kidney.[267] The LBO property described in Chapter 3 as a basic property of the IO generally also applies to the present matter. In studying the physiology and diseases of the macrorepair mechanism, that mechanism *must* be considered as an integrated and sequential *system* that organizes quite different subprocesses, regulatory and enabling factors, and tissue elements in time and space to achieve its ultimate goal.

Recall that the following common plan characterizes all functional entities of the lower-level IO, where $F(Sk)$ signifies any skeletal function including the above-described biological processes.

$$F(Sk) = \begin{array}{c} S \\ \downarrow \\ \downarrow \quad \downarrow L \\ \downarrow \quad \downarrow \\ \hline C:P:D:O:A:M \end{array} \quad\quad (5)$$

With suitable modifications and qualifications, that plan also applies to L_2- and L_3-level IO entities, including those that concern this chapter. The preceding paragraphs pointed out that some problems involve impaired "P" cell response which can impair the formation of granulation tissue, or callus, or BMU-based remodeling as in Figure 8. Other problems involve maldifferentiation or "D" as in the production of scar instead of woven bone and cartilage as in Figure 9. Yet others involve malfunctions of already differentiated osteoblasts and chrondroblasts, or "A" cells as in the mineralization defect associated with vitamin D-deficiency rickets and osteomalacias. While no presently known malfunctions are attributed to capillary or IO-intrinsic "O" malfunctions, such malfunctions have not been specifically sought so they would not likely be recognized if they do exist. Experience suggests that some such examples should occur, for almost everything possible in the way of disease does occur in actuality. In particular, the elaboration of new capillaries is one of the essential activities in producing the granulation tissue, so some biological failures characterized by insufficient production of new granulation tissue may represent examples of impaired capillary proliferation. The following paragraphs will look in somewhat more detail at some of those malfunctions.

B. The Abnormal RAP

An *obtunded RAP* fails to accelerate otherwise normal healing processes in normal fashion.[261] It accounts for many of the rate problems mentioned earlier. It could reflect an "S" effect that arises extrinsic to the IO as in diabetes mellitus, malnutrition, vascular insufficiency, or in denervated tissue. Some drugs may also obtund the RAP. Likely candidates include certain disphosphonates and fluoride, and some immunosuppressive, antiarthritic, anti-inflammatory, and oncologic chemotherapeutic agents. An inadequate RAP can also reflect an IO intrinsic effect, as in some types of neurofibromatosis, and still unexplained cases occur. In any case, an impaired RAP impairs all stages of healing without altering their kind or sequence.[261] Clinically it becomes apparent as a delayed but ultimately successful union. To repeat, perhaps 50% of clinically delayed unions reflect such an effect, wholly or in part.[267]

Hyperactive or prolonged or otherwise pathological RAPs can also occur, but they do not appear to impair the bone repair mechanism.[1] A few of them were discussed in Chapter 11.

C. Abnormal "P" Response

Precursor cell activities directly provide the new cells that form the new capillaries and supporting cells of the granulation tissue, as well as the cells that form the bony and chondral matrixes of the callus and the resorption and formation activities of the subsequent remodeling and modeling processes. Thus, without that "P" activity no healing at all is possible. Probably somewhat different signals with differing MES values activate or enable the "P" activity involved in the different stages of bone repair. Inadequate "P" responses could reflect an inadequate IO-extrinsic stimulus to the "P" activity (including an inadequate RAP), or an extrinsically depressed "P" response as in Didronil[208,408] or fluoride intoxication or in radiation-damaged tissue, or some systemic conditioning effects as in the panhypopituitary state, in primary hyperparathyroidism, or in Addison's disease.

The author suspects that a significant fraction of the clinical examples of an inadequate healing response to some injury reflects the fact that at the time of injury the local tissues were in an "Off" state with respect to their ability or competence to respond to that injury, and that IO-extrinsic factors of both "L" and "S" classes determined the locations, durations, and depth of that "Off" state. IO-intrinsic "deafness" of the "P" modality to an otherwise adequate stimulus can also cause an inadequate "P" response, again as in some cases of neurofibromatosis and congenital pseudoarthrosis of the tibia. An inadequate "P" response limits the amount of new tissue produced during healing, and it is reflected clinically by inadequate amounts of callus. While in that regard an obtunded RAP has an apparently similar effect, certain clinical evidence can usually distinguish the two effects. In the author's experience and at least in part, between 30 and 50% of established clinical nonunions appear to derive from inadequate "P" activity.

Excessive "P" responses can lead to excessive amounts of callus as in some cases of osteogenesis imperfecta and myositis ossificans traumatica, and to added periosteal bone in pulmonary hyperthrophic osteoarthropathy, or to pathological thinning of a normal callus due to accelerated and augmented second stage processes as in some cases of throtoxicosis and completely resorbed bone grafts.

D. Abnormal "D"

Production of scar or related soft tissue elements (e.g., fatty tissue) instead of bone and cartilage could follow an abnormal differentiation of the daughters of precursor cell activity, and equally for the first (the commoner case) and second healing stages (an uncommon case). It too could arise from extrinsic "S" effects (e.g., in severe primary hyperparathyroidism) or from an intrinsically disordered response of those daughter cells (e.g., some forms of neurofibromatosis and congenital pseudarthroses). Perhaps 10 to 20% of established biological failures derive in part from "D" malfunctions, and embryologic data suggest that malfunctioning enabling mechanisms cause them, rather than the regulation of continua.

E. Abnormal "A"

IO-intrinsic defects might make the effector "A" cells of concern here (osteoblasts and chondroblasts) form an abnormal matrix as in Baker's disease (fibrogenesis imperfecta ossium) and osteogenesis imperfecta, and/or fail to initiate its mineralization.[233,242] However, such IO-intrinsic defects in bone repair occur rarely in clinical practice. IO-extrinsic disorders could also cause inadequate matrix synthesis as in starvation, or inadequate mineralization of the matrix as in vitamin D-deficiency osteomalacias, 1° and 2° hyperparathyroidism[45,729] and rickets, and in intoxications with drugs such as β-aminopropionitrile[285,702] or aluminum.[242] This kind of malfunction underlies most Looser's zones in osteomalacic skeletons but it is not common in clinical practice.[527,579]

F. Defective "M"

Local properties inherent in normal woven and lamellar bone normally guide their replacement and/or turnover by BMU-based remodeling throughout human life and the modeling of bones during growth. Consequently, bone that is defective in its composition, organization, and/or mechanical properties, could also cause malfunctions of the repair mechanism. Certainly, macromodeling malfunctions occur for that reason in osteogenesis imperfecta,[326,432,653,671] in lead intoxication and fluorism, and in certain rare aseptic necroses of major epiphyses. These "M" defects also appear rarely in clinical practice. Examples would include all known forms of osteogenesis imperfecta and fibrogenesis imperfecta ossium.

G. Abnormal "O" and "C"

To repeat, little is known about the possible roles of malfunctions of the capillaries that supply the basic L_2-level units of fracture healing, and of the intrinsic organizing or "O" potential of those units, and for all stages of the intact repair process. That ignorance derives in part from a lack of systematic histopathologic and biochemical study of biological failures of the bone repair mechanism, and in part probably because such malfunctions might be quite rare or even incompatible with life.

H. Temporal Incoherence

As the foregoing text implies (the percents add up to more than 100%), a given clinical biological failure or malfunction of the repair mechanism often combines several of the malfunctions noted above, which has concealed the separate determinants of those fractures from most observers. In part, that may represent an analog of the ubiquitous but only recently recognized MCN effect that was described in Chapter 4. To understand this, take note first of the fact that the above-described stages of healing could be written as this sequence: a:b:c:d:e:f, where the letters have the following meaning:

Letter	Meaning
a	Injury
b	Granulation tissue production
c	Produce organic matrix of the callus
d	Mineralize the callus
e	Remodel the callus
f	Model the callus

The similarity of the above situation with that diagrammed in Figure 16 in Chapter 5 should be apparent. To complete it is the fact that for a variety of reasons, in any typical fracture the initiating stimuli of each stage occur at different times in different regions of the fracture, which is why in a given fracture one can find in various parts of it at the same time, granulation tissue, scar, woven bone, and lamellar bone. Thus, the situation in the large bones of a healing human fracture could look like this:

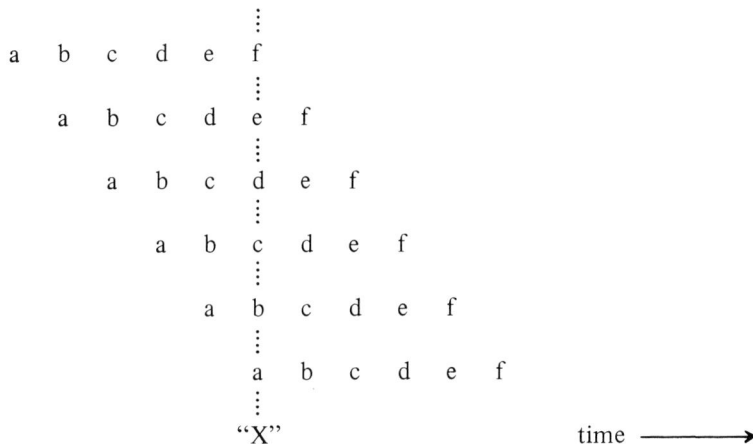

Accordingly, a state of temporal incoherence exists and the basic situation needed for an MCN effect also exists. It follows that small and large doses of drugs and brief and prolonged periods of treatment, and properly timed sequential-intermittent as opposed to continuous treatment should have different effects on this system. It also follows that both coherence research and coherence treatment may solve some otherwise refractory problems of this system.

V. THERAPEUTIC PROSPECTS

We will consider here some factors that relate both to the prospects of future treatment and to the research that could improve it.

A. Electrical Stimulation (ES)

This treatment technique resulted from a biologically naive interpretation of the original work of Fukada and Yasuda,[268] who discovered that microvolt-range electrical potentials can arise on the surfaces of dry mature bone when it is strained in flexure. The possibility that such potentials governed bone modeling attracted many subsequent investigators. When some of them introduced electrodes to pass currents through bones in an effort to cause what is termed modeling in this book, instead of modeling new woven bone often appeared around the negative electrode. That suggested that such electrical treatment might help fractures to heal. The matter has since become an active field of study, but for various reasons it remains somewhat confused. However, electrical currents can now be applied to a healing fracture at least by two means: via electrodes introduced surgically into the bone fragments and/or the fracture itself,[88] or by induction via external coils[62] or fields[795] applied close to but outside of the skin. It is not the purpose here to discuss the problems of that technique and research. Rather, the published ES results suggest to the writer that the modality has at least two separate and reasonably obvious effects plus a third more speculative one.

First, the invasive forms of ES in particular can evoke the local production of new woven bone adjacent to the negative electrode, just as normally follows an ordinary traumatic injury. In other words, in some cases ES can initiate the histogenesis of new woven bone. That woven bone then sets in motion anew the remaining links in the chain of repair events encoded in Relations 1 to 3 and in Figure 5. Consequently, when an intrinsically normal IO responds to obtunded stimuli following an initial injury, ES can sometimes supplement those stimuli and accelerate healing by potentiating the production of a callus. The possible direct effects of ES on granulation tissue production, angiogenesis, precursor cell activity, differentiation of new cells, BMU-based remodeling, and modeling remain unknown so far, in no small part because of the anholistic concepts underlying many of the experiments directed to the problem. However, when the "P" activity becomes intrinsically deaf to its extrinsic regulation and/or when some factor forces differentiation from hard to soft tissue producing cells, ES can fail to evoke woven bone production and bone union. Such phenomena may cause ~70% of the failures of such treatment. When it succeeds, ES probably works by evoking special local physical, cellular, and biochemical effects (see Section V.D), and it should become more predictable and effective when we learn how to potentiate and/or supplement it effectively by biological and biochemical means, and at what stages during the overall healing process it will prove most effective.

Second, as Takahashi et al.[721] have shown, ES also stimulates a new RAP that can accelerate all phases of any already ongoing healing. A significant fraction of the clinical successes of ES treatment seems to the author to relate to this effect, which probably differs from the histogenetic effect on woven bone production noted above. Unfortunately, no ES research to date has studied the effects of ES on the separate IO terms and repair subprocesses and the equally important transitions from one process to the next that are shown in Relations 3 and 5.

Third, the author suspects that somehow and in some patients ES can cause an "Off" state of the regional tissues, speaking with regard to their competence to respond appropriately to injury, to switch over to the "On" state, whereupon the hoped for normal healing process becomes enabled and ongoing.

B. Bone Grafting

The surgical procedure involved in performing an onlay or inlay bone graft causes a major new local injury that evokes a new local RAP on its own account. That accelerates subsequent healing. Where the original injury could fail to evoke an adequate RAP, the second injury caused by the operation often succeeds in evoking one. That potentiation of healing is known as the "second injury phenomenon".[140,156] Those phenomena would seem to account for some of the successes of bone grafting procedures.

Fresh autogenous bone grafts also add some relatively labile biological/biochemical factors that can further potentiate healing, in part by fostering the differentiation of "P" daughter cells towards osteoblasts and chondroblasts rather than towards fibroblasts and fat storage cells. Local biochemical factors that might participate in that potentiation would include prostaglandins, leukines, leukotrienes, Urist's BMP, angiogenic factors, several "growth factors", and fibronectin (see Section V.D). This area would seem to have great therapeutic promise. It represents a potentially fruitful area of research.

C. By-Passing the Callus Phase

To repeat, about 15% of the clinical nonunions seen in the U.S. reflect an inability of injured local bony tissues to produce fracture callus, either in adequate amounts or at all, and for reasons alluded to earlier and as illustrated in several figures in this chapter. Primarily Swiss workers have shown that one can still obtain union in such cases, and without depending on callus production, because the bone BMU-based remodeling potential usually remains normal. As Rahn[613] and Schenk[666] have described, if one provides rigid mechanical fixation of closely apposed and well-fitted fragments, that will allow that remodeling process to transfix the fracture plane with numerous new secondary osteons, a process accelerated some 10 to 50× above normal by the RAP evoked by the operation done to apply the fixation. Several types of implants, as well as sliding bone grafts, can provide such fixation. In the past 30 years, the author has treated about 100 established tibial biological failures of union by securely fixed sliding bone grafts without a single failure.

D. Direct Biological Stimulation

The current literature lists a growing number of biologically active biochemical substances that can somehow directly stimulate and/or potentiate naturally evoked new woven bone production, the essential hard tissue element in any first-stage healing. Those substances include certain prostaglandins,[329] Urist's BMP,[742] fibronectin,[724] one or more "bone growth" factors, and "X". They usually arise and act in the "L"-domain of the IO and promise to provide simpler, more economical, more effective, and biologically more sophisticated solutions to many malfunctions of the bone repair process in the future. However, much less is known about those substances that must also exist in nature to enable and regulate angiogenesis during all healing stages, the underlying precursor cell activities, and the differentiation of new cells.

Some systemically circulating or "S" agents can also affect the responses of the bone repair processes to local signals. Such "S" agents include, but are not confined to, growth hormone, some somatomedins, adrenalcorticalsteroids, immunosuppressive agents, and parathyroid hormone. However, the male and female sex hormones have proven ineffective in that regard, and may even slightly depress bone repair.

Some enabling and/or regulatory substances may act by blocking a normally present inhibition of an otherwise inherent tendency for uninjured cells to proliferate. Consequently, some supposed "stimuli" to repair may in fact and instead remove or disable a preexisting inhibition of cell proliferation. In effect, they may release a brake rather than floor an accelerator. Also, the systemic and local factors that initiate and regulate precursor cell activities, the differentiation and activities of their daughter cells, and the sequences of the normal bone repair processes will probably prove numerous, and to have important interactions and overlap, and to have major differences as well. For example, some prostaglandins may affect only some parts of the 1° healing phase and play no role in controlling the continuum phases of remodeling and modeling, while adrenalcorticalsteroids seem necessary in physiologic concentrations to enable a normal 1° healing stage, but pharmacologic concentrations can actually depress the 2° stage activities. Finally, the hypophysectomized animal usually cannot develop either a normal 1° stage or 2° stage, or an adequate RAP. As research on such matters continues, further such differences and effects should be found. Currently they represent an information vegetable soup that mixes analogs of the earlier-described MCN phenomenon with parallel and series gating, and with the enabling, saturation, and continuum effects described in Note 2 in Chapter 2, and with the different effects of drugs that relate to their dosage, their mode of administration, and the duration of treatment that are referred to in Note 1 in Chapter 11. The author suspects this system will prove much more complex than the above text suggests, somewhat as the mechanisms underlying a blood clot or the immune response[434] have proven complex. Consequently, much work lies ahead before we can understand it well enough to satisfy the "Frost criterion" (e.g., you do not understand a system until you can make it dance to your drum rather than that of fate at least 95 times out of 100).

E. Future Research

As already suggested, the idea of coherence research described in Chapter 5 would seem to be applicable to studying the problems of the bone repair process (and chondral, tendon, and ligament healing), and of its responses to drugs, hormones, and other factors. Like BMU-based remodeling, the normal bone repair process involves an ordered sequence of subprocesses, each couples in some way to its successor and, in most fractures, in a state of temporal incoherence. Thus, the abstract features exist for tem-

Table 4
SOME APPARENT FUNCTIONS OF THE
MACROREPAIR PROCESS

Domain	Functions
L_3	To heal gross injury to bone, fibrous tissue, and cartilage
	To wall off infection and infestation
	To wall off metastases
	To defend against mechanical abuse

poral coherence and incoherence, for the action of varied enabling and regulating agents, and for an operational MCN analog that must somehow be by-passed. Understanding the complex details of the natural controls of each subprocess and the coupling of one subprocess to the next could and almost certainly would benefit by inducing and/or making investigative use of the properties of temporal coherence as described in Chapter 5. A critical factor in any such research will lie in working out the σ values for the various subprocesses, and, in that respect, in also establishing standard experimental models of fracture healing in laboratory animals. Considered in that light, most research on bone, chondral, and fibrous tissue healing that has been published to date is at par in its scope and depth with the research on human bone remodeling that was done before its quantized BMU basis was generally accepted. What has been missing in the fracture healing studies has been an accounting of the operational features of the skeletal IO. That healing has been generally conceived up to now as a matter of cell → organ, when in fact and indisputably, it is a matter of cell → IO → organ, and the IO here includes the features indicated previously in Figure 5.

F. Summary

The current literature on this subject is descriptive and often confusing, partly because we lacked an adequate conceptual framework for organizing and interpreting its factual content. That lack can be remedied easily, so accelerated progress and more effective means for treating bone repair problems should materialize in the coming years. First, however, there is much to learn, ponder, and understand. Table 4 lists some of the apparent purposes of the macrorepair mechanisms in skeletal physiology.

VI. ILLUSTRATIVE QUESTIONS

Interest in human fracture healing problems runs high in the world at the moment. Vast sums have been invested in studying some of its features since about 1970. However, most of that money aims at exploitation of something observed or supposedly known, rather than at basic research. The exploitation has its uses, as any orthopedic surgeon who treats fractures knows, but the field has now reached a point where the next really major advances should come from newer and better understanding provided by basic research rather than by exploitation. After all, when conventional minds analyze problems they tend to see conventional explanations and solutions for them rather than fundamentally different and better ones. Those who do basic research have plenty of problems to study, as the following list of questions will illustrate.

What specific agents — physical, biochemical, biological — enable and then regulate the continuum phase of granulation tissue production? And the second phase of callus production? And the third phase of remodeling and modeling?[1,95] What are the real physical, chemical, biological, and biochemical details of those activities, their enablements, and their regulation (see Figure 10)? And the overriding RAP? What governs

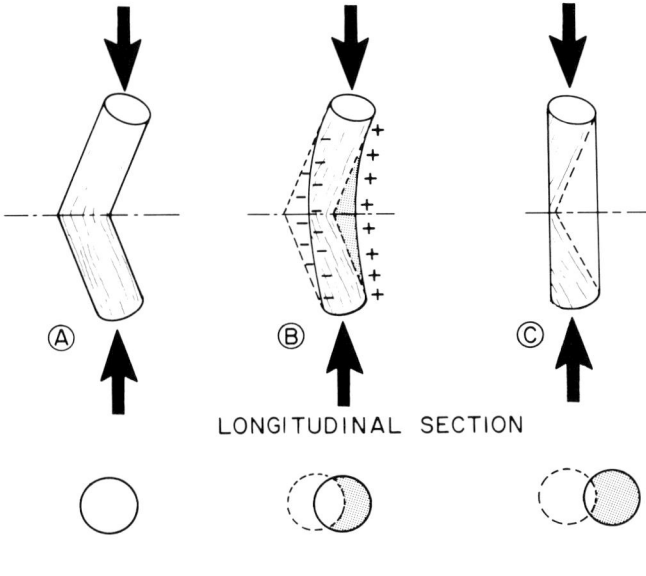

FIGURE 10. The flexure-drift relation. Since it has proven difficult for many people to understand at first, and since it is one of the later phases of normal fracture healing, this matter deserves the emphasis of repetition. The malunited bone on the left has a significant abnormal angulation, as shown, much like the proximal or upper tibial fracture in Figure 1. Vertical loads on it increase its angulation, and that increase constitutes dynamic flexure, the concave-tending facing of which points to the reader's right. If it exceeds the MES, then formation drifts will arise on the right or concave-tending surface, and resorption drifts on the left or convex-tending surface. Both drifts will then proceed to reshape the structure to that configuration on the right and as shown in the cross-sectional views below. That modeling process stops when dynamic flexure ≯ MES stops, so under this loading mode it is also a flexural neutralization or limiting activity. To repeat, it is not the original angulation itself that makes those modeling drifts appear, but rather the dynamic *change in flexure* under load; if in the left-hand situation a loading mode were contrived that tended repeatedly to straighten the already curved bone then that would evoke drift patterns and directions exactly the opposite of those shown on the right. (From Frost, H. M., *The Physiology of Cartilaginous, Fibrous and Bony Tissue*, Charles C Thomas, Springfield, Ill., 1972. With permission.)

the associated precursor cell activities, the "A" cell activities, the capillary growth, and the "D" phenomena? What brings any one of those processes to a stop? How does it know when to quit?

How does ES work? And on each healing stage, and within each, on each of the terms in the basic-state equations of Chapter 3? What sets the natural σ values of each stage? What can impair, e.g., prolong them? How do commonly used drugs affect each stage, and each activity within each stage?

What specifically comprise the histopathologic, biochemical, biological, and radiographic features of various kinds of nonunions and delayed unions? How do commonly used drugs enter the picture? And endocrine, nutritional, and systemic biochemical factors?

What significant modifying effects on the healing process, *in toto* and by parts, associate with systemic diseases, commonly used drugs, different modes of fracture treatment, age, sex, the nature of the "L" environment of the fracture region, and the recent dietary and physical exercise histories?

Note: Many of the above questions could apply equally well to problems of fibrous tissue healing.

What precise roles do "On-Off" states of the regional tissues play in normal healing and in biological failures of healing, if any?

162   *Intermediary Organization of the Skeleton*

What else of importance that pertains to the domain of this chapter do we still need to know? What are the errors in the chapter? What are its important omissions, whether because we have not recognized their relevance or discovered them?

Note 1

ON THE ACCEPTANCE OF IO-ORIENTED SKELETAL RESEARCH

Basic facts had exposed the operational importance of the skeletal IO for the case of lamellar bone by 1966[229,230] and for some others of the basic skeletal tissues by 1973.[236,241] What follows is part of the record of its diffusion within the field at large.

A. Program Time

Between 1962 to 1981 inclusive, well over 10,000 skeletal papers, abstracts, and posters were accepted by program committees and given at the meetings of the American Academy of Orthopedic Surgeons, the American Orthopedic Association, the Orthopedic Research Society, the Association for Bone and Mineral Research, the Gordon Research Conferences, the International Calcified Tissue and International Calcium Regulating Hormone meetings, and the annual meetings of varied other Western and Oriental societies of pathology, dentistry, internal medicine, anatomy, nutrition, pharmacology, endocrinology, pediatrics, biomechanics, orthopedics, rheumatology, cell biology, and molecular biology, including their veterinary medical counterparts. However, only 14 of those presentations dealt with the IO.

B. Access to Publication

Between 1963 to 1983 inclusive, well over 9000 skeletally oriented articles were accepted and published by the editorial staffs of the American and British *Journals of Bone and Joint Surgery, Science, Acta Scandinavia Orthopaedica, Laboratory Investigation, Calcified Tissue International,* and numerous other indexed and refereed journals in the developed nations that deal with physiology, pathology, dentistry, veterinary medicine, internal medicine, pediatrics, anatomy, orthopedics, nutrition, biomechanics, pharmacology, metabolic bone disease, and skeletal growth disorders (certain exceptions are noted below). Only 13 of those publications dealt with the skeletal IO.

Over 1000 abstracts were accepted and published by the Orthopedic Research Society alone for the years 1982 to 1984 inclusive. Only two dealt with the IO, none did in 1984, and the program committee rejected the report of the first successful ADFR treatment of osteoporosis by Colin Anderson of London, Ontario.[24]

C. Access to Funding

From 1966, when the influence of the static histomorphometrists began to ride high in most organizations that funded medical research, to 1981 inclusive, over 2500 skeletally oriented grants were funded by the NIH, NASA, the AEC, the Batelle, Ford, Kroc, rheumatologic, Osteogenesis Imperfecta, Paget's Disease, and other foundations in the U.S., the British M.R.C., their Scandinavian, German, French, Belgian, and Latin-American equivalents, and numerous other agencies that support medical research in the developed nations. Hundreds of those funded projects used the anholistic static histomorphometry, but only two grants were funded in the U.S. during those years for projects that depended on dynamic histomorphometry, meaning WSDH as defined in Note 1 in Chapter 4, and only four such projects were funded in other nations. Because of that drought, only the writer's laboratory could do dynamic histomorphometry in 1967. By 1975 the laboratories of Takahashi, Jee, Meunier, Melsen, Jaworski, and Recker had joined it, but private funds supported their initial dynamic work.

D. Awards

From 1960 to 1984 inclusive, various kinds of skeletal research gained over 300 awards in Europe, Scandinavia, and the North American continent. Not one related to the skeletal IO.

E. Summary

Whatever the reasons, the granting, journal referee, program committee, and lecture hall mechanisms frowned on the new skeletal holism until the middle 1980s, when it seems to approach its season. George Herbert said it in one way: "Good and quickly seldom meet." John Stuart Mill put it differently: "Wrong opinions and practices gradually yield to fact and argument; but facts and argument, to produce any effect on the mind, must be brought before it."

F. Exceptions

The main sources of IO-oriented publications up through 1983 have been the *Henry Ford Hospital Medical Bulletin* (now the Journal), plus *Clinical Orthopaedics* under the aegis of a far-sighted editor (Urist), plus four other instruments: the *Journal of Metabolic Bone Disease and Related Research,* the *Presse Lyon Medicale,* the proceedings of four international workshops on bone histomorphometry, and a series of monographs published by Charles C Thomas. Those vehicles were receptive to the new holism because various σ bone group members influenced or conceived them.

Chapter 13

CHOOSING APPROPRIATE MODEL SYSTEMS FOR SKELETAL RESEARCH

"Because the knowledge of our predecessors has placed us part way up the ladder of knowledge, we go even farther with less effort than they to discover what they could not see themselves."

(Paraphrased, Blaise Pascal, about 1647)

ABSTRACT

Some fundamental differences between the lamellar bone modeling and remodeling systems as defined in the new bone lexicon explain why their malfunctions can cause different groups of diseases, why efforts to find cures for diseases of one system by studying the properties of the other usually fail, and why efforts to extrapolate to the intact steady-state skeleton the responses that can appear reproducibly in in vitro or in subsigma in vivo bone systems have failed too. That accumulated evidence fits the IO theoretical projections that modeling systems should not reliably model remodeling systems, or conversely, and that transients cannot model steady states in intact skeletons. Valid model systems should duplicate the physiological system(s) that cause the human disease, including the envelopes and any other special anatomical localizations, and the transient or steady states involved.

I. INTRODUCTION

Other chapters and notes in this book refer to some of the problems that have arisen in skeletal research by using model systems that do not provide valid analogs of the human disease that led to an experiment or series of experiments.

This chapter offers specific information concerning appropriate experimental models for human bone diseases with the understanding that analogous approaches can apply to chondral, dental, and fibrous tissues and to synovial disorders. The bone disorders have created the most numerous examples of choosing inappropriate model systems and Note 1 in Chapter 2, Note 2 in Chapter 6, and Note 1 in Chapter 9 provide examples. The problem is widespread and touches many areas of contemporary skeletal research and thought.

The discussion begins with the origins of the terms modeling and remodeling and reviews their meanings according to the new skeletal lexicon.

A. Background to the Terminology

As long ago as 2700 B.C., the Egyptian Imhotep[351] probably knew that when a long bone fracture heals with an obvious deformity, in a young child it usually corrects during subsequent growth, but in an adult it persists. Before 1900 A.D. it had become known that that child's bone "sculpting" involved coordinated motions in tissue space of the periosteal and cortical-endosteal surfaces of the malunited bone, that formation of bone on some surfaces and resorption on others caused those surface motions as described in Chapter 7, and that analogs of them in healthy growing bones maintain their basic shapes and proportions as they grow in length and diameter from the tiny fetal model to adult size.[422,452,592,623] Naturally, then, by the late 19th century both the normal and postfracture sculpting phenomena had become known as "bone remodeling". Unaided vision can perceive the effects of those activities on intact bones on ordinary X-ray films, and as shown in the Figures in Chapter 7.

Another activity can also occur in bone. Semimicroscopic and not readily seen on routine X-rays or by gross anatomical study, Leeuwenhoek[447,448] and Havers[317] first described one of its signatures, now termed the secondary osteon or haversian system. Later workers added further signatures, including the reversal and arrest lines,[343] osteoid seams, and Howship's lacunae.[182,273,343] That internal turnover of compacta by secondary osteons had also become known as remodeling by the 19th century. Some earlier authorities may have suspected that the sculpting and haversian phenomena differ somehow,[490,751] but that remained obscure until 1962 to 1966, when new histomorphometric techniques using tissue time markers led to the discovery that bone turnover occurs in semimicroscopic packets on all envelopes of the adult human bony skeleton (see Note 2 in Chapter 4).[193,229,230,315] By 1966 it had become clear that it follows its own game rules, that its enablement, regulation, and effects differ in basic respects from those of the sculpting phenomenon, and that haversian remodeling was only one case of it.[379,414] It had also become clear that each activity causes its own group of diseases, that the former controls shape and configuration while the latter can change or more often reduce the amount of bone independently of its configuration, and that one seldom if ever provided a valid laboratory model of the problems of the other.

That created a problem. What terminology should distinguish the sculpting from the turnover phenomena, when both were commonly known as remodeling? At first they were termed surface remodeling and internal remodeling and so described at the Bone Biodynamics Symposium at Henry Ford Hospital in 1964,[231] and those terms are still used by some investigators.[822] However, they were too similar, so around 1966 the writer began to call the former activity modeling (macromodeling in this book, for micromodeling was not recognized in 1966), as in modeling in clay, and redefined the remodeling term to designate only the semimicroscopic BMU-based turnover of bone.[230,236,237] Those meanings begin to appear in standard texts and they serve in this book as well.[H,V,Y,379] Still, in retrospect a more apt term for the packet-like remodeling activity would be useful, in part because many clinicians unfamiliar with the IO and its jargon continue to use the remodeling term to signify both phenomena, and that usage has a certain logic.[222] Therefore, in this book:

$$\text{modeling} = \text{sculpting}$$

$$\text{remodeling} = \text{turnover in packets}$$

Figure 1 illustrates both of those activities as they occur in humans.

Comparing the activities those two terms stand for in the new lexicon requires reviewing their origins in the IO next, beginning with L_2-space-time.

II. THE SKELETAL IO (AND GROWTH, MACROMODELING, AND REMODELING)

A. General Remarks

In reading the remainder of this chapter, the reader should keep in mind that IO-intrinsic and IO-extrinsic malfunctions can each cause disease in their own right, and that a combined group also exists. Table 3 lists some examples of all three types.

L_1-Domain IO entities serve as the bricks that assemble in various ways to create the more specialized L_2-domain entities, where the clinically known activities of growth, macromodeling, and remodeling can be said to have their origins. Macromodeling occurs primarily during growth in all bony vertebrates, and in lamellar bone two different L_2-level entities produce it, while remodeling occurs throughout life in man and in other long-lived animals and a single kind of L_2-level entity produces it. In L_2-space-

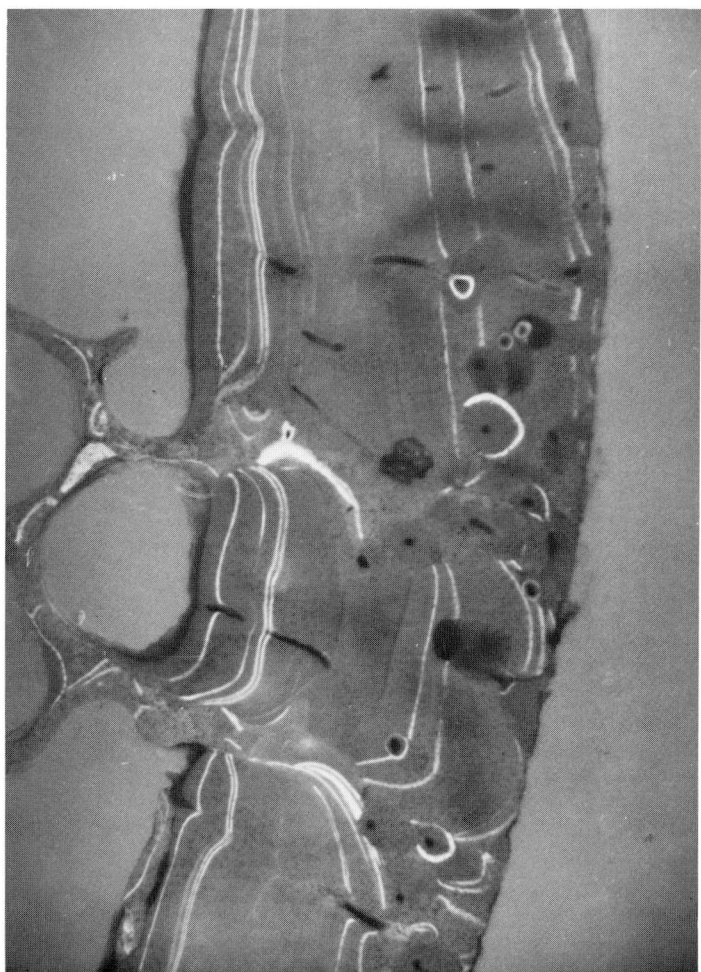

FIGURE 1. Cross-section of the pleural cortex of the middle third of a sixth rib from a young adolescent girl. Undecalcified, blue-light fluorescence microscopy, about 24×. Periosteum to the right, marrow cavity to the left. The bright bands represent fluorescing tetracycline bone labels deposited incidentally over the previous 8 years when she received the drug for intermittent respiratory infections. The section and its labels illustrate eight phenomena. First, some small amount of quantized BMU-based lamellar bone remodeling occurred with the cortex, revealed by the circular labels therein. Second, bone modeling formation drifts occurred repeatedly on the cortical-endosteal surface. Third, a corresponding series of resorption drifts occurred on the periosteal surface, and like a carpenter's plane or spokeshave, they paid little attention to the intrinsic architecture of the underlying bone. Rather, they were guided by the nature of the dynamic bone flexural strains under normal usage. Fourth, and accordingly, this active cortex was drifting through tissue space towards the reader's left, which was towards the skin and away from the lung. The middle third of growing ribs normally are moved laterally or away from the lung by appropriately patterned drift systems on the periosteal and cortical-endosteal surfaces. One biomechanical reason for that directional drift of the rib has been described elsewhere.[238] Fifth, during the approximately 8-year span covered by these labels, the macromodeling mechanism turned over considerably more of this compact bone than the remodeling mechanism did. Sixth, the mean age of the bone tissue in this cortex is half or less than half of the child's chronological age of ~12 years because of the constant addition of new bone and removal of old bone caused by both the modeling and remodeling processes. Seventh, most of the secondary haversian remodeling occurred in the older bone on the right rather than in the younger bone on the left. At least part of the reason for this is that that remodeling arises to remove microdamaged bone and replace it with new, undamaged bone of the same lamellar type. Microdamage occurs as a function of increasing time, so older bone will develop more of it than young bone. Eighth, in L_2-space-time there is no connection or coupling between formation drifts (see also the next figure), but in making a new secondary osteon or haversian system the resorption and formation activities occur in the same place, sequentially, always in the order R → F, and the F always directly follows the R in time.

time both modeling and remodeling act in the approximate domains of the millimeter and their steady-state responses to regulation tend to take 2 months to develop.

B. Growth

As noted in Chapters 3 to 5, in broad terms this activity increases the number of cells and the amount of intercellular material, both of which then increase the size of a tissue or organ. They make an accountant from an ovum.

As they grow, skeletal tissues can also acquire an external form in relation to their surroundings, whether those surroundings comprise other IO units of the same kind (e.g., adjoining columns of cartilage cells in an epiphyseal plate) or of a different kind (e.g., bone next to periosteum). Factors external to the tissue, such as a mechanical force and/or a biochemical or other influence of an adjoining tissue, cause that external form or architecture by locally allowing or potentiating growth of the tissue in some places and directions and restraining it in others. Such extrinsically determined L_2-level structures include secondary osteons, circumferential lamellae, and trabeculae, all made of the same simple L_1-level tissue, lamellar bone. Note here a topological quirk: marrow lies within bone in the geographic anatomical sense, but external to it in the IO functional sense. That extrinsic control of the bulk shape of growing tissues and organs represents macromodeling, as described in earlier chapters.

Growth-dependent cellular proliferation ceases at maturity; growth hormone and somatomedin(s) (and perhaps other systemic agents) control it directly and it increases the number of cells and the size and amount of tissue in an organ.[379] It also appears to enable the macromodeling activity.[263] The growth-independent class of cell proliferation occurs throughout life, including after maturity, it appears in wound and fracture healing, it normally replaces exhausted or dead cells in all endothelia, epithelia, connective tissues, and the hematopoietic system, and it does not change the total number of cells or the amount of tissue present in an organ, or its size either.[263] It also supports the remodeling activity.

Since in the L_2-domain the proliferation or "P" term in Relations 3 and 4 in Chapter 3 gates the endocrine regulation or "S_h" of growth into the IO, the state equation would look like this:

$$F(G) = (S_h:L) : \overrightarrow{(P:C:O:A:M)} \qquad (1)$$

The programmed relation could look like this:

$$F(G) = \begin{array}{c} S_h \\ \downarrow \end{array} \begin{array}{c} L \\ \uparrow \end{array} \longrightarrow \qquad (2)$$
$$C:P:D:O:A:M$$

C. Lamellar Bone Macromodeling

To repeat, macromodeling molds the macroarchitecture or configuration of growing tissues and bones in all bony vertebrates regardless of their size, and the faster the growth the faster the macromodeling.[235,379] In bone, and excepting the effects of chondral activity, it establishes the outer and marrow cavity bone diameters and the cross-sectional and longitudinal shapes.

1. Control

In lamellar bone at least two factors extrinsic to an IO unit can regulate its modeling activities after growth and any other factors have enabled them. The mechanical loads carried by the whole bone can both enable and guide them, at least in part,[393] and the adjacent juxtaosseous tissues can also guide them.[263] The macromodeling system is enabled by increasing mechanical usage. Apparently, it is not enabled by acute mechanical disuse but rather it remains dormant under those conditions. The mechanical loads carried by a bone depend in their turn directly on growth-dependent increases in body mass, muscle strength, and usage[393] and on bone length (which affects the flexural moments and strains generated by normal usage of a bone).[238,249,267] Depress, arrest, or increase either growth and/or mechanical loads, no matter how, and bone modeling activity typically changes likewise, although usually after certain time delays and transient changes that reflect the sequential features of the underlying IO modeling units, which is encoded in Relation 4.

2. Age Incidence

Bone macromodeling subsides to trivial levels after skeletal maturity in species that do mature, and its effects appear on ordinary X-rays in species as disparate in size as the sparrow and elephant.

3. Distribution

It affects primarily the periosteal and cortical-endosteal envelopes, and Figure 2 shows an example.[379] As defined here, it does not affect the haversian envelope or spongiosa, although poorly studied microscopic analogs of it do occur there, including the "waltzing osteon". Due to its properties, modeling *must* change the size, architecture or shape, and the amount of any affected tissue in the skeleton, including bone.

4. Operational Features

Organ-level lamellar bone macromodeling sums the separate activities of two kinds of L_2-level multicellular units, each of which arises physically, biologically, and locally independently of all others in L_2-space-time. As described in Chapter 7, those L_2-domain units are the osteoblastic or formation drift, and the osteoclastic or resorption drift. Each has its own capillary, and as far as we know at present each performs its own function of either R or F without any sequential changes of the R → F nature in L_2-space-time such as do occur in BMU-based lamellar bone remodeling.

The programmed relation for a mechanically controlled formation drift, F(FD), could be written thus, where "S_m" signifies an S-mode mechanical agent, M_1 signifies the preexisting bone, and M_2 any new bone that the unit might make as the result of a formation drift:

$$F(FD) = \begin{array}{c} S_m \downarrow \\ \rightarrow M_1 \leftarrow \\ \downarrow = = = \Rightarrow \\ C{:}P{:}D{:}O{:}A{:}M_2 \end{array} \quad (3)$$

The dotted arrow represents the feedback loop, meaning that new bone, "M_2" deposits on old bone, "M_1", and changes its architecture, which then changes how M_1 will respond to strains under subsequent mechanical loads, S_m.

A simple state equation for a formation drift could look like this:

$$F(FD) = (S_m) : (M_1{:}L) : (P{:}C{:}D{:}O{:}A{:}M_2) \quad (4)$$

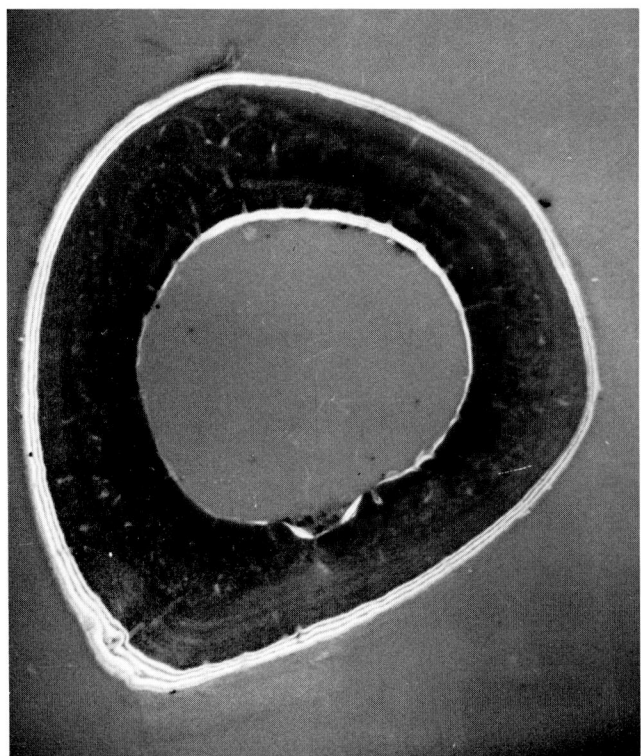

FIGURE 2. Undecalcified cross-section through the middle third of the diaphysis or shaft of the femur of an experimental growing rat. Anterior quadrant above, posterior quadrant and linea aspera below. This control animal received three intraperitoneal tetracycline bone labels a few days apart before sacrifice. The labels show the following facts pertinent to this chapter. First, the entire outside diameter of the bone was expanding due to periosteal formation drifts. Second, that expansion was most rapid towards 7 o'clock and least towards 2 o'clock, so the whole shaft was being moved down and to the left in tissue space by those drifts. Third, in the 2 o'clock quadrant the formation drift was in the "Off" state when the middle of the three labels was given. Similar phenomena can be seen in some of the adolescent's labeled formation drifts in the previous figure. Fourth, the cortical-endosteal surface showed a formation drift at 1 o'clock and either a resorption drift (likely) or an "Off" state (possible but less likely) at 7 o'clock. Fifth, there is no secondary haversian remodeling inside the cortex at all, so this bone provides what Frostian skeletal authorities term a "modeling-only system", excepting of course the events that occur beneath its growing epiphyseal plates. (Specimens courtesy Dr. H. Roth, Southfield, Mich., and R. Hattner, San Francisco.)

Observe that while the "P" or precursor cell activity represents the primary gate for endocrine control of the growth of these IO entities, the M_1 term serves that gating function for the regulation of growth by concurrently acting S-domain mechanical forces applied to the bone.

Resorption drifts would have no M term below the bus since they do not make new solid intercellular materials, and the clasts resorb bone that was made by other IO units in the past. So the state equation might look like this:

$$F(RD) = \begin{array}{c} S_m \\ \downarrow \\ M_1 \leftarrow \\ \downarrow \\ L \\ \hline \Rightarrow \\ C:P:D:O:A \end{array} \qquad (5)$$

and the simple state equation would appear thus:

$$F(RD) = (S_m) : (M_1 : L) : (P : C : D : O : A) \qquad (6)$$

with a feedback loop from A back to M.

The dotted line indicates the feedback loop in this system where the effects of "A" by modifying the structure of M modify how it responds to the subsequent loads, S_m. The same kinds of mechanical factors, i.e., dynamic strains, control both formation and resorption drifts as described in Chapter 7.

Note Bene: In the operational terms of A, R, and F, bone formation drifts comprise A-F entities with no R involved. Resorption drifts comprise A-R entities with no F involved.

5. More on Distribution and Control

In L_o- or organ-level space-time, drifts distribute in patterns over the periosteal and cortical-endosteal surfaces of the compacta — patterns that prove strongly stereotyped in a given bone of a given species.[238,478] As noted in 1963,[824] that stereotypism happens because both the neurologic control of the musculature and the anatomical details of muscles, e.g., their sizes, insertions, and origins, are similarly stereotyped,[241,824] and they provide the major loads carried by the bone.[824,708] As described later and still in the direct sense, bone drifts appear to respond poorly at best to known manipulations of the diet, biochemical factors, drugs, or the endocrine system.[267] Table 1 summarizes salient properties of bone macromodeling as understood at present.

As earlier chapters described, the A cells of bone drifts, the osteoclasts and osteoblasts, have limited individual functional lifespans, on the order of 2 weeks for clast nuclei and 1 month for blasts.[374,376,572] Consequently, continuing bone modeling for the 18 or so years the human needs to mature, or for the 1 year a rat needs for the same purpose, requires repeated creations of new macromodeling units, and thus continual precursor cell and related activities. As noted in 1973,[238] it follows that the P rather than the A activities represent the ultimate rate-limiting factor for steady-state bone macromodeling, for they determine if, where, how much, and what kind of macromodeling will occur, and when. If for any reason that precursor cell activity stops, then modeling will stop too, and totally both in vitro and in vivo, but after a time delay of some 2 to 3 weeks in a rat and perhaps twice or thrice that in a child.[263] Because of the sequences inherent in bone drifts and encoded in the above relations, a steady-state treatment effect on bone macromodeling needs 1 or 2 months to develop.[267] Accordingly, effects noted earlier than that, both in vivo and in vitro, should represent transients that cannot persist in the intact subject and that alone should neither cause nor cure disease, as Parfitt, the author, and others have noted on several occasions.

In a 10- to 15-year-old human, a completed drift can deposit or remove from less than 1 mm³ to over 100 mm³ of bone, so that volumetric property can vary over a wide range, approximating 1 to 1000,[267] and the labeled formation drifts shown in Figures 1 and 2 reveal that property clearly.

6. Distributive Properties

Referring back to the "andistributive property" described in Chapter 2, the various elements of L_2-domain macromodeling activity involve in their proper order, growth → IO-extrinsic macromodeling factors → bone drifts. Two kinds of drifts, the resorption (r) and formation (f), can be activated (a), and the amount or volume of tissue resorbed or formed by each (v), can vary approximately 1000-fold. A relation that

172   *Intermediary Organization of the Skeleton*

Table 1
BONE MODELING PROPERTIES (AT L_2-LEVEL)

1. Two different kinds of bone drifts: formation and resorption.
2. Each occurs independently of others in L_2-space-time.
3. Each has its own capillary, and its control gates primarily via "M".
4. Each requires L_o-domain growth, and each occurs in all bony vertebrates.
5. Neither normally occurs after maturity in man.
6. Each occurs primarily on periosteal and cortical-endosteal envelopes, so the greatest effects are confined to compacta.
7. Neither has major effects on spongiosa or nongrowing bone.
8. Causes over 90% of annual compact bone turnover in an 8-year-old child.
9. Must alter architecture, size, and quantity of compact bone.
10. Directly regulated by growth, muscle forces, and skeletal mechanics.
11. No so far proven *direct* steady-state hormone, nutritional, biochemical effects.
12. Value of σ approximately 1 month in the rat, 3 months in man.
13. Volume of tissue moved per L_2-domain drift can vary more than 1000-fold.
14. Each drift requires a capillary and growth-dependent precursor cell proliferation.

describes the distributive properties of those events could be written thus, where (g) refers to growth:

Distributive properties of bone macromodeling (L_2-domain):

$$(g:m)\overset{\rightarrow}{:}(a:r)$$

and

$$(g:m)\overset{\rightarrow}{:}(a:f) \tag{7}$$

where these constraints apply:

$$\begin{aligned} g &= F(L_2\text{-mode growth}) & V(a:r) &\neq k \\ m &= F(S\text{-mode factors}) & V(a:f) &\neq k \\ a &= F(g) & k &\neq F(f) \\ a &= F(m) & F &\neq F(r) \end{aligned}$$

By way of explanation, "(g:m)" means that growth (g) must occur before modeling (m) can arise, and they represent an interaction (:) that can then somehow enable (a) a resorption drift (r) to form (a:r). The volume of bone (V) resorbed by that drift (a:r) is not a predetermined fixed quantity, it is not a constant ($\neq$k). In terms of their distributive properties then, g:m does not equal m:g, for the latter term would mean that modeling enabled growth, and that is clearly not true. So while g.m = m.g in ordinary arithmetic it does not in the biological systems considered here because those functions lack the distributive property, i.e., they are andistributive. Finally, in modeling systems the resorption drift (r) is not a biological function (F) of a formation drift (f), or r $\neq$ F(f), and also conversely, or f $\neq$ F(r).

Note: This chapter discusses macromodeling so the material does not apply to micromodeling, except where so stated.

7. Functions of Macromodeling

The apparent purposes of this activity seem at present to be two.[238,379,608] One purpose consists of adapting architecture (which includes size, configuration, and location in tissue space) to the mechanical demands made of the structures of the skeleton by usage. The other purpose appears to be that of assisting the growth processes in constructing the size of the adult from that of the infant. In essence then: macromodeling is an architectural game.

D. Lamellar Bone Remodeling

In the new bone lexicon this redefined term signifies exclusively a semimicroscopic, packaged form of replacement of older lamellar bone by new lamellar bone, and thus a special kind of bone turnover.[232] Figure 1 showed an example of it.

1. Age Incidence, Distribution, and Control

It occurs in significant amounts and throughout life only in large and long-lived animals including man, for only trivial amounts appear in the compacta of small, short-lived species as well as in infants and in the fetus where modeling is especially active.[237,379] (See Table 5 and also the discussion of spongiosa below.) In man it occurs on all four bone envelopes,[230] but it does not appear to respond directly to growth. As for bone strains and mechanical usage, increasing usage actually seems to inhibit remodeling; under conditions of acute disuse that inhibition appears to be removed, allowing a period of increased remodeling to ensue. It also responds, and sensitively so, to varied dietary, endocrine, pharmaceutical, and systemic biochemical factors, to some local ones, and probably to certain indirect effects of mechanical loading, including fatigue-like bone microdamage.[134,237,241,572]

2. Operational Composition

As described in Chapter 4, a single L_2-level entity, the BMU produces bone remodeling when some activating (A) event marshalls a batch of osteoclasts that resorb (R) a packet of bone. The properties of that activating agent suggest one of the switches described in Note 2 in Chapter 2, rather than one of the continua. After the resorption phase ends, new osteoblasts replace the clasts and form (F) new bone in the previous defect. That ARF sequence consumes a time period named σ, normally some 3 to 4 months in duration in man, and it creates a new bone structural unit (BSU) lying on a reversal line (see the figures in Note 2 in Chapter 4). In L_2-space-time the remodeling R and F always occur in the same place, always in the order R → F, the same capillary nourishes both, and, under most conditions including aging, they occur in closely similar and minimally variable amounts, some 0.05 to 0.1 mm³ of bone turned over per BMU.[154,199,232] The evidence indicates that μ, the term used by histomorphometrists to signify the activation event, acts as the primary gate for the system, although good evidence also indicates that some agents can gate into the system via the "L" term. A simple state equation for lamellar bone remodeling could be written thus:

$$F(BR) = (S):(\mu:C):[\overrightarrow{(L:P:D:O:A)}:\overrightarrow{(L:P:D:O:A:M)}] \qquad (8)$$

3. Distributive Properties

As for the distributive properties of BMU-based remodeling, the L_2-domain activity does not require growth (g) or any modeling (m) influences in the same sense that modeling does. In fact, in the human newborn, macromodeling is intense but little BMU-based remodeling occurs.[190,765] Accordingly, the (g) and (m) terms of Relations 7 do not apply here. Rather, the activity requires some local initiating or enabling event termed activation (a), followed by volumetrically and spatially stereotyped packets of resorption and formation. In symbols then:

Distributive properties of bone remodeling (L_2-domain):

$$(a):(r:f)$$

where these constraints apply: (9)

$$\begin{aligned} a &\neq F(g) & r &= F(f) \\ a &\neq F(m) & f &= F(r) \\ V(r{:}f) &\sim k \end{aligned}$$

Note that in the remodeling system the resorption packet (r) is a function (F) of the formation packet because, as the writer discovered before 1964, they are tethered or coupled,[229,824] so what happens to one usually happens to the other, although not necessarily equally (the Frost-Heaney observation; see Note 2 in Chapter 4 and Note 1 in Chapter 16). Therein lies a major operational difference from the drifts that cause bone remodeling.

The histological or ARF phases of a typical BMU finish after 3 or 4 months in healthy people, so continuing remodeling requires continuing activation of new BMU. If activation ceases, then in normal subjects and within 3 months or so all remodeling also ceases.

4. The ΔB · BMU

The remodeling BMU uses a special, slowly acting mechanism to change the amount of bone in the skeleton or on an envelope. Created as a new property by the L_2-domain cellular associations that form a BMU, and one that is unique to that BMU, it seems to underlie most — but not all — lamellar bone losses and gains in adult-acquired osteopenias and other diseases of the adult skeleton. Unlike modeling, it usually adds or removes net amounts of bone without changing the shape or external size of a bone. As described in Chapters 4 to 6, if the R phase removes more bone than usual, the F phase usually deposits more too, and conversely, but not necessarily equally, as shown in the figures in Chapter 5 and Note 1 in Chapter 2, Note 2 in Chapter 6, and Note 1 in Chapter 10. When they differ then an incremental or quantized bone deficit or excess will result. The ΔB · BMU parameter signifies such quantized changes per completed BMU and as noted in earlier chapters, it normally has a positive value on the adult human periosteal envelope, a zero value on the haversian and a negative value on both the cortical-endostial and trabecular envelopes. Those typical balances cause the slow changes in envelope sizes that occur in the adult during normal aging.

5. Remodeling Variants in the Spongiosa of Growing Animals

The above description applied to BMU-based lamellar bone turnover in human adults, and a misconception has arisen in some quarters that that is the only place where BMU-based bone remodeling occurs and that all other bone resorption-formation phenomena represent modeling. However, that is incorrect because at least two other behaviorally analogous ARF activities can occur beneath epiphyseal plates in all growing mammals, both large and small.[413,414]

First, the mineralized cartilage at the floor of that plate becomes resorbed in packets and then partly replaced by packets of new woven bone to form the primary spongiosa.[379] Hence, an ARF sequence occurs which is analogous to that described earlier for lamellar bone, so here too one must infer that something couples or tethers the R packet to the F packet. Control of that activity seems to arise locally, and also to depend directly on the growth of the overlying epiphyseal plate, and more on that growth than on the effects of systemic agents, drugs, hormones, etc.[379,413] Beyond that we know little about the details of its regulation.

Second, then a further ARF-type remodeling activity replaces that primary spongiosa with the secondary or permanent spongiosa composed of lamellar bone. Due to their typically negative $\Delta B \cdot BMU$ value, a net loss of hard tissue substance accompanies each of those ARF processes, so the spongiosa progressively reduces in amount with increasing distance from the epiphyseal (or apophyseal) plate. At least in part, the signals that evoke each of the above ARF processes depend on the inherent properties of the preexisting tissue, which comprises mineralized cartilage for the first, and the primary spongiosa itself for the second process. That suggests that both processes may depend on one or more of the switch-like mechanisms described in Note 2 in Chapter 2. Indirectly, therefore, but nevertheless closely, both processes depend on longitudinal growth. If growth slows or accelerates then both ARF processes usually do likewise. However, the details of their responses to most hormones, drugs, and nutritional factors, as well as local ones, remain unknown, in large part because extant studies have failed to discriminate properly between the separate $\Delta B \cdot BMU$ effects and processes that are involved, and failed too to understand and/or measure all of the relevant dynamic relationships in the normal and diseased systems that served as experimental models. Kimmel and Jee[413] have described some of the requirements for obtaining accurate and valid information from this system, which many investigators have used to model various aspects of adult human metabolic bone disease. Figures 3, 4, and 5 illustrate some of the features of the above processes.

Accordingly, enough is not yet known about the properties, enablement, regulation, and diseases of this form of BMU-based subepiphyseal trabecular remodeling in growing mammals to determine whether it can provide a valid model of the remodeling disorders of the intact human adult. As far as any available hard evidence is concerned, it may or it may not. However, in principle its distributive properties suggest that it does not, because it depends directly on L_3-mode and higher-order longitudinal bone growth, although it does show the volumetric stereotypism associated with BMU-based lamellar bone remodeling.[267] The distributive relations for this system and its constraints might be written thus:

Distributive properties of subephiphyseal remodeling of spongiosa (L_2-domain):

$$(g):(a):(r:f)$$

where

(10)

$$\begin{aligned} g &= F(L_3\text{-mode growth}) & a &\neq F(m) \\ a &= F(g) & r &= F(f) \\ V(r:f) &\sim k & f &= F(r) \end{aligned}$$

Table 2 summarizes some salient properties of bone remodeling and Table 3 lists certain differences between bone modeling and remodeling as currently understood.

6. The Functions of Remodeling

Four major functions or purposes of the remodeling mechanism seem apparent at present. During growth it replaces one kind of tissue with another. From that one may infer that the latter tissue provides more necessary and enduring physical and behavioral properties than the former and also that something about the properties of the latter makes it impossible to by-pass the bother of replacement. That replacement purpose relates to growth by endochondral ossification. The second purpose of remodeling consists of maintaining structures that already exist by removing packets of dam-

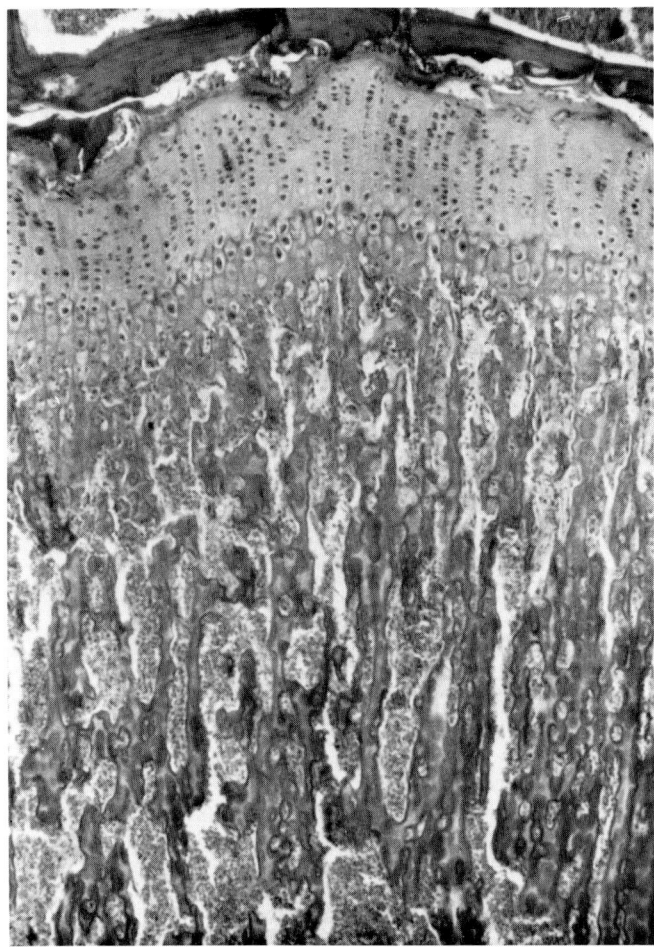

FIGURE 3. Longitudinal section through the proximal tibia and epiphyseal plate of a control normal rat. Decalcified, H & E stain, about 40×. The ossification center of the epiphysis lies above, the proliferative zone of the growth plate next, the hypertrophic zone below that, and the primary spongiosa below that. Direction of growth of this plate is upwards. The individual columns of chondral cells can be seen; each represents a clone. The gradual increase of marrow at the expense of spongiosa as one moves down and away from the plate can be seen and is due to the normally negative value of the $\Delta B \cdot BMU$ of the remodeling processes that are involved. The longitudinal grain of the spongiosa is determined by local modeling factors.

aged or otherwise defective material and replacing it with new material of the same kind. The third function or purpose consists of providing one of the three systems that can control homeostatic blood-bone interchange, as described in Chapter 10. The fourth function appears during repair as noted in the previous chapter, for remodeling replaces an established callus with mature and biomechanically competent tissue in both bone and fibrous tissue. In essence then, remodeling serves the needs of replacement, maintenance, and homeostasis.

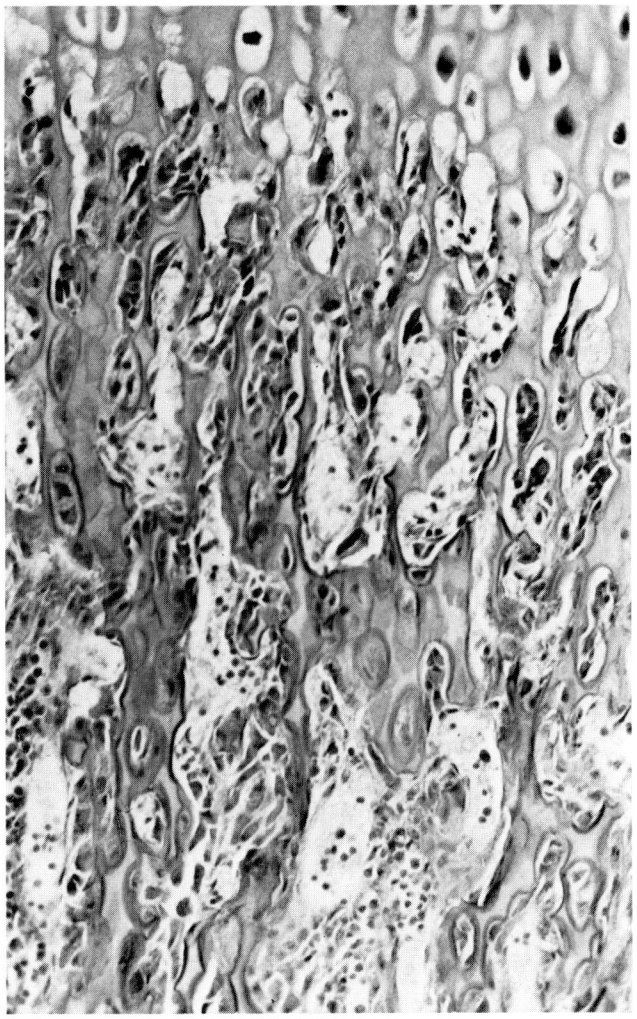

A

FIGURE 4. Decalcified, longitudinal, H & E stained sections through the proximal tibial epiphyseal plate (above) and primary spongiosa (below) of rats. About 130×. (A) The normal, control animal. Note the vertical and parallel alignment of the chondral cells in the growth plate and the underlying spongiosa, and the proportion of the marrow cavity occupied by trabeculae. In this field the mineralized cartilage above undergoes partial resorption, following which woven bone deposits alongside the vertical bars of unresorbed mineralized spongiosa. In the center of this field then lies primary spongiosa. (B) A corticosteroid-treated animal, showing a reduction in the amount of spongiosa and loss of the clear, longitudinal alignment of the chondral cell columns and the primary spongiosa beneath them. This appearance has been widely interpreted to reflect the effect of an absolute increase in the rate of bone resorption. However, other factors must be measured and corrected for in order to justify that conclusion and they include the following. (Photomicrographs courtesy S. Stanisavljevic and A. R. Villanueva.) First, the steroids decrease longitudinal chondral growth, so the mean age of the chondral and trabecular tissue in the rat on the right has increased significantly. That has allowed more chondral creep to occur and is part of the reason for the loss of longitudinal order in the chondral cells in the plate. Second, the decreased longitudinal growth reduced the amount of new spongiosa added beneath the plate, while at the same time leading to an increased mean age of the spongiosa that is already there. Consequently, the latter has had a longer time than normal to reduce in amount under the influence of the normally negative $\Delta B \cdot$ BMU that characterizes the local remodeling processes so the reduced amount of spongiosa could reflect those factors as well as any absolute increase in resorption or a more negative $\Delta B \cdot$ BMU value. Third, the formation of new bone in this field has virtually ceased so even a greatly reduced absolute rate of resorption could accompany an increased rate of net bone loss.

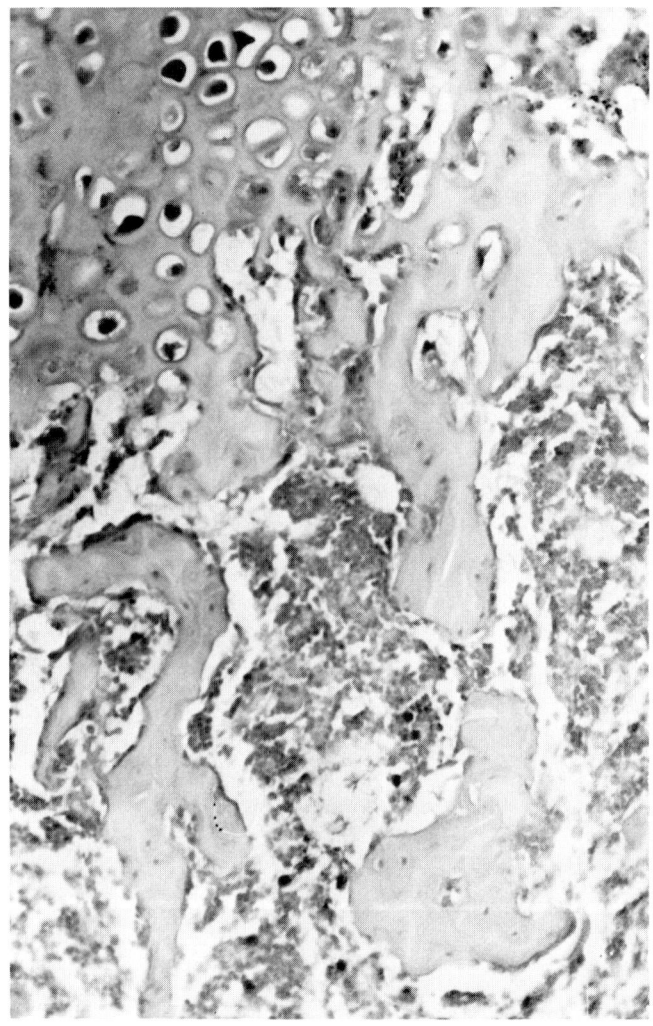

FIGURE 4B

III. CONCERNING INTRASKELETAL MECHANISM OF BONE DISEASE

When any important function deranges then a disease results, and the malfunctions of bone remodeling and modeling cause two different disease groups that could be called the growth modeling-dependent and the remodeling-dependent groups.[o]

A. Growth Modeling-Dependent Lamellar Bone Disease

To recapitulate, modeling determines the outer and marrow cavity diameters and cortical thickness of growing bones, and therefore the amount of compacta contained in them and the amounts of resorption and formation on their periosteal and cortical-endosteal surfaces. It also shapes their transverse and longitudinal architecture to fit their mechanical usage, so modeling malfunctions affect compact bone primarily and mostly during growth. They can cause insufficient or excessive accumulations and/or mechanically inappropriate architecture. Certain biochemical defects (i.e., vitamin D-

Normal

Rickets

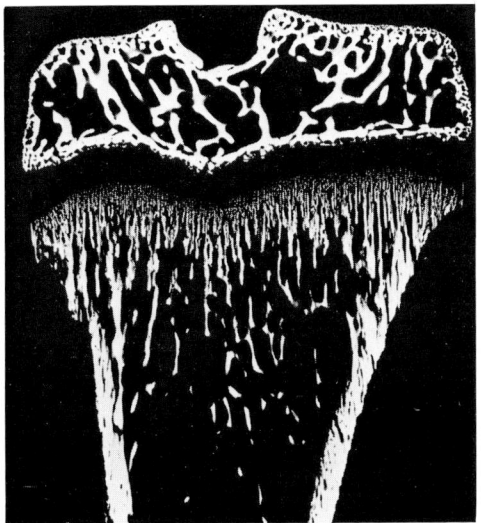
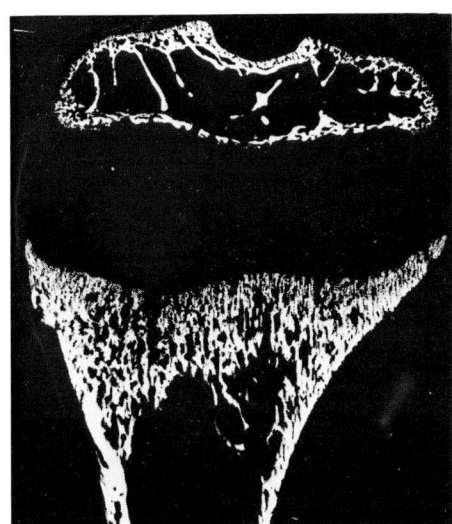

Normal

Rickets

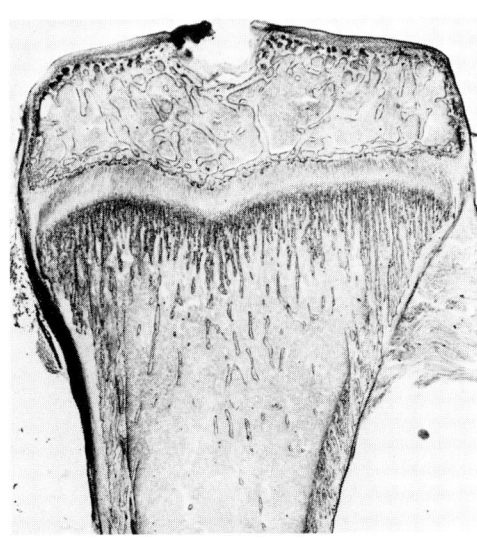
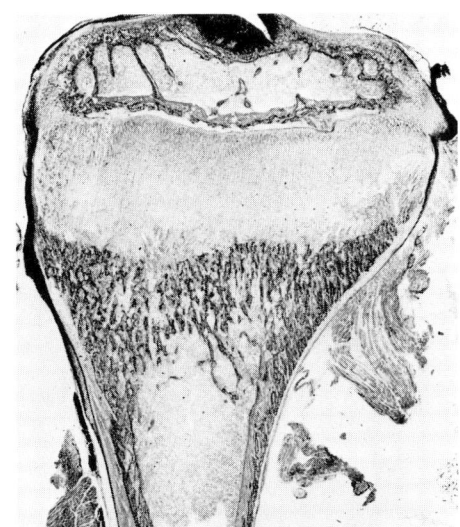

FIGURE 5. Undecalcified, longitudinal sections of the proximal tibia of young rats, microradiographs above and transmitted light photomicrographs below. About 15×. The left column illustrates the appearance of the normal state, the right column that of rickets induced by a diet deficient in vitamin D and calcium. Compare the normals to the previous two figures, remembering that the microradiograph above reveals the actual distribution of calcium in the tissues. Note also a significant but little remarked phenomenon: while the total amount of spongiosa decreases steadily with increasing downwards distance from the plate, the trabeculae that remain become progressively thicker due to a locally positive $\Delta B \cdot BMU$. The phenomenon proves that a normal mechanism for thickening already established trabeculae exists in nature. The rachitic state shows a number of abnormalities. The most striking one is a great depression of mineralization of cartilage in the epiphyseal plate, a depression much greater than the mild retardation in its longitudinal growth. As a result of that imbalance the plate thickens and due to chondral creep its cellular columns became intertwined, which makes them appear disorganized in thin sections. Also, the compacta is thinned, the secondary spongiosa is coarser and spatially somewhat disorganized, and the ossification center is osteopenic. At any given distance below the plate, the mean age of the spongiosa in the rachitic animal exceeds that of the normal. (Photomicrographs courtesy of Drs. J. Inoue, T. Haba, and H. Takahashi, Niigata.)

Table 2
BONE REMODELING PROPERTIES (AT L_2-LEVEL)

1. Single kind of unit, combining a resorption (R) and formation (F) phase.
2. R always precedes F but both occur in the same L_2-domain-space or location.
3. One capillary supplies both phases, and their control gates primarily via μ.
4. Continues throughout life in man; does not require growth.
5. Occurs in trivial amounts in compacta of small animals.
6. Occurs on all four bone envelopes in all long-lived animals and man.
7. Greatest effects occur in spongiosa, in part due to surface to volume ratio properties.
8. Causes over 95% of the annual bone turnover in the human adult.
9. Need not alter architecture, size, and quantity of bone, but may, and if so the special $\Delta B \cdot BMU$ mechanism accounts for it.
10. Directly affected by numerous endocrine, biochemical, nutritional, and pharmacologic factors, and by bone microdamage.
11. Not directly influenced to any great extent by growth, muscle forces, or bone mechanics (except that vigorous mechanical use can inhibit it).
12. Value of σ approximates 3 months in man, 2 months in dogs.
13. Volume of tissue turned over per unit varies little, ±30% around a mean value.
14. Requires a capillary and nongrowth-dependent precursor cell proliferation.

Table 3
COMPARISON OF BONE MODELING TO REMODELING

Property	Modeling	Remodeling
Growth	Dependent	Independent
Growth-dependent cell proliferation	Dependent	Independent
Growth-independent cell proliferation	Independent	Dependent
R and F	Independent	Coupled
$k r \sigma$ applies	No (?)	Yes
Envelopes	On C-E and P only	On all four
Before skeletal maturity	Occurs	Occurs
After skeletal maturity	Usually absent	Continues

Table 4
EXAMPLES OF HUMAN PATHOLOGY IN WHICH REACTIVATED SKELETAL MODELING CAN OCCUR IN ADULTS

Paget's disease	Acromegaly
Osteophytes, juxta-articular	Ribbing's disease
Pulmonary hypertrophic osteoarthropathy	Alveolar ridge recession
Fibrous dysplasia, polyostotic	Osteomalacic bowing of long bone
Osteogenesis imperfecta tarda	Cooley's anemia
RAP-dependent	Chronic osteomyelitis
Osteoid osteoma	

deficiency rickets) can also cause formation drifts to deposit circumferential lamellae of abnormal quality. Pathologically reactivated modeling can arise in some adult diseases, usually mixed with remodeling phenomena, and usually enabled in part at least by an accompanying RAP.[267] Table 4 lists some examples of that phenomenon and Figure 6 provides examples of some of the phenomena discussed here and below.

As for most IO-extrinsic steady-state modeling disorders in children, e.g., the osteopenias of muscular dystrophy, renal failure, paralysis, or biliary stenosis,[404] it seems to be primarily the direct effects of the disease on growth, muscle strength, and activity that cause the bone modeling disorder.[166,262,708,783] Any effects of the disease itself di-

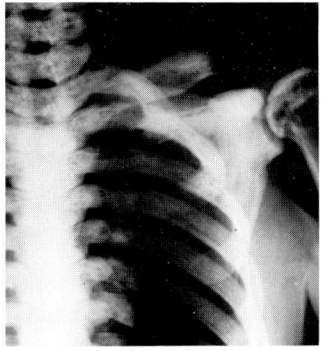
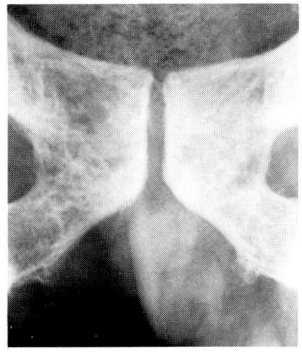

A   B

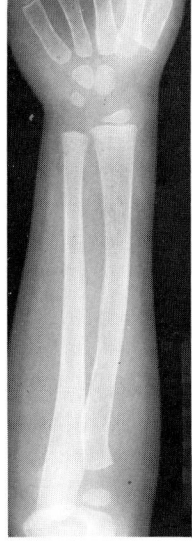

C

FIGURE 6A to C. (A) AP X-ray of an adolescent with osteopetrosis, a group of IO-intrinsic diseases that has at least one known L_c-domain cause; osteoclasts unable to resorb calcified cartilage, woven bone, and lamellar bone at normal speeds. Thus, those IO entities that depend in some way on resorption of hard tissue to function normally do not do so in this disease and it can affect all four envelopes and both modeling and remodeling of bone. Bone marrow transfusions and transplants have supplied normal precursor cell populations and led to a cure of this disease in some laboratory and human trials. (X-rays courtesy Dr. M. Castle, M.D., Detroit.) (B) X-rays of the symphysis pubis of a patient with a rare affection named axial osteomalacia by the author in 1962,[262] showing the coarse trabeculation characteristic of the disorder. It is an endosteal envelope-specific phenomenon which has not been well studied so far (although Teitelbaum et al.[775] have recently reported interesting studies of two subjects). It is still not known whether the increased amounts of spongiosa found only in the axial skeleton of these patients arose during adult life as the consequence of a lamellar bone remodeling malfunction, or if they reflect a persisting endowment of subepiphyseal remodeling that acted primarily during the person's skeletal growth. (C) X-ray of the forearm of a child with Gaucher's disease, and an IO-extrinsic osteopenia due to net losses of cortical-endosteal and trabecular bone, due in turn to an "L"-domain abnormality, a hyperplastic bone marrow. Here the osteopenia of cortical bone represents at least a partly disordered modeling modality, while the trabecular bone tissue volume deficit reflects disordered remodeling. Like the preceding case, the specific abnormality here consists of an abnormal interaction between bone and marrow, and until it is found and defined, studying either tissue alone ex vivo is unlikely to shed much light on the details of the pathogenesis of these two diseases. (X-rays courtesy Dr. Wm. Eyler, Henry Ford Hospital.)

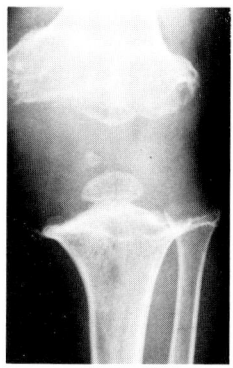
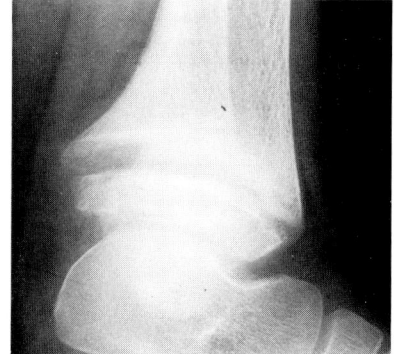

D   E

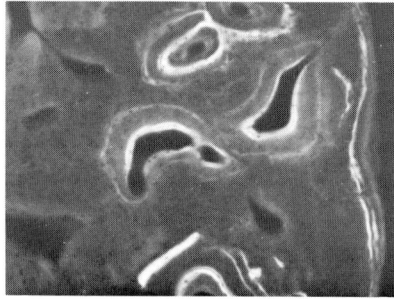

F

FIGURE 6D to F. (D) AP X-ray of the knee of a young child with Morquio's disease, a form of dwarfism. Here an IO-intrinsic disorder of chondral growth and modeling exists. Growth retards and mineralization of cartilage likewise, so the mean age of the tissue increased significantly. Partly for that reason and perhaps partly too because of an abnormal quality of the cartilage, its bulk creep increases greatly, which causes a major part of the resulting abnormalities in articular and epiphyseal shape and proportions. Here too L_2-mode growth seems to be relatively greater than L_1-mode growth, so the epiphyses appear wide in relation to their height. While study of the biochemistry of this cartilage in ex vivo systems would be useful, the specific behavioral factors that cause the clinical problems relate to growth, creep, and mechanically controlled modeling. The abnormal bony configuration simply reflects the adaptation by an intrinsically normal bone modeling modality to abnormal instructions delivered by the diseased chondral "conductor". The intrinsically normal bone modeling potential reacts normally here, but due to the chondral disorder it receives unusual modeling "instructions". Hence the flared bony metaphyses, which simply fit the bone to the pathologically expanded epiphyses. (E) Lateral X-ray, ankle, of a young lad with renal failure, secondary hyperparathyroidism, and renal rickets. That thickened and widened the distal tibial epiphyseal plate by retarding chondral mineralization relative to longitudinal chondral growth. That also increased the mean age of the unmineralized cartilage and allowed increased amounts of both unit and bulk creep to occur in the growth plate. This IO-extrinsic form of rickets is accompanied histodynamically by an IO-extrinsic renal osteodystrophy. (X-rays courtesy Dr. R. Haliburton, Windsor, Ontario.) (F) Cross-section, undecalcified, fluorescence microscopy, of compacta of an 11th rib biopsy of a woman with osteogenesis imperfecta. She received two tetracycline bone labels 10 days apart prior to the biopsy, and took the same antibiotic at more remote times in the past for other reasons. Periosteum to right, marrow to left. The bright bands represent glowing tetracycline deposited during active bone formation. This was the first case of osteogenesis imperfecta ever to be so studied, and to our considerable surprise, dynamic histomorphometry revealed her tissue-level bone turnover proceeded about three times faster than normal for her age, while her cell-level bone formation proceeded at two thirds of normal. Similar findings have been reported since in other victims of this disease by the author and by other investigators. That meant that the antecedent explanation for the features of this disease, "an inability to make bone as fast as needed," was untrue, so their explanation had to lie elsewhere. This was only one of many instances in which IO insights and dynamic histomorphometry revealed that antecedent explanations of skeletal phenomena were incorrect. In fairness, many earlier workers drew correct inferences in other matters from only fragmentary evidence, the proof of which had to await IO insights and the development of dynamic histomorphometry.

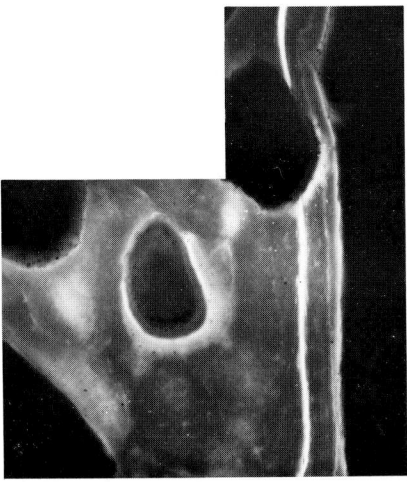

FIGURE 6G. Another human adult cross-section of a 10th rib, biopsied after double tetracycline bone labeling to study a metabolic bone disease. Periosteum to the right, marrow to the left. Due to the patient's osteomalacia, the appositional rate was so markedly reduced that the two labeled, forming haversian systems show only single labels, and the other two show none because they were in the "Off" state mentioned in Chapter 4. The vertical bright band is a subperiosteal tetracycline bone label, deposited several years previously and incidentally during the treatment of an unrelated infection. That label shows that a periosteal osteoblastic or formation drift was occurring on this cutaneous cortex of the rib at the time. Such drifts are unique to bone modeling, not remodeling, so while this adult no longer grew in any accepted clinical sense, and was not modeling either as far as the naked eye could see, the microscope and the labels revealed that he was modeling very slowly. Accordingly, some small amount of bone modeling can occur in adults, at

rectly on the IO are probably minor, noting however the exception shown in Figure 6, upper right. Two facts can illustrate that phenomenon. First, a skeleton that cannot satisfy its typical mechanical usage will fracture under that usage, for as described in Chapter 7, at least one special and certainly major function of modeling consists of fitting bones to meet their mechanical usage. Second, and excepting uncommon IO-intrinsic diseases such as osteopetrosis, hyperphosphatasia, or osteogenesis imperfecta, IO-extrinsic disorders rarely cause bone mechanical incompetence in pediatric pathophysiology, and then only in extreme states. For example, the profound osteopenia of the lower limbs of a child with myelodysplasia is certainly abnormal when compared to a normal child of similar age and stature, but it fits quite well the equally profoundly reduced *typical* mechanical usage of those limbs.

As far as is known at present (but no systematic studies are known to the author), lamellar bone macromodeling does not significantly affect normal spongiosa, so unless a defect affects both macromodeling and remodeling (as in osteopetrosis), the one system probably cannot yield valid information about the behavior of the other.

Note Bene: To study any direct effects on any bone resorption and formation activities of any challenge whatsoever in intact growth modeling-dependent systems, one must first define, measure, and correct for the effects of that challenge on growth and all processes that depend on it.

To extrapolate a response of such systems to the steady-state human skeleton requires ensuring that the experimental response is also a steady-state one.

B. Remodeling-Dependent Bone Disease

Malfunctions of this system divorced from growth and modeling effects occur most

184   *Intermediary Organization of the Skeleton*

Table 5
MODELING/REMODELING CLASSIFICATION OF THE LAMELLAR BONE
CHANGES IN SOME HUMAN SKELETAL DISEASES

Growth/Modeling-Dependent

Gigantism, osteogenesis imperfecta,* and Marfan's syndrome; the osteopenias of muscular dystrophy, Gaucher's disease, biliary stenosis, myelodysplasia, metabolic acidosis, postpolia and other childhood-onset paralyses, malnutrition and Turner's syndrome; pseudohypoparathyroidism, Pyle's disease,* and hypophosphatisia.*

Remodeling-Dependent

The adult-acquired osteoporoses, including senile, postmenopausal, Cushing's, corticosteroid-induced, postparalytic, migratory osteoporosis, hyperparathyroidism, thyrotoxicosis, the RAP; all adult-onset deficiency and malabsorption osteomalacias, spontaneous fractures, fibrogenesis imperfecta ossium.*

Combined

Acromegaly, osteopetrosis,* Paget's disease of bone,* Ribbing's disease, all vitamin D-deficiency forms of rickets, familial vitamin D-resistant rickets,* idiopathic juvenile osteoporosis.

Note: An asterisk signifies an IO-intrinsic bone disease as defined in the text. Chondral disorders are not classified here.

---

clearly in human adults and they certainly can make the skeleton mechanically incompetent under normal usage, e.g., in the osteoporoses.[1,739,740] Remodeling system disorders can also cause pathological losses, surfeits, and/or maldistributions of lamellar bone,⁰ and as recorded in Chapter 6, they too can deposit bone of abnormal quality. While an adult's remodeling occurs on all four envelopes so it can affect both compacta and spongiosa, the approximately 10 to 20 times larger surface-to-volume ratio of the latter causes most IO-extrinsic remodeling disorders to affect it more than compacta, excepting some envelope-specific disorders that have been discussed elsewhere.

Note Bene: To extrapolate a response of an experimental remodeling system to the steady-state adult human requires that the experimental response likewise be a steady-state one that is not influenced by, or that is corrected for the effects of, growth.

Table 5 classifies some clinical disorders that affect bone primarily, and Table 6 some that affect chondral tissues primarily, according to the above concepts.

C. In Vitro Systems: Their IO-Oriented Properties

Since cell, tissue, and organ culture systems are widely used to study the regulation of bone cells and bone cell associations, a brief look at certain of their basic properties seems in order here. A bone cell culture system contains only the effector cells or A, and a local but always artificial environment or L, so its state equation could be written thus, letting $f(BC)$ signify the bone cell output of the system:

$$f(BC) = (L){:}(A) \qquad (11)$$

A tissue culture system such as the fetal rat calvarium adds the bone so it could be written thus, $f(TC)$ signifying the output of the system:

$$f(TC) = (L){:}(A{:}M) \qquad (12)$$

Table 6
GROWTH/MODELING CLASSIFICATION OF THE CHONDRAL CHANGES IN HUMAN SKELETAL DISEASE

Pure Growth Dependent

Pituitary gigantism, pituitary dwarf, short stature of paralysis, progeiria,* Marfan's syndrome*

Pure Modeling Dependent

Blount's disease, coxa valga, coxa vara, tibial torsion, metatarsus adductus, club foot, congenital hip dysplasia

Combined

Achondroplasia,* Morquio's disease,* Jansen's disease,* vitamin D-deficiency rickets, familial vitamin-D resistant rickets*

Note: The asterisk signifies an IO-intrinsic-origin disorder. All other disorders are IO-extrinsic in origin. The table concerns only the chondral tissues and not the bony or other skeletal tissues.

Organ culture systems such as a fetal rat femur contain neither functioning capillaries nor the ongoing P:D:O or S elements of the intact system, so while some authorities have considered them fundamentally different from tissue culture systems, in the terminology and contexts of this text their state equations would look exactly like the above one. Some authorities have also assumed that since the intact fetal rat femur in organ culture was an organ, the essential functions of the organ remained intact after it was dissected out and placed in the culture medium. However, that is surely incorrect.

It follows that the above systems can model the responses of only already existing effector cells to L or extrinsic regulation, *and only for select matters that occur in* L_1-*space-time.* They lack the steady-state C:P:D:O and S functions that form essential parts of the intact IO at all levels in the intact subject. As long as those systems do not supply those missing activities and relationships, they can model transient phenomena of the intact system, but only those transients that can arise from modifying the behavior of the A cells alone, and only phenomena that occur in L_c- and L_1-space-time. That becomes clear, as an example, by comparing the projections to the intact subject of the results of studies by Goldhaber et al.,[284] Holtrop,[337] and Raisz et al.[614-616] with the true effects shown in Takahashi's experiment in Figure 1 in Chapter 10.

As models of human bone disease, the concern of these paragraphs lies in three assumptions made by some who use such systems. If investigators or their audiences extrapolate such data directly to steady-state disease then they expose those assumptions, which are as follows.

First, since in vitro systems lack the aforesaid further essential activities and relationships of the intact IO, and since their A cells do have limited functional lifespans, then extrapolating their inherently transient responses directly to a steady-state human necessarily assumes that only the L and A activities matter in vivo and all others do not. Second, any such extrapolation also assumes that the transient and steady-state distinction in the intact organism has no merit. Third, it assumes that no operationally important difference exists between the growth modeling-dependent skeletal activities of a rodent or avian fetus and shown in Figure 2, and the BMU-based bone remodeling of the intact, adult human.

The in vitro systems run down in 3 weeks or so but the intact living skeleton cannot develop a steady-state response of its growth, modeling, or remodeling activities to some challenge in less than the σ values for those processes, or about 3 to 8 weeks in healthy rats and 3 to 4 months in healthy humans. Furthermore, in many of the human skeletal diseases that interest the research community at present, the σ values of the modeling and remodeling systems can become ten times or more longer than normal, so to observe steady-state responses of those diseased systems to various agents one must treat them for equal lengths of time, something that is impossible at present with all of the ex vivo systems.

To be blunt then, compelling evidence assigns the same merit to the above assumptions as assuming that if you release your pen at high noon, then because the sun lies directly overhead and because its gravitational field immensely exceeds that of the earth, the pen will fall upwards. The basic facts are true but the conclusion is patently false because that statement of true facts omitted others that are both relevant and crucial to the proposition. Figure 7 provides an illustration of the discrepancies that usually result when one extrapolates tissue culture data directly to the intact subject.

Note Bene: Current cell, tissue, and organ culture systems model L_c- and L_1-domain phenomena. They do not and cannot model the intact L_2-domain and higher-order skeletal systems, so their responses represent possible transients in L_1-space-time of the intact subject, but not steady-state ones.

D. Envelope-Specific Phenomena

As was first pointed out in 1966,[230] the bone deficits found in, e.g., senile, postmenopausal, and corticosteroid-induced osteopenias, arise primarily where marrow touches bone, and therefore on the trabecular and cortical-endosteal envelopes. Such adult-acquired envelope-specific deficits of bony tissue should represent some effect of the underlying systemic disorder directly on the juxtaosseous marrow, the response of which then changes the bone balance behavior, e.g., the $\Delta B \cdot BMU$ of those BMU that touch it physically, meaning those BMU on the two envelopes mentioned above. To model such disease in vitro or in vivo would require a system that exposes marrow to the proposed systemic agent, and also to a bone that has a viable BMU-based remodeling capability or C:P:D:O:A:X, and that also adjoins marrow tissue.º For a corticosteroid-induced osteopenia in a woman the essential experimental relationships would look like this:

$$\text{corticosteroids} \longrightarrow \text{marrow} \longrightarrow \text{endosteal } \Delta B \cdot BMU \longrightarrow \text{endosteal bone loss} \quad (13)$$

$$\underset{\text{------?------}}{\uparrow}$$

The X-ray of the foot in Figure 20 in Chapter 7 illustrates that exact effect of corticosteroids.

The question-marked by-pass means the drug might also affect the remodeling system directly by parallel gating, so its observed effect of shedding endosteal bone could sum two different kinds of differently routed effects that occurred simultaneously. Varied other envelope-specific patterns of the features of bone disease are now known, including involvement primarily of compacta, or of the periosteal envelope, or of the axial skeleton.

Consequently, any model system that lacks those elements of the human disease under study in effect seems to try to understand walnuts by studying watermelons, or cardiac striated muscle by studying jejunal smooth muscle. The present in vitro systems do lack those elements, and in addition they usually study periosteal rather than cortical-endosteal or trabecular bone surface activity (see Note 1 in Chapter 8).

Note Bene: Experimental systems that do not duplicate the observed gating features of a human disease do not model that disease.

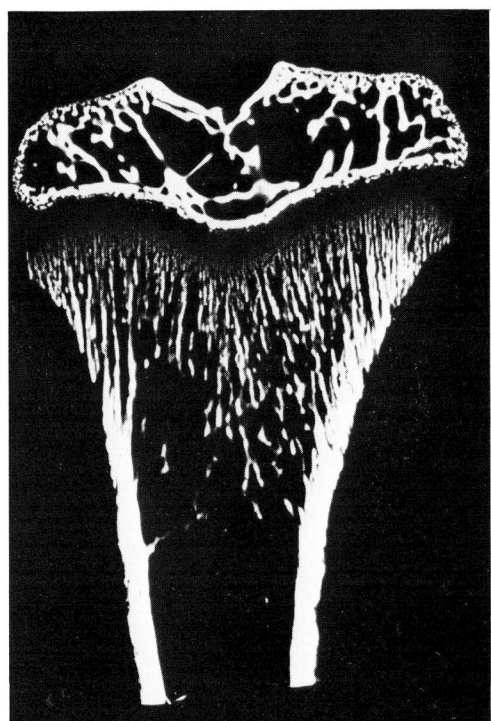 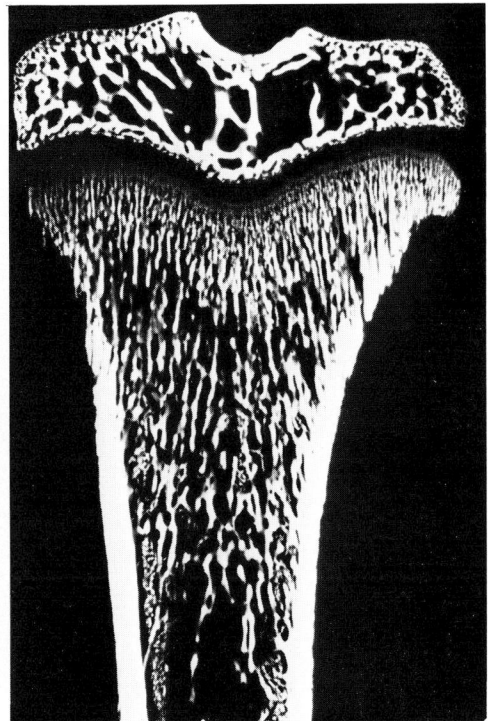

FIGURE 7. Microradiographs of longitudinal undecalcified sections of the proximal tibia of rats, about 50×. The left one is a normal control animal while the right one received daily injections of a prostaglandin, PGE_2. In bone-resorbing tissue culture systems that agent consistently causes increased bone resorption, which led many physiologists to conclude that the agent causes net bone loss in the intact mammalian skeleton. Due to their prevalent atomism none of the tissue culture workers who studied the bone effects of the drug before 1981 bothered to check its effects on bone formation in their systems, so the idea became prevalent that prostaglandins affected resorption exclusively and caused many or most pathological bone destructive processes as well as the osteopenia of many or most osteoporoses. As can be seen on the right, when given to the intact animal and for longer than the σ value of the system, it leads instead to a considerable accumulation of extra bone in the marrow cavity of the metaphysis and epiphysis, and to some cortical thickening too. This effect on the intact animal was first observed by High and reported to the skeletal research community with the permission of the Upjohn Co. at the 1981 Sun Valley Bone Workshop, sponsored by the University of Utah. For reasons that need not be detailed here, she was not allowed to publish those findings but several nonplussed and astonished tissue culture authorities in the 90 scientists who attended that meeting and who heard and saw that evidence immediately returned to their laboratories and began to study PGE effects on bone formation in their tissue culture system tissues. They soon began to report that it increased bone formation but they omitted acknowledging the source of their original inspiration to study its previously ignored and, in fact, discounted effects on that activity. (This field can provide many subtle amusements and other kinds of entertainment to an informed observer, and many frustrations to a young discoverer who lacks the protection of established stature and influence.) Again then, clear evidence that tissue culture systems operate by L_c-domain rules, that the intact system operates by the additional rules that apply to L_1- and higher domains, and that the latter rules, not the former, dominate and determine what happens in the intact animal. (Microradiographs courtesy of Drs. W. S. S. Jee, Salt Lake City, and T. Haba, Niigata.)

E. Skeletal Turnover Mechanisms

Many research investigations study the turnover of the mineral and organic components of skeletal tissues, using gravimetric, histomorphometric,[641] chemical, or radiobiological methods alone or in some combination.[77,292,405,406,641,785,786] To interpret such studies accurately requires a knowledge of the different intraskeletal mechanisms that can affect that turnover, as well as of the σ value for each mechanism and its particular

188 Intermediary Organization of the Skeleton

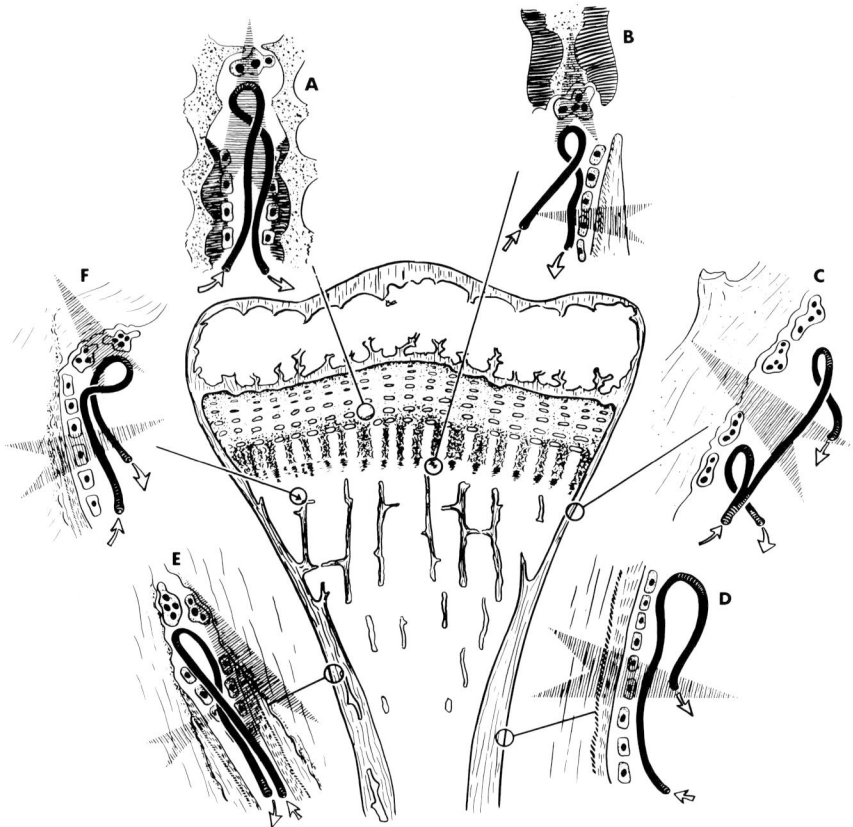

FIGURE 8. This sketch of the proximal tibia of a growing child, seen in longitudinal section, identifies some of the L_2-domain or middle-IO systems that participate in its growth and maintenance. The shaded arrows indicate the directions of the processes, directions determined by local micromodeling and macromodeling influences. The bold lines represent the capillaries that nourish the systems, the "C" term of Relations 3 and 4. (A) Longitudinal erosion by chondroclasts of mineralized cartilage, and centrifugal deposition of new woven bone on the surfaces of the remaining mineralized chondral bars or plates, to form the primary spongiosa. (B) Erosion of the primary spongiosa by osteoclasts and deposition on the remaining material of new lamellar bone to form the secondary spongiosa. (F) BMU-based remodeling or turnover of that secondary spongiosa. Due to the negative $\Delta B \cdot BMU$, the older regions of spongiosa (which are the farthest from the growth plate and have undergone more cycles of the above remodeling process) contain less bone than the younger regions. (C) An osteoclastic drift on the periosteal metaphyseal surface reduces its outside diameter. A matching cortical-endosteal formation drift also occurs but is not shown, so the cortex not only does not thin and vanish, it actually thickens with time because the formation drift exceeds the resorption drift in amount. (D) On the diaphyseal periosteal surface a formation drift deposits circumferential lamellae which increase the outside diameter of the bone. (E) BMU-based remodeling inside the compacta gradually replaces the circumferential lamellae with secondary osteons. The fragments of circumferential lamellae lying between those osteons are then named interstitial lamellae. (From Frost, H. M., *The Physiology of Cartilaginous Fibrous and Bony Tissue*, Charles C Thomas, Springfield, Ill., 1972. With permission.)

effects on hard tissue balance (i.e., on net gains or losses of tissue or of mineral), and of various "push-pull" modes of regulation that can affect each of them (see Note 1 in Chapter 10).

The turnover mechanisms will be listed below, first for the growing skeleton, then for the adult, understanding that in adolescence the two classes overlap. Figure 8 illustrates some of the activities described below. The term "turnover" here will mean removing some of the old and replacing it with new. Two different forms of skeletal tissue turnover occur: that due to histological combinations of tissue resorption and formation by specialized cells such as osteoclasts, chondroclasts, and osteoblasts, and a molecular-level turnover produced by the activities of cells already existing inside the tissue, such as osteocytes, chondrocytes, fibrocytes, and odontocytes, a turnover that does not involve mass resorption or formation of all components of the tissue at essentially the same time. In a telegraphic style, the histological mechanisms will be described first.

1. Turnover in the Growing Skeleton[379,413,414,606]

Formation of the primary spongiosa — Mineralized cartilage removed, woven bone deposited by a BMU-based packet-like L_2-domain mechanism. The σ value is 1 to 3 weeks, normally negative tissue balance.

Formation of secondary spongiosa[379] — Primary spongiosa removed by a BMU-based L_2-domain packet-like mechanism; lamellar bone deposited. The σ value 2 to 6 weeks, normally negative tissue balance.

BMU-based remodeling of secondary spongiosa[379] — Older lamellar bone removed, same kind of new bone deposited by a BMU-based packet-like mechanism of L_2-domain. The σ value 5 to 12 weeks, normally negative tissue balance.

Cortical-endosteal modeling — Compact bone removed, circumferential lamellar bone deposited by two separate and unrelated L_2-domain mechanisms, the activities of which sum at the organ level. The σ value is 2 to 8 weeks, normally negative tissue balance.

Periosteal modeling — Same properties as in cortical-endosteal modeling except normally positive tissue balance.

Collective spongiosa processes[379,413,414] — During normal growth the total amount of spongiosa in the skeleton steadily increases when expressed in ABV referent. Since none of the processes that affect spongiosa and that are listed above have normally positive tissue balances, an astute reader might sense a contradiction. The problem lies in an omission. The remodeling mechanism that creates primary spongiosa out of mineralized epiphyseal plate cartilage has a negative value when the amount of mineralized cartilage removed is compared to the amount of primary spongiosa that replaces it. Since that primary spongiosa did not exist before, it becomes a new deposit in the spongiosa "bank", and at a rate that depends directly on the longitudinal growth of the overlying growth plate. Because that creation of new spongiosa proceeds faster than the remodeling processes that follow behind it, the total amount of spongiosa in the skeleton normally continues to increase during growth and a net loss begins only after growth stops at skeletal maturity. Figure 9 diagrams those relationships.

Collective compact bone processes — Here too L_o-domain factors so distribute the individual modeling drifts over the cortical surfaces, and so affect their relative magnitudes, that during growth the net amount of compacta in a skeleton increases steadily, in part due to net periosteal bone additions that exceed the net cortical-endosteal losses, and in part due to increases in the length of the bone by longitudinal growth.

The surface and percolation homeostatic mechanisms — As described in Chapter 10, these two systems cause special kinds of turnover of the mineral deposits in the skeleton without significant accompanying turnover of the organic components of the matrix. They can cause net gains and losses in the mineral ions that are involved, including relatively large proportions of calcium, magnesium, sodium, phosphate, and carbonate, plus smaller proportions of many other anions and cations.

Excepting the last named mechanisms, each of the above systems turns over both the mineral and the organic components of skeletal tissue. To a great extent, each follows the regulatory dictates of growth, although other factors can apparently modify those

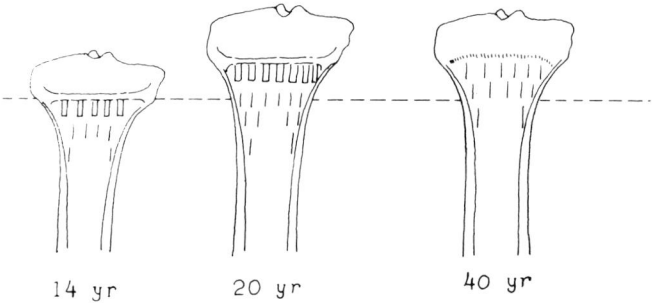

FIGURE 9. As the upper 14-year-old tibia on the left grows, the conversion of mineralized cartilage at the bottom of the epiphyseal plate into primary spongiosa (open bars) leaves a new supply of spongiosa behind the growing plate. The horizontal dashed line cuts all three figures at the same time age level, here 14 years, meaning for example that where it cuts across the tibia of the 20 year old, the spongiosa at that level initially formed at age 14 years. Between ages 14 and 20 years the epiphysis grows upwards, in the process creating even newer supplies of primary spongiosa above the original supplies found at age 14. The spongiosa formed at age 14 has since undergone a number of remodeling cycles and been replaced by secondary spongiosa (vertical solid lines), and due to the negative $\Delta B \cdot BMU$ value of the remodeling BMU that caused that turnover, the original amount of that spongiosa has progressively reduced. The total amount of spongiosa in this metaphysis is greater than at age 14 because the growth process added more than the remodeling processes removed. This forms one of the balance situations discussed in Note 1 in Chapter 10. The phenomenon continues through age 40 but now with no offsetting addition of new material by growth, so a general net loss has become evident.

growth-dependent responses to some degree. Each also depends in the direct sense on different kinds of things. For example, modeling depends primarily on growth and mechanical usage, while creation of primary spongiosa requires that mineralized cartilage exist first, and that depends directly on growth. Creation of the secondary spongiosa requires that the primary spongiosa exist first, and that also depends on growth.

Note Bene: Measuring the net amount of mineral gained or lost by the whole skeleton or any single intact bone of an intact subject combines the separate contributions of 8 distinct intraskeletal mechanisms, each of which sums a separate R and F or balance subprocess, so such data alone cannot be interpreted in terms of any single one of those 8 mechanisms, or any single one of their 16 subprocesses.

2. Turnover in the Adult Skeleton

Here the situation becomes simpler because skeletal maturation halts the growth process, and along with it those other processes that growth somehow enables, including the macromodeling of hard tissue. That leaves only two basic mechanisms operating in quantitatively significant amounts in the healthy, skeletally mature individual.

BMU-based remodeling — This turns bone over on all four envelopes. Normally, the special $\Delta B \cdot BMU$ mechanism determines the tissue balance for a given envelope,[245] and those envelopes have the typical balances recorded in earlier chapters. The σ value for this system approximates 3 months or more in man, and about 2 months in an adult rabbit; its bone balance properties seldom reflect unilateral changes in only resorption, or only formation. They always do reflect changes in the $\Delta B \cdot BMU$ param-

Table 7
CLASSIFICATION OF SOME ANIMAL MODEL
BONE SYSTEMS

Growth/modeling dependent		Remodeling dependent	
Rat	Quail	Adult human	Adult sheep
Mouse	Young rabbit	Adult dog	Adult cow
Shrew	Puppy	Adult primate	Adult deer
Hamster	Kitten	Adult pig	Adult goat
Chick	Infant	Adult horse	Adult cat

eter, and directly so; observed changes in net bone balance directly reveal only the changes in that parameter or in the activation frequency (μ).[232,267] Envelope-specific and other regional-specific responses to varied challenges of the $\Delta B \cdot$ BMU mechanism occur frequently, and similar features probably apply to the activation of new BMU too.

The surface and percolation homeostatic mechanisms — These continue to operate in adults as they do in children, although perhaps more slowly in a unit amount of adult bone than in an equal amount of a child's bone. As in the child, both mechanisms can turn over the mineral phase of the bone but not to any significant degree its organic part.

Note Bene: Experimental systems and protocols that do not provide, define, and measure separately the behavior of the particular physiological system that applies to steady-state human disease or physiology will simply mix the responses of that system with those of the other ongoing systems in the intact subject to provide a collective result that cannot be interpreted in terms of the behavior of any one system.

IV. PROPOSITIONS AND POSTULATES

In choosing a laboratory model of some human skeletal disorder or other problem, one should acknowledge three propositions and try to satisfy five postulates. The propositions are

1. Most skeletal diseases represent steady-state, L_o level malfunctions that arise in an intact IO.
2. Many bone diseases exhibit regional-specific, tissue-specific, and/or activity-specific features.
3. Experimental models that do not duplicate such features have limited uses.

The postulates are (analogs of Koch's)[421]

1. Characterize the disease or other feature under consideration as growth-modeling or remodeling dependent, or both, or as something else, (e.g., myelofibrosis, fracture healing).
2. Characterize its envelope-specific and any other geographically specific features.
3. Characterize it as IO-intrinsic, IO-extrinsic or some combination.
4. Characterize it as a transient or steady-state problem.
5. Then choose or devise an experimental system that duplicates those features.

Tables 7 to 10 summarize some of the information needed to satisfy those postulates for the case of human bone problems.

Table 8
FACTORS TO CONSIDER IN CLASSIFYING A HUMAN BONE DISEASE

Organizational domain	Physical activity
IO-intrinsic	Steady state
IO-extrinsic	Transient
Growth	Modeling
Lamellar bone remodeling	Mineralized cartilage remodeling
Primary spongiosa remodeling	Homeostatic surface system
Homeostatic percolation system	Envelopes
Axial skeleton	Appendicular skeleton
Compacta	Spongiosa
Age	Sex
Skeletal maturity	Nature of the MES

Table 9

In the intact growing bone or skeleton, calcium and related cations and anions can come from or enter into
    Mineralized cartilage (B)
    Secondary spongiosa (B)
    Periosteal compacta (B)
    Percolation homeostatic system (B)
    Primary spongiosa (B)
    Cortical-endosteal compacta (B)
    Surface homeostatic system (B)
    Teeth (U + B)
In the intact animal such ions can accumulate via
    All of the above, plus
    Soft tissues (B)
Such ions can leave the intact animal via
    Sweat (U)
    Urine (U)
    Hair (U)
    Feces (U)

Note: B = A bidirectional process so a balance of some kind determines net gains or losses. U = a unidirectional process.

Table 10
DISTRIBUTIVE PROPERTIES OF LAMELLAR BONE MODELING AND REMODELING

Distributive relation	Conventional name of the activity
g:m:(p,c):a:r	Resorption drift
g:m:(p,c):a:f	Formation drift
(p,h,c,t):a:(kro):r:f	Bone remodeling BMU

Note: Let a = activation; g = growth; f = formation; r = resorption; m = modeling; (kro) = applies to the intact IO system only; p = periosteal envelope; h = haversian; c = cortical-endosteal; and t = trabecular envelope.

Note that choosing "modeling-only" systems, such as actively growing rodents or birds, to model a remodeling disorder becomes particularly inappropriate in view of the observation that systemic disorders seldom directly and seriously affect an intrinsically normal modeling process, which nearly always produces a bone suited to its sometimes abnormal but typical mechanical usage, and under the control of the growth process. The same systemic factors can exert major effects on BMU-based lamellar bone remodeling.

Three matters merit a brief final comment. First, studying the properties and behavior of the IO requires special techniques, insights, and strategies that are not widely known or even fully developed so far. They are described in several recent symposia,[c-F] but those matters lie peripheral to the major concerns of this text. Second, since knowledge of the skeletal IO is still incomplete, room remains for differences in judgment on the above matters so this text should not be chiseled in stone. Third, this chapter focused on some human adult bone problems, but a similar approach to the choice of model systems applies to the problems of growth and aging, and of cartilage, fibrous tissue, dental tissues, and synovia.

V. ILLUSTRATIVE QUESTIONS

Most fields of science pass through a series of fads as one topic attains a more general interest than it really deserves when put in perspective with the whole, only to be replaced in time by another, and so on. In the long run, of course, things tend to balance out. The use of modeling-only animal systems to model the remodeling-only diseases of adult humans has been such a fad since about 1950, which makes this chapter appropriate. Those modeling-only systems have seen extensive use but also inappropriate analyses and extrapolations of their properties. A great deal of information must still be obtained about their dynamic properties, the factors that enable them, and those that regulate their continuum phases. A sample of the questions that need answers in order to make better use of such systems follows.

Are the detailed histological and biochemical effects and metabolism of vitamin D and steroid hormones the same in rodents, birds, and man? If not, how do they differ? And with respect to other effects on all skeletal tissue? What are the answers to the same generic questions for all other vitamins, nutrients, hormones, and drugs?

What agents enable, and what others then regulate, the continuum phases of BMU-based remodeling of primary and secondary spongiosa, of mineralized cartilage, lamellar bone, and fibrous tissue? How does man differ in those respects from popular animal model systems (see Table 5)? What animal models can provide valid models of BMU-based remodeling in man? And for fibrous tissue remodeling? How do the absolute and relative differences in ages of man and the animal models affect the use of one to study the problems of the other? What are the functional lifespans of the typical clast nucleus and the typical blast? And how are they affected by age, sex, species, drugs, nutrition, endocrines, activity, and anatomical location?

Why is BMU-based remodeling usually less active, and in both surface and volume referent, in epiphyseal than in metaphyseal spongiosa? One could speculate that the greater, more focused mechanical loads in the ossification center than in the wider mass of spongiosa beneath the epiphyseal plate more effectively inhibits the marrow-derived tendency described in Chapter 5 that seems to accentuate activation and make the ΔB-BMU more negative on trabecular and cortical endosteal surfaces; but is that the real explanation? Or one of them?

Does that same inhibition of remodeling in the epiphyseal ossification center also inhibit the BMU-based repair of microdamage sufficiently to account for disorders such as osteochondritis dissecans, Perthes' disease in children, and idiopathic aseptic

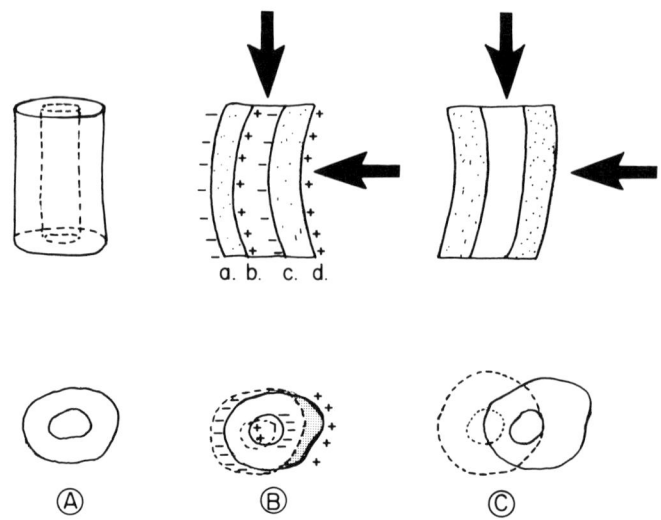

FIGURE 10. Much, if not most, of the periosteal and cortical-endosteal bone formation and resorption in growing mammals represents bone resorption and formation drifts that local flexural strains directly enable and regulate. In other words, it represents mechanically controlled macromodeling. Some influence of growth probably also exists as suggested by Figure 2, and if so it would act via the "L"-gate and in the "L"-mode described in Chapters 4 and 5 (periosteum would represent "L" on the periosteal envelope, and marrow on the cortical-endosteal and trabecular envelopes) rather than — or as well as — directly on the local cells as an "S"-mode effect. This figure illustrates again the mechanical basis of that control of cortical macromodeling. When, to the bone on the left, one adds a compression end-load, and a flexural side load as in the middle, that causes a dynamic flexural strain as shown. When that strain equals or exceeds (i.e., $\geqslant$) the MES then formation drifts (+) arise on the concave-tending surfaces and resorption drifts (−) arise on the convex-tending ones; they recontour the bone as at the right, where the cantilever flexure caused by end-loading a now curved bone neutralizes the nontrivial direct flexure, so net flexural strains no longer equal or exceed the MES. That is at least one of the basic mechanisms described in Chapter 7 that creates resorption drifts on the metaphyseal periosteal surface, and formation drifts on the diaphyseal periosteal surface. To repeat discussions elsewhere in this book, that is a truth of this system. It is probably not the whole truth, however. If some real-life situations should seem to violate it or be unexplained by it that does not mean this truth should be thrown out; rather, it means we lack another part of this story and we should spend our energies seeking it rather than trying to bury the truth we already have. As for the concerns of this chapter, the message here is twofold. The changes induced in this macromodeling process by hormones, drugs, and other circulating agents relate much more to growth, bone architecture, and mechanical usage than to the modeling or homeostatic mechanisms, and transient changes (less ∼3 to 5 weeks after challenge in a rat) do not reveal the nature of steady-state treatment effects on the intact system.

necrosis of the femoral head in adults? Or is it a contributing factor with other so far unidentified ones?

Exactly what temporal and ABV relationships characterize the remodeling sequences beneath an epiphyseal plate? And the responses of those and the bone modeling system to longitudinal growth? How do systemic agencies — hormones, sex, diet, nutrition, climate, activity — modify those responses? And in man and animal models both? Figure 10 suggests some pertinent questions and food for thought here.

How might one study such phenomena in both animals and man? What are the proper referents for making valid comparisons across the lines of species, age, sex, activity, climate, body size, and metabolic and growth rates?

What are the important omissions from this chapter? What else about this problem do we need to know that remains unperceived or not yet discovered?

Note 1

ON TISSUE AND ORGAN CULTURE TECHNIQUES

The main text may seem to belittle cell, tissue, and organ culture techniques so often that a reader might believe the writer held them of little value. *Nothing could be further from the truth.*

Those powerful and sophisticated techniques have demonstrated their usefulness in too many ways to deny or lessen their value. The problem lies elsewhere. The "bone" worried here constitutes the fact that many people who use those methods and/or who try to interpret and apply their results fail to perceive a limitation affecting the relevance and clinical meaning of those results. That is, and as applies to renal problems, the results rarely extrapolate directly to the intact steady-state organ, animal or man. Quite simply, the strategy of cell → organ is clearly naive. Because drug "Y" enhances osteoclast vigor in cell or tissue culture does not mean it will cause a steady progressive bone loss in the intact skeleton. Because a drug depresses osteoclast vigor in those experimental systems does not mean it will cause a steady progressive accumulation of bone in an intact skeleton. Cells play by one set of rules and the IO by different ones. The IO rather than isolated cells directly determines what the intact skeleton can and will do under a given treatment. The cell does so only indirectly and, at present and in skeletons, for the most part unpredictably. When the results of in vitro experiments are interpreted according to the basic postulate of cell → organ, then such interpretations run into the LBO and MCN properties of the IO that place the probability of making such a prediction correctly somewhere below 1/362,000, as described in Chapter 3. Evidence that the conclusion suggested by that line of reasoning is correct, whether the road that led to it contains flaws or not, lies in a fact mentioned elsewhere in this book: direct extrapolation to the intact, steady-state mammalian skeleton of cell, tissue, and organ culture data alone has not once correctly predicted the responses of an intact skeleton. Alfred Nobel once made a similar prediction, i.e., a prediction of the ultimate effect of a primary cause without accounting for all of the other truths that lay between the two. Rendered in the sense that by making war too horrible, dynamite and its chemical relatives would eliminate war, he predicted: "On the day two army corps can annihilate each other in one second all civilized nations will recoil from war in horror."

Note Bene: Cell and tissue culture techniques excel all other currently available ones for studying the properties of living cells and their interactions with L-domain regulating agents, and the results of those methods can extrapolate directly to L_c-space-time and to L_1-domain properties of the IO, provided appropriate properties of the intact system are accounted for in the process. At present, however, L_2- and higher-domain properties of the skeleton cannot be studied with those techniques, so they must be studied in the intact, living system by other means, including dynamic histomorphometry. In witness thereof, not a single one of the basic IO phenomena described in Note 2 in Chapter 4, Note 1 in Chapter 5, and Note 2 in Chapter 7 was discovered or even suspected by investigators using cell and tissue culture methods, nor by any biochemist, molecular biologist, histologist, or pathologist. That fact is recorded here, not to belittle those methods or those who used them, but to support the contention that one needs *all* of the pertinent information to understand the skeleton, and to the degree that any important part of it is missing, knowledge falls short of that goal. The various chapters of this work should leave little doubt in any unbiased mind that the IO of the skeleton contains a great deal of that pertinent information.

Failing to perceive such facts, or perhaps denying them as Chapter 14 points out (some molecular and cellular biologists believe their disciplines alone deal with the big truths, and they tend to defend that idea when it is threatened), many people continue to extrapolate from the above laboratory models directly to humans, which continues to raise false hopes in physicians and pharmaceutical houses. A prayer of a wise surgeon comes to mind here: "Lord, if I make mistakes today, let them at least be new ones . . . " In that respect, trying to extrapolate directly from cell to man constitutes a very old mistake.

As Parfitt recently summarized,[572] current cell, tissue, and organ culture methods do allow one to study many L_c- and some L_1-domain phenomena under controlled conditions. When their results are interpreted in direct relation to the properties of the next higher level of the IO they can reveal many new and useful things. An increasing number of authorities who use those techniques agree with that view.

Chapter 14

AN OPTIMIZING STRATEGY FOR SKELETAL RESEARCH

"I wish I understood everything I know about this patient's problem."

(R. Dwyer, a surgeon, 1974)

"Serendipity solves problems only slowly and at great cost. Contrariwise, knowing the principles of action of a system allows one to solve its problems with seemingly magical economy of thought and time."

(H. M. F., 1978)

ABSTRACT

Because of certain statistical properties of the IO and because of the nature of its natural controls, a downwards analysis through the organizational ladder will lead to an understanding of disease and to cures for it more rapidly and assuredly than an upwards analysis. The argument depends on special definitions of disease, function, property, randomness, and the research move.

I. INTRODUCTION

About 1969 at one of the Sun Valley Bone Workshops organized by Jee, the author gave a lecture that men such as Arnold, Bartley, DeLuca, and Talmage still remember vividly. The innocuous-sounding subject: how to maximize skeletal research productivity given the available support, talent, and facilities. The lecture lasted only some 30 min, but the ensuing 2 hr and more of discussion were marked by vigor, strong emotions, and violently divergent views. The rest of the sessions in the morning had to be rescheduled.

What caused the heat? A simple idea that addressed concepts only and did not point even indirectly at any person or group. The reasons why an impersonal idea could make a roomful of mature and experienced scientists engage in a 2-hr shouting match contain a message for anyone planning or beginning a career in skeletal research or medicine.

The simple basic idea has two parts. First, it contends that the organ, the IO, and the cell have about the same importance in the intact human body. Second, it proposes that the most efficient way to find real biological functions so one can study them and their controls rather than the overwhelmingly more numerous and irrelevant trivia defined in Chapter 2 consists of starting at the top of the biological organizational ladder and tracing the functions down it layer by layer to their sources.[235,255,267] Appropriate facts behind those proposals make it hard to contest them on the basis of reason alone.

The cell and molecular biologists in that audience perceived exactly that message of the main argument. They realized immediately that it and its author threatened their cherished and generous "mistress" with a treacherous stab in the back, out of the dark and without warning. The following analysis might help to understand their reactions. Begin by observing that that mistress is both a concept and a perspective. As a relation one might define her thus, where $I(x)$ will signify "the fundamental importance of "x" in the overall scheme of human physiology", and $F(x)$ will mean the functions of "x" as the Glossary (see Appendix I) defines a function for use in this book.

Table 1
DEFINITIONS OF CERTAIN TERMS IN THIS CHAPTER

Term	Definition
Function	See the Glossary and Chapter 2
Disease	See the Glossary and Chapter 2
Downwards path	Start the study of a problem at some level of organization and proceed downwards from there, level by level
Property	See the Glossary and Chapter 2
Random choice	Any feature chosen for study that has not first been proven to cause the disease in question
Move	Any research project intended to yield information about a disease, its diagnosis, or its treatment
Serendipity	The actions of pure chance, of Dame Fortune
Upwards path	Start the study of a disease at a lower level on the organizational ladder and then extrapolate the results to the next higher or even higher levels

The mistress:

$$I(\text{macromolecules}) > I(\text{cell}) \gg I(\text{all else}) \tag{1}$$

$$F(\text{cell}) \to F(\text{organ})$$

One should understand that the cellular and molecular biologists who heard this lecture had already invested a major part of their careers in contingent scientific postures, in the charms and rewards of the above mistress, as described in Note 2 in Chapter 14.

The stab in the back, on the other hand, might be written thus:

The threat:

$$I(\text{macromolecule}) \sim I(\text{cell}) \sim I(\text{IO}) \sim I(\text{organ}) \tag{2}$$

$$F(\text{macromolecule}) \to F(\text{cell}) \to F(\text{IO}) \to F(\text{organ})$$

As elsewhere, the one-way arrows signify the one-way street defined in Chapter 2, so, for example, $F(\text{IO}) \to F(\text{organ})$ would read thus in words: "the functional properties of an organ depend directly on the functions supplied to it by the IO, which in turn depend directly on . . . ", and so on.

The above digression is provided here because that mistress has seduced many scientists and rewarded them generously with grants, equipment, students, honors, page space, prestige, respect, and influence. Accordingly, some such scientists might react as strongly as some people in the Sun Valley group to any serious threat to her preeminent status. Whatever else this book does, it certainly threatens to raise to the level of that mistress the somewhat dishevelled commoners who, up to now, stood humbly and in respectful silence at the foot of her pedestal (the reader should find in the above paragraphs a bit of impish humor but none of ill intent).

With those ideas and observations in mind, the argument itself can begin by considering some reasons for medical research. Table 1 supplies definitions for some of the special terms used in this chapter.

II. THE ALGORITHM

A. Some Goals of Medical Research

The people who actually do research have varied motives and goals: curiosity, or their chief ordered it, or they truly want to understand, or they want it as a lever to obtain status, prestige, or power, or under the drive of ambition or of a compulsion of obscure origin, or perhaps no other opportunity existed when a pressing need for employment arose. Few people who do research have as their true main goal trying to find cures for disease, and many of those few may be excellent clinicians but indifferent investigators. The converse holds true as well, for most good investigators are indifferent clinicians. The reason lies in the fact that the mental traits required to excel at each are somewhat antithetical, as pointed out in Note 1 in Chapter 15. Thus, the goals of investigators are varied and mixed.

However, the goals of those who support that research are simpler and fewer. They usually intend that the studies they support will produce the kinds of new information that will improve our knowledge of disease and our abilities to foresee, diagnose, and cure it, regardless of the motives of the persons who actually do the work. In that regard, the supporters of research can back one of two routes.

B. Two Ways to Find Cures

The following two antipodal strategies may oversimplify the problem, but they also reveal the major alternatives involved in the process.

1. The Random Path

At one extreme, one could support trials of randomly chosen drugs and other agents, alone and in combinations, until success finally comes, or one could support studies of randomly chosen cellular and subcellular properties until the causative one(s) finally surfaces, noting that "random" in this chapter means any choice made without advance *proof* that the studied property causes or relates directly to the cause of the disease.

That random path is the approach of serendipity mentioned in the aphorism at the beginning of this chapter. It represents betting in a roulette game where the odds against the player approximate $> 10^6$ to 1. Some 2000 years of experience expose the win-lose properties of that game, and it is phenomenally wasteful of time, talent, resources, and labor. It is slow. It is a poor investment for anyone looking for a good return in less than 50 years.

2. The Downwards Search

At the other extreme, one could study the properties of the disease in the intact subject and then trace its causes and roots downwards layer by layer through the biological organization, and entity by entity across each layer, concentrating on functions rather than on randomly chosen properties. In that process, one would first identify the causative malfunctions, then find out what caused them.[255] Basically that repeats the major steps in the scientific revolution described in Chapter 1: first, find and understand the basic mechanisms that control the system, then find and understand what controls those mechanisms in nature. The track record of that strategy in all fields of science indicates that it eventually usually reveals how to make those mechanisms dance to our drum instead of Dame Fortune's, meaning how to cure their malfunctions.

This strategy may seem reasonable as a statement of faith, but to invest time, talent, and resources in it requires some convincing reasons.

One collection of reasons follows. It depends on known properties of the skeleton and also on the special meanings assigned in this book to the terms disease, function, random, and control.

C. A Basis for an Optimizing Algorithm

Chapter 2 described some basic properties of biological systems that pertain to the

Table 2

Level	Minimum number of roots
(L_{is}) Whole man	$5 = 5$
(L_{os}) Articulated skeleton	$5^2 = 25$
(L_o) Organ	$5^3 = 125$
(L_3) Upper IO	$5^4 = 625$
(L_2) Middle IO	$5^5 = 3,125$
(L_1) Lower IO	$5^6 = 15,625$
(L_c) Cell	$5^7 = 78,125$
(L_{org}) Organelle	$5^8 = 390,625$
(L_m) Macromolecule	$5^9 = 1,953,125$

present matter, provided the research has the goal of finding the root causes of disease and cures for them. That problem represents a kind of game in which one must make a series of research "moves" to win, and in which in principle anything from an infinite number of moves to a single move could lead to a win. The problem is to find the particular kinds and sequence of moves that will lead to a probable win in the fewest moves. The pertinent facts follow.

1. The Function/Property Ratio

As noted in Chapter 2, every biological system, whether a cell, an IO entity, or an organ, has >10^6 enumerable properties, less than 20 or so of which will represent functions that are supplied directly to and that are needed by the next-higher level of organization. Keep in mind that a function differs from its control.

2. The Function/Disease Relation

Given the special definitions of function and disease in this book, and excepting trauma, infection, and other catastrophic challenges from the exterior environment of the subject, then only functions can cause disease and only when they malfunction, regardless of what makes them do so.

3. The Odds of Random Choices

It follows that if one picks a particular property of a system to study without advance *proof* that it is the, or one of the, deranged function(s) that cause the disease, there is less than 1 chance in 1 million that it does cause the disease.

4. Functions Evolve Upwards

Table 2 shows how the skeletal organization evolves from the macromolecule to the man. Table 3 shows how structures can evolve upwards as one climbs the organizational ladder from bottom to top, here for an osteopenic spine. Table 4 illustrates how functions might evolve upwards through the ladder, here for the mechanical competence of the spine, meaning freedom from pain and structural failure under normal mechanical demand. Each of those tables diagrams nine separate steps between and including the macromolecule and the man, and each level should associate at least five of its separate structures and/or functions to provide one new function, entity, or structure to the next higher level. That statement seems conservative, for further functions and structures could be defined at each level. At a minimum, therefore, the disease-relevant properties of the intact subject could sit on top of at least 5^9 separate macromolecule-level features, which equals 1,953,125 of them, and also on top of 5^7 or 78,125 separate cell-level features.

Consequently, even if one correctly chose a particular cell-level function rather than

Table 3
EVOLUTION OF A STRUCTURAL DISORDER (SPINAL OSTEOPOROSIS)

Level	Entities	Minimum number of roots
L_{is}	Osteoporotic woman	5
L_{os}	Spine, GI tract, cardiovascular, GU, others	5^2
L_o	Bone, joint, gonad, pituitary, others	5^3
L_3	Envelopes, spongiosa, compacta, disc, others	5^4
L_2	Circumferential lamellae, osteons, trabeculae, marrow	5^5
L_1	Lamellar bone, periosteum, endosteum, capillary, others	5^6
L_c	Precursors, 'blasts, 'clasts, endothelium, others	5^7
L_{org}	Nucleus, Golgi, mitochondrion, membrane, others	5^8
L_m	Macromolecule (1), (2), (3), (4), others	5^9

Table 4
THE EVOLUTION OF ONE FUNCTION IN THE BIOLOGICAL ORGANIZATION OF THE SKELETAL SYSTEM

L_{is} Mechanical competence of the spine

L_{os} Maintain alignment, collective stability, painless joint and bone function, motion, collective strength, protect neural contents, allow free egress of spinal nerves; growth of spine as a unit

L_o Bone shape, size, content of spongiosa and compacta; provide painless bone function, strength, endurance, stiffness; provide space for hematopoietic marrow; provide intervertebral disc functions; facet functions; L_o-mode growth

L_3 Appropriate bone modeling on two envelopes, remodeling on four envelopes, chondral and ligament modeling; ligament remodeling; macrorepair ability for bone, ligament, marrow; homeostatic mechanisms

L_2 Basic mechanisms for L_2-mode bone, fibrous tissue and chondral modeling, remodeling, growth; bone homeostatic function; innervation; L_2-mode functions of marrow, stroma, and hematopoiesis; microrepair mechanism for bone and ligament; lymphatics

L_1 Histogenesis and L_1-mode growth of woven and lamellar bone, ligament, disc material, hyaline cartilage, marrow; provision of rigid, strong bone tissue, compliant chondral tissue

L_c Osteoblasts and osteocytes for lamellar and woven bone; chondrocytes and chondroblasts for hyaline and fibrocartilage; fibroblasts and fibrocytes; marrow cells, innervation

L_{org} Cell nucleus, mitochondria, endoplasmic reticulum, vacuoles, Golgi apparatus, cell membrane, centriole, nucleolus

L_m The typical human cell is said to have >30,000 genes that can encode the production of an equal number of proteins, polypeptides, and other macromolecules

a trivial property to study, but did not first prove that it also is the function that causes the disease in question, the odds that it would not cause it so its study would be a waste approximate 78,125/1, or $p > 0.999$. Under those circumstances, one would have to study $\sim 78,125/2$ or 39,062 such functions to achieve a probability of $p \sim 0.5$ of having studied the causative malfunction. This, then, is another poor investment.

5. Identifying the Causative Malfunctions

It follows that consistent success in playing this game requires separating functions from trivial properties, and also separating the far more numerous irrelevant functions from the few errant ones.[267] That observation has considerable weight because each research "move" on this game board, each study of a particular feature, would represent committing one laboratory and 12 or so people to the project for a period of 2 to 5 or more years. Consequently, searching blind alleys would waste enormous amounts of time, material, and support. The project itself might focus on the role of anaerobic metabolism in cartilage growth, or of alkaline phosphatase or parathyroid function in senile osteoporosis, or of running on the orientation of osteons in the femur, or of calcitonin or estrogen on σ for bone formation in secondary hyperparathyroidism, or of a globulin in rheumatoid arthritis.

Fortunately, the cell-level malfunctions that cause a disease can typically be found, and with a probability ~0.9, in at the most seven such moves, and the relevant L_2-level IO malfunctions can be found in at the most five such moves, and often fewer and with a similar probability of a "win".

By way of explanation, to identify a function directly and in only one move, study the interactions between any two adjacent levels of organization and determine what the lower supplies to the higher that the latter cannot do without.

With regard to disease and malfunction, one can find the abnormal features that characterize disease at any level of organization by the simple strategem of comparing the diseased system to the healthy one. Then one finds what malfunctions at the next lower !evel caused those at the higher one, and so on downwards.

6. The Upwards Search

The reverse cannot be done with acceptable odds of success, i.e., one cannot predict the organ-level effects of a given lower-level malfunction or function, unless he has *complete* information about the structure, functions, relationships, and other features of the entire system from the cell to the woman.[235,263,267] Such complete information does not yet exist for any structure or organ of the skeleton, and in that respect the major gap in the chain of skeletal knowledge constitutes the IO. Accordingly, the upwards search strategy encounters the odds and random choice problems described earlier.

7. Recapitulation

The malfunctions that cause a disease can be found in only a few moves by starting at the top of the organizational ladder, identifying the features of the disease by comparison with healthy subjects, then identifying the malfunctions at the next lower level that caused those features, and repeating that process of *downwards search and identify* to the level of the cell and below. By that means, in, at the most, six moves one should find the cellular malfunctions that cause the disease, and which kinds in which tissues and organs, at what age and in which sex, and with a probability much greater than the odds of success by playing the game randomly.

As for the work involved, many of the above moves have already been made, for since about 1870, clinicians, biochemists, and pathologists have cataloged fairly good gross microscopic and chemical descriptions of disease at the organ level, and in many instances at lower levels as well. Accordingly, the real research in following this algorithm would often begin by seeking the L_3-domain roots of already known organ-level abnormalities.

In any such search it becomes important to know the nature of the functions and related properties that each level of organization supplies to the next higher one, and that each entity within each level supplies to the next higher one. That requires knowing the purpose as well as the nature of those functions, as described in the closing part of the text in Chapter 2. Earlier chapters in this book tried to identify some of those

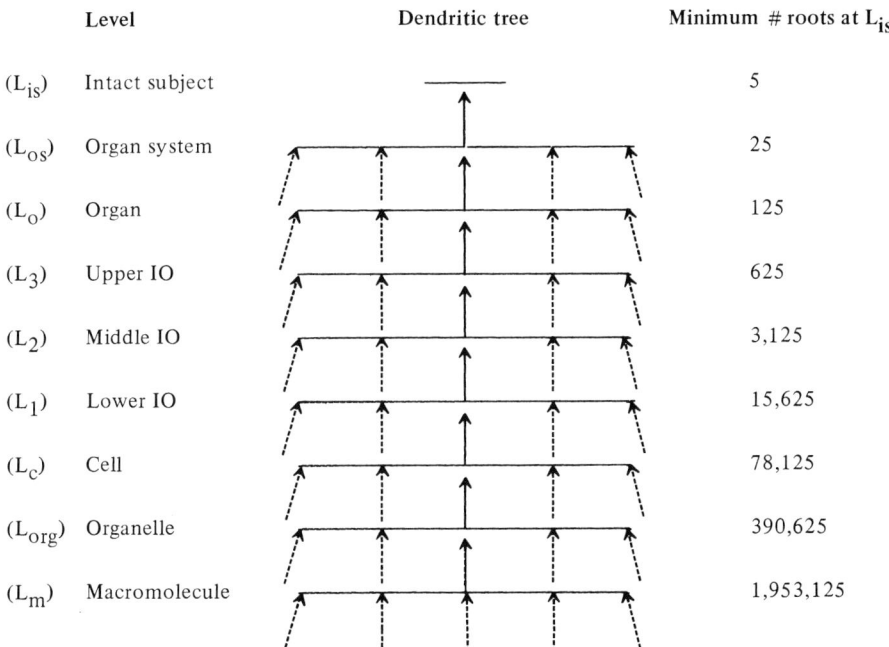

FIGURE 1. The "dendritic property". The levels of biological organization are listed on the left. In the middle it is assumed that each essential function at each level from top to bottom depends on no fewer than five functions provided by the next lower level. The roots at the next lower level are shown by the continuous line arrow for one of those five roots; the other four are indicated by the dotted arrows and the five roots of each of those four are not shown. The right-hand column indicates the number of roots, in principle and at the next lower level, that underlie a skeletal function at the L_{is}-level, the intact, living subject. Thus, if a single root malfunction should cause a particular disease of the intact subject, it could be one of 3125 possible ones at the middle IO level and of 78,125 possible ones at the cell level. If combinations of two or more root malfunctions should cause the disease, then the number of the above possibilities increases enormously. It follows that lacking proof that some function at some lower level causes a given disease, then these odds make it overwhelmingly likely that it does not. Consequently, some means of separating the pertinent functions from the irrelevant ones would be highly useful.

purposes, but many more must await perception. What is already known of such matters could err in some respects.

8. The Control of the Causative Malfunction

Simply finding the malfunction(s) that causes a disease does not alone provide the information needed to cure it. That requires one more move in this game: find out what enables and regulates that mechanism in nature, and what kind of error(s) in those controls made it malfunction. Now that particular matter can be studied with the certainty that it is not some irrelevant or trivial matter. Lacking that certainty, then the probability problems defined earlier would apply.

Figure 1 diagrams the dendritic property that underlies the above strategy. It was described in 1971[235] and 1973[238] before some of the IO features described in this book were perceived. That dendritic property constitutes a fact of nature and of the mammalian skeleton, and therefore a valid way, even if an unconventional one, to describe the organization and functions of the body. Figure 2 provides an example of a matter that was incorrectly explained on the basis of tissue culture work.

Table 5 summarizes the main features of the above algorithm and Figures 3 to 7 illustrate how a downwards search and identify analysis might work. This part of this

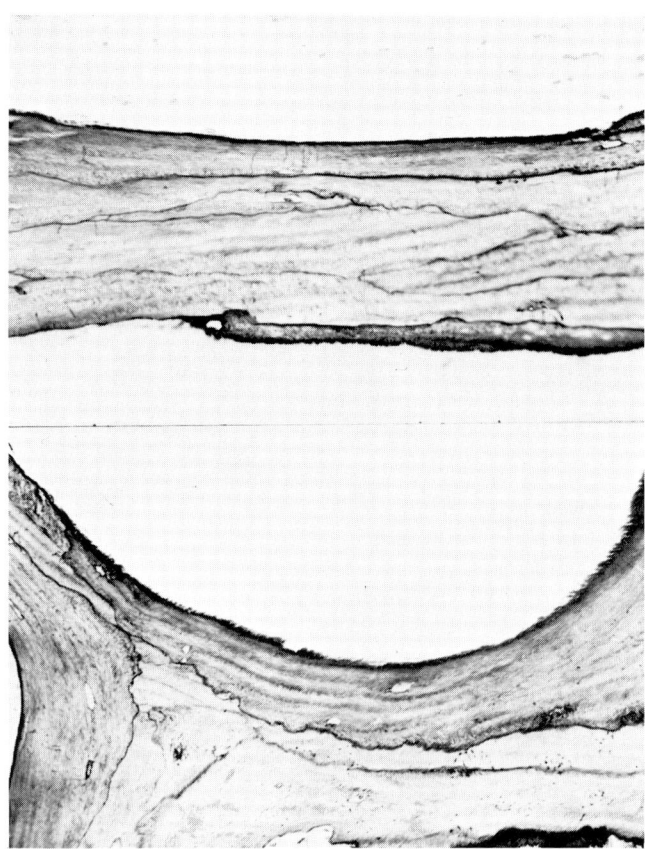

FIGURE 2. Undecalcified sections of spongiosa from transilial biopsies of a patient who received treatment with NaF for 2 years for her osteoporosis. Pretreatment biopsy above, posttreatment below, stained to show cement lines. About 140×. The sections illustrate the observation that to understand complex systems one must study them as a whole, rather than in arbitrarily selected parts. The thickness of the bone packet or BSU in the posttreatment biopsy exceeds any of those in the pretreatment biopsy, and systematic measurements of the biopsies of the many patients in this study showed that was a true treatment effect. But how did it occur? Was new lamellar bone formation initiated *de novo* as some have proposed on the basis of tissue culture work? In that respect, the packet rests on a scalloped cement line rather than on a smooth arrest line such as shown in the figures in Note 2 in Chapter 4. That means that a BMU activation event occurred first and a resorption stage followed; the formation stage followed that. Thus, the treatment increased trabecular thickness by acting on complete BMU-based remodeling units, and the resulting questions become some of those discussed in Chapter 6. Obtaining the answers will require studying this system in the living subject as a whole. Note that none of the factors that directly determine the treatment effects shown here can be studied in current tissue or organ culture systems, in part for reasons described in Chapter 13. Those direct factors include mean wall thickness and mean surface extent of the finished formation packet, and the mean erosion depth and surface extent of the finished, preceding resorption packet, which together determine the value of the $\Delta B \cdot BMU$ parameter. Note too the relatively thin bone packets between cement lines in the pretreatment biopsy. That implies reduced resorption and formation per completed BMU in the untreated disease and illustrates features referred to in Note 2 in Chapter 6 and Note 1 in Chapter 10. (Photomicrographs courtesy of Dr. A. J. Olah, Bern, Switzerland, who reported this study at the Fourth International Histomorphometry Workshop at Arhus, Denmark, in 1964.)

chapter will close with a quote from an astute and perceptive observer of over 330 years ago, who also pondered the matter of how to find out how things work, and in a systematic and logical way. Thus: " . . . it appears advisable to me to look back from the perfect animal, and to inquire by what process it has arisen and grown to maturity, to retrace our steps, as it were, from the goal to the starting place, so that at last when

Table 5
SUMMARY OF THE OPTIMIZING ALGORITHM

1. Start at the top of the organizational ladder.
2. Compare the diseased to the healthy intact system to identify the characteristic L_{is}-level abnormalities of the disease.
3. Find the malfunctions at the next lower level of organization that caused the disease-related abnormalities at the upper level.
4. Repeat that downwards search, level by level, until the root cause(s) of the disease has been found.
5. Find the malfunctions of the enabling and regulating agents that account for the root malfunctions, keeping in mind that those malfunctioning agents may disturb several different functions, and at several different levels on the organizational ladder, at the same time.

we can retreat no farther, we shall feel assured that we have attained to the principles . . . '' So spoke William Harvey in 1651.

III. COMMENT

The preceding sections of this chapter were laconic to minimize misunderstanding of their message. The message rests on facts and the history of all research supports it, even though certain nuances of the problem are not discussed here. For example, an affection such as periodontal disease, postmenopausal osteoporosis, a stress fracture, an arthritis, achondroplasia, a fracture nonunion, or an osteomalacia may have more than one root cause; some of those causes may be IO-intrinsic and others IO-extrinsic, and abnormal ''S'' and ''L'' agents may directly disturb the functions of more than one IO entity and more than one level of organization.[673,686,687] Granted those possibilities, the above algorithm still provides the essence of a way to circumvent the constraints and penalties associated with directing skeletal research randomly, by relying on serendipity. Certain other facts apply to any acceptance and implementation of the above algorithm and they deserve some brief comments.

Few people in the medical care complex of the world have the traits that suit them for doing the kinds of research that can improve medical care in fundamental ways, noting that making available what can already be done differs from discovering how to diagnose, treat, and predict disease better than can already be done. Also, only a small fraction of the resources of a medical care complex and of a nation will support such fundamental research. The imperatives of day to day medical and surgical needs make it thus.

Such facts make both the talent and the support for that kind of research scarce, and they will remain so. It behooves people who can influence the course of basic research to try to see that it deploys both wisely and effectively.

Medical research sums up the contributions of at least three different kinds of activities. One exploits what is already known, as in improving the design of an existing endoprosthesis or the chemical structure of a drug. Another discovers new phenomena, such as the action of penicillin on some bacteria or the action of vitamin D on growing epiphyseal plates.[207] The third innovates, as in Paul Ehrlich's idea of the chemical bullet[183] or the ideas of evolution by genetic mutation, or the concept of plate tectonics, or the concept of feedback in control systems.[778]

All three kinds of research are necessary to the whole even though a dedicated practitioner of any of the three may disparage the value of the others. Granting that, most fundamental advances in the life sciences relate to the basic mechanisms that govern the nature and behavior of tissues and organs in bulk. In understanding those basic mechanisms, some fields advanced farther and faster than others, and some are laggards. Two obvious laggards in human physiology constitute the study of man's mental, emotional, and social functions, and skeletal physiology (those observations reflect

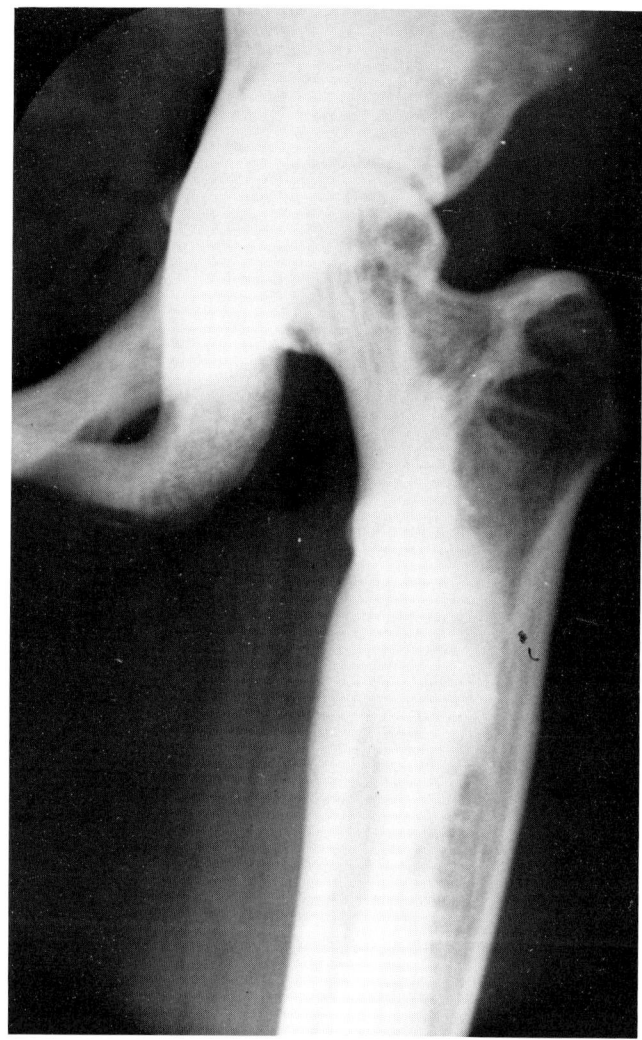

FIGURE 3. AP X-ray of the hip and upper femur of a patient with a form of Albers-Schonberg disease or osteopetrosis, in which clasts supposedly cannot resorb mineralized tissue. Indeed, the original mineralized cartilage core of this femur remains visible inside the medullary cavity. Part of the femoral head is medullarized and cancellized, as is the greater trochanter and upper femoral neck, while the large marrow cavity of the upper femur reveals that cortical-endosteal resorption drifts occurred. The supra-acetabular bone remains extremely dense. Here then there is a resorption defect, but one that varies in different places and at different times of life.[476] It affects all varieties of the hard tissue remodeling activity more than the bone growth and modeling activities, and than the chondral growth and modeling activities (the shapes of the hip joint and the greater trochanter reflect the effects of both those activities). One must explain those features, and a simple and sole defect in the ability of clasts to resorb bone or mineralized cartilage cannot explain it, for that would make it everywhere the same. That does not mean such a defect does not or cannot exist. Rather, it means that it can only be part of the whole truth in this patient. Something peculiar to L_2- and L_3-space-time and the properties of the IO is (are) also involved, perhaps in the form of defects in some kinds of local MES signals and helper cell systems as suggested in Note 1 in Chapter 7.

the complexity of the systems rather than any incompetence on the part of those who study them).

Perceiving the operational IO in our skeletons forms a fundamental advance in knowledge, and evidence begins to appear in print that others agree.[522,822] The added

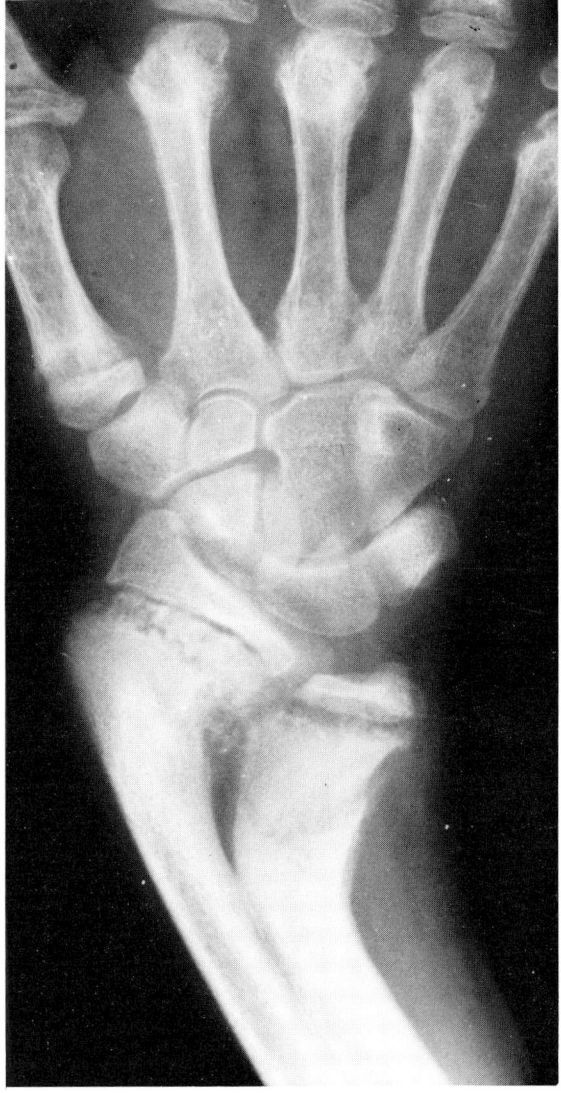

FIGURE 4. This adolescent boy had severe chronic renal failure and he was not maintained on dialysis. As one result, he had the syndrome known as renal rickets, which, however, is more complicated than a mere impairment of cartilage mineralization. Note the following. The distal ulnar epiphyseal plate is thickened and irregular, as is that at the base of the first metacarpal on the left, but the distal radial epiphyseal plate is not, nor are the intercarpal joint and carpometacarpal cartilage spaces, where growth of articular cartilage occurs. While both distal radial and ulnar epiphyses are displaced ulnarward, the displacement did not occur in the epiphyseal plates. Rather, it occurred in the spongiosa beneath the plates, which was probably disorganized like that in the rachitic rat shown in a figure in Chapter 13 and poorly mineralized, too, so it was unusually compliant and susceptible to creep in shear, and the major part of the displacement shown here occurred there. One would suspect a combination of microdamage and creep in the primary spongiosa. Bone modeling activity has been struggling manfully to recontour the radial and ulnar shafts to fit what the displaced epiphyses — which are covered on all sides by hyaline cartilage at this age — demand of them. One thing stands out: the major abnormalities occur where local growth and/or turnover are fastest.

holism the IO brings to the field is based partly on still arcane facts and it occurred relatively recently without any blare of trumpets to announce it to a large audience. Also, it threatens the preeminent status of cellular and molecular biology so it has

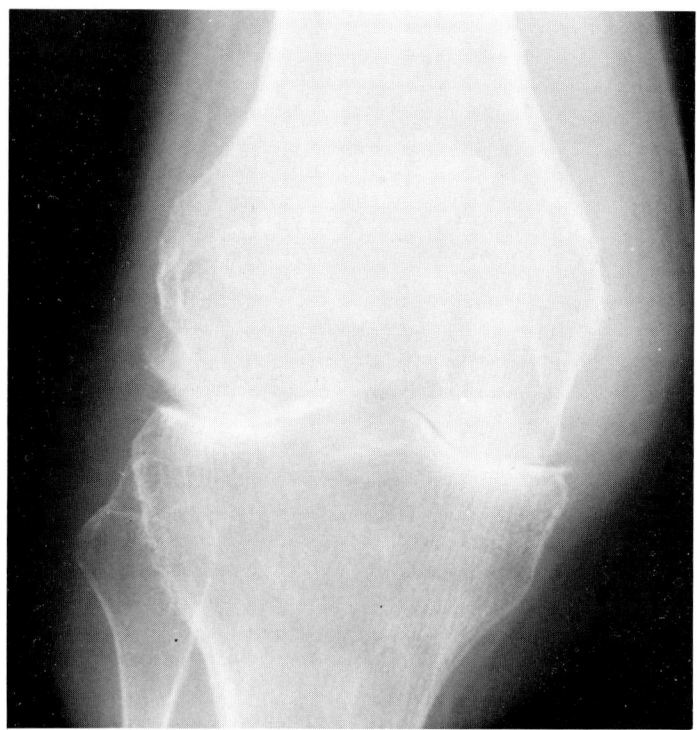

FIGURE 5. An AP X-ray of the knee of an adult with rheumatoid arthritis. Note that the cartilage space between the femur and tibia has thinned and virtually disappeared, meaning the articular cartilage is gone. A downwards search and identify analysis of this matter might proceed as follows. The pannus associated with rheumatoid disease does not directly cause the cartilage space narrowing because it arises at the margins of the joint, it is very fragile, and it would be crushed and destroyed instantly between the two joint surfaces under any load. It can only cause marginal erosions. So the cartilage either wore away from usage, or in the underlying bone some endochondral ossification became enabled and replaced the cartilage with bone from below. If the former, something happened that reduced the synthesis of new chondral matrix components below the rate of wear (which could have accelerated due to some change in the composition or structure of the matrix). If the latter, something should have served as a switch to enable further mineralization at the chondral tidemark. Of course both things could have occurred. That analysis raises immediate and simple questions that pertain to L_2- and L_3-domain matters of the IO of joints, and that could be answered relatively easily by appropriate dynamic biochemical and histomorphometric studies.

already evoked certain reactions that probably represent territorial protectiveness and defense. (Marcus Cicero put that neatly over 2000 years ago: "When you have no logical basis for an argument, then abuse the plaintiff.") For those and other reasons, some defensive reactions may arise to the algorithm in this chapter, and also to the nature and roles of the IO, the properties of which form the pedestal on which the algorithm rests.

Only partly for reasons given above, the downwards path strategy of deploying skeletal research offers more success in fewer moves at less expense and more quickly than the random path that characterized such research up to now, and than the upwards path that many cellular and molecular biologists currently favor.

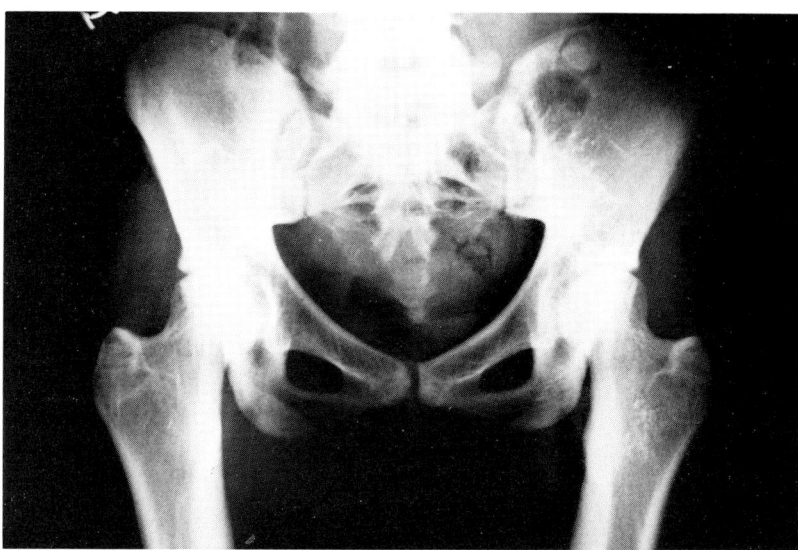

FIGURE 6. This and the next figure illustrate an unusual bone problem in an adult woman in her 40s on renal dialysis for chronic renal failure. She has no significant musculoskeletal complaints and the X-rays were obtained during routine screening for evidence of renal osteodystrophy, which this is not. Analyze the problem downwards with that limited information. The gross configurations of the hip joint, greater trochanters, pelvis, sacrum, and upper femurs are normal, as are the articular cartilage spaces, implying that chondral growth and modeling are also normal. The bony configuration of femur and tibia are abnormal, however, in that the diaphyses are too wide and the cortexes too thick, especially the medial cortex of the upper femur; the femoral heads are also quite dense, implying an increased trabecular bone volume. The thickened and dense regions all occur where bone carries large compression loads (while the long bones also carry major flexural loads, the spongiosas in the caput and the epiphyses about the knee do not), which implies that this bone is unusually compliant as a material, so more bone was needed to reduce strains $\leqslant$ MES, or for some reason probably not related to her renal disease the MES was set smaller than usual in this patient so modeling had to supply more than the normal amount of bone to attain greater than normal bulk stiffness. Of course, "X": something goes on here that differs from the above factors. The author finds it tempting to consider that uniaxial compression loading and strain can also control bone macromodeling, that flexure is only one of the important mechanical factors in that activity. No detailed studies of bone histology and dynamics are available in this woman. (X-rays courtesy Dr. C. Wehling, Southern Colorado Clinic, Pueblo.)

The author believes it would be a mistake to let the direction of future research depend on serendipity or the upwards path because of the poor "win" probabilities of those two strategies. There may be statistical merit in the idea that random studies of enough things will eventually provide a mastery of disease. There is equal merit in the idea that when enough monkeys sit before enough typewriters for enough millions of years, one of them will finally type the text of Tolstoy's *War and Peace*. Likewise, if one is hungry, then if he sits in a chair long enough with his mouth open, some day a roast duck should fly into it. Theophile Gautier spoke of such belated gifts of Dame Fortune, serendipity's paramour, thus: "Fortune loves to give bedroom slippers to people with wooden legs and gloves to those who have no hands." One could add: cures to those who have already succumbed to their disease.

A strategy that could increase the useful output of skeletal research by one or more orders of magnitude would yield great benefits for which a pressing need exists, mean-

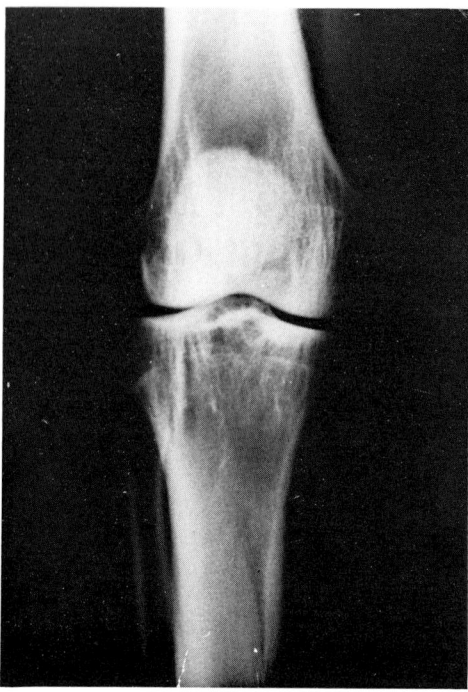

FIGURE 7. AP X-rays of the left knee of the patient in the previous figure.

ing cures for the diseases of our parents and children while they are still here rather than after they pass on, provided of course that the potentials and constraints of contemporary expertise and support would allow it.

In this man's judgment they would. The potential impediment here is not what *could* be done today, but what people *would* do or not do, and why.

Note 1

THEORY: SOME OF ITS VALUE AND USES

"It's difficult to force fools from the chains they revere."

(Voltaire, about 1770)

Many years ago, Dr. B. Frame of Henry Ford Hospital invited the author to tell an audience of clinicians interested in metabolic bone disease about some of the then recent findings concerning the human bone remodeling process. Some of the material found in Chapters 4 and 5 was trotted out with accompanying photomicrographs and headstrong enthusiasm. Suddenly an internist in the audience muttered: "But that's just theory!" He nodded in his chair the rest of the hour.

In one way he was right. As a clinician he had to care for his patients, his schedule was overloaded, and with respect to diagnosis and treatment he wanted to know what he could do tomorrow that bettered what he had done today. He needed reliable recipes rather than the information that might help to make better ones so an hour spent on then unproven ideas of how to do it better, maybe 10 or 15 years in the future, seemed to waste his time.

From another point of view he lacked insight into the realities of medical progress. It begins with something new that is applied to something old and then refined and exploited. That work usually litters the trail behind it with the mistakes and failures that eventually lead to a success. Thus, the arsenicals that Ehrlich devised to treat lues led to modern antibiotics, and with further input from Fischer, Pasteur,[578] and others, to modern bacteriology, biochemistry, and immunology. Those developments plus the persistence of Mendel,[502] led to the synthesis by Jacob and Monod[360] of the basic interactions between DNA, RNA, and ribosomes, and thence to a nascent genetic engineering.

The aforesaid something new can begin as an accidental observation such as the recognition of the effect of quinine on malaria. It can also begin as an innovative idea that is then worked on, tested, improved, revised, and finally brought to fruition as a proven new field of research and/or practice. One could recall here Jenner's idea that vaccinia can immunize against smallpox,[382] Pasteur's that all life stems from life, at least since Paleozoic times, Newton's that a gravitational force attracts matter to matter, and Einstein's that matter and energy form different states of the same thing, e.g., $e = mc^2$.

The physical sciences have evolved remarkably and explosively in the past 200 years. A large factor in that evolution constituted the injection into those fields of innovative theories of how things work, and, also, of innovative adaptations of the already known to solve previously refractory problems. Hence, Maxwell's field equations, and Bell's, Edison's, Marconi's, and Tesla's practical applications of phenomena modeled by those equations. Numerous other examples exist, recalled by names such as Ampere,[18] Darwin,[155] Banting and Best, Dalton,[150] Watt, Fraunhofer, Lister,[461] Semmelweiss,[692] Goddard, Bohr, Shapley, Gauss, Virchow,[752] and Fleming,[207] and, on a smaller scale, the visionary concept of temporal coherence in IO entities, and then its practical adaptation by Anderson[24] to the treatment of osteopenias.

In most physical scientists of today those factors instilled respect for the few who specialized in theory, the business of organizing existing knowledge and ideas into new shapes that lead to new ways of perceiving and using things. So when a physical scientist accumulates a track record of success in that respect then he usually receives the respect and encouragement of most of his colleagues.

Some of the medical sciences lag in respecting and encouraging their own young theorists. This is particularly true of the skeletal community. The reasons need not be discussed here, but the phenomenon is revealed by an automatic suspicion of innovative theory and its creator, and an automatic attack and destroy reaction to both. It is real and it holds this field back. It delayed by nearly 20 years the general acceptance of the indisputable evidence that bone remodeling is a quantized activity, while Note 3 in Chapter 7 revealed how it delayed progress in a biomechanical subfield. Young people embarking on a career of skeletal research or its application to clinical problems could learn a little about it as an aid to dealing with it constructively. The author can offer advice there, based on a modest track record of theorizing that began to bear fruit in 1960 to 1964,[823,824] and on experience with counterproductive reactions to "that's only theory" in its varied forms and nuances. That advice follows.

Innovation is the essence of fundamental improvement in any field of science, so the long-range and basic interests of the field require valuing and fostering rather than disparaging it.

Such innovation forms a small corner of the total medical endeavor. The major effort of the medical care complex of the world lies in using and distributing knowledge and expertise that already exists, and most of its imperatives override investigative concerns and needs. It follows that insofar as he works with theory, a man stands apart from the mainstream of clinical practice, and few clinicians will understand his problems and concerns. Oppenheimer put that matter thus: " . . . anyone working at the frontier of . . . science is . . . a long way from home and . . . the practical arts that were its matrix and origin . . . "

No theorist is always right and even the best ones are wrong more often than right. Clinicians do not understand that, but good scientists do and do not disparage such errors when they become evident and are corrected. Since a few incompetent and/or envious people (and the strange species, *aves mirabilis ansapiens,* discussed shortly) may hold up such errors as evidence of the incompetence of the originator, a young theorist in this field can benefit from a thick hide. On that matter the author speaks from experience, having published in earlier days a fair number of speculations that ultimately proved absolutely wrong. In part those errors reflected inexperience and lack of a competent foil.

The inexperience lay in a blurred perception of the great difference between facts and their meaning, so some of the latter were confused with the former (see Note 3 in Chapter 7). Like the squirrel in the Preface, a fact is a fact whether one understands it or not, but its meaning can be subtle and one always infers it at risk of error, so he should usually do so with caution. Only experience can teach that caution.

The competent foil is another mind familiar enough with the subject to criticize ideas free of their creator's biases, thereby discerning errors of judgment, comission, and omission and ambiguities of expression. Example: a young dental student once answered with complete confidence in the accuracy of his knowledge, in response to the author's question "what does compression do to bone?" Answer: "It stimulates bone resorption of course." The author's next question, as to what he was sitting on at that moment, eventually extracted the desired answer: the ischial tuberosities. Were they under compression then? Yes. In sitting for a lifetime did they resorb? "Oh." The student was bright (and is now an internationally respected surgeon) and he got the point: there was clearly more to the problem than his initial answer contained and assumed, and than his dental basic science education revealed, so until he understood those further factors he put the first answer in temporary storage as maybe right, but maybe wrong too, or maybe even unrelated to the truth.

In this effort, as in others, experience teaches respect for the difference between facts and their meaning, and many young men trying on various faces in search of their own will err before they learn that. Apologies for such errors are unnecessary if one accepts them as such. In testimony, the verified IO-related discoveries listed in Notes 2 in Chapter 7, Note 1 in Chapter 5, and Note 3 in Chapter 7, and Table 7 in Chapter 7 were distilled and separated by time, experience, and data from the shorter list of errors that originally accompanied some of them. In the process the verified was preserved, the errors forgotten, and the former comprises a substantial contribution to the field. Others could have made those contributions had they not feared making errors in the process that their colleagues would ridicule. In point of fact, the historical roots of this field instill that fear in many contemporary skeletal workers. It mixes with a warranted one of publishing erroneous facts.

The above means that progress always requires a venture into the unknown, so it *always* risks error. Accordingly, the cost of freedom from error is progress itself. Try not to confuse fact with theory, learn to use and build on theory, and when an error occurs when trying to do so then take comfort in an aphorism of a Scot named Burns: "The best ground bears weeds as well as flowers."

The small fraction of good theory is the goal, the gold, and the diamond. It put the atom to work, man on the moon, and his fingers inside his living heart and brain. It justifies the frustrations, labor, polemics, and failures that usually accompany it.

New theory requires tests of its merit. Good theorists test their creations themselves, using whatever means and help they can obtain from other sources. A theorist may fear testing his brain child, especially when others do it and in public, but he must not shirk it and cannot avoid it. On that matter the English physicist, John Tyndall, wrote thus in 1871: "The brightest flashes in the world of thought are incomplete until they have been proved to have their counterparts in the world of fact."

Initially, most new theory will either be rejected or ignored by most people in the field (e.g., plate tectonics in geology, coherence treatment in metabolic bone disease, the treatment of Semmelweiss from the medical profession and the reaction of 1960 to 1970-era biomechanicians to the macromodeling rules, and bone physiologists of the same era to the quantum concept of bone remodeling). There is usually nothing personal in that rejection, but if the theorist takes it as a personal affront and reacts to it as such he may make enemies who will use their influence to belittle the theory, its creator, and its supporters, to restrict access to publication, program time, and grants, and to deny the creator the credit for his achievements. One should expect that initial rejection and continue to diffuse the theory by any means available. If the theory has merit then an occasional young person, usually a stranger, will become intrigued by it and follow it up. Then, two instead of one may preach a new message into some wilderness, and then four, then eight, and so on. One clue to success in diffusing an innovative idea or theory may surface sooner or later. It will be when others begin crediting the matter to themselves or to other people than the true originator.

One should not protect theory as he would children or wealth.

New theory deserves publication for two major reasons. First, to see if facts will destroy it or reveal needed revisions. In that regard, many with momentarily unoccupied shotguns will enjoy taking casual shots at it, so there should be no lack of testing or need to seek it from others. The Greek, Pericles, put it thus nearly 2500 years ago: "Envy will exert itself against a competitor while he lives, and will applaud him only after his death." The envious are not noted for passing up opportunities to demean the worth and works of those whom they envy. Second, if the theory has merit then it is a tool for doing better in better ways, and by others.

Most people find hard to understand at first a theory that contains a really new way of looking at things, and that affects older more than young people (see Note 2 in Chapter 1, Note 1 in Chapter 2, Note 1 in Chapter 9, Note 2 in Chapter 14, and Note 2 in Chapter 3). Francis Bacon put his experience with that phenomenon thus, paraphrased somewhat: "Young men invent better than they judge, they achieve better than they advise, and they are more suited to new projects than to established ones, for the experience of age can direct and counsel well in matters already known to age. But in really new things age can mislead youth."

One solution to that problem lies in presenting the idea again and again from differing points of view. Like the concept of the limit in calculus or the ARF idea described in Note 1 in Chapter 9, repeated exposure can succeed where one exposure fails. The solution does not lie in talking louder or in demeaning the intelligence or knowledge of critics, actions which make some listeners suspect a deception rather than a genuine effort at instruction.

Once satisfied his brain child has real merit, a theorist must turn to the separate problem of conveying it to others, a problem complicated by the fact that many good theorists do not communicate well. In that effort, obtaining page space in a refereed journal or in some standard text will not suffice. The problem requires finding conceptual stratagems, analogies, and situations already familiar to the audience that can serve as vehicles to convey the basic ideas of the theory, and doing so in clear and pithy language. For that reason, a theorist should try to understand how his audience thinks and what it already knows that might bridge its knowledge to what the theorist would teach. Samuel Johnson put it rather succinctly: "There is no other method of teaching that of which anyone is ignorant but by means of something already known." Thus, the group of people the theorist would instruct or inform should dictate his prose, his vocabulary, and his analogies. It would waste everybody's time, for example, to use the jargon and detailed knowledge of thermodynamics or group theory in an article or lecture addressed to busy endocrinologists or orthopedic surgeons, and conversely.

So, "That's just theory" does not represent an intelligent, informed and perceptive observation. Rather, it betrays a lack of scientific experience and maturity.

A special kind of bird lurks in most scientific audiences and lecture halls with a preprogrammed mental and emotional reaction to any real innovation. That innovation presses the switch that starts the program running, and with variations its basic elements are these. Whatever experiments the theorist did with result A, this bird did twice as many years before with exactly opposite results. Whatever meaning the theorist perceived in the data of others, this *aves* reports either than a Transylvanian scientist reported it first in 1753, or a German scientist proved it false in 1919. Whatever provided the critical facts that supported the theorist's concepts, this feathered creature states categorically are untrue. Whatever the theorist has learned from direct experience, this being claims to have studied, too, and twice as much and thrice as long, and found no such thing.

In 40 years of listening to such "program" content emerge from the beaks of this species of *aves mirabilis ansapiens,* the author has found only one way to deal with it: ignore it. One cannot conduct a useful scientific discourse with this species for it categorically rejects truth as false, it invents facile falsities and then proclaims them to be immutable truths, and it regards all true heuristic and holistic thought as intolerable and mandating a verbal *auto da fe.* The species does have one use, however. Since its program is never switched from "Off" to "On" by something completely devoid of merit, when it does start running its target can take comfort in the probability that he does have something of value to say.

Most clinicians will not be interested in new theory. They practice an honored and demanding craft, the imperatives of which place high value on proven things, recipes basically, that work consistently or with known side effects and variability regardless of whether one really understands why they work or not. Most of them become interested in new theory only when it has developed to the point that it has or soon will have useful clinical applications. That does not criticize clinicians or their craft; it merely records a fact that a theorist will live with whether he likes it or not.

The typical young theorist in this field will be employed in some capacity (research fellow, instructor, junior assistant professor, or scientist) by others, which exposes him to certain risks that few people have acknowledged, let alone discussed. He depends for continued employment, job security, and promotion on the opinions his superiors have of his work. Typically those superiors will not really understand his work if it is pioneering and innovative, so they tend to form their opinions of it from the evaluations of others in the field, and particularly of other already established authorities. The latter people usually react negatively to a truly innovative insight into the meanings of facts, so if the young theorist in his enthusiasm over a new insight begins to try to diffuse it aggressively, and also to show that it invalidates prevailing views, he will be judged unreliable, erratic, as overinterpreting the data. The next time his employment renewal comes up he may find himself denied promotion, downgraded, fired, or asked to resign. Thus, his job security is ultimately controlled by the opinions and advice of older people he may not even know personally, and who typically if not always will view unproven innovation as a threat and a flaw rather than as potential gold, and who cannot see any more clearly into the future than the author or the reader of these lines.

In a way, that happened to the author. When the quantum concept of bone remodeling, the modeling-remodeling distinction, and the core macromodeling concepts were formulated between 1960 and 1966,[229,230,823,824] their evidence was still mostly unpublished and thus arcane to any outside critics. The new ideas also directly threatened the scientific postures of most of their critics, which included their own publications, and the strategies behind their own research, teaching, grant evaluations, and journal refereeing. They reacted by judging the new ideas as scientifically ludicrous. Collectively, they rejected them and, in 1967, finally stopped all outside grant support for the research that had led to them. Had the writer been employed as a full-time scientist that would have terminated that work and all subsequent publications right then. However, the writer earned his living as an orthopedic surgeon and teacher; research was simply a hobby. Since orthopedic colleagues held the caliber of the surgery and teaching to be of high quality, loss of grants was an annoyance that had no effect on job security and was corrected by having the institution (Henry Ford Hospital) support the work out of the surplus of the orthopedic department, so the work and publications continued. In simple words, the writer's job security was quite independent of other's evaluation of his research. That provided the additional time that allowed a younger generation of investigators, not crippled by the conceptual chains and postures of their own predecessors, to discover for themselves that those new ideas were based on hard fact. Their work and publications then verified and successfully diffused the new ideas among the scientific community, and in a way led to this book. Since the course of subsequent events finally showed that the above negative judgments and the withdrawal of all outside support of the research in question was a blunder rather than an act of true acumen, some of those responsible have fallen back on Marcus Cicero's advice. Being unable to justify their stand with the force of logic, they resort to abusing the plaintiff (as noted elsewhere in this book, the human factor in this field can provide various kinds of entertainment to the informed observer, some of it humorous, some not).

The message in the above scenario for the young theorist becomes plain and simple. Soft-pedal any radically new meaning(s) perceived in data for a while, and instead spend time and energy on producing and publishing those collections of data that will, in sum, make the new meaning at least a necessary possibility if not the proof of it, even in the mind of critics. In that process, exercise some care to avoid creating open confrontations with established dogma, because there will always be established authorities who cherish and protect that dogma as zealously as a religious fanatic might protect his religion from any imagined heresy. In skeletal science such zealots often have influence, and will not be averse to wielding it to eliminate what they conceive as heresy.

Note 2

SOME EFFECTS OF AN INVESTMENT IN A SCIENTIFIC POSTURE

"When men are easy in their circumstances they are naturally enemies to innovations."

(Joseph Addison)

This note considers an aspect of that observation which applies to the skeletal field. It has caused some once promising careers in skeletal research and disease to fizzle, and, in frustration, loss of respect and loss of credibility. The following episode will introduce it, after which it will be explained.

Some years ago (before 1975) the author witnessed — but did not at the time understand — the following events at a large skeletal meeting before an audience that exceeded 400 people. A young foreign scientist had been invited to report some findings in the bones of osteoporotic patients which were obtained with a significantly modified existing technique. Because of that modification, no other observer had done similar work so no one could verify or refute the findings in question. It was that investigator's first visit to the U.S. and his 3rd year in skeletal research, and the few publications that issued from that work were unknown to most of the U.S. research community.

The investigator was invited by an individual who is currently an effective, positive force in skeletal research and communication, and who came across and sensed in this new work a potentially useful innovation. Two discussors, A and B, were appointed by the program committee to comment on the new work. Both had used the unmodified technique in their own research, so from one point of view the visitor represented a competitor. Also, for reasons that stemmed from pecking order strife, but not otherwise explained here, B was an avowed enemy of the person who invited the young visitor.

The new work was duly reported and discussion time came.

Prof. A thereupon noted that because the technique was new, nobody else could confirm or contest the new findings. He applauded the work and urged that it continue and that others try it. All in all, he gave a balanced and courteous discussion, and also in effect welcomed a newcomer to the field, a stranger to our own land, and another youth trying on faces in search of his or her own.

Then Prof. B came to the podium and savagely skinned the visitor alive for having the gall to botch up a soundly established technique, for suggesting that certain "well-known" facts about osteoporosis were fallacious, and for insulting the audience with phenomena that were spurious because nobody else had ever seen them.

It was a cruel public execution, not unlike stealing a blind man's pencils and then clubbing him in the groin to boot. That investigator never revisited the U.S. and dropped out of sight several years later. With that in mind, on now to "posture".

In school and during postgraduate training, a future scientist or physician must absorb much knowledge that is new to him or her, much like a sponge, and without the experience that would allow evaluating the accuracy and limitations of that knowledge. Then the student graduates, begins to work and to lecture, teach, and publish. During that process and the years it consumes, he develops particular ways of looking at his field, its problems, and of studying them. Some years of this make him known for that work and those views, which his own students will continue when they begin their careers. Finally, he will have acquired mechanisms that support his work and his views, and agree that his work is important. In short, the former student has become an authority. Those developments represent a serious investment in a scientific posture that combines elements of prestige, ego, anxiety, labor, risk, recognition, performance, publications, ambition, credibility, and influence, and a significant fraction of a lifespan as well. Of course, that investment is not necessarily bad.

Now something unforeseen may happen. Some new idea or person may threaten that posture by suggesting it lies on fundamental errors or inadequacies. People who have a major investment in such a posture can react to that threat in different ways, depending on their character, ethics, maturity, and intelligence. A few can evaluate such a threat dispassionately and, if reason requires it, then accept any flaws in their own posture and go about correcting them. Hence, Prof. A above.

A majority of such people react by defending their posture, by digging in their heels. They may do so, but out of fear of what others may think if the once respected authority admits he gave them flawed advice in the past, or if another publically exposes it as flawed. They may do it because they cannot comprehend the merit of the threat. They may also fear loss of their support, or their access to employment or grants or ready publication, or invitations to write and lecture, or of young fellows who want to study and train under them and enhance their output (and ego). Whatever the particular reasons, most authorities react both defensively and negatively to any threat to their scientific posture, although with varying degrees of vigor and directness. Hence, Prof. B above.

An illustrative fact: while dynamic histomorphometry has provided the chief and most powerful tool for studying the skeletal IO from its inception from about 1964 through 1986, none of the several dozen investigators who currently use it had reached 40 years of age when he or she decided to make it a tool of his or her own career. Their older mentors all had the same option, but they would not or could not pay the price. Figure 8 illustrates another situation in which the above defense reaction led to a more than 20-year delay in zeroing in on the true pathogenesis and pathology of a well-known hip problem. A comment of Mark Twain seems appropriate here: "Loyalty to petrified opinion never yet broke a chain or freed a human soul." One could add, nor freed a mind from a cage of petrified ideas.

The above properties have one more important effect on communication between science and clinic. Because the clinician has to face, live with, and manage the disastrous consequences of overt errors, he also learns to distrust other clinicians who make unusual numbers of errors and to discount their judgment and, let it be added, properly so. Accordingly, when such a clinician becomes involved in some kind of basic research he usually carries the same attitude into the laboratory, the program committee, the referring of articles, and the grant evaluation. Most such clinicians are truly unaware that productive and innovative research, the basic stuff out of which the clinical miracles of tomorrow will grow, inherently requires making many errors as the price of one solid new contribution. Both the basic scientist and the clinician must understand that situation if they are to communicate effectively.

Since new people in a field usually have not had enough time to develop a major investment in the above-described posture, they have little to lose, so they receive new ideas more readily than their own teachers who have invested in such a posture. Therein lies at least one factor behind the controversy that surrounded the skeletal IO and its roles in health and disease until quite recently, and which still surrounds some aspects of it (see Note 3 in Chapter 7). It will take time for that remaining reservoir of resistance to dissipate. The historical record suggests that that dissipation will not come about by the existing and resisting authorities changing their minds and reevaluating their own previous postures. Rather, it should come about as they retire from active work and become replaced by younger men who busily develop different and newer postures. Thus it has usually been; it will probably remain.

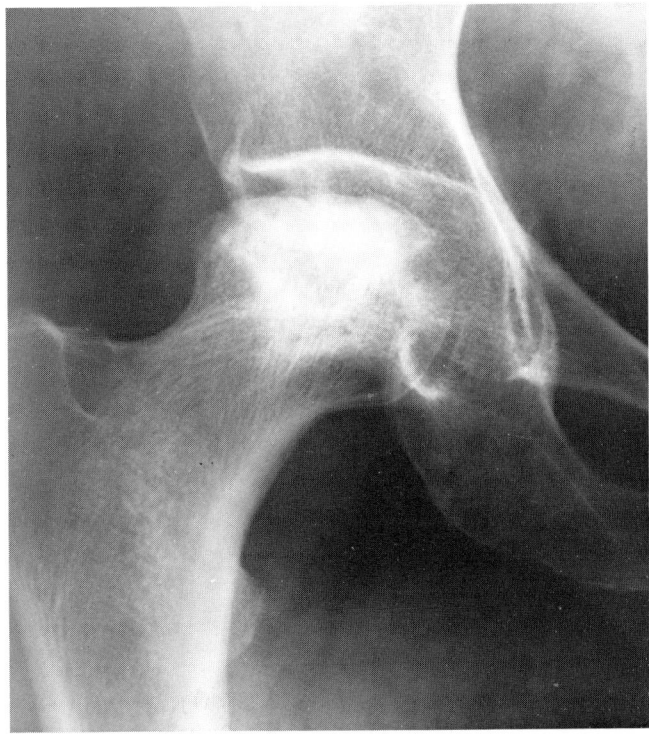

FIGURE 8. A microdamage microcosm. AP X-ray of the hip of a young adult male with a syndrome called "idiopathic aseptic necrosis of the femoral head". It has been widely assumed, but totally without proof, that the common and fundamental initial cause of this condition constitutes small-vessel vascular embolism that leads to an ischemic infarct of the femoral head.[6,25,394] That assumption grew out of earlier clinical experience — 1928 to 1965 — with femoral neck fractures, where an infarct certainly can occur that can lead to a subsequent aseptic necrosis of the femoral head that looks exactly like this one on X-ray examination, and that behaves subsequently in the same way. As one result, since about 1958 a considerable number of experiments have attempted to reproduce this lesion in lower animals by varied experimentally caused embolisms and other mechanisms of infarction without significant success. As another result, most discussions of the pathogenesis of the lesion report the above assumed fundamental cause as though it were proven fact, and their authors reject out of hand any other view.[158,394] The writer's clinical experience with and pathologic studies of the condition and its analogs eventually led ca. 1964 to a quite different view of its pathogenesis that was described in 1964 (*Workshop on Aseptic Necrosis of the Femoral Head*, NIH, 1964) and again in 1973,[241] but that was rejected by most others at the time. An exception was the Swiss rheumatologist, Zinn, who had the advantage of reviewing personally some of the writer's pathological material at Henry Ford Hospital around 1964. Since 1980, however, several other investigators, including Glimcher, have published independent findings that have rejuvenated that view (but without citing the earlier reports). It reads about like this. In this condition the primary common denominator is microdamage, which begins to accumulate in the spongiosa in the dome of the femoral head, usually along the plane of maximum shearing strains. That ultimately separates a roughly conical portion of the head, base up and point down as shown here, from the surrounding and underlying bone. That separation allows vertical pistoning of the conical portion, which can disrupt its fragile capillary blood supply and infarct it. Meanwhile, much repair activity arises in the spongiosa adjacent to the separated part, as well as vascular invasion of it. The repair and invasion activities include new bone deposition, both woven and lamellar, which has made both the conically separated part and the living distal bone next to it radiodense in this particular X-ray. Some of the author's specimens also revealed clear evidence of a series of revascularizations and reinfarctions of that part. The overlying articular cartilage remains viable since it obtains its nutrition from the synovial fluid rather than from the underlying bone. The subsequent clinical course is well known and not pertinent here, nor in dispute. Different things apparently can cause microdamage to begin to accumulate enough to initiate this syndrome. They include true infarction of the head after fracture of the femoral neck or after a traumatic hip dislocation, biological suppression of BMU-based microdamage repair by corticosteroids, immunosuppressive agents, or radiation damage, or, as here, cases of still unknown cause occur. However, and significantly, the condition does not occur in situations where true capillary or arteriolar embolism or is-

chemia is known beyond doubt to occur, including subacute bacterial endocarditis, occlusion of the major local arterial supply by trauma, thrombosis or arteriosclerosis, Buerger's disease, and the clinical syndrome named fat embolism that can follow multiple severe trauma. In the U.S., the cases in which a true original ischemia was not the initiating cause outnumber those postdislocation and postfracture cases where a gross capital infarct did form the primary cause. Hence, the somewhat stereotyped clinical and X-ray features of this syndrome comprise the final result of a variety of primary causes, exactly as the writer proposed in 1964 and 1973. The idea that embolism was the common primary event was simply another default judgment, made because before 1964 no other cause or explanation was even imagined, and it was also influenced by the work of the German pathologist and thinker, Axhausen[50] and by an American orthopedist, Phemister. While the microdamage thesis did provide a different explanation that, to repeat, has accumulated substantial independent support since 1980,[31,120,495,798] in 1964 it met the attack and destroy reaction, and, with another of Dame Fortune's minor ironies, from some of the same individuals who now support the idea, although for whatever reason they do not cite the 1964 to 1973 descriptions by its original proponent. Parenthetically, one likely reason the human syndrome has not been duplicated in animal experiments relates to a comment offered in the discussions in Chapters 7 and 9: it probably takes ~1 to 3 years for a given mass of lamellar bone to begin to develop significant amounts of microdamage and therefore to become at risk of a gross "stress fracture" if something impairs its repair. Exactly such time lags separate the initial fracture or hip dislocation from the initial clinical evidence of its onset in the typical human postfracture or postdislocation case of aseptic necrosis of the hip. The laboratory experiments referred to above seldom lasted longer than 6 months, which is insufficient time for fatigue-induced collapse of the spongiosa in the caput to begin to occur. Furthermore, to demonstrate that microdamage in its early forms requires the use of a simple but very specific technique devised by the author,[217,219] which no subsequent author discussing this problem has yet employed (while Wilde, Meunier, and Vignon have done such studies and shared their material with the writer, they have not yet published them). Accordingly, they could not have perceived any early forms of microdamage that may actually have existed in their material. As with the Charcot joints discussed in Chapter 11, if the above rationale is correct these lesions should cease to progress clinically if new microdamage production is reduced below its rate of repair. Deloading the hip by going totally nonweight bearing on crutches for some 6 to 9 months can do that, and since 1964 about 40 such hips in the author's hands all did arrest and heal. Some required subsequent arthroplasties because of the effects of loss of sphericity of the head, but none because of continued or recurring collapse.

Chapter 15

SOME DESIGN PROBLEMS IN HIP REPLACEMENT PROSTHESES:
AN ANALYSIS

"To solve biomechanical problems, begin by finding how nature did it."

(H. M. F., 1969)

ABSTRACT

Many failures associated with the femoral component of the current generation of hip endoprostheses used in humans result from a failure to duplicate the shear locking, load defocusing, and longitudinal support mechanisms provided by the intact living upper femur, and also from a lack of awareness of the role of the MES phenomenon and strain per se in setting the biologically tolerable limits of strain and microdamage in living lamellar bone. As a result, many current designs create load shunting and focusing mechanisms that concentrate local overloads in the bone, and that in turn lead to failure of the bone in mechanical fatigue. Those designs can also cause expansile bone drifts around the upper stem that gradually loosen a mechanical bond between bone and the device.

I. INTRODUCTION

In many fields of knowledge, a particular mode of progress occurred repeatedly. For a time, sometimes a long time, data and ideas of the most diverse types accumulate but seem to have little or no connection. The data and ideas can come from different levels of organization of the system or field, and from different times, places, and materials. Then one day, flashes of insight reveal some of the connections between them, and a new unifying concept is born (survival of the fittest[155]) or one previously known in another field is carried over to explain the problems of the field in question.[245,778]

Some of the fundamental unifying concepts of physics, chemistry, biology, astronomy, mathematics, evolution, economics, and psychiatry developed in that way, and only as examples.

So it is with the subject of this chapter, which weaves a partly finished fabric out of separate threads that come from static and dynamic mechanics and structural engineering, from the materials properties, composition, and structure of bone from the ultramicroscopic to the organ levels, from the growth, modeling, remodeling, and micro- and macrorepair mechanisms of that tissue in the domains that embrace both the IO and the organ, and from clinical observations accumulated over a span of 4 decades. This fabric is still incomplete (*a* truth is not *the* truth), but it betters its antecedents. It is another step towards even higher ones.

This chapter also contains another example of an IO-oriented mode of analysis of a contemporary problem, and of one use of a bit of biological wisdom:

The mode of analysis follows that outlined in Chapter 14.
The bit of wisdom appeared in the above aphorism.
The problem lies in clinical failures associated with the femoral component of past and present "generations" of hip replacement endoprostheses.

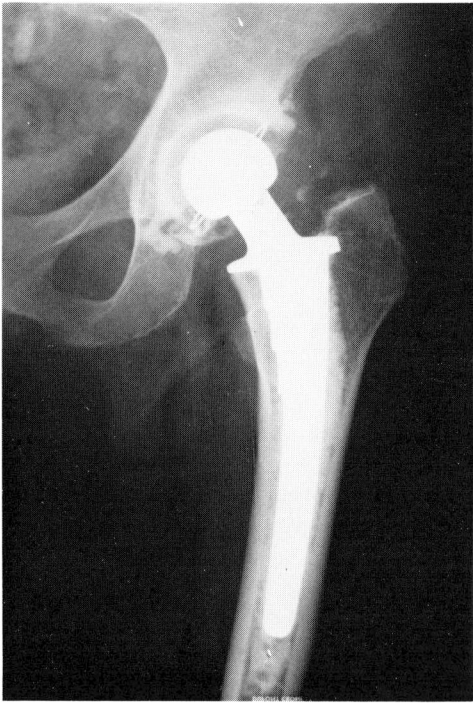

FIGURE 1. An AP X-ray of a total hip replacement arthroplasty done by the author in 1980, showing the placement and orientation of the acetabular and femoral components. The clear zone around the metal head or ball is the radiolucent polyethylene socket. The white material around it and the stem of the femoral component is the special cement, a polymethylmethacrylate mixed with a barium salt to make it visible on the X-ray.

The manifestations of the problem will be described first, then analyzed with the aid of material already presented in earlier chapters, and then some possible solutions will be described. It should be understood at the outset, however, that while this chapter outlines a trail of errors that were committed collectively in designing and using such devices, the errors could not have been foreseen by any reasonable means known to the author and that applies to the current problems as well as to the earlier ones that have become a part of medical history.

II. THE PROBLEM

A. Brief History

Figure 1 illustrates one of the current generation of total hip replacement arthroplasty systems in use throughout the world as a treatment for various kinds of painful and disabling arthritis of the human hip joint. It represents a stage of an evolution that began in the late 1940s when a French surgeon, Judet (d'Aubigne[46] has reviewed that development) caught the attention of the orthopedic community by replacing the femoral head in elderly patients who had sustained a fracture of the femoral neck with a ball made of plastic from which a stem protruded into the marrow cavity of the distal fragment, somewhat as in Figure 2. The patient began to walk immediately afterwards,

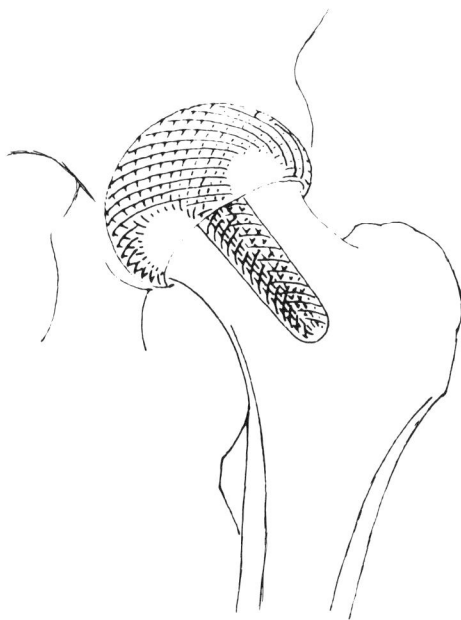

FIGURE 2. The original Judet hip prosthesis made of acrylic plastic replaced only the femoral head and left the socket intact, and it looked about like this.

whereas such hip fracture patients previously had to refrain from bearing weight on the injured extremity for months, and often the fracture did not heal even when it was pinned, leading to further problems if the patient survived the ordeal; many elderly patients did not survive it.

While the procedure helped many patients, in use it developed problems, too, and there ensued world-wide a phase of redesign and re-redesign as each newer design solved earlier problems but exposed different ones, often simply by lasting longer. In the 1960s an English surgeon, John Charnley, conceived and brought to reality the idea of replacing both the ball and the socket, and of fixing them in place with a special cement to obtain a good mechanical bond and transfer between device and bone of the mechanical loads that crossed the artificial joint. The procedure was usually done for various kinds of arthritis of the hip; it was a great advance and its descendants benefit tens of thousands of people annually around the world. Again, redesigns and re-redesigns followed and still go on, for while the present generation of prostheses provides a mean service life of ∼10 years compared to ∼1 year for the original ones, it also has revealed problems.

The original prostheses soon developed catastrophic wear and/or broke due to the materials of which they were made, their basic design, and to manufacturing methods. The current generation of devices do not wear out and rarely break. Rather, failures of the bone now provide the major problem, and failures on the femoral side outnumber those of the acetabular side. The current situation looks like this, in three parts.

Mechanically speaking, a typical femoral component implanted at operation has a ball, an angled neck, a base plate, and a stem. The latter fits down inside the marrow cavity of the upper femur, which is usually reamed to fit it. The base plate sits on the edges of the trimmed base of the femoral neck. A special plastic cement is often inserted into the marrow cavity of the femur before inserting the femoral component,

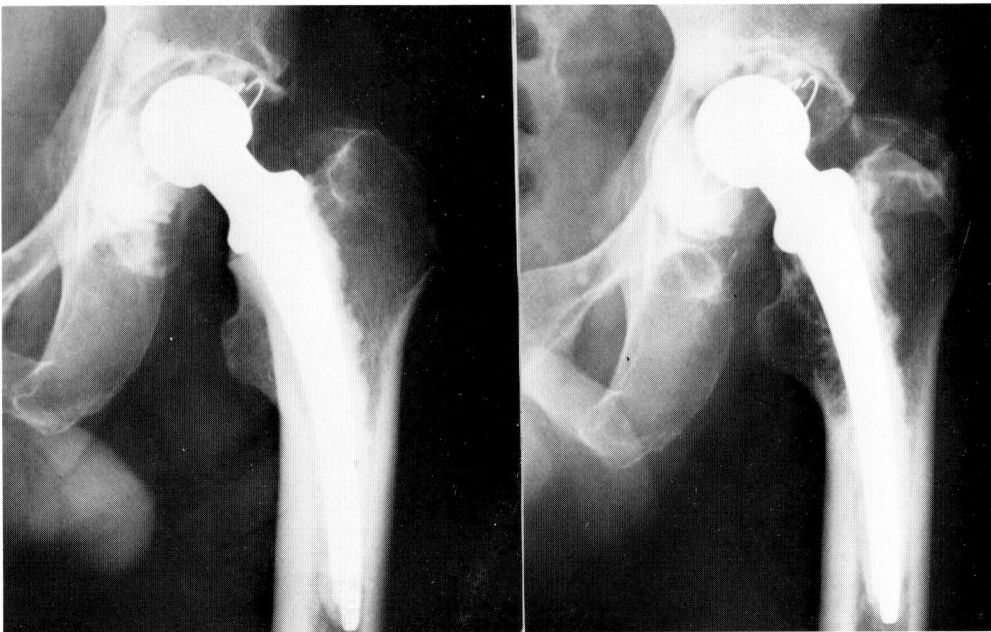

FIGURE 3. AP X-rays of a total hip replacement arthroplasty done on a man at age 67. *Left:* the X-ray taken 3 months postoperative. Pain was still present on activity. The base plate plane of this device normally inclines about 40° away from horizontal, and since it was inserted in some varus the base plate plane here is even steeper than that. *Right:* 3 years later the acetabular component has become loose. The stem of the femoral component has settled slightly into the medullary cavity of the femur due to resorption of the bone under the base plate. Much of the bone in the femoral calcar medially and beneath the base plate has disappeared and the lesser trochanter has become quite osteopenic. These problems progressed and led to a revision of this hip later. One can see here the shear skid, battering ram, and load focusing effects on the inner wall of the calcar that are described in the text. (X-rays courtesy Dr. Muryl Laman, Southern Colorado Clinic, Pueblo.)

and when it sets or hardens it provides a mechanical bond between the component and the bone over a large surface area. The acetabular component or socket is also inserted in a prepared bony bed and cemented in place in a way that duplicates the location and facing of the anatomical socket. The ball is then placed in the new socket, the wound is closed, and the patient is sent to the recovery room and then to the ward. Figure 1 illustrated those features.

Postoperatively, the patient begins to move the new hip and to walk on it within 1 day or so after the operation, and leaves the hospital walking with some support (crutches or a walker) within 1 to 2 weeks. Full weight bearing usually resumes within 1 to 3 months. The relief of preoperative pain provided by the artificial joint usually proves dramatic; previously stiff joints regain nearly normal mobility and the usual patient feels he or she has received a new lease on life, and in fact has.

B. Subsequent Course

In 2 to 4 years, X-rays of the new hip begin to show some loss of bone in the femoral calcar[303] beneath the base plate of the prosthesis, and evidence of loosening between the cement enveloping the stem and the surrounding femur, as well as between the cement and the stem of the device. That loosening progresses gradually but a bit more rapidly with each passing year. Minor discomfort associated with activity also appears and progresses gradually. Within 1 decade or more, about half of these hips will need to be revised, i.e., redone because of problems related to that loosening. Figure 3 illustrates such a problem. While many different things can go wrong with these devices and their use, some occur more often than others; the above sequence abstracts the most common mode of failure of the current generation of THR systems. So, there, is a problem.

Solving any problem can begin by trying to define it by a series of increasingly accurate approximations, each of which becomes tested and then revised and/or improved. When the problem is finally defined accurately and completely, the knowledge needed to do so usually also reveals how to solve it.

The present problem will be analyzed with the aids of hindsight, personal experience with it and related matters, and a knowledge of the skeletal IO. The analysis begins by trying to see how nature solved the same problem, which may sound odd since nature does not do total hip replacement (THR) arthroplasties. However, two matters require a comment first.

First, the previous generations of these devices had serious enough problems in the devices themselves that appeared soon enough after the operation that the typical clinician or engineer had little opportunity to see if the living bone objected to what the device asked of it, and, if so, how. As a result, experts in the field did not need to know much about the biological and biomechanical properties of that bone, and in fact (if in retrospect) did not account for them in their redesigns. That situation has changed, for the current failures reveal that current designs demand things of living bone that it cannot endure. It follows that bony failures will continue until someone identifies and corrects those unacceptable demands.

Second, in the following pages the reader will find some new terms that signify both real phenomena and useful, real, and even essential concepts. The terms include load defocusing and focusing, load shunting, the skid effect, shear locking, the stiffness lag, and unit load multiplication.

III. THE NATURAL SOLUTION

The following paragraphs will consider, in order, what the problem is not, what the normal hip achieves, and then how it does so.

A. The Problem is Not

First, the major problem does not lie in the manufacture and usage in vivo of an artificial ball and socket bearing. The metal on high-density polyethylene-bearing mechanism, buried in and lubricated by the normal body fluids, has proven surprisingly satisfactory. The bearing surfaces per se do not wear out in the time periods that have been involved so far, and their observed rates of wear suggest that they could probably endure for twice the length of time that they do currently, or even longer.

Second, the major problem does not lie in the chemical and immunologic acceptability by the living body of the materials used to make the components, nor in tumorogenesis. Some materials used in the past did create such problems, but extant methods of testing the biocompatibility of materials intended for implantation in the body have proven effective in detecting questionable materials before they are implanted in humans.

Third, the major problem does not lie in the strength and fatigue life of the materials in the components. It does form a minor problem (it formed a serious one before 1960), and if the present service life of these devices should somehow be doubled then fatigue-life problems in them could appear again.

Very simply, the major problem is that the bone cannot endure the demands posed by a normal lifestyle and the inanimate parts put into it.[631]

226 *Intermediary Organization of the Skeleton*

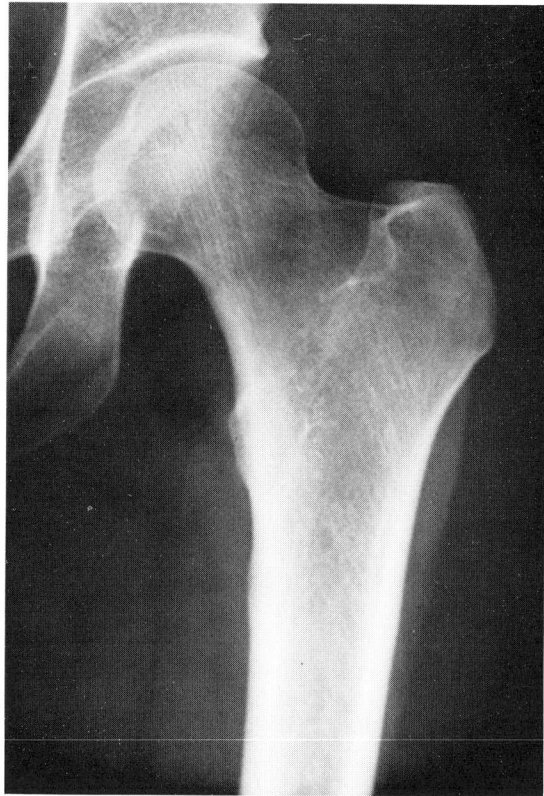

FIGURE 4. An AP X-ray of the normal left human adult hip of a male in his late 40s. Use it for comparison with the drawings in subsequent illustrations. (Courtesy Dr. Nick Alsever, Southern Colorado Clinic, Pueblo.)

B. What the Normal Hip Achieves

The natural hip shown in Figure 4 provides three major features.

The bearing — The normal bearing mechanism endures under normal mechanical usage for the human lifespan of some 75+ years, and with a probability of failure over that time span, and of any individual hip, of $p<0.001$, excluding of course the consequences of injury, infection, and neoplasms.

The loads — That bearing mechanism carries large loads that vary from moment to moment in their orientation, magnitude, and loading rate. Momentarily and not infrequently they can exceed five times the body weight. Those loads pass downwards from the socket, across the joint space, into and through the femoral head and neck, and into the intertrochanteric region, then the upper diaphysis, and finally the midshaft. Figure 5 diagrams a normal hip and certain features of its loading.

The support — The living bony structures beneath and supporting the actual joint surface also endure under those loads for the same lifespan, and with a probability of failure of the intact bone under normal usage and over the lifespan of $p \sim 0.0001$, again excluding the consequences of injury, infection, and neoplasm. That provision affects and includes the bony parts of the femoral head, the neck, the intertrochanteric region, and the femoral shaft.

Note Bene: Since the bony parts of a normal hip that endure reliably for the whole human lifespan nevertheless loosen and fail within ~ 10 years with $p \sim 0.5$ when they

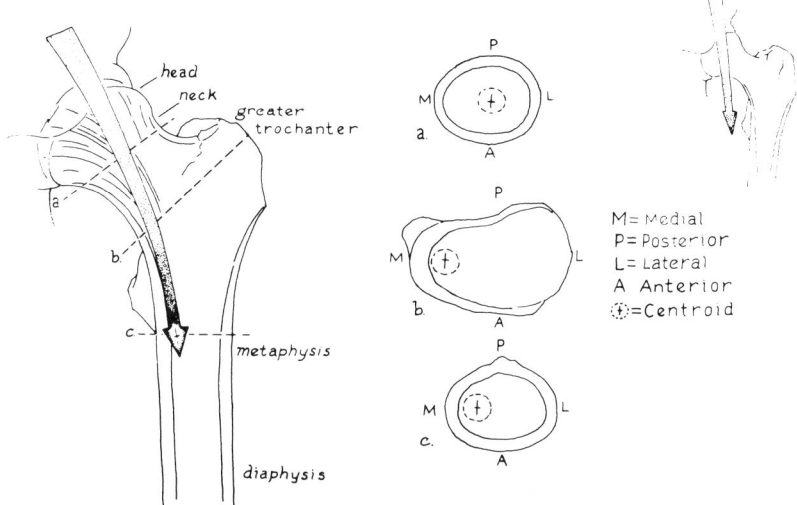

FIGURE 5. A diagram of the human left hip and upper femur as seen in AP view, with the major anatomical parts labeled. Compare to the previous figure. The large, slightly oblique, and slightly curved arrow represents the time-averaged load resultant transferred through the joint itself down into the upper femur. The dotted lines a, b, and c indicate the cross-sectional levels shown in the middle of the figure. Level b indicates approximately where the base plate of a typical hip arthroplasty component would lie. The upper right drawing illustrates the essentially vertical load resultant associated with standing and leisurely walking.

support artificial heads and necks, the designs of the latter *must* deliver challenges to living bone that saturate or disable its normal maintenance and adaptive mechanisms, and its SOS mechanisms too.

Up to the present, few students of these matters would find anything novel in the argument, but what follows could change that. From characterizing briefly *what* was the natural way to transfer loads from the natural hip to the midfemur, attention turns next to *how* the author conceives that that was done. The text will then compare certain details of the typical artificial hip to the natural one, and that comparison (see Note 3 in Chapter 2) will expose some phenomena that can explain at least some of the reasons for the problem, and at the same time suggest how to solve it.

C. How the Normal Hip Achieves It

The average load resultant — Figure 5 diagrammed a normal human hip as seen on an AP X-ray. The large arrow represents the orientation and location in tissue space of the compression load resultant carried by the hip and upper femur during a vigorous activity over 1 month. That resultant averages over time or integrates the orientations and magnitudes of all of the different loads, major and minor, carried by the hip joint during all of the sitting, walking, climbing, running, and jumping activities over 1 month. The inset on the right indicates the different and basically vertical load resultant that can occur during standing and ordinary walking. It ignores the effects of medially pulling muscles that act mostly during more strenuous activities and that the therefore more oblique left-hand resultant includes.

The left-hand figure diagrams the greater loading situation (GLS) mode henceforth, and the right-hand one the lesser loading situation (LLS) mode.

The normal load-transfer situation — As the vertical load passes down from the joint into and through the bone, spongiosa carries most of it in the femoral head, both compacta and spongiosa share it in midneck at *a* (with respect to Figure 5), and at the base of the neck, section *b*, mostly compacta carries it as it does entirely at section *c* through the upper part of the shaft and at all lower sections.[303] Thus, the major structural support to the vertical loads on the hip constitutes spongiosa within the head and compacta at the base of the neck, and that load transfers from spongiosa to compacta in the intervening neck and the lower part of the head.

Amount and trajectory of vertical load approximately match the amount and distribution of compacta at the base of the neck — Comparison of the location of the load resultant at Figure 5b with the distribution of the amount of compacta around the cross-section at that level reveals that the much thicker medial cortex or calcar lies where most of the vertical load passes down through the section (M), while the lateral cortex (L) carries a small part of it or even a tension load, in which case a *neutral axis* for flexure will exist within the section.[241,272]

The inwaisting function — The inwaisting of the femoral neck and of the proximal femoral shaft in Figure 5 mean the vertical compression loads on the compacta in those regions generate an inwaisting force of cantilever origin as described in Chapter 7, since vertical loading on an inwaisted shape always generates a cantilever-origin inwaisting force. As described in that chapter, that inwaisting force implies some kind of neutralizing outwaisting or bursting force.

The normal shearing loads — A compression load on a structure shortens it slightly which represents a compression strain, and that generates a resisting compression stress in the material of the structure, as shown by the vertical arrows in Figure 6.

However, that load also always generates a shearing strain and a resisting shearing stress in the material that becomes maximum at a 45° angle from the line of action of the compression load.[202,241,599,822] Those shearing stresses and strains result from a corresponding shearing load within the material that comes from the external vertical compression load. If the material is relatively weaker in shear than in compression, then a compression overload will make it fail in shear at 45° to the line of action of the compression load, as in the middle inset of the figure.

If the material has a grain, i.e., if it has anisotropic materials properties as lamellar bone does, and if the line of action of the compression load parallels that grain, then the orientation of the above maximum shearing strain and stress tilts from the 45° angle to align more closely to the line of action of the compression load, as in the right-hand inset of the figure.

Note Bene: Those shearing strains and stresses and the shearing loads they imply always arise whenever a rigid structure carries a compression load. Shearing strains and stresses also arise under tension, flexural, and torsional loads, but excepting tension they tend to align differently than when uniaxial compression causes them. Consequently, under both GLS and LLS loading modes the vertical compression loads carried by the normal hip also generate shearing loads and corresponding shearing strains and stresses in the cortex at the base of the femoral neck, as the shearing arrows in Figure 7 show for both loading modes. As a relation one might write the above thus:

$$\text{compression load} \rightarrow \text{corresponding shearing load} \qquad (1)$$

The normal shear-locking and load defocusing mechanisms[267] — A normal adult bone is "biomechanically adapted", meaning that during growth the operation of the bone modeling laws described in Chapter 7 so adapted the configuration of the bone to its typical mechanical usage that over all of its surfaces and under all of its major usages its strains seldom attained the lower limit of the MES for bone strain, and for any kind and orientation of nontrivial loading that it had to carry with any frequency.[249,267]

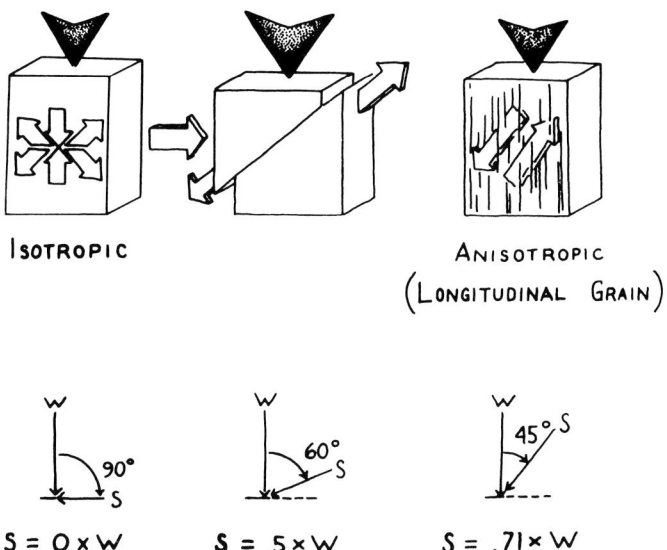

FIGURE 6. *Upper row:* at the upper left a vertical compression load descends on a structural block. It causes strains that generate a resisting internal compression stress (vertical facing arrows), and associated shearing stresses aligned at 45° to the line of action of the vertical load. Increasing the vertical load as in the middle inset can cause it to fail in shear, as shown. When the structural material has a longitudinal grain as on the right, then the shearing strain and stress tend to align a bit more parallel to the grain than the 45° shown on the left. *Lower row:* given a vertical load, w, applied to some structure along the line of action shown by the vertical arrow, then the shearing load it generates equals w cos θ, so at a right angle or 90° to its line of action the shearing load is zero, at 60° to its line of action it is half the vertical load, and at 45° to its line of action it is 71% of the vertical load. The base plates of most hip femoral components incline at 40 to 50° from the vertical load on the human hip so vertical compression loads create large shearing loads, the line of action of which parallels those base plates.

That MES limit applies not only to tension and compression strains, it applies to the corresponding shearing strains also. Under a vertical compression load the base of a normal femoral neck normally carries a shearing load that corresponds to the compression load resultant shown in Figure 7 and that aligns it about 45° to it. Normally that shearing load diffuses or defocuses over the entire cortical cross-section shown in Figure 5b. Accordingly, each square millimeter of cortical cross-section area at the base of the neck carries its own share of that load. It does so because each unit amount of the bony material bonds firmly to all of its neighbors, and that bond in shear makes the material a rigid solid rather than a fluid. In effect, that bond "locks" each layer of the material that parallels one of the shearing arrows in Figure 7 to its neighbor in the adjacent plane, so the two cannot slip past each other in shear. The effect resembles gluing a throw rug to the underlying floor. Hence, a shear locking mechanism that provides rigidity and stiffness. It is the particular and basic materials property that distinguishes the liquid from the solid state that is a normal feature of bone. In this situation, while most of the vertical load on the hip is carried by the compacta in the calcar region, the corresponding shearing load is carried by or defocused over the compacta in whole section shown at Figure 5b, including any part of it that might be loaded

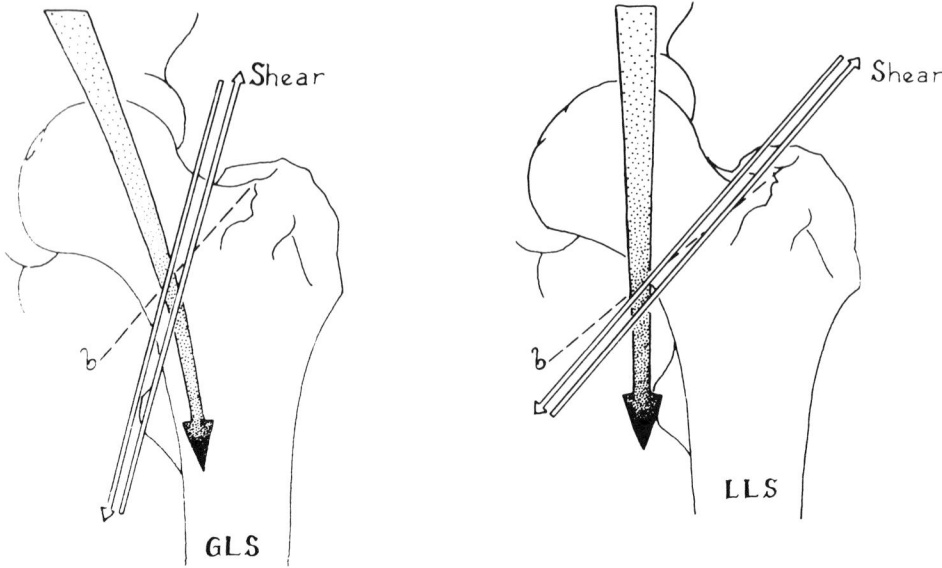

FIGURE 7. The left drawing shows the approximate orientation of the maximum shearing load, strain and stress in a normal human hip under the greater loading situation, while the right drawing shows those same shearing features for the lesser loading situation. As the text indicates, those shearing features *always* accompany vertical compression loading, and normally the loads defocus over all 100 of the cortical cross-section area units shown in Figure 5b. As the previous figure indicates, those loads are large. For a 400-lb vertical load on the hip on the left, the shearing load in the plane of the shearing arrows would approximate 71% of that, or ~ 285 lb.

in tension due to flexure. One could well say it is "defocused". That might be written thus:

$$\text{load} \;\rightarrow\rightarrow\; \text{defocused shearing load} \quad\quad (2)$$
$$\uparrow$$
$$\text{shear lock}$$

where defocusing means spreading a load out over as much of the supporting structural material as necessary to keep strains everywhere below the MES.

For convenience, say that the cross-section area of the cortex at the base of the neck that carries that shearing load, which is the cortical area shown in Figure 5b, contains 100 area units (in life it might approximate 6 cm²). Then the shear-locking mechanism causes each of those 100 units of area to carry about 1/100 of the total shearing load at the base of the neck, so none of those units would develop shearing strains equal to or larger than the MES for lamellar bone shear. The total number of such cross-section area units in a particular femur would depend of course on the diameter of the cross-section and the cortical thickness; the particular value of the MES for shear would depend on the relative orientations of the shearing load and the local grain of the bone.

That describes a "load defocusing mechanism" that, to repeat, spreads a particular load over as many bony domains or unit areas as needed to reduce the strains in each and all of those domains below the MES. The mechanism exists because each unit amount of that compacta locks rigidly to its immediate neighbors in a way that provides ultimate strength in shear of some 8000 psi, and in compression of some 18,000 psi. (Table 3 in Appendix I provides conversion factors for English and metric units.) As a result, to make the normal structure fail in shear one must make every one of those 100 area units in any one layer of the material slip past its neighbor in the adjacent layer, and all at the same time. That summation of the strengths of each small domain in the structure provides the great bulk strength of the intact bone.

In sum, the normal hip defocuses its superimposed joint loads, including particularly the shearing loads, through sufficiently numerous area units of bone on cross-section to keep the three principal strains in each unit or cubic millimeter of compacta below their respective MES values. The shear lock provides most of that load diffusion by bonding adjacent bony domains together so that each unit volume of the bone shares any load carried by its neighbors. Later on it will be shown what can happen when that shear lock is broken.

A similar defocusing mechanism will spread any transverse shearing loads due to torque of the neck on its base over a comparable number of bony unit areas and domains, so the shearing loads of any torque (and of flexure too) are also defocused in the bulk unit formed by the normal femoral neck, intertrochanteric region, and proximal femoral shaft.

Repair of microdamage — Chapter 7 reported that a major purpose of the mechanical MES in lamellar bone appears to lie in helping to place enough structural material in every part of every bone to keep the strains that can produce fatigue phenomena, i.e., microdamage, below the MES during normal usage. In time, however, some bone microdamage should always occur under normal mechanical usage and even in biomechanically well-adapted bones. As noted in 1963,[824] in 1966,[230] and 1973,[237] one function of the BMU-based remodeling mechanism should consist of repairing such microdamage before it can accumulate enough to threaten the structural integrity of an intact bone.[252] In the human hip and the supporting femur, if that mechanism becomes incompetent then complete fatigue fractures can and do occur. They can follow local treatment by ionizing radiation,[587] treatment by certain drugs such as the diphosphonates,[208,209] and immunosuppressive agents, infarction, and in many kinds of osteomalacia.[804] Idiopathic cases also exist. Further, when some repeating event sufficiently overloads a perfectly healthy bone often enough, then it can fail completely in fatigue and within less than 1 month, as Rutishauser demonstrated more than 30 years ago in an in vivo situation[657] and as Carter has more recently reviewed in laboratory situations.[822]

Structural grain and load orientation — In all known species of bony vertebrates, the grain of lamellar bone always aligns parallel to the major compression and tension loads carried by the bone, which means that it also aligns parallel to the major corresponding strains. As one result, that grain also aligns perpendicular to the major transverse shearing loads and strains caused by torque and flexure. Due to that grain, lamellar bone has much greater compression and tension strength and stiffness parallel to its grain than across it. Since that grain is established and fixed by micromodeling during the initial deposition of any bone, and probably under the control of the local dynamic strains during its deposition, one should infer that that fit of grain orientation to strain orientation is design-purposeful, that it represents the optimal arrangement for the provision of adequate stiffness and fatigue life by the living lamellar bone as a structural material. In short then, lamellar bone grain fits the orientation of the loads it carried during its deposition.

Finally, in healthy bony skeletons one does not observe the following situations:

1. Significant (i.e., >MES) tension or compression loads or strains perpendicular to the grain (which would represent tangential or hoop loads).
2. Significant (i.e., >MES) shearing loads or strains parallel to the grain (which usually occur under flexure, and some under torque).
3. Unneutralized, significant (i.e., >MES) bursting loads arising in the marrow

Table 1
MAJOR FEATURES OF THE NORMAL HIP

1. All loads on it produce corresponding shearing loads.
2. The materials shear-lock produces the state of rigidity in bone.
3. The shear-lock defocuses otherwise harmful load concentrations.
4. An MES value exists for each of the principal strains.
5. The biomechanically adapted femur nowhere strains >MES.
6. Strains <MES produce minimal amounts of microdamage.
7. The remodeling mechanism repairs microdamage in normal amounts.
8. The normal grain aligns so as to best use the anisotropic mechanical properties of bone in relation to the orientations of its loads.

space, with their corresponding resulting tangential or hoop tension loads and strains.
4. Finally, recall from Chapters 3 to 5 and 7 that the structural changes that adapt the configuration and size of living bone to any changes in its mechanical usage cannot occur efficiently in normal human adults because, with some few and usually RAP-induced exceptions, effective structural macromodeling occurs only in growing skeletons.

Table 1 summarizes some of the above material.

IV. MAN'S SOLUTION

This analysis will consider the problem from the viewpoints of strain and fatigue rather than stress and strength, so it may take some readers time to understand it. The strain-fatigue viewpoint exposes some kernels of the problem and also suggests some solutions that stress-strength analyses have not.

Figure 8 diagrams a THR arthroplasty as seen in AP view, e.g., from the front. The following facts apply to it, although they may arise in a different order than the following description, depending on circumstances that will become clear.

A. Base Plate Fit

The base plate sits atop the base of the neck, which the surgeon cuts off to remove the natural head and neck at the time of surgery, using a power saw or similar means. The cross-section area of bone lying under the base plate has the approximate size and shape shown in Figure 5b, so most of the earlier-mentioned 100 area units of bone should still exist to carry the superimposed vertical load. The surgical procedures are still relatively crude, and the actual fit at the end of the operation more likely resembles situations shown in Figure 9, where high spots or asperities with small individual areas may prop the base plate slightly above the plane of the surface cut and/or where small malalignments or tilts of the base plate plane on the underlying bone surface can concentrate the whole vertical load on very small fractions of the total bone surface. The black dots within circles in the figure indicate such regional overloads, and typically their contact area with the base plate will total less than 1% (1/100) of the whole bone cortical cross-section area. Such mechanical misfits are probably the rule rather than the exception. In other words, poor mechanical fitting → local unit-load overloads.

B. Effects of Local Unit Load Multiplication

When the above factors apply, then the whole vertical load on the hip will descend through the base plate and focus onto those drastically reduced areas of supporting bone, to multiply their unit loads in proportion, i.e., to make them 100 or more times larger than normal. Since fatigue damage apparently begins to accumulate catastroph-

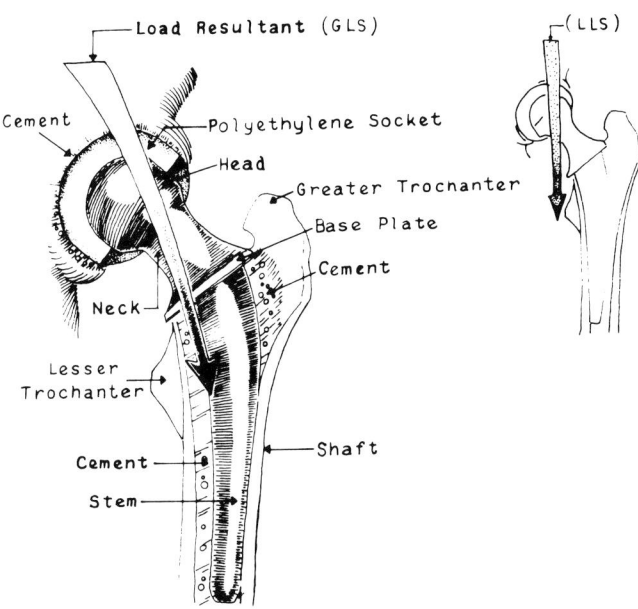

FIGURE 8. Drawing of a cemented-in THR arthroplasty from the front, showing the essential relationships and the nomenclature.

ically in fresh wet human bone when its loading cycles within a range of only about two times the upper limit of the MES,[110-113, 260] those overloaded small bone regions under the base plate should certainly develop rapidly increasing burdens of microdamage. If that bone is alive, then the local microdamage will enable intense BMU-based local remodeling, which always begins with resorption (remember, ARF). Bone formation will begin only 1 month or so later, and for straightforward reasons it will need a further 2 to 6 months to become structurally effective (to become effective it must mineralize sufficiently to become approximately as stiff as the surrounding bone and that "stiffness lag" takes months to resolve even in normal bone).

If the patient receives medications and/or has diseases that prolong the σ for remodeling and/or for completion of mineralization to 1 year or more, which certainly can occur in many of the patients who need these operations,[503] then the formation phase may not become structurally effective and the stiffness lag may not disappear until 6 to 18 months after the resorption phase begins. In that case, considerable net local loss of bulk structural stiffness and strength will accumulate due to uncompensated or sluggishly compensated resorption. Accordingly, that weakened region collapses, and that transfers the part of the base plate load it carried to adjacent asperities or regions, where the above processes may simply repeat themselves. That can create a migrating locus of chief support of the device.

A competent RAP that develops after the operation will accelerate the above compensatory processes, but it can also increase the remodeling space, which further weakens the involved local bone. If a RAP does not develop, then the microdamage still continues to accumulate unrepaired. In summary, and as first described in 1966[230] and again in 1973,[241] the following applies, where MDx signifies microdamage:

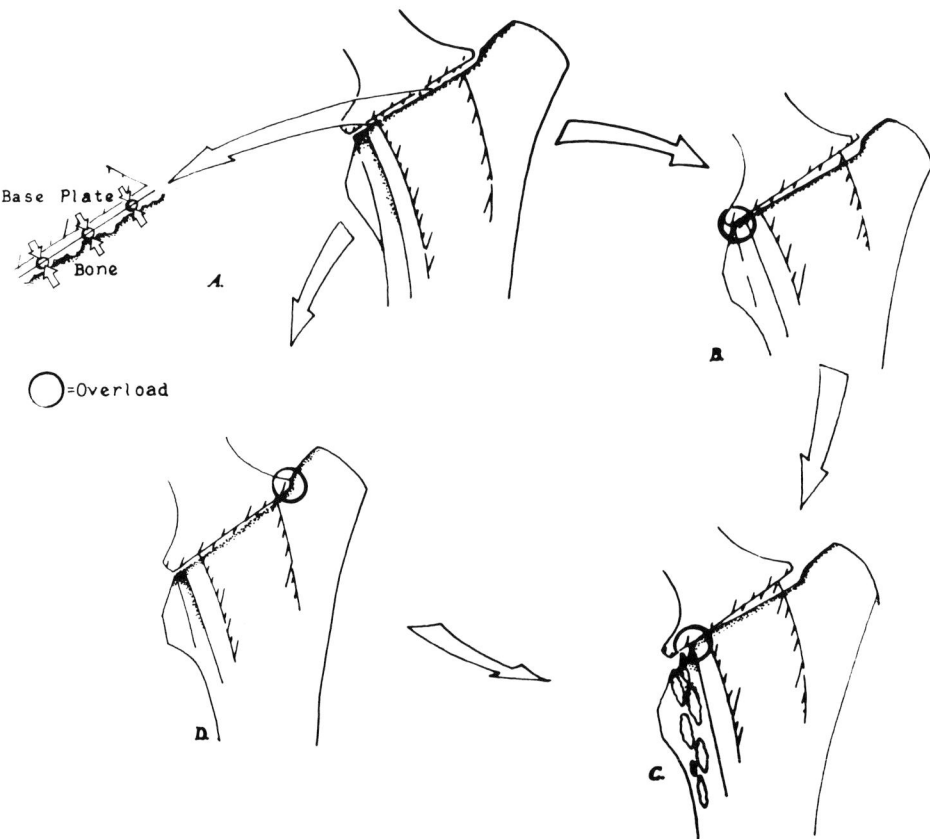

FIGURE 9. On the left, asperities on the bone surface beneath the base plate can prop it up slightly so they can carry the whole load on the hip. That greatly multiplies their unit loading. At B and D, even slight malalignments of the base plate plane relative to the cut femoral neck plane can focus the total vertical compression hip load onto small regions of the underlying supporting bone, which multiplies their unit loadings. At C, when microdamage accumulates rapidly under a calcar that is locally overloaded by a base plate malalignment or other means, the resulting intense BMU-based bone remodeling makes it very porous, which further weakens it, which in turn accelerates the production of further microdamage and allows a certain amount of crumbling. Figure 3, right, illustrated an actual example of that sequence. The main locus of vertical base plate bony support then moves laterally to the neighboring uninvolved bone, where the process then repeats itself.

local unit overloads → increased local MDx

increased MDx → increased remodeling

→ more fragile bone

→ increased remodeling space   (3)

→ decreased structural support

→ crumbling, collapse of overloaded bone

→ migrating locus of chief support

Partly because of the above processes, within some months after the operation the location of the major transfer of the vertical mechanical loads from device to femur can change from the bone under the base plate to the inner wall of the femur that envelops the stem, so that situation is discussed next. For any designs that lack an effective base plate support, whether by intent or otherwise, the following features can appear immediately.

C. Differential Longitudinal Stiffness

Cementing the stem of the prosthesis into the marrow cavity as shown in Figure 8 provides an initial mechanical bond between the quite irregular adjacent cortical-endosteal bone surface and the cement, and between the latter and the stem of the prosthesis. As one result, during weight bearing a sizeable fraction of the vertical hip load, rather than passing from the base plate to the underlying cortex at the base of the neck, travels instead down the stem of the device, though the surrounding cement and thence into the cortical-endosteal bone of the femoral shaft, much as shown in Figure 10. A variety of strain gauge studies have shown this, examples including those by Crowinshield, et al.,[142,143] Harris,[352] Radin,[442] and colleagues. The metal stem is usually some tenfold stiffer than the surrounding compact bone, so under a superimposed compression load as in the left-hand insets of Figure 10, and in the lengthwise interval between the base plate and the tip of the stem, *the bone shortens more than the stem,* as the dashed horizontal index marks indicate. While those motions or strains are small, they are real. At first, elastic deformation of the enveloping bone and cement probably accommodates them but they overstrain the mechanical bond between stem and cement in vertical shear so it gradually cleaves or separates. As that up and down sliding motion recurs at each step the person takes, that cleavage propagates upwards towards the base plate, loosening the bond between stem and cement and/or cement and bone. In summary and briefly: different stiffness of bone and device → increased shearing loads at their interface.

The above developments plus those at the base plate mentioned earlier can now bring to the fore a third effect (again, in particular cases these events may occur in different order).

D. The Cone-Piston Effect

The bulk shape of the stem-cement unit (or of the stem alone in some more recent cementless designs) is conical as Figure 10 suggests. When the above processes sufficiently loosen the mechanical bond between the component and the bone, the device pistons up and down even more inside the bone during normal activity, which in effect drives a cone into the tapered marrow cavity of the upper femur, as shown in the right-hand insets of Figure 10. That flares the upper part of the cortex outwards relative to the lower part, also as shown in the figure. That outflare occurs because no effective hoop constricting force exists around the perimeter of the bone immediately under the base plate to limit it, somewhat as shown in Figure 11.[267] The natural hoop constriction, the lower part of the neck, was removed at operation by cutting the neck across at its base, a process that should also create local microcracks that run vertically down into the bone from the cut surface, that break the circumferential tension bonding in the bone that normally exists there,[658] and that the cone-piston effect tends to accentuate.

If the operation and/or the stimulus of the earlier-mentioned events should create an effective local RAP, that can enable a certain amount of local bone modeling, even in elderly adults.[267] As Chapter 7 indicated, compacta responds to the kind of outflaring flexural strains shown in Figure 11 by developing periosteal formation drifts and endosteal resorption drifts. That drift combination increases the outside and marrow cavity diameters of the bone, as shown in Figure 10E, and it occurs in many of the clinical bone failures of these devices. The tangential hoop tension strains that must

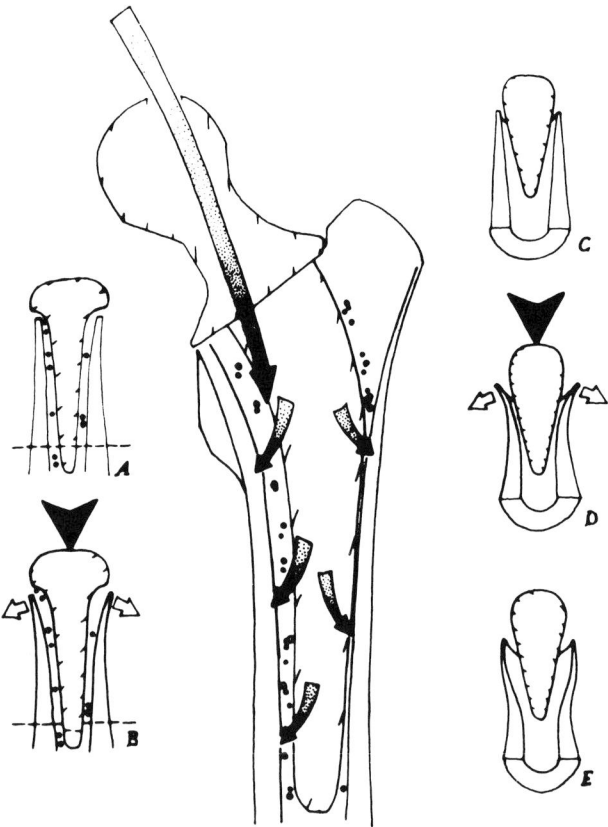

FIGURE 10. *Center:* a typical cemented-in hip femoral component transfers much of the vertical compression load on it through the stem and obliquely down into the surrounding cortical-endosteal bone. Because the stem is much stiffer than the bone, under vertical loads a vertical shearing strain arises in the plane of the interface between the stem and the cement, and in time that leads to a cleavage between stem and cement. That cleavage then allows increased vertical piston-like motions of the very stiff stem within the less stiff surrounding bone and cement. The left insets illustrate the effect, via the dashed horizontal index lines. When that situation develops, then as in the right-hand insets, the conical shape of the physical unit formed by the stem-cement combination causes an outflaring flexure of the upper cortex, meaning the concave-tending facing is centrifugal rather than centripetal. If a RAP also exists locally, that flexure can evoke modeling drifts that expand the outer and marrow cavity diameters of the upper femur, as in E, which loosens the upper part of the device. The next figure explains that outflaring flexure in more detail.

accompany that outflaring cortical flexure have also been measured directly both in living subjects and in the laboratory.[352,442,489] That expansion of inner and outer diameters completes the loosening of any mechanical bond between the proximal stem and the adjacent cortical-endosteal bone.

The lower of the left-hand insets in Figure 10 also diagrams this dynamic expansile strain of the upper femur when the conical prosthesis is driven down into it.

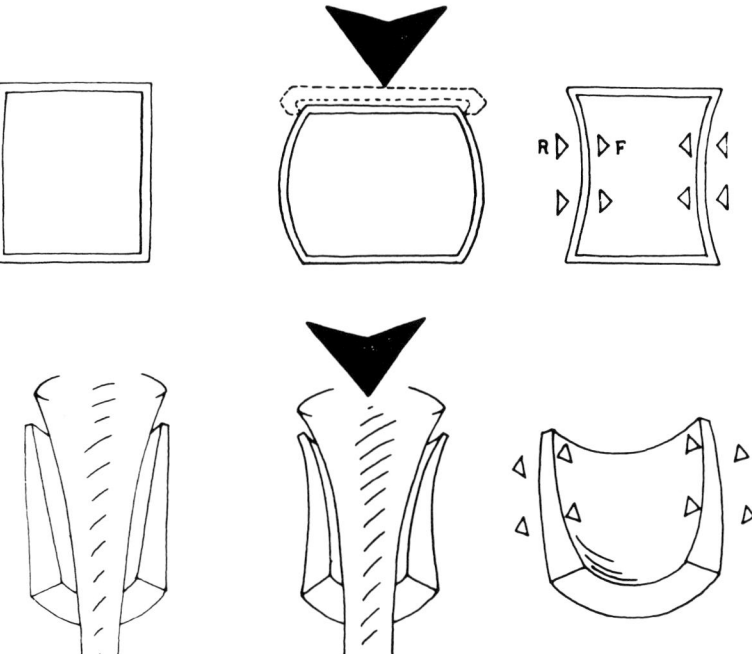

FIGURE 11. *Top row:* a vertebral body seen from the front and filled with spongiosa as on the left will shorten under a vertical load as in the middle. Since the vertebral endplates are stiffer and stronger than the spongiosa in the middle of the vertebra, they act like a constricting hoop or clamp, as the dotted part of the figure suggests, that prevents lateral expansion of the endplates. No such hoop effect restrains the compacta at midvertebra, however, so under the pressure of the compressed spongiosa within, it bulges outwards as the middle diagram indicates. The concave-tending facing of that flexure is centripetal, e.g., it points to the marrow cavity, so surface drifts arise that move the cortex towards the marrow cavity to inwaist it as the small arrows in the right-hand drawing indicate and as was described in Chapter 7. *Bottom row:* when that upper hoop-like constricting restraint is removed, as it is at the base of the femoral neck when the surgeon cuts it off, then because the thick compacta below provides the same hoop-like restraining function below, the same vertical piston-like load on the internal contents of this bone now causes an outflaring flexure of the upper cortex, e.g., one in which the concave-tending facing of dynamic flexure now faces outwards or centrifugally rather than inwards. Accordingly, if that evokes any bone surface drifts, they will be ones that expand the outside and marrow cavity diameters, as at the lower right.

E. The Shear-Shunt and Skid Mechanisms

In all but one of the THR femoral designs through 1983 that had one, the baseplate plane inclines from the horizontal plane by about 45°, as shown in Figure 12, and the underside of the base plates are smooth and flat, as shown in the upper left diagram in Figure 13. Referring back to Figures 6 and 8, it becomes clear that a 45° base plate plane closely parallels the orientation of the major shearing load caused by the vertical load on the hip.[241,267] Because the underside of the base plate is smooth rather than shear-locked to the underlying bone, it becomes a skate or skid that performs a highly injurious shunting of the total normal shearing load into another load carried by another location in the bone.[267]

In this particular "shear-shunt mechanism", the vertical load-induced shearing load tends to drive the prosthesis down and to the left, and that load focuses on the upper

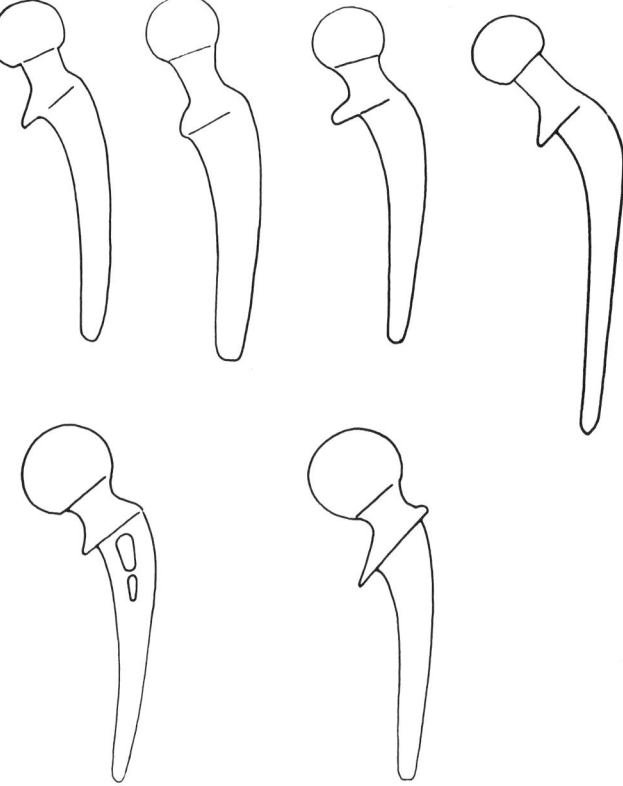

FIGURE 12. These diagrams illustrate (to scale) four current total hip replacement femoral components in the top row, and two often-used femoral head-neck replacement prostheses below. The latter are used to replace the femoral head but are sized to fit properly within the normal socket. Note the inclinations of the base plates, all of which are smooth and flat on their undersides. Compare also to Figures 1 and 3.

inner surface of the calcar, as a leftwards-directed compression load, somewhat like a battering ram, as shown in the longitudinal and cross-sectional views in Figure 14.[267] As a result, only a few unit areas of that endosteal calcareal bone have to carry a compression load intended in nature to be a shearing load shared by 100 unit areas of cortical cross-section at the base of the neck. That "unit load multiplication" catastrophically overloads those few calcareal areas so microdamage should arise in amounts that overwhelm the vital L_2-domain remodeling mechanism that normally repairs it. That would cause that part of the calcar to begin to resorb and to collapse, as shown earlier in Figure 3. With full respect to colleagues who attribute this resorption to an absence of a normal vertical compression load on the part, the author believes that at least in part (and likely for the most part) it is caused by the above medial battering ram effect of the medial stem, driven sideways and medially against the inner wall of the calcar, the natural grain orientation of which is not suited to endure it. The essence of the above might be written thus:

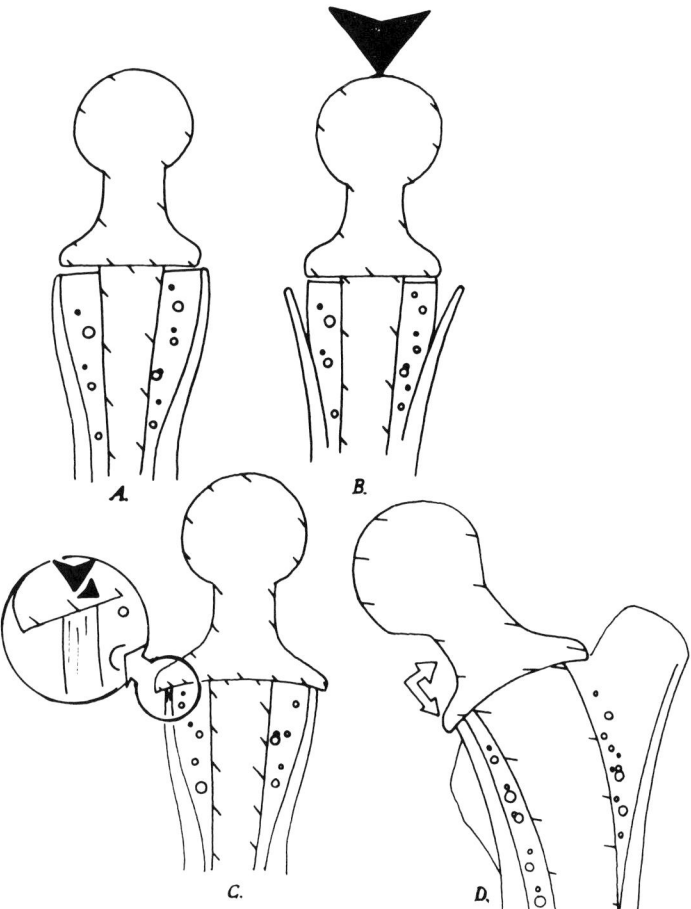

FIGURE 13. *Top row:* from a lateral view this again shows the cone-piston effect and the outflaring cortical flexure it causes under a vertical compression load. *Bottom row:* if the base plate is wide enough to overlap the femoral cortex, and is also concave on its underside to generate a necking-in or inwards or centripetal force on the cortex when under a vertical compression load, that can reduce or prevent the outflaring cortical strains, and the centrifugal drifts that can loosen the device.

shear skid → shear shunt

shear shunt → unit overloads on inner upper calcar (the battering ram effect) (4)

unit overloads → increased MDx production

increased MDx → local structural collapse

To recapitulate, the lack of a shear lock between the underside of the base plate and the femoral compacta on which it rests shunts and focuses the total shearing load that parallels the inclined base plate plane — and normally that load is large, about 70% of the vertical load — onto a small region of the upper inner femoral calcar as a direct compression load directed outwards, not unlike a battering ram. That local unit load multiplication overloads the small calcareal region supporting it, so it begins to accumulate microdamage, to crumble, and to become slowly resorbed. Since the calcar

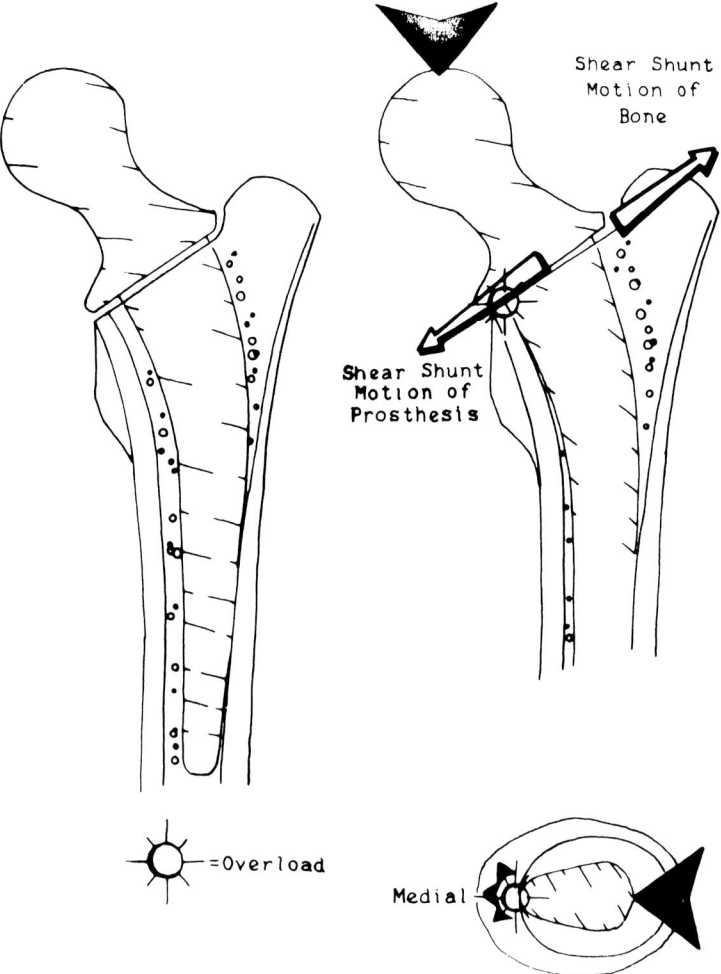

FIGURE 14. When the device on the left carries a vertical load as on the right and when the underside of the base plate is a smooth skid, then the 45° shearing load generated by the vertical comparison load essentially parallels the plane of the base plate. That shearing load equals about 71% of the vertical load here, so a 300-lb vertical load on the head will cause a 210-lb shearing load on the base of the neck that parallels the plane of the base plate. That tends to move the femoral component down and to the left on the underlying bone. The chief thing that restrains the device from moving down and to the left is the contact between the upper inner surface of the calcar and the upper medial surface of the stem, as indicated by the black circle and the cross-section at lower right. That represents a region of shunted and focused load where the unit loading becomes multiplied many times over its norm, and where that load now acts in a direction that the grain of the calcareal bone is not fitted to accept.

normally carries a large fraction of the vertical load on the hip, even with a slightly loose stem, when it collapses under the base plate the adjacent, but more lateral, supporting bone will carry it as well as its own share, which overloads it, while the cone-pistoning effect increases. As the upper regions of the calcar crumble under the battering ram effect, the lower ones assume that load, and so on. Hence, two causes of a migrating locus of chief structural support. The above phenomena also increase toggling, as shown later.

F. The Outflaring Cortical Flexure

When made to undergo repeated dynamic flexural strain equal to or greater than the MES, the surfaces of compact bone will drift towards the concave-tending facing of that flexure, as described in Chapter 7. In most adults, a bone cannot readily change its macroarchitecture in response to some new, threatening situation, while what modeling adaptations do occur around the femoral component of a prosthesis usually make the situation worse instead of better, for as just described they react in a positive feedback mode.[267] That occurs because in a normal femur the compacta at the base of the neck is "closed". In the presence of a hip prosthesis, however, the operation opened the upper end of the cortex, as in Figures 11 and 13, so lacking an effective hoop restraint or constriction at its open end, under the pistoning effect of the femoral component it develops exactly the opposite flexural patterns from those that occurred in the vertebra discussed in Chapter 7 and shown in the upper insets of Figure 9. That outflaring flexural pattern in the upper femur evokes the earlier-described outwaisting drifts that tend to loosen any mechanical bond between prosthesis and bone.[267] In essence then, when centrifugally concave-tending drifts arise: if concave-tending flexure faces centrifugally, if it equals or exceeds the MES for modeling, and something enables local modeling, then cortexes drift centrifugally and the stem of device loosens.

If nature cannot remedy that problem then man should, and that solution seems direct. Making the underside of the base plate concave, as shown in Figure 13C and D, will allow the vertical load to generate an inwaisting force component to neutralize or limit the outflaring local flexure of the compacta, and the tangential hoop expansion strain that accompanies it as shown in B.[267] Some manufacturers of another kind of hip prosthesis, the Austin-Moore design, provide exactly such a concavity that usually does work satisfactorily in that respect.

G. Load Focusing can Multiply Trivial Total Loads into Catastrophic Local Ones

The total loads on a normal femur during normal activities, such as walking and standing, remain biomechanically trivial simply because the earlier-described shear-locking mechanism defocuses them over whole cross-sections at any and all levels of the bone. As a result, they should never cause strains anywhere in the bone that reach or exceed the lower MES. The same loads on the femoral component design shown in Figures 8 and 14 can allow the base plate skid and shear-shunt mechanisms to shunt and focus that total shearing load onto a very small region of the inner upper calcar, as a locally expansile battering ram effect. In principle, if that region equals only 1 of the 100 area units referred to earlier, then its unit loading multiplies by a factor of 100, which overloads it with consequences already described. The same load focusing mechanism applies to the cases of asperities and base plate tilts. Consequently, an individual's perfectly normal leisurely activities can cause small regions of overload in his hip arthroplasty that tend to loosen it and make it begin to migrate. Table 2 summarizes some of the above material.

In a particular case the sequence of the above events may differ. For example, the differential longitudinal stiffness can become the initial problem in some cases, the base plate skid can become the initial problem in other cases, and the outwaisting drifts in still others. The preceding paragraphs describe what does happen in many of these failures, but the order in which they occur probably varies in different patients. One of the reasons for that uncertainty lies in the fact that the in vivo strain studies published so far have dealt with early postoperative situations, when some of the above described effects may not have had enough time to develop.

Table 2
MAJOR FEATURES OF THE ARTIFICIAL FEMORAL COMPONENT

1. Underside of the base plate is flat and smooth. No shear lock exists. The base plate becomes a skid. Skid effect shunts shearing load into internal battering ram. Skid effect focuses shearing load onto upper inner calcar.
2. Different stiffness of stem and bone cleaves any mechanical bond of stem to bone.
3. Vertical pistoning results which can evoke expansile bone drifts around the upper femur that loosen the component.
4. Poor initial fitting of bone to device causes local unit load focusing and multiplication.
5. Grain of the calcar is poorly oriented to withstand an internal battering ram impulsive loading.
6. Any local overloads cause increased microdamage production.
7. Increased microdamage production causes increased remodeling and remodeling space. New bone needs months to become stiff enough to carry its share of load.

V. POSSIBLE SOLUTIONS

The reader will understand that the following paragraphs present the author's thoughts and they require proof before they merit general acceptance.

A. Transmission of the Vertical Load

Nature, after 1 billion or so years of experimentation, settled on transferring the hip loads completely from the femoral head to the compacta at the base of the neck, rather than down and obliquely into the cortical-endosteal surface of the proximal shaft as many current THR designs arrange it. Therefore, the femoral component of THR systems should do the same via the base plate. That requires the base plate to extend far enough anteriorly, medially, and posteriorly to completely cover the compacta at the base of the neck, as suggested by the lower two insets in Figure 13. Other authorities have recently come to the same conclusion, e.g., Lewis et al.[456] In simple words: transfer vertical loads to compacta via a base plate.

B. The Shear Skid and Shunt

The shear-shunt mechanism is real, injurious, and common although it has received little specific attention so far. Considerable experience with failed prosthetic designs exposes it as a major factor in the failures of living bone, and it also affects some current designs for artificial knees, shoulders, elbows, ankles, and digits.

To reduce the skid effect the base plate angle could lie more nearly parallel to the ground, as shown in Figure 1. To prevent it, the underside of the base plate should be rough, perhaps on a 0.5 mm scale, with small but numerous and appropriately shaped asperities that can indent into each unit area of the supporting underlying compacta, and thereby provide an effective shear lock to defocus the shearing load over all of the underlying cross-sectional bony support, as nature intended it to be.[267] That would also prevent excessive torque of the base plate on the underlying bone and around the long axis of the stem. Figure 15A illustrates the idea. Such a surface could even be produced at the time of casting the component, which would eliminate a currently expensive manufacturing operation that produces a rough surface into which the surrounding living tissue — hopefully bone — is supposed to grow to provide a mechanical bond that does not involve the use of any cement.

To draw again on the biological wisdom that opened this chapter, nature has solved the problem of attaching one rigid but separate bony part to another while large loads of various kinds transferred from one to the other. That system constitutes the epiphysis lying on top of the metaphysis but separated from it by the epiphyseal plate. As noted in 1973,[241] numerous small hills and hollows in the underside of the epiphysis fit

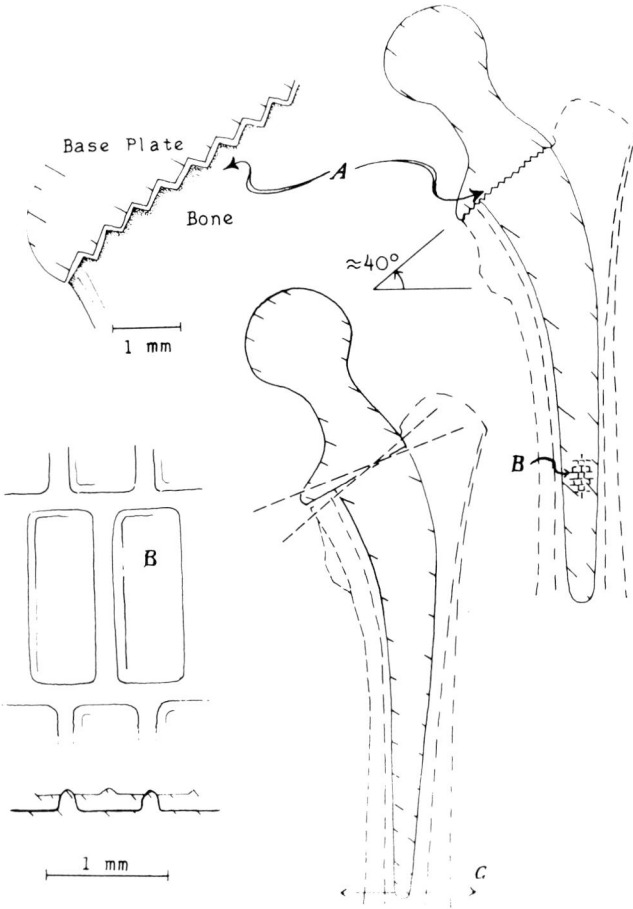

FIGURE 15. (A) Making the underside of the base plate rough, somewhat as shown here, would provide a shear-lock of the base plate to the whole bone cross-section supporting the base plate, one in which the greater the vertical load, the greater the locking effect too. At C on the lower right the effect of toggling of the stem on base plate and neck-cut alignment is shown. Part B illustrates a possible geometry of the surface pores in the stem that could allow an ingrowth of tissue that would provide a transverse shear lock, but would allow slight vertical motions of stem relative to bone without shearing off any bony ingrowth. Plan view above, cross-section below. The pores and ridges on the underside of the base plate could have a somewhat similar rectangular geometry. Reducing the base plate angle from the usual 40 to 45° angle towards a more horizontal plane would also reduce the shearing load that parallels the base plate plane, as in Figure 1.

into matching hollows and hills on the top of the metaphysis to provide a shear-locking mechanism that increases in effectiveness as the uniaxial compression loads on the part increase.

In simple words then, provide a shear lock that defocuses total transverse shearing loads over the whole bony cross-section.

C. The Inwaisting Factor

The underside of the base plate should be made concave rather than flat to create the necking-in or inwaisting or circumferential constricting force mentioned earlier.[267] That would neutralize the outflaring piston-wedge effect shown in Figures 10 and 13. Figure 13C and D illustrate such concave shapes.

D. Toggling

In such a design, the major function of the stem would change from that of a major carrier of the superimposed load to the more limited one of minimizing toggle motion around the base plate fulcrum. Even though slight, such motion would cause the base plate to tilt back and forth, and a large enough tilt could overload small regions of the underlying bone as shown in earlier figures. To minimize that toggling, one might make the lower end of the stem fit the marrow cavity more closely, either by enlarging its cross-section or by surrounding it with a close-fitting collar. The stem might also be made longer, or means might be provided to allow small flexural motions between the base plate and the stem so that stem toggling need not also tilt the base plate itself. Some other possibilities exist too.

E. Mechanical Fit and Tolerances

The above suggestions imply that these procedures should achieve a precision of fit between bone and base plate and shaft and stem at the time of surgery that is not possible with current techniques, so current surgical procedures and tooling may need improvement to provide the necessary fine mechanical tolerances.

F. The Cement

As many experts have suspected, implementing the above procedures might permit discontinuing the use of the cement that serves currently in many designs to provide a mechanical bond. The present cement at least poses its own problems mechanically and biologically, and because of them it will have to go sooner or later. In that regard, a new generation of knee prostheses depends on rough surfaces of the components in contact with the bone, plus precise fitting of the two, to provide the necessary shear-locking mechanism, and the early results look quite promising.

G. Biologic Fixation

Until the femoral component can be made to have nearly the same bulk longitudinal stiffness as the bulk femur around it, it will probably prove futile to try to solve the loosening problem by making at least the distal half or two thirds of the stem rough on a 100-μm scale so that woven bone — actually, callus — can grow into its cavities.[267] The relative vertical motions of the stem and the surrounding bone may be too large near the tip of the stem of present designs to allow a good metal-to-bone bond to endure in pits of such small dimensions. One can better visualize the problem by thinking of the bone in Figure 10 as made of pliable gum rubber and the femoral component as made of stiff plaster. Then fit them together as shown and repeatedly push down on the head and then release it. The tip of the stiff stem will move up and down inside the rubber marrow cavity as the upper portions of the rubber femur shorten under vertical loading and then elongate again upon deloading, but the stiff plaster stem does not. That relative motion will be greatest near the tip of the stem and zero at the base plate, and at the tip of the stem it could easily exceed half of the 50- to 100-μm diameter of one of the pits on the surface of the stem, so that could shear off any ingrowing bony nodules. A third solution might control the longitudinal compliance of the stem by innovative design and use of structural materials to make its bulk compliance approach that of the enveloping bone.

H. The Biology

This problem has a strictly biological side too, as well as the primarily biomechanical one that has occupied us so far. For example, bone macromodeling cannot compensate for most design errors in adults, so those errors must be avoided. To repeat, in the future those who design these devices will have to add to their analytical repertoire the concepts of the newer bone biology and the IO, and cultivate the habit of thinking in terms of dynamic strain and fatigue, rather than in terms of strength and static stress. Note 1 in Chapter 7 and Note 1 in Chapter 9 go into those matters further.

Because a competent microdamage repair mechanism can probably salvage some situations that otherwise would lead to failure, it is necessary to know how that mechanism actually works adjacent to these arthroplasties, what can be expected of it in terms of kind, magnitude, and time, and what cannot. A significant but minor group of failures arises more from disordered bone repair, remodeling, and RAPs,[503] than from errors in design of the components or in their insertion. Again, it is necessary to know more about them in the future.

I. Generalization to Other Situations

It has probably become clear by now that the above general approach to the analysis of the hip problem, and especially the shearing matters, should extrapolate with appropriate modifications to analogous problems involving other joints. That will require learning to look at dynamic strain rather than static stress, at fatigue rather than ultimate strength, at shear locking, skid effects, and load shunting, focusing and defocusing mechanisms, and at unit load multiplication mechanisms regardless of whether the terminology used to signify those mechanisms in this text survives intact or not.

VI. ILLUSTRATIVE QUESTIONS

Total joint replacement arthroplasties have been a craze in orthopedic surgery since about 1964, and a bewildering number of things have been tried to improve their durability, function, and applicability. Yet they continue to fail, although after longer and longer periods of usage. As noted earlier, failures of the tissues rather than of the components provide most such problems today. Those failures relate to an interaction between the design of the devices and their materials properties on the one hand, and the properties and capacities of the basic biological and biomechanical functions of the skeletal IO that earlier chapters discussed on the other. One central fact should be kept in mind in pondering such matters, and that is that the structural support provided by the bony structures of an arthritic joint will last the patient the rest of his days if left in place. Accordingly, if they fail under artificial components that proves the design of the components is at fault, not that of the bone. Those basic, essential functions certainly include macromodeling, remodeling, micro- and macrorepair, and the RAP, plus their IO roots; their failures can arise as IO-intrinsic or IO-extrinsic problems or as combinations. The latter can arise from some diseases and some drug effects or from unwitting errors in design, emplacement, and usage of the components themselves.

It seems remarkable therefore that no IO-oriented study of such important matters as they apply to these arthroplasties has yet been published. That seems like trying to treat pneumonia without learning anything first about the pathophysiology and bacteriology of both the disease and the organ it affects; medicine was last in that position for that disease 130 years ago.

Some sample questions follow that require answers. The questions will not go away until the answers have been supplied.

How much bone is devitalized by a THR operation? And where? For how long? How is it revitalized and how quickly? What added effects does the cement have? How do the remodeling and RAP mechanisms respond to the operation? And to subsequent use? And to the inanimate materials used, the drugs taken, the diet, and the level and kinds of activity?

What is the saturation point for the repair of microdamage? Why? What physical, biologic, and biochemical agents affect it, and how, when, and where? How do commonly used drugs affect it? Where does microdamage occur most and where the least? How do age, disease, medication, sex, diet, and activity affect its development and its repair? How much modeling can the operated adult hip produce? Where and when? For how long? What physical, biological, and biochemical agents affect it, how and when? What are some optimum geometries and dimensions for producing an artificial shear-locking mechanism? What is the pattern of mechanical forces across and strains in the normal and operated hip (note that extant analyses usually omit consideration of the effects of the hip adductors, obturators, and hamstrings, yet these powerful muscles were provided for good purpose, they function consistently in life and if ignored in theoretical analyses those analyses thereby become unrealistic)? How do age, sex, activity, diet, and race affect them? And in painless and in painful hips?

What is the actual distribution of compliance around the various regions of major joints, and in health and disease? How does a replacement arthroplasty affect the adjacent bone compliance in both the short and long term? Do commonly used drugs affect that compliance? If so, how? How does activity affect that compliance? And its component parts: loading magnitude, range, frequency, and rate? And durations thereof?

What are the biological limits of unit loading for living lamellar bone when loaded parallel, tangentially, and radially with respect to its grain, and in tension, compression, and shear? And in torque and flexure? And for bone ex vivo? How large an expansile intramedullary load can a living bone accept?

What are the real MES values for strain in compression, tension, and shear, and along and across the grain of lamellar bone? What is the relative importance of ultimate strain magnitude and strain rate in those MES values? How do drugs, age, sex, and disease affect those values?

What else that is important in these matters have we failed so far either to perceive or to stumble across? What has this chapter overlooked and what more must be known?

Note 1

COMMUNICATION BETWEEN LABORATORY AND CLINIC

The problems medical investigators and clinicians experience in communicating with each other are well known to both camps and are rife with misunderstandings, both humorous and contentious. In the balance, the instances of ineffective and even counterproductive communication outnumber effective ones and the situation presented in Note 3 (in Chapter 7) is only one of many examples. Many of those problems stem from the different mental traits and occupational experiences and hazards that characterize the competent investigator and clinician. The author has worn both the research and clinical hats productively for over 40 years, at least in the judgment of many colleagues in each camp, and since this troublesome but nontrivial matter has not been formally discussed by others, a final note concerning it seems appropriate.

Consider then the personal qualities needed to excel in each field and some contingent matters.

The investigator — Effective and productive exploratory research (excluded from this category are the development activities, such as measurements of already known features to further decimals, in further material and with better or different tools) requires an eagerness to question and put to the test any idea, no matter how innovative or untried, no matter how unorthodox, no matter how deeply it might threaten or run contrary to conventional ideas, wisdom, dogma, or knowledge. While most such innovations fail, all successful ones come from that process, and the inevitable price of any success is the failures. Also, the investigator's worth is measured against the number and value of his successes rather than of his failures. In effect, the system within which he functions punishes a lack of success rather than the perpetration of overt errors, so that is his chief occupational hazard. Also, his successes are published to make them available to others, but his failures are seldom made known beyond the walls of his laboratory. Most important, any failures or errors will affect test tubes, reagents, materials, expendable cultures, and laboratory animals, so they cause no harm to human beings or the investigator himself. Since those attitudes and facts are the essence of productive and innovative research, most successful investigators will have them, and they use them both habitually and instinctively. They seldom analyze the relationship between their attitudes and the kind of work that forms their career. That instinctive, automatic, and even eager willingness to try the untried but seemingly logical innovation usually also colors correspondingly an investigator's politics and social philosophy. In one phrase, successful innovative investigators react to an exciting and logical new idea or prospect thus: "Let's try it; it might work!" They do not have to deal with any troublesome personal, moral, emotional, ethical, or legal consequences of any failures of their innovations so they do not worry about or concern themselves with such. Nor for that reason do they usually understand others who must.

The clinician — Effective and competent medical and surgical practice occurs on a different stage populated by patients who have concerned relatives and ready access to not overly busy attorneys, all three of whom place such a high value on human life and health that when either suffers from a physician's acts or lack thereof, no matter how inexplicably or how carefully the physician tried to avoid it, they may seek justice via the courts or vengeance out of the muzzle of a gun. Physicians themselves have been known to become so troubled and distraught over such happenings that they retired from practice or committed suicide. Furthermore, patients bring to their doctors diseases and other problems only dimly understood — so dimly in fact that most new and exciting developments in laboratories fail when tried in the clinic for the first time, and usually have unexpected side effects[320] or may actually do harm to or even kill the patient.

In the past, that observation has offended some nonclinical people, but 40+ years of clinical experience reveal it as a blunt and inescapable reality. The offended ones in effect sat behind a TV set in some safe rear area, monitoring front line soldiers in a battle and criticizing them for ducking the enemy's gunfire. No clinician can practice his art for more than 15 years without forcibly confronting that reality. One might sum it up like this: the overwhelming majority of departures from the time-proven clinical recipes that provide the real basis of contemporary medical and surgical practice failed when they were finally tried in patients, and they also often caused unexpected side effects, and some of them did harm. Furthermore, while the patient and his relatives hope the physician will be successful in dealing with a patient they also understand that even doctors (!) cannot correct every health problem. Consequently, a physician is rarely punished for his failures to succeed but he can be devastatingly punished morally, legally, and even corporally for doing overt harm to his patient in the process of trying to help him or her regardless of whether that harm is done with the scalpel or the prescription pad. Thus, in various ways the chief hazard the clinician faces is the consequences of committing errors, and not the consequences of failures to succeed. As for publication, while the game rules of research diffuse successes among that community and ignore overt failures, in clinical practice exactly the opposite occurs. The successes are largely ignored but overt failures are made much of.

It follows that the experienced and competent physician of necessity acquires a deep and instinctive distrust of innovations, of the untried and unproven, and that attitude correspondingly colors his politics and social philosophy. Further, any physician who cannot learn those facts and attitudes is truly an incompetent menace and is usually ejected from the clinical practice arena in one or another way. In one phrase then, the competent physician's reaction to a proposed innovation is this: "First prove it will not harm my patient".

The above differences in attitudes do not stem from differences in the basic intelligence of the people in each camp. Each camp has its share of low, middle, and high intelligence. They do explain why the investigator often perceives clinicians as hidebound old reactionaries who oppose any real progress, and why the clinician perceives the investigator as a young, wooley-headed, if often articulate, radical who would cause worse problems with his unproven cures than the diseases he would treat with them.

A recent situation illustrates one facet of the matter. Certain research discoveries raise the prospect that a particular class of compounds might effectively potentiate the bone healing process. Putting on the investigator hat for the task, the author helped to design an initial animal experiment to test the idea, and with the intention that a success would justify the further development work and expense needed to provide a safe and proven new therapeutic process for human bone healing problems, many of which were described in Chapter 12.

The experiment went well and the compounds worked, but at that point the author had to switch hats, to put on the clinical one and, speaking now as an orthopedist directly responsible for what he asks his patient to accept or submit to, to insist on the aforesaid development work. However, the scientists (and perhaps its management level, nonmedical people) of the institution balked and decided instead to try to move directly into the clinic. They began polling medical centers where that might be done and everywhere physicians asked the same questions they had already been told needed answers before trying the material in humans. They were asked for their proof that the material would do no harm, and that it would also help. Of course, without the development work they could not provide those proofs. The management in that institution still remains unwilling to commit the funds needed to provide it so the matter currently rests in limbo. As an aside, publication of the above experiment is also being held up on the grounds that it might give other institutions a commercial advantage if it were published (Joseph Heller's Catch 22 in a new disguise?).

One more example can illustrate what can happen when competent clinicians become involved in research and bring to it their clinical conservatism. It is exhumed from the archives in order not to offend any currently active people, but the phenomenon it illustrates remains very much a current one.

Between 1957 and 1960 the author sent three manuscripts to a prestigious orthopedic journal, the menu of which focused on clinical matters but was also intended, according to its staff, to contain examples of productive and innovative skeletal research. However, they were evaluated by clinicians imbued with the same reservations they would apply to a new operation or drug that might be used on children tomorrow. The reviewer's comments were that the methods used were obviously impossible and the findings were obviously artifact. In response to the author's steaming rejoinder to the editor (who in retrospect made gracious allowances for another young man trying on faces in search of his own), he was informed politely and courteously, but firmly, that it would not be necessary to send the slides to their reviewers for their own evaluation of the material, findings, and methods, and that because of the prestigious nature of the journal, it could not afford to publish unproven or controversial material; after all, any new operation printed in its pages would be performed the next month in most of the nations of the world.

Accordingly, the original descriptions of the author's now standard technique for making and staining undecalcified sections,[217,218] of the now well-known fact that osteoid occurs regularly in perfectly healthy adults as well as children,[804] and of the increasingly important fact that actual mechanical microdamage occurs normally in human bones[220] were published elsewhere.

When the author wants reliable clinical information today he still reads that journal, for in that respect it was and remains excellent. However, an interesting additional fact is that since 1955, at least, it has never published the original report of a single new fundamental basic science discovery or concept. Accordingly, when the author and most others in the research end of skeletal medicine want to find out what is really new in that field, this prestigious journal is the last one consulted, if it is consulted at all.

In sum, it behooves investigators who interface with clinicians to try to put on the latter's moccasins for a few leagues to understand their problems and situations, and of course the converse applies too. Investigators must understand that the consequences of doing harm to patients can be devastating to both patients and doctor, while the investigator behind some unhappy trial walks away from it unharmed. The company may be sued but not the investigator. The clinician needs to understand that progress requires new ventures, and new ventures into poorly understood systems virtually guarantee that mistakes will occur. The two camps have to devise *modi operandi* that satisfy the imperatives of each and violate the needs of neither. As proof it can be done, witness: penicillin, aspirin, blood transfusion, appendectomy, and internal fixation of fractures.

Since the attitudes of the productive investigator are actually harmful, even disastrous, when carried over into the clinical practice area (which includes teaching, refereeing, and program committee work), while the attitudes of the competent clinician would be equally disastrous and counterproductive if carried over into the laboratory area, few individuals can do both well.

Chapter 16

A SUMMATION

"There is a time for everything and a season for every purpose under the sun."

(John Ridlon, surgeon, about 1880)

"Knowledge is proud that he has learned so much; wisdom is humble that he (understands) no more."

(William Cowper, about 1790)

I. A PERSPECTIVE

A graying teacher close to retirement listens to a student enthuse over the course material of the final semester. The student is eager, he worked hard, did well, and enjoyed his subjects. Briefly, he approaches the threshold of his career overwhelmed by the wealth of knowledge in his chosen field and doubting that it will see further major changes because his predecessors already did the most important work and made the major discoveries.

That amuses this teacher, for experience lends eyes and ears beyond youth's ken that hear himself some 40 years before voicing similar thoughts to classmates over a beer after the final final exams. Yet in the following decades, undreamed of developments so transformed his field that a student of 40 years ago put in a contemporary operating room or laboratory could only gape in bewilderment at what happened there, somewhat as American Indians must have done when they first saw the white man's thunderstick kill game invisibly at 50 paces or more. The eye of experience also saw vanished generations of students standing in the same awe of their knowledge and equally innocent of what lay ahead, generations receding unbroken to the ancient Indus valley and prepharaonic Egypt and that should extend unbroken into the future too. Yet, and to quote William Harvey, discoverer of the circulation of the blood:" . . . true philosophers . . . never regard themselves as already . . . thoroughly informed . . . " observing that in his time a philosopher signified what we call a scientist today.

So the teacher listens, but amused and with half his mind while the other half wonders what new miracles that boy would have seen when he neared his own retirement, what firm knowledge of the present would prove flawed by then. The teacher hopes too that the dice of mortality might leave him enough time to see and relish some of those developments himself.

Above then a perspective that seems appropriate at this point in this book and at this time in this field.

II. A METAMORPHOSIS

It can be exciting to live in a time of a constructive metamorphosis and a privilege to contribute to it, two observations that need a bit of background to define their present context.

Note 2 in Chapter 1, Note 1 in Chapter 2, Note 1 in Chapter 3, Note 2 in Chapter 6, Note 1 in Chapter 9, Notes 1 and 2 in Chapter 14, and Note 1 in Chapter 15 discuss some mental traits that can hobble physical and biological research. One of those traits consists of depending on time-proven algorithms to seek copies of already-known mechanisms in any new systems or problems that come into the focus of research. Of

Table 1
THE CHRONOLOGY OF SOME CONCEPTS AND PERCEPTIONS THAT ARE BASIC TO THE SKELETAL IO

Pre-1963	Cell → organ; skeletal physiology depended on the varied potentials of already existing differentiated cells and the systemic agents that regulate them; singularity and independence assumptions
1963	Coupling of resorption to formation; the LOBO effect
1964	An operational IO for bone; physiological quanta based on special multicellular units; the $k\sigma$ property; "On-Off" states; transient and steady-state behavior; time-averaging properties in macromodeling; dynamic, repeated strain as a modeling determinant; the flexure-drift and neutralization mechanisms
1966	The growth-modeling distinction; the RAP; envelope-specific dynamic distinctions and local, e.g., "L"-gated or site-specific control
1972	Definitions of function, disease, enablement, regulation, stretch-hypertrophy relation for fibrous tissue; chondral growth-force response characteristic; streaming potentials; operationally important IO for all skeletal tissues; the MES for bone.
1979	Temporal coherence and incoherence
1979	Growth-force response curve for cartilage corrected
1980	MCN, LBO properties; the MES for all skeletal tissues; mechanical and biomechanical competence; gating; micro- and macromodeling distinctions; fatigue endurance as a skeletal tissue design goal;
1982	Load partition, focusing, defocusing, and shear-locking mechanisms; first formal statement of general IO properties.

course, such proven algorithms can save time and effort, and more often than not they work well in daily life and work.

Those algorithms tend to fail when applied to a new kind of problem, for their design uniquely fits them only to the already understood. Adhering to them virtually guarantees a failure to master fundamentally different systems or problems or to devise fundamentally better ways to synthesize and organize existing facts. In times past, such proven algorithms showed that neither man nor bee can fly, that the locomotive will never replace the horse, and that the secrets of the living brain and cell will remain known only to God. Such analyses usually ignored "X", the still unknown but pertinent factor(s) that beckons irresistably to some men from beyond the charted perimeters of the known and understood.

The domain of physiology and disease that contains the skeletal IO presents many kinds of problems that were never even imagined. They were discovered by accident, and then efforts had to be made to understand them and to seek any denominators they might have in common with other areas of skeletal and general physiology.

Success in those efforts required learning to observe, think, and hypothesize outside the bounds of most previously successful skeletal algorithms (John Dewey put it thus: "Every great advance in science has issued from a new audacity in imagination."). That kind of success led to the content of this book, which distils the contributions of many minds and which contains a fabric of perceptions that was woven from many threads, or from the many terms in a large set of simultaneous equations that might model the form, function, and development of the skeleton. Table 1 lists some of those perceptions, the basic concepts and facts of the IO that set its properties apart from those of unassociated cells and of intact people, and that set it apart from the ideas of skeletal physiology and disease that prevailed in 1963 when recognition of an operational as opposed to a merely anatomical IO began to unfold. As outlined in Note 1 in Chapter 16, no single one of those perceptions was known, suspected, or even imagined when the author graduated from medical school in 1945.

Those perceptions grew out of the recognition of some of the intraskeletal mechanisms provided by the IO. As some of the results of those perceptions, it is now possible to predict major features of the configurations of bones, joints, tendons, ligaments, and fascia from first causes and initial conditions, and to predict their adaptations to varied mechanical usage. It is also now possible to predict from first causes most of the clinically known forms of osteoporoses, osteomalacias, dwarfism, chondrodystrophies, and rickets, plus many kinds of malfunctions of the healing and homeostatic functions of the skeleton. The successes of the above predictions greatly outnumber the failures so far. Those new perceptions include the IO itself, plus the realization that the following three-part strategy mentioned in earlier chapters underlies the construction and physiology of the adult vertebrate skeleton.

First, the fetus is constructed as a marvelously intricate, integrated living machine that faithfully and blindly materializes a long-since completed master blueprint that sums up what was learned from innumerable bioengineering, biochemical, and other kinds of "experiments" done over geologic spans of time by the processes that evolved human life from the primordial soups and climate of this planet.

Second, as one part of that blueprint the fetal organs and tissues receive special action principles that dictate how they respond to subsequent biochemical and mechanical usages.

Third, the fetus enters the world and a different environment with new demands, as an independent, lustily squalling infant. As it grows, its structures and physiological systems continually adapt to their new usage according to the above-mentioned action principles.

To a significant degree then, major architectural and functional features of the infant skeleton are fluid, open. Another collection of facts hid that simple truth for a long time. Not only do the initial fetal conditions remain stereotyped from one infant to another, the normal postnatal usage of the normal infant skeleton proves equally stereotyped, and precisely that combination of pre- and postnatal stereotypism makes adult skeletal architecture so similar from one person to another. In its turn, and as the writer noted in 1963,[824] that stereotyped usage reflects the stereotypisms of the initial fetal muscle, bone, and joint anatomy, mass of the body, the gravitational field of the earth, and how the nervous system controls muscle activity and its demands on the skeleton.

Note Bene: Another bit of biological wisdom follows from the foregoing. The changes in structure and function that occur in disease reveal exactly how fluid the postnatal skeletal design truly is, meaning how, where, when, and how much it can change its usual configuration in response to unusual usage. In a real sense, disease can serve as a special kind of analytical microscope. In the laboratories of the clinic and operating room it provides diverse natural experiments that offer and beg understanding by inquisitive and sometimes unorthodox minds in order to reveal how to manage disease better than it is done now.

As one example and concerning skeletal mechanics, man's experience with designing and constructing his own machines led him to consider them in terms of stress and strength, and to make them fully ready and guaranteed — in writing — to perform their tasks before they left the factory. It was collectively assumed that the construction of the skeleton followed a similar strategy. Earnest searches for the game rules of such a strategy have produced relatively little understanding, if a plethora of data. Only recently have some reasons become apparent. One reason lies in the fact that real skeletal design appears to focus more on fatigue and dynamic strain than on static stress and strength (and it may have even more concerns and purposes than are now apparent). Another reason relates to a matter referred to earlier: the infant skeletal design is not necessarily the adult one. A third relates to a fundamental but still poorly understood general fact: the systems of the body often distinguish trivial from nontrivial, and respond only to the latter. Thus, to study what governs such a system one should first find out the thresholds that separate its trivia from its nontrivia, and then address the behavior of the latter, for there will lie the true meanings and purposes of the system. Studies of its behavior in trivial circumstances will usually only reveal random, enigmatic, and design-meaningless noise,[267] as illustrated in Note 3 in Chapter 7.

Such basic truths have their most direct roots in the skeletal IO rather than in the properties of unassociated cells, but because that IO became apparent as an operational reality only recently, many of its facts and the inputs of the innovative minds that will lead to a better understanding lie ahead rather than behind. When they arrive, they should lead to understanding of the same depth and confidence that apply as examples to renal, gastrointestinal, pulmonary, and cardiologic matters today. In the past, an ignorance of the IO made the field rely on serendipity and the clinical recipes it provided. Ambrose Pare put the matter of an empirically based clinical practice quite succinctly: "I dressed his wounds; God healed them."

The above developments have initiated a metamorphosis in thought in both kind and scope that must lead in due time to a profound metamorphosis in clinical practice in this field. Knowledge and understanding, including of the IO, have developed to the point that they begin to allow predictions of new effects that can serve the needs of research, diagnosis, and treatment and that have reasonable chances of being correct. In different words, theory begins to implement serendipity usefully, and in time it could relegate the latter to a supporting role rather than the dominant one it has held for the past 6000 years of recorded human history. The time is not far off now when, rather than having to *allow* wounds to heal, they can be *made* to heal.

III. SOME CONCLUSIONS

A. Some Lessons of the Recent Past

The development of renal physiology offers eight basic lessons that bear on any strategy for studying and analyzing skeletal physiology and disease. It will take the field time to digest those lessons, which form a part of the messages of this book, and which also form seeds of future clinical expertise. They are summarized below.

The nature and organization of macromolecules and organelles directly determine the properties of intact cells.

The nature and organization of cells and their intercellular materials directly determine the properties of an intact IO.

The nature and organization of an IO directly determine the properties of intact organs and organ systems.

The nature and organization of organs and organ systems directly determine the properties of the intact vertebrate organism.

Those four links can be said to comprise a fundamental chain of physiologic knowledge, and that idea merits encoding it as a final relation of the main text. Thus:

$$\text{The chain of knowledge:} \quad \text{(macromolecules and cells)} \rightarrow \text{(IO)} \rightarrow \text{(organs)} \rightarrow \text{(the intact human)} \tag{1}$$

A major feature of the system encoded by that relationship consists of the fact that when all of its four links are fully formed and joined, then one can predict from his understanding of the system the effect of a given but previously unobserved malfunction at any level, including the cellular level, on all subsequent levels, including man. That property also allows one to design treatment from theory, as in coherence treatment.[24] If any intermediary link is broken, such as the IO, such a prediction cannot be made, and the L_{is}-level effects of drugs, hormones, disease, and genetic malfunctions

Table 2
DOMAIN OF ORGAN SYSTEMS

(L_{os})

Domain of organs (L_o) — Bones, Joints, Ligaments, Tendons, Teeth, Fascia

Domain of the IO (L_n) — Homeostasis, Macrorepair, Remodeling, Macromodeling, Micromodeling, Growth, Histogenesis

Domain of the cell (L_c) — Specialized cell functions, Molecular level turnover, Differentiation, Proliferation, "X"

Domain of macromolecules

must then simply be observed and cataloged and somehow remembered without being understood. Also in that circumstance, the effects of treatment cannot be predicted in advance, so they must be found by observing them, which means that treatment then is arrived at by trial and error. To repeat, to understand how its molecular basis bears on a disease in the whole man requires fully forming and joining all four links.

For such reasons, one cannot predict the organ-level effects of cellular responses to drugs, hormones, and other agents until he knows how those cellular responses affect the IO.

Likewise, one cannot predict the effects of the IO on the intact human until the nature of that IO and its functions, its purposes, are known and understood. Since the operational importance of the skeletal IO was only recently perceived, much should still remain unknown about it.

Table 2 arrays the above relationships to illustrate the basic framework of the structural and functional relationships of the skeletal systems. Each of the domains in that table functions and relates to its neighbors according to its own collection of special rules, and because of the statistical and other properties of this system that were outlined in Chapter 2, those rules cannot be predicted. They must be found by observation. All presently known rules of this system were found by that means alone.

If the messages of Relations 2 in Chapter 14 as well as in the above paragraphs are essentially correct, as the author believes, then the IO could be said to contain about one quarter of all the knowledge that is needed to master skeletal disease, but relatively little is known about it today, compared to what is known about cell-level and lower matters as well as the whole man. In evidence and as Note 1 in Chapter 1 and Note 1 in Chapter 12 document, less than 1/1000 of past and current research and analytical efforts relate to the IO.

To repeat, mastery of disease requires that each of the four links of skeletal knowledge (L_m-L_c, L_n, L_o, L_{os}) and (L_{is}) be formed and connected to its neighbor, because a break in any link breaks that chain and prevents the mastery that only it will allow.

Viewed in that light, contemporary ignorance of the IO reveals a previously unsuspected imbalance in the emphasis of skeletal research and the support and teaching activities behind it. Prudence and wisdom suggest that efforts should be made to correct that imbalance, to form, join and close that currently defective link.

The above facts and syntheses suggest that the atomistic reasoning inherited from the past should depart the future functions of the classroom, the journal referee, standard texts, and the deliberations of agencies that support skeletal research. It belongs in the archives of medical history along with laudable pus, leeches, and humors as another — but former — step in medical progress. It is a revealing fact that no standard 1984 text of general human medicine, physiology, anatomy, or pathology fails to consider the renal IO, the nephron, yet none of them, and likewise no standard orthopedic text, considers the skeletal IO.

B. The Future

The next decade or two in this field should prove fascinating as well as productive, turbulent, and spiced by features of the human pageant as emotion, fads, opinion, territorial jealousies, protective reactions to threatened scientific postures, and polemics interface with reason, burgeoning knowledge, and scientific progress. It was written some 2500 years ago in the Book of Daniel that many will run to and fro and knowledge will increase. Understanding should increase too as the seasons of man and medicine, and of learning and teaching, flow one into the next and as the IO enters its season wherein its properties are bent to the needs and purposes of man and medicine.

During those developments, any would-be "builder" or heurist might keep in mind that any of the great leaps upwards of today usually becomes another of the small steps of tomorrow. In the words of the Roman emperor Marcus Aurelius Antoninus: "Everything that exists is in a manner the seed of that which will be." Everyone knows that seeds are smaller than the trees that grow from them. Now let us put this seed in the ground, and watch and tend what grows.

Note 1

PRE-IO CONCEPTS OF SKELETAL PHYSIOLOGY

IO-related knowledge only began to appear in print and in volume about 1964. It may help to understand the nature and magnitude of its impact on this field to outline some of the basic ideas and knowledge that preceded it, and which in many respects it threatened and in fact has since overturned. This note summarizes briefly some of that pre-IO knowledge and body of concepts, which was completely unaware of an IO, of any quantum-like behavioral properties above the level of the cell, of the transient steady-state distinction, and of the true reasons for the existence and nature of those states in all five basic skeletal tissues. The basic scheme of skeletal organization was assumed unwittingly and generally to be cell → organ, as encoded in Relation 1 in Chapter 1. It will become apparent that good progress has been made since 1963, regardless of how much farther remains to go. However, it must be understood that the material that follows in no sense belittles the contributions of earlier investigators. The contributions of each generation stand on those of earlier generations, without which things like growth, modeling, repair, sexuality, and reproduction would still be explained as proof of magic or the intervention of the hand of Zeus, and joint lubrication as an influx of marrow fat into the joint cavity. Also, the nascent understanding of the IO does not demean, replace, or criticize contemporary knowledge of molecular or cellular biology or the ongoing research in those fields. Rather it complements that knowledge and research by adding a missing link to the chain of skeletal knowledge and understanding that will ultimately lead to mastery of skeletal health and disease.

As a brief overview, knowledge of the IO revealed the pathogenetic and anatomical-pathological types of osteoporoses and osteomalacias, exposed their malfunctioning causative mechanisms, and led to current research that will eventually reveal how to correct them. IO knowledge and perspective finally revealed some of the basic mechanisms that control skeletal architecture so that a basis exists for a quantitative theory of macromodeling activity. Many of the skeletal mechanisms that respond to endocrines, nutrition, activity, and other controlling agents have been identified and some are now under active study. The field abounds with pertinent basic questions and problems that were not even imagined by its authorities in 1960, and a few signal successes have begun to sprout from that knowledge and perspective that could not even have been understood by 1960-era knowledge and understanding. Now to what was "known" in 1963.[A,B,332,490,527,579]

It was known that osteoblasts made the new organic matrix of bone, but it was believed that its subsequent mineralization depended directly and only on the milligram per deciliter-based calcium-phosphate ion product. When that product exceeded about 25, then mineralization occurred, but when it fell below that value, mineralization defects of undefined nature resulted. The nature and different kinds of such defects known today were unknown and unsuspected. The vigor of already existing osteoblasts supposedly directly and solely determined how much bone was made in the skeleton and where and how fast. Hormones, drugs, and other agents supposedly acted directly on those cells to modify how much new bone they made and how quickly. No serious thought was given by internists, orthopedists, pediatricians, and physiologists to how those osteoblasts got there or to the possibility that malfunctions of the mechanisms that produced and supported them rather than of the blasts themselves might underlie some diseases. The developments of autoradiography and tritiated thymidine labeling of dividing cells in the 1940s and 1950s[34,446] soon led to their application to some skeletal matters, so by 1964 it was probably apparent to many skeletal analysts that cell population dynamic factors could underlie some bone disease. It was the writer's misfortune to express that in print first, and to begin to produce evidence that the idea was correct.[229,230,232,385,386] Therefore, he attracted most of the gunfire from then established authorities and their students, who sensed a radical threat to the comfortable and orthodox views of what today can be regarded as yesteryear. Also, the physiology of lamellar and woven bone were assumed unwittingly to be similar, and therefore their responses to hormones, drugs, and other regulatory agents likewise.[221] Organ-level bone formation was seen as the direct and unmodified summation of that of individual osteoblasts. Similar assumptions applied to chondroblasts and cartilage, synovial cells and synovia, fibroblasts and fibrous tissues, cementoblasts and cementum, odontoblasts and dentin, and ameloblasts and enamel.

Osteoid seams were believed to occur only in osteomalacia, and sometimes in rickets and infants.[220,700]

It was known that osteoclasts resorbed bone, but it was believed their direct responses to hormones and drugs determined how much bone was resorbed in the whole skeleton and how fast. Again, no serious thought was given to the mechanisms that produced and supported the clasts, nor to the idea that those mechanisms could play an important role of their own in disease. The same applied to chondroclasts and cementoclasts. The cellularly based resorption of fibrous tissues was unknown and it was believed that injured cartilage was incapable of effective repair.

It was assumed as another default judgment that osteoclasts responded in the same way to circulating agents regardless of their skeletal location, and similarly for osteoblasts, so that each formed a functionally single cell population. That was named the *singularity assumption* in 1973.[o,237] From that it was further assumed as a third default judgment that any net bone loss reflected either an increased resorption, or a decreased formation, while a net gain reflected either a decreased resorption or an increased formation. Hence, the *either-or assumption* described in 1981.[o] Those three assumptions persist today in varied forms with varied nuances but they only expose the atomistic views of the speaker or writer involved.

In 1963 there was no knowledge, nor even a suspicion, of the bone remodeling BMU and the coupling of R to F that occurred in it, excepting in the author's laboratory as described in Note 2 in Chapter 4. The ΔB·BMU was unknown, as renal clearance was in 1920. The skeletal role in homeostasis was assumed to reflect solely the responses of existing osteoclasts to parathyroid hormone. The coupling that appears between resorption and formation in the lamellar bone remodeling BMU has only been generally acknowledged since about 1975, so while it forms the subject of several current research projects it was not generally known in 1963 and it still is not generally realized that analogous forms of coupling occur in other IO entities and activities and in other skeletal tissues and structures, even though when it is specifically sought the evidence for such couplings is abundant and clear. Table 3 provides a partial list of them; they still await systematic study.

Bone disease was assumed to sum directly the malfunctions of individual osteoblasts and osteoclasts, so all answers to disease would presumably be found in and explained by the continuous malfunctions of unassociated but differentiated cells. Therefore, all cures would occur by correcting the disorders of such unassociated bone cells and by continuous treatment, and those ideas applied to all five basic skeletal tissues, not merely to bone. It was not even imagined that intermittent-sequential treatment — or research strategies — could add a new dimension to the treatment of skeletal disease, or to its study, or to its understanding.

There was no knowledge or even suspicion of the MCN property, the LOBO, krσ, or LBO properties, or of the fundamental transient and steady-state distinctions described in Chapters 3 to 6. Temporal incoherence and coherence were unknown, as was also the ubiquitous MES property of all regulated systems, inanimate as well as living

Table 3
EXAMPLES OF COUPLED PHYSIOLOGIC SKELETAL PROCESSES

This activity	Is coupled to	In this domain
Resorption of mineralized cartilage	Deposition of woven bone	L_2
Resorption of primary spongiosa	Deposition of secondary spongiosa	L_2
Resorption of lamellar bone	Deposition of lamellar bone	L_2
Resorption of cementum	Deposition of cementum	L_2
Deposition of enamel	Deposition of dentin	L_2
Mineralization of cartilage	Chondral growth	L_2
Production of synovial fluid	Absorption of synovial fluid	L_2
Net global bone tissue loss	Renal calcium and phosphate excretion	L_{is}
Growth of hyaline cartilage	Mineralization of hyaline cartilage	L_2
Growth of a joint	Mechanical usage	L_o
Tendon length	General body growth	L_3
Tendon diameter	Muscle strength and usage	L_3
Bone diameter	Muscle strength and usage	L_3
Skeletal macroarchitecture	Mechanical usage	L_o
Micromodeling	Mechanical strain	L_1
Microdamage	Remodeling	L_2
Macrorepair	Injury	L_3

(it is one thing to remember a mass of facts; it is another to perceive in them some of their common causative and behavioral roots). Growth was imprecisely defined; macromodeling, micromodeling, and remodeling were not distinguished; the dependence of the first on growth and the independence of the latter on it were unrecognized. The existence of at least two operational classes of cell proliferation and of L_1 and L_2 growth modes were not perceived, although the evidence was there. The mechanically important facts in skeletal architecture and physiology were assumed to be static stress and ultimate strength, and to orthopedists, supposedly compression stress stimulated the formation of hard tissue directly, and tension stress stimulated its resorption directly. Orthodontists, on the other hand, believed exactly the opposite. The responses of bony, chondral, fibrous tissue, and dental structures and architecture to mechanical usage were completely baffling and some people even disputed the idea that such responses occurred.

There was a complete lack of awareness of the role and absolute necessity in skeletal health, function, and design for the "back up" activities encoded in Relations 3 and 4 in Chapter 3, and that lack applied to all five basic skeletal tissues. Accordingly, and to repeat, all forms of regulation were assumed to act directly and only on already existing differentiated cells. Standard texts on skeletal physiology, metabolic bone disease, pathology, and orthopedic surgery tried to account for skeletal clinical-level phenomena solely in those atomistic 19th century terms. The basic knowledge, the facts, had been known for over 50 years, but their significance with respect to postnatal skeletal physiology and disease remained unperceived until 1964 and afterwards.

No forms of mechanical fatigue damage of skeletal tissues were known. Apparently, many observers suspected on theoretical grounds that such damage should occur, and Rutishauser had published some pioneering in vivo experiments that dealt with that matter,[657] but the initial response to the first published evidence of microdamage that arose during life in human material was a widespread and flat rejection. A published discussion by Milch in 1964, a watered-down version of the actual one, is illustrative (at the conference on Aseptic Necrosis of the Femoral Head, sponsored by the surgical study sections of the NIH in 1964). As with many other IO features described in this book, the evidence was later rediscovered by other investigators who were unaware of the earlier publications. The legend to Figure 8, Chapter 14 provides further information on this matter. Only recently has an awareness grown that all structural skeletal tissues must have vital mechanisms that repair such microdamage, so that disease related to that damage must reflect the influence and interplay of both the factors that cause it and the factors that repair it. In truth, such diseases reflect a disturbance in one of the physiologic balances described in Note 1 in Chapter 10.

In 1963 it was thought that osteoporoses were essentially one disease with respect to the intraskeletal mechanisms that were involved, so any major differences lay in the nature of their medical causes.[773] Thus, the osteopenia of most osteoporoses supposedly reflected either an accentuated resorption of normal bone by hypervigorous osteoclasts or, in the minority view, depressed bone formation by enervated osteoblasts. It was not realized that, for example, an osteopenia could present in a 60-year-old woman because she accumulated too little bone during growth, so a normal age-related loss could leave her with too little bone by age 60 years. It was vigorously denied and considered preposterous that increased net bone loss could occur in the presence of a decrease in the absolute rate of bone resorption. The heterogeneity of the causative intraskeletal mechanisms of osteoporoses already suggested by the writer in 1964 on the basis of early histomorphometric findings was unknown but it is now well documented although the initial descriptions and projections have become forgotten.[231]

Also in 1963, it was thought that osteomalacias resulted from a single malfunctioning intraskeletal mechanism, loosely termed and hazily conceived as a mineralization defect. Not until the nature of the bone remodeling BMU was defined did it become possible to both predict and understand how different and numerous the causative intraskeletal mechanisms and histomorphometric and physical features could be, and that severe osteomalacia could occur in the presence of normal conventional serum chemistry values. Examples of most of those situations have now been reported. They are so diverse, but individually many of them are so uncommon, that so far no successful effort has been made to classify all osteomalacias on that basis, according full respect to several colleagues who made an effort to do that in the past decade. It seems clear, however, that such a classification can now be conceived and that it will require using what is known about the IO as its framework.

In 1963 another default judgment assumed the design of the adult skeleton and its physiology reflected the blind and faithful execution of a predrawn, master genetic blueprint for all of its cellular, biochemical, and structural functions and properties. It was not even suspected that skeletal design followed the three-part strategy described in Chapters 7 and 16, and that during growth the infant's skeleton can adapt to a significant degree to its usage so the mature skeleton sums up the separate contributions of initial fetal conditions plus the subsequent postnatal adaptations, and according to principles of action that govern the behavior of the intraskeletal mechanisms of the IO. Most anatomists, of course, favored the expression, "form matches function", but when specific game rules were proposed for that relationship they usually sought their flaws rather than the limits of their truths. Note 3 in Chapter 7 describes one example of that reaction to new ideas. Again, the basic facts were there; one can identify them in retrospect, but their meanings were not perceived. Those postnatal adaptations apply to biochemical and endocrinologic activities and potentials as well as to strictly structural ones. As one consequence, great efforts have been made to define the characteristics of particular diseases and to find cures for them; little use has been made of the idea that each disease is also an experiment performed by nature that can reveal the general functions of the normal IO and of the agents that normally regulate it. That latter road holds the promise of leading to cures for whole groups of diseases.[267] One example is the ADFR treatment of osteoporoses,[24] a cure that can be applied, in principle at least and regardless of their particular causes, to most osteoporoses rather than to only one of them. In this area, a great potential lies in collaborative research that involves both human and veterinary medical scientists, for many potentially useful animal models of human disease exist and their study could help to understand the IO controls and functions that underlie skeletal disease. However, that potential remains mostly untapped even though some productive human-oriented work has been done (e.g., by High, Norrdin, Sumner-Smith, Black, and VanSickle). A recently formed CORE group of young veterinarians interested in skeletal pathophysiology has the potential for exactly such collaboration. It is modeled after the σ bone group formed by the author around 1972.

No effective accounting was made of the important and ubiquitous effects of geographically locally gated regulation (described in 1966[230] and defined here as the "L" gate of the basic IO state equations) on the control of cellular- and supracellular-level activities in skeletal tissues. Although the first unequivocal evidence of "L"-gated regulation in the skeleton was described and identified as such in 1964, and although the mechanism was also described in 1966,[230] and although numerous clinical examples of it were actually known before 1963 in other adult organ systems and in embryology, not until the late 1970s did it begin to receive the acknowledgment and attention of a handful of other authorities and only after 1980 has it begun to gain wide acceptance.

In 1963 it was assumed that only continuous treatment was needed to cure skeletal disease, and of course continuous treatment has earned a firm place in medical therapeutics. The kind of brief, intermittent and sequential treatment involved in coherence treatment and research exposes an additional dimension and potential in therapeutics and research that was not even imagined in 1963, even though analogs of the idea were well known by then in physics, chemistry, electronics, and embryology.

In one sense, the above ignorance of IO-related features seems remarkable, because in retrospect one can see that evidence of their existence was known and not in serious dispute by 1963. The "retrospectroscope" reveals that that evidence included innumerable firm clinical, gross anatomical, pathological, and histological as well as embryological facts. For example, evidence for the functional independence of bone balance on the various skeletal envelopes was already known during the career of John Hunter in the 1700s, evidence for the bone remodeling BMU acting on all four skeletal envelopes was known by 1875, and evidence for the remodeling BMU involved in endochondral ossification and for the micromodeling activity in all skeletal tissues was known before 1900, as was also gross anatomical evidence for the flexure-drift relation and the flexural neutralization mechanism. Indeed, the meaning of those phenomena must have occurred to at least some other observers before the author entered the scene, but if so they experienced and succumbed to the "attack and destroy" reaction from others, as described in the Preface to this book. This field still does not have many builders, many true heurists. It needs more, and to some degree the older heads in it are responsible for that lack. Unwittingly, they may also have sown the seeds of a revolution. Throughout human history, rigid and self-protective dogma and authority have been the possessions and weapons of older men, and they have generated in youth the forces that ultimately overturned them and replaced them with something new, and often even better. Then the young revolutionaries became in time the possessors of another dogma and self-protective authority, and that has continued from the time of the legendary Gilgamesh to today. The author senses such an upheaval beginning in skeletal science in this decade.

As for the effect of dogma, many young people today may not realize how disastrously can the ridicule of one's colleagues affect the career of a young anatomist, histologist, physiologist, or other full-time scientist. When someone makes a living doing that kind of work, then survival closely and inseparably couples to the respect and influence of colleagues, which includes their opinions of novel interpretations of otherwise undisputed facts. In the skeletal field particularly, that has been a factor in preserving orthodox ideas and methodologies, and in discouraging innovative basic ideas and lines of work. This field usually does accept enthusiastically a new *tool* for studying the skeleton. Examples would include the electron microscope and electron emission spectroscopy, the varied uses of radioactive tracers, tissue time markers, and chromatography, the microradiograph, strain gauges, the CAT scan, the mass spectrometer, varied forms of spectroscopy, the computer, statistical analysis, X-ray diffraction, histomorphometry, the ultracentrifuge, automated amino acid analyzers, and fluorescent antibody techniques. The field still disparages unorthodox proposals of the biological meanings of the information those tools provide. That must and will change and perhaps it is doing so as this is written, but it is a change that still meets strong resistance.

As suggested above, in 1963 the features of normal health and disease were conceived as the direct result of the functions and malfunctions of already existing differentiated cells. Those features included what are now known, as examples only, to consist of modeling, remodeling, the L_1- and L_2-modes of growth, repair, and the diseases known as osteoporoses, osteomalacias, dwarfism, and other children's deformities, arthritis, and periodontal disease. It was not realized that the factors that control precursor cell activity, including its enablement and regulation, dictate *where* growth, modeling, and remodeling occur, *when* they occur, how *quickly* they occur, their *durations*, and *how much* of them occur, so that many major features of skeletal disease (of course, not all) reflect various malfunctions of precursor cell enablement and regulation and/or of supporting cells, including blood vessels, rather than malfunctions of the differentiated daughter cells produced by "P" cells. In a sense, the factors that control precursor cells are the building contractor who hires, disposes, and controls the output of the workmen who actually assemble the walls, floors, plumbing, and electrical system of a house. Considered a bizarre idea when it was stated between 1964 and 1966,[229,230] since about 1978 other authorities have perceived its validity and begun to act on it.[582,583,616]

As one result of the above, little was known in 1963 of the agents that control postnatal precursor cell activities and the differentiation of daughter cells in postnatal life, and even today there is minimal hard information about those matters, but some of them are now under study.

In 1963 it was not generally realized that enabling an activity could be a different process with different rules than controlling it once it was running. Similarly, starting an automobile differs from driving it, and conception differs from the subsequent creation of an embryo and fetus. Accordingly, about 1963 most disease-oriented skeletal research and thought focused on what this text terms the regulation of already ongoing physiologic continua, and virtually ignored the enabling functions which, as indicated above, determine when, where, how quickly, and how long most tissue-level skeletal activities will occur.

Parenthetically, in terms of disease there is a big difference between a malfunction of the enablement of conception, and a malfunction of the enablement of postnatal growth, modeling, remodeling, or repair. If conception does not occur no disease results, because no baby will be born regardless of whether the male-female couple involved is disappointed or delighted over that failure. When in an otherwise healthy *postnatal* growing or adult body the enablement of the above-named basic skeletal functions malfunctions, then disease can and usually does result in a living person, it may cause much suffering and hardship, and it must be dealt with. Table 4 provides a sample listing of such diseases that involve various tissues and organs, understanding that many of them also involve malfunctions of regulation as well as of enablement, as Note 2 in Chapter 2 defined those terms.

Table 4
SOME DISEASES THAT PROBABLY INVOLVE
MALFUNCTIONS OF ENABLEMENT MECHANISMS IN
VARIOUS TISSUES

Congenital limb deficiencies	Low-turnover osteoporoses
Type I osteogenesis imperfecta	Low-turnover osteomalacias
Fracture nonunions	Spontaneous fractures
March fractures	Supernumerary digits
Syndactyly	Most forms of dwarfism
Gigantism	Acromegaly
Pseudohypoparathyroidism	Thyroxtoxic osteopathy
Paget's disease of bone	Rickets (most forms)
Spastic cerebral palsy	Athetosis, congenital
Club foot	Marfan's syndrome
Hyperparathyroid osteopathy	Dentinogenesis imperfecta
Rheumatoid arthritis	Osteoarthritis
Slipped capital femoral epiphysis	Blount's disease (tibia vara)
Pathologic microdamage accumulations	Spontaneous tendon ruptures

In the over 20-year span that separates the above described 1963 views and knowledge from the initial inscription of these words on scratch paper, the field has indeed come a long way. Its numbers, ideas, and vocabulary have each changed and grown. One must also infer that in the next 20 years the field will go even further proportionally, that the changes ahead will be even more numerous, varied, and basic, and will grow even more in scope, depth, subtlety, and sophistication. However, it seems safe to predict that the IO concept will endure and grow rather than wither and be replaced.

Note 2

SCIENCE AND RECIPES: A POSTFACE

Skeletal clinical expertise (medical and surgical) developed largely by trial and error. That is, from an immense variety of empirical treatments tried for skeletal problems in the past 2500 years, the expertise of today retained the good and forgot the bad, so it comprises a collection of time-tested recipes used by craftsmen rather than by scientists. Like the great wall of China, it was assembled stone by stone over generations of man to the accompaniment of joy and sorrow, birth and death, courage and cowardice, and always, pain and labor.

Physicians and biochemists today try to explain why those recipes work, but the explanations are often naive and flawed. To illustrate, when the ken of experience and mature perception reaches back over the centuries, it sees that physicians of today find much of what physicians thought, told their patients, and studied in 1900 was naive and flawed, even humorous, and those physicians felt the same with equal justice about their predecessors of 1800, and so on back to antiquity and the Egyptian, Imhotep. That same ken realizes that in the year 2060 physicians will perceive us similarly (a sobering realization to one who, early in his career, was so sure he knew it all, or at least the important parts).

If some hold the medical expertise of today in high regard as a true science, that reveals a basic ignorance and provincialism, the unwitting intellectual arrogance mentioned in the Preface to this work. The imperatives of medical practice and its patients' needs so preoccupy physicians that they can fail to see their expertise as only a pebble on a beach that stretches beyond view into the future, an observation that concerns perspective. People tend to measure current achievements and knowledge against those of the past, and on that scale the achievements and knowledge of today occupy most of the length of the measure that goes back to the time of Gilgamesh. On the meter of total knowledge and mastery of the skeletal system, the expertise of today occupies only a few of its thousand millimeters.

Those remarks do not disparage the recipes or the craft, or those who practice it. When one does not understand an organ system well enough to predict most of its diseases and their cures, then he must use and accept what chance, experience, incomplete knowledge, and flawed understanding allow. A craftsman uses what works whether he understands why or not. In metaphor, not all Stradivarii make violins.

However, great changes are near in how physicians conceive, diagnose, and treat skeletal disease. Growing understanding of the system finally allows predictions of disease and treatment from theory that have reasonable chances of proving correct. A significant factor in those changes is the IO. Ignorance of that link in the basic chain of skeletal knowledge has prevented predicting correctly the effects on people of innumerable cell-level malfunctions and related matters. Consequently, those effects had to be found and cataloged without being understood. Then, somewhat as happened when Watt produced a working steam engine, efforts were undertaken to understand how cell- and molecular-level matters caused those effects. Because of the above changes, a future generation of clinicians can apply a true, or at least a much better, science to skeletal disease and can understand why it works.

Since a lead time separates advances in basic understanding from translation into clinical expertise, for this writer's further active years as an orthopedic surgeon, proven recipes should remain the mainstay of clinical work, and in a not too remote future other physicians may view that mainstay as we view that of the U.S. civil war: interesting history — and thank God progress occurred afterwards.

Now, somewhat as a Nobel Laureate put it some years ago and to revert to the first person singular: while trying to plant a seed and learn to speak a new tongue of science that my elders found strange and knew not, I grew old, and suddenly I find my future is behind me. So others will tend that seed as I learn to watch another future begin to unfold at the hands of younger men speaking an even stranger tongue.

Welcome! to it, and them, and all other builders . . .

APPENDIX I

"We cannot improve a science without improving the language or nomeclature which belongs to it."

(Antoine Lavoisier, 1789)

I. INTRODUCTION

At the large meeting in Dearborn, Mich. in May of 1983 (*Clinical Disorders of Bone and Mineral Metabolism*) organized by Drs. B. Frame and J. T. Potts, data were presented on the incidence of spinal "crush fractures", mostly asymptomatic, in untreated and treated osteoporotic people, based on longitudinal X-ray studies. The estimated frequency of such fractures ranged from some 300 to 800 separate incidents per 1000 person years in people over age 60 years. Like other orthopedists in the audience, the author was incredulous because if 3 to 8 out of every 10 people over age 60 developed a new spinal compression fracture every year, it would take 10 times more orthopedic surgeons than already practice in the U.S. simply to care for them, and no time would be left over to care for club feet, torn ligaments, long bone fractures, and dislocated joints.

The real problem related to definition rather than fact. To orthopedic surgeons, a spinal crush fracture means a gross compression fracture, usually after an obvious injury, and its usually severe symptoms will put the typical patient in bed for 1 or 2 weeks. It takes some 2 months to heal properly, and the sudden change in posture it causes can cause activity-related lumbosacral-level backache for another 9 months or so.

However, the scientists in question classified as crush fractures any radiographically apparent change in vertebral body height, endplate concavity or configuration of the anterior cortex of the centrum. Most such changes occur without obvious injury, symptoms, or disability. X-ray studies of older people do show that they occur, but the "crush fracture" term chosen to signify them meant one thing to the speakers and quite another to many of their listeners.

The confusion that has its origin in semantics appears in any rapidly evolving field, particularly when it borrows terms and ideas from other disciplines and particularly when the affairs of once distinct disciplines begin to fuse into new and larger ones. It can cause communication and credibility problems, so defining the terms that convey complex ideas and phenomena becomes essential. Such problems led to Table 1 which lists terms that still have varied meanings in the skeletal field in 1984, particularly when their usage crosses interdisciplinary lines.

Such problems also led to the Glossary herein. It provides the meanings assigned in this text to many special terms. Note also the following codes attached to its entries. A single asterisk (*) identifies terms borrowed from nonmedical fields that retain their orthodox definition in the donor field. A double asterisk (**) identifies new terms coined to characterize certain previously unnamed features of the skeletal IO. A triple asterisk (***) identifies terms that have been redefined or restricted in meaning when they appear in this book, whether borrowed from other fields or belonging to orthodox medical terminology. Since the effects of the IO touch many disciplines, its vocabulary constantly grows and borrows from diverse sources, and as Lavoisier well knew. The absence of an asterisk identifies a standard medical term used in its standard meaning.

Table 2 collects and lists the meanings of certain abbreviations and acronyms that appear frequently in the main text or in the Notes.

Finally, when a new but fundamental and important property or relationship becomes perceived or devised, then it becomes a necessity of convenience and brevity to name it, hopefully choosing a short name that will convey the essence of what it signifies. As examples of the space-saving property of such names, review in the following Glossary the definitions of enablement, function, σ, coupling, ΔB·BMU, and MES. It would become impossibly clumsy and time-consuming to substitute the whole paragraphs that define them wherever they appeared in this book. Jargon inescapably accompanies progress, as Lavoisier knew. However, the creator of new terms is at a disadvantage, for often others can perceive more apt names. Thus, what the writer originally perceived and named a "phase lag pool", Jaworski aptly renamed the remodeling space, while the "tethering" between R and F in the remodeling BMU originally discovered and described by the author between 1964 to 1966 later was more aptly renamed coupling by Harris and Heaney, and the delay between osteoid formation and its mineralization first measured and described in 1963 by the author was later renamed the osteoid or mineralization lag time by Baylink. On the other hand, terms such as BMU, σ, activation, and mean wall thickness, also coined by the author, have survived intact, at least so far. The important thing here lies in the ideas signified by the doubly and triply asterisked terms. They survive and in fact expand. Accordingly, some doubly and triply asterisked terms in the following Glossary may yield to better ones in years to come, but the basic ideas and facts they signify will probably remain (although not unmodified).

Since the terms of the state equations appear frequently throughout this work, they are again listed and defined briefly below and the Glossary of terms will follow directly afterwards.

II. DEFINITIONS OF SYMBOLS IN THE STATE EQUATIONS

A   *Activity* of existing effector cells and the cells themselves, i.e., osteoblasts, odontoblasts, fibroblasts, chondrocytes, endothelial cell, synovial cell, etc.

C   *Capillary(s)* of any multicellular functional unit

D   *Differentiation* of new daughter cells into specialized effector cells, i.e., "D" → "A"

L   *Local* environmental sources of extrinsic regulation of functional units

M   *Materials* produced by differentiated "A" cells, such as bone matrix, fibrous tissue, dentin, cartilage, synovial fluid; thus, "A" → "M"

O   *Organization* of differentiated cells, intercellular materials, and capillary into intrinsically structured functional units

P   *Progenitor* or precursor cell activity that, upon command, makes new daughter cells, so "P" → "D" → "A" → "M"

S   *Systemic* agents extrinsic to the IO that can regulate its functional units, such as hormones, drugs, electrolytes, and some mechanical influences

X   Important but not yet named or discovered factor(s)

IO   The intermediary organization that occupies the tissue level and bridges the cell to the organ

III. GLOSSARY

Absolute bone volume**   (ABV) The volume of bone tissue left after subtracting the vascular and marrow spaces, and regardless of its degree of mineralization.

Activation**   The process of making something begin. See enablement below.

Activation frequency**	The writer originally coined and defined this histomorphometric term as the number of new remodeling units that begin in unit time in a unit amount of tissue or on a unit of its surface. Therefore, a kind of birth rate. The term can generalize to the enablement of any entity and/or activity of the IO, and thus it can apply to all five basic skeletal tissues.
ADFR	Activate-depress-free-repeat. A form of coherence treatment described in Chapter 5.
Andistributive*,***	In mathematics, a property such that the arrangement or order of terms is important. In the book it means that the sequence or order of the parts of a biological system in time or/and in space is crucial to its nature, its functions, and its control by enabling and regulating agents. As a result, changing that order either creates a different system with quite different properties or creates an artificial system that does not exist in nature. See Chapter 2.
Anholistic	The property of not accounting for the whole system, of omitting essential parts of it. The antonym is holistic.
Appositional rate**	In histomorphometry the mean thickness of new bone deposited in unit time on bone-forming surfaces. It provides a useful index of the cell-level vigor of osteoblasts. Increasingly it is subdivided into the matrix and mineral apposition rate although the differences between the two are rarely large.
Arcane*	Assume a world of blind people into which comes one person who can see. None of the blind people ever saw day and night follow each other so they have no knowledge of it, yet the sighted person perceives that after the first day has passed. His knowledge is firm and indisputable on the basis of the evidence that is there, but it is arcane in that none of the other people in that world have ever seen it themselves and consequently they could choose to believe that the sighted person tries to deceive them or himself, and go on believing there is no such thing as night and day.
Association	Any combination of tissue elements or organs that provides special behavioral and structural properties that are lacking in the uncombined elements themselves.
Biomechanical competence**	The ability of a tissue to model appropriately in response to excessive strains, and also correct its internal microdamage. It is not the same thing as the mechanical incompetence described later.
Blasts	Differentiated cells that actively engage in synthesizing intercellular organic matrixes, e.g., osteoblasts, odontoblasts, fibroblasts.
BMU**	Basic multicellular unit, usually referring to BMU-based lamellar bone remodeling. The quantized discrete "packet" of bone remodeling. The term is coming to signify the histologically apparent activities of the unit, its R and F. The new bone moiety it makes is becoming known as the BSU or basic structural unit.

Bone mass	In clinical medicine an imprecise term which, when used by clinicians, usually has the particular meaning of ABV as defined above. In physics and chemistry, however, the true physical or gravimetric mass of a bone or sample.
Bone balance	Net gains (positive values) or net losses (negative values) of bone tissue as a result of the summarized effects of new bone formation and the resorption of older bone. Strictly speaking, the bone balance should be a derivative, that is a rate of gain or loss, meaning a first or higher order derivative, but often also used in the sense of delta (Δ), meaning a net gain or loss over some period of time.
Brittle*	In materials properties, breaking or rupturing with minimal prior deformation. A brittle material may be also a weak one (chalk) or a strong one (glass or high carbon steel).
BSU**	Basic structural unit. The new bone moiety formed by a BMU, and that serves structural and homeostatic functions (and possibly others) until replaced by later remodeling by even newer bone.
Cantilever flexure***	Bending caused by applying an end load to a curved column or other structure such as a bone or part of the cortex of a bone as described in Chapter 7. The combination of its curved shape and the end load causes its curvature to tend to increase and that increase represents the flexure in question here. The same effect can be achieved by applying the end load offset from the center of the cross-section at the end. See Figures 7 and 8 in Chapter 7. In bones, cantilever flexure usually limits and/or neutralizes supra-MES strains caused by direct flexural loads.
CGFRC**	In chondral macromodeling, the chondral growth/force response characteristic described in Chapter 8.
Coherence***	Referring to an IO population and to time, coherence is the state in which multiphase entities begin and evolve together in step, e.g., in register in time or in synchrony.
Coherence treatment**	Inducing a state of temporal coherence so that the proper timing of the delivery and withdrawal of an agent can confine its effects to any one of the separate, sequential internal stages of an activity that is provided by a biological entity, IO or otherwise. The stratagem circumvents the nullifying effects of the MCN property on the continuously treated entity or on a temporally incoherent population of them.
Collagen	The fibrillar protein that forms the basic structural organic component of fibrous tissues, bone, cartilage, dentin, and cementum.
Compliance*	In mechanics, the opposite of stiffness. Therefore, the ability to yield or deform under an applied load. Rubber is compliant, steel is not.
Component*	In mechanics, a force, usually a load, that is part of a collection of forces that combine to form a single equivalent force named the resultant. A component of a resultant may itself be the resultant of two or more subforces. See resultant below and Figure 2.

Compression*	In mechanics, a force that presses two things together, and therefore the opposite of one that pulls them apart. Also, the strain caused by such a force, and the stress that resists it. See Figure 3 in Chapter 7.
Continua*	The ongoing phase of any process that began earlier, whether or not starting it involved different controlling factors.
Control***	In this text, the combination of factors that enable and then regulate any and all kinds of skeletal activities, thus control = enablement + regulation.
Coupling**	This term has several different meanings at present, depending on who uses it, but when used by others it usually refers only to bone turnover. Current confusion about its meanings in the literature justify an effort to explain it here, and they also illustrate the value of using space-time domain referents accurately. First, in this book it names the L_2-domain mechanism(s) that connects or tethers an initial packet of resorption to a directly following one of formation, e.g., as in the lamellar bone remodeling BMU. The mechanism(s) responsible for that kind of coupling remain unknown. Second, the L_{is}-domain/ Frost-Heaney observation described in Note 1 in Chapter 16 notes that in the intact human, total bone resorbing activity connects or couples somehow to total bone-forming activity, so that in steady and nonmalignant states limitless, generalized losses or gains of bone do not occur, either in disease or in response to hormones, drugs, or other agents. Of course, that means that something connects total formation to resorption without of itself revealing how. It represents a fact that once looked for an explanation. We now do have three explanations for that fact, and others may become apparent in the future. The known ones include the BMU-level coupling named above, plus the MCN regulatory effect described in Chapter 3 that applies to L_1, and its analogs to L_2- and L_3-domain bone remodeling, plus the L_o-domain growth and macromodeling mechanisms described in Chapters 4, 6, and 13. Third, then, from one point of view the growth-dependent, L_o- and higher-level macromodeling determinants couple or connect periosteal and cortical-endosteal resorption and formation drifts in stereotyped patterns. As a result, when formation drifts increase in one part of a bone, resorption drifts usually increase in others, and the conductors in that orchestra include growth, mechanical usage and "L"-mode bone controlling effects such as that found in Gaucher's disease, leukemia, and in sickle-cell anemia. However, that coupling arises uniquely in L_3-space-time and does not exist in L_2-space-time. In sum then, depending on who uses the term "coupling", it can mean bone remodeling BMU (L_2-domain origin), bone macromodeling (L_3-domain origin), Frost-Heaney observation (L_{os}-domain origin), or any other systems that cou-

Creep** — In materials science a gradual, time- and load-dependent and irreversible deformation of an otherwise rigid, solid material or structure. Fair amounts occur in normal cartilage, less in tendon and ligament, little in normal bone, a bit more in osteomalacic bone, and very little indeed in enamel and dentin. Increased amounts can occur in diseased tissue.

ple their responses (varied origins). To repeat, this book advocates an even more general use of the coupling term, as described in the last entry above.

Cytes — Differentiated cells that reside in already elaborated intercellular organic matrixes, e.g., osteocytes, chondrocytes, synovial cells, cementocytes, fibrocytes, odontocytes.

Diaphysis — The shaft of a long bone. See Chapter 4.

Direct flexure*** — Bending of a structure caused by applying a compression or tension load directly to its side, as in Figure 7 in Chapter 7.

Discrete* — The property of having finite and measurable dimensions and limits in time and space.

Disease*** — Any condition that impairs the ability of the organism to function properly in its natural environment. See Note 3 in Chapter 2.

Domain — In this book, the term usually signifies a particular class of referents that apply to the level of biological organization, and/or to the space and time dimensions appropriate to the properties and activities of a particular level or organization and/or to the different IO or other entities in it and/or to the activities and functions themselves that are unique to such a level or entity.

Drift — A unidirectional motion through tissue space of a bone surface, an intact cortex, an intact diaphysis, or a tooth socket. It usually results from bone macromodeling as described in Chapter 7.

Effector cell*** — Any differentiated cell, such as an osteoblast or osteocyte, chondroblast, or chondrocyte. Signified by "A" in the IO notation used in this work.

Elasticity* — The ability of a structure or material to return itself to its original shape and dimensions after being deformed by a load, and then deloaded. See also resiliance and viscoelasticity.

Enablement** — The process of initiating or starting a biological activity that previously represented a potential activity rather than an already ongoing one.

Endoprosthesis — An artificial device implanted in the body to replace some part of it and its functions, such as an artificial joint, artery, tooth, bone, or ligament.

Endosteum — The flattened sparse layer of cells on cortical-endosteal bone surfaces that lies between the bone and the marrow. Sometimes also signifies the comparable layer on trabecular surfaces. See Chapter 5.

Envelope** — The functionally discrete bony surfaces of the skeleton, including periosteal, haversian, cortical-endosteal, and trabecular. See Figure 1 in Chapter 5.

Epiphyseal plate	Also called the growth plate, at which growth in length occurs in long bones. A major anatomical difference between an epiphyseal plate and an apophyseal plate is that the former has a joint adjacent to one side but the latter does not. Some anatomists argue heatedly over this matter. See Figures 21 and 22 in Chapter 5.
Epiphysis	That part of a growing bone that lies between the joint surface and the epiphyseal plate. See Figures 21 and 22 in Chapter 5.
Equilibrium*	The state in which a system's input equals its output, its synthesis equals its degradation, total resorption equals total formation, and so on. Therefore, a special case of the steady state.
Fatigue*	In mechanics, structural failure due to cyclic loading well within the ultimate strength limits of the structure or material. The ultimate strength equals the load needed to break or rupture the structure or material under a single application. Fatigue damage begins as one or more of several kinds of microscopic damage that progresses slightly at each load-induced strain until finally so little material in the cross-section of the structure remains undamaged that a normal load causes what little remains intact to fail.
Feathering**	A defect in bone mineralization that arises independently of histological bone-forming activity. The author has recognized several different patterns, which suggests several different causative mechanisms are involved. Still a poorly studied matter.
Feedback*	In cybernetics, an arrangement in which a system senses its current output status or behavior by some means and converts that output into a signal that regulates its output status or behavior in the immediate future. The game rules of the system determine if the responses occur in the negative or positive feedback mode.
Flexure*	In mechanics, bending, as in bending a pipe, pencil, or paper clip. It is a kind of strain or deformation that combines in specific patterns all three principal strains. See Figure 1, and also cantilever and direct flexure.
Force*	In physics and as Newton said, that which can accelerate matter. In biomechanics, two special categories of forces are recognized. Those forces applied to a material or structure from without it are named loads, and the internal elastic forces of the strained or deformed intermolecular bonds within the material are named stresses.
Fragility*	In mechanics, breaking or rupturing with little energy absorption. Thus, the opposite of toughness.
Function***	Any property, activity, or combination of them that one entity in a biological organization provides directly to the next-higher level of organization and that the latter level cannot function normally without. In the relations in this book such functions are signified by F. This is a special restriction of the conventional meaning of the word. In

	this book the mathematical general meaning of a function is signified by *f*. See Note 4 in Chapter 2, and also Chapter 2. Functions form a very small subset of the properties of any biological system.
Gate***	Any biological entity or other feature that serves as a funnel to direct the entrance into a biological organization of the factors that control its operation. In doing so, any such gate affects how that organization responds to its enablement and/or to its subsequent regulation. The meaning resembles the meaning of a gate in electronic digital logic, and as noted in Chapter 3, serial and parallel gating and combinations of them occur in the skeletal IO.
Halo volume**	The perilacunar bone surrounding osteocytes which can demonstrate defects in mineralization and/or in matrix composition and structure independently of the more distant, surrounding bone. Halo volume abnormalities occur regularly in familial vitamin D-resistant hypophosphatemic rickets, in chronic renal failure, and in certain osteomalacias.
Haversian	Refers to osteons, either primary or secondary as desired.
Histogenesis	The matter of initiating the formation of any type of tissue in a place where its active formation does not already occur.
Histomorphometry	The methodology that provides information about the quantitative static and dynamic properties of bulk tissues from measurements made on thin sections of that tissue. See Note 1 in Chapter 3, Note 1 in Chapter 4, and Note 1 in Chapter 6.
Holistic*	The matter of accounting for the whole system, and each of its various parts in their proper perspective in relation to the whole. The antonym therefore would be "anholistic".
Insertion	In anatomy, the attachment of a muscle or its tendon to the bone or other structure that it acts on. The other end is named the origin.
Intermediary organization***	The tissue level that bridges unassociated cells to intact organs and that contains the supracellular systems, mechanisms, and game rules that directly underlie the health and diseases of the organ.
kro**	The property of some IO systems and/or activities in which the amount of work performed by the entity is somehow predetermined at the time it is enabled, so it continues to function until that amount of work is completed and then it stops functioning, and in essence regardless of how slowly or quickly the work was done.
LBO**	The losing betting odds property. See Chapters 2 and 3.
LOBO**	The lamellar-on-bone-only property. See Chapter 2.
Load*	In mechanics, the outside force that, when applied to a structure, then deforms it to create its internal stresses.
MCN**	Multiple, chained, neutralizing effects that protect the intact system from the otherwise harmful effects that various agents can have on any of its unassociated internal parts.

MWT	In histomorphometry, mean wall thickness, the mean thickness of the layer of bone deposited by a bone remodeling packet. Therefore, an index of the amount of bone in the completed packet. A complementary MED or mean erosion depth has been suggested recently by Jaworski and used by Melsen and Mosekilde and their students as an index of the depth of the resorption cavity or bay that precedes the bone forming phase of a BMU.
Macromodeling**	The process by which agents external to an actively growing tissue or structure shape its directions and rates of growth to cause the naked eye level features of its architecture.
MES**	The minimum effective signal, the lowest signal strength that will evoke the appropriate responses of a feedback system. Signals equal to or larger than the MES will control the system, while those smaller than the MES will not. Usually not a step function but, rather, a range of gradual change.
Metaphysis	The part of a bone between the shaft and the epiphysis or epiphyseal plate. Lately often called the physis. See Figures 21 and 22 in Chapter 5.
Mechanical incompetence**	The state of fracturing, collapsing, rupturing, stretching, and/or giving rise to pain while under normal mechanical usage. Often, but not always, the result of an underlying biomechanical incompetence.
Mesenchymal cell	See precursor cell, which is a better term for what mesenchymal cell meant to most people around 1963.
Microdamage**	Microscopic forms of physical damage to a structure or material which, if allowed to accumulate, will ultimately lead to complete fracture or rupture under perfectly normal loads. An early kind of fatigue damage in biological structural materials.
Micromodeling**	The process by which agents external to an actively forming tissue determine its fiber grain and the orientation of its cells and capillary as it deposits and in L_1-space-time. It is a new if basic idea and term, so further insights may expand on this definition.
Micropetrosis**	Domains of dead bone and usually quite stiff bone surrounded by domains of living and less stiff bone and in physical continuity with them. The canaliculae and even the osteocyte lacunae in micropetrotic bone can become completely filled with solid mineral deposits. The condition can arise independently of infection, microembolism, and microthromboses in the regional blood supply.
Modeling**	The process by which agents external to a growing tissue influence its growth speed and direction and its fiber grain in ways that create its microscopic and gross architectures. Subdivided into micromodeling and macromodeling mechanisms. See Chapters 7 to 9.
Moments*	The system of principal internal stresses in any structural member subjected to flexural loads. See Figure 1.

Mu**	The histomorphometrist's term for the activation frequency, the number of new IO units that begin in unit time in a unit amount of tissue. Therefore, a kind of birth rate.
Negative feedback*	In cybernetics, a control system mode in which an error in a regulated property creates a signal that makes the system then reduce that error.
On-Off**	The ability of a multicellular system to alternate between periods of full or continuous activity and periods of relative or total inactivity. The speed during the "On" state can vary somewhat like a car that is moving.
Origin	In anatomy, the base attachment of a muscle, usually proximal to its insertion.
Osteoid seam	In histology, the as yet unmineralized organic matrix of bone, usually newly synthesized and usually forming a layer on preexisting mineralized tissue that can vary in thickness from $\sim$3 to $\sim$30 μm. Also sometimes simply called osteoid. Excessive amounts of osteoid have been named hyperosteoidosis by Meunier and colleagues in Lyon. See Figure 2.
Osteomalacia	In medicine, one of a varied group of metabolic bone diseases in which defects in the mineralization of bone tissue arise, defects in its repair of microdamage and macrodamage may arise, and an excessive amount of osteoid accumulates in the skeleton due, in part at least, to a retardation in its mineralization in L_2-space-time. See Chapter 6.
Osteon	In histology, a haversian system, either primary or secondary. It is the best known kind of BSU, and is found within compact bone.
Osteopenia***	In medicine, a skeleton that contains less bone tissue than normal when expressed in ABV terms. See Chapter 6.
Osteoporosis***	In medicine, a disease that combines mechanical incompetence, an osteopenia, and an underlying medical cause(s). See Chapter 6.
Packet	See quantum, below.
Parallel gating**	In digital logic as well as in IO physiology, the situation in which different regulatory agents reach an IO entity simultaneously but by different gates to exert their particular effects on the functions of the intact IO unit. See Chapters 3 and 6.
Partition Property**	In engineering analysis and with respect to the natural structural mechanical design of the skeleton, this means that its architecture meets the demands of all of the various major loads that it carried during its postnatal growth. Since those loads are usually varied in their lines of action as well as in magnitude and kind, and also at the times they occur, there will usually be an excess of structural material when the actual architecture is tested under any single real load, or under any single resultant of many loads. To understand the fit of architecture to mechanics, one must partition or "resolve" or subdivide

	all of the loads it carries over some time period into a collection of appropriate, representative, and different subresultants or parts.
Perichondral ring	The layer of cartilage enveloping the lateral side of an epiphysis. See Figures 21 and 22 in Chapter 5.
Periosteum	In anatomy, the soft tissue layer adhering to the outside surface of a bone.
Precursor cell	In histology, any cell that can give rise to a differentiated effector or "A" cell, regardless of the precise mechanism involved (e.g., by cell division or modulation and directly or via several intermediary steps).
Positive feedback*	In cybernetics, a control system mode in which an error in some regulated property creates a signal that then makes the system increase the error. Often called the "vicious circle".
Principal strain*	In mechanics, any of the three fundamental strains considered alone; tension, compression and shear. The bulk strains of solid materials and structures can always be expressed as particular combinations of the three principal strains.
Principal stress*	In mechanics, any of the three fundamental stresses: tension, compression, and shear. The stress always corresponds predictably to the strain in kind but not necessarily in magnitude. A property called Young's modulus of elasticity or simply the elastic modulus defines the stiffness of a material and therefore relates the amount of stress associated with a particular strain of that material. An increasing modulus means increasing stiffness.
Property***	In this book, any enumerable or stateable feature of any entity in a biological organization, whether physical or behavioral, and whether important or trivial.
Quantum***	Herein, the property of forming some kind of packet or natural unit that has measurable limits in time and space. When one breaks such a quantum up into its component parts then the properties and functions that it provides to the intact subject vanish, so in that sense it provides activities and structures that cannot be further divided. They are created by the very association that is characterized as a biological quantum in this text. The packet or quantum could refer to an anatomical structure such as an osteon, a trabeculum, a chondral clone, or an epiphyseal plate, or to the amount of physiologic work supplied by any such packet, or to a combination of both. Usually in the intact organ, and due to their normal state of temporal incoherence, the effects of the many quantized packets of its IO tend to sum up in a way that conceals the quantized or incremental basis of its functions and other properties.
RAP**	The regional acceleratory phenomenon (see Chapter 11).
Referent**	That to which any datum or other observation is referred. It might include time and/or space dimensions, anatomical location, level of biological organization, sex, anatomical structure, age, a physiological activity, or any

Regulation***	other specifiable feature. The organizational domains referred to in the main text form a subset of the general class of referents.
	That which controls the continuum phase of a biological activity. It is distinguished in this book from that which enables or initiates a previously quiescent or absent biological activity or other feature. Starting an automobile engine would be an enablement. Then driving it would be regulation.
Remodeling***	In this book, a turnover of tissue in discrete, semimicroscopic packets, and usually ones in which an R phase precedes an F phase and some activating or enabling event initiates the process. Hence, ARF. The remodeling mechanism can replace one type of tissue with another, or replace older tissue with newer tissue of the same kind. The older scientific literature and the current clinical literature also use the term to signify what the writer has named modeling or sculpting and that is admittedly confusing. See Chapter 13.
Repair	Healing of injury. The microrepair mechanisms that act on microdamage should be distinguished from the macrorepair mechanisms that act on grosser forms of damage, for they differ in their nature, their scale, their physiologic controls, and in their apparent purposes.
Resilience*	A kind of elastic behavior in which a deformed material or structure returns nearly as much mechanical work, and as quickly, to its environment in returning to its original shape and dimensions as was exerted on it by a load in deforming it. A tennis ball or a high-quality spring are good examples. Bone is fairly resilient while cartilage is not.
Resorption	The removal of an already elaborated tissue such as bone, mineralized cartilage, or ligament by special and usually multinucleated cells named clasts. The term refers to the activities of those cells, individually or collectively as the case may be. The term does not, repeat not, refer to a negative bone tissue balance, meaning any net loss of tissue. The practice of using the word resorption to signify a negative bone balance or a negative delta should stop, for it has caused needless confusion and misunderstanding.
Resultant*	In mechanics, the single force, including its line of action in tissue space and its magnitude, that represents the summed up effects of two or more separate forces called components of the resultant. When engineers break a resultant down into its individual components they say they "resolve" it. This text speaks of "partitioning" it into its component subresultants or parts. See Figure 3.
Rigidity*	In mechanics, the property of retaining shape and dimensions under an applied load. When a material lacks rigidity in shear, but is still rigid in tension and compression it is a fluid and it will flow under an external force,

Saturation***	whether slowly or rapidly. When it is rigid in shear too, then it is a solid. In physiology, the maximum level of response of a physiologic mechanism in a biological system to a challenge or controlling agent, so that stronger stimuli do not lead to further increases in the same kind of response of the mechanism in question.
Shear*	In mechanics and as a strain, the kind of motion that occurs between a rug and the floor when the former is pulled over the latter. Also, the resistance to that motion (shearing stress) and also any load that tends to cause it (a shearing load).
Serial gating***	In digital logic as well as in IO physiology, the situation in which the action of some agent on a biological system passes sequentially through two or more kinds of cells or other biological compartments before reaching its ultimate target.
Sigma**	The natural time period of a system, the lead time that must elapse between the delivery of a sharply pulsed challenge and the final and characteristic response of the system to that challenge. Also frequently referred to as a latent period. All biological systems possess the property.
Steady state*	In cybernetics, chemistry, and physiology, any state or mode of behavior of a system that it can be made to maintain indefinitely by indefinitely continued action of one or more controlling agents.
Stiffness*	In mechanics, the property of resisting deformation under an applied external load. It is the inverse or opposite of mechanical compliance, and it is not the same thing as strength. There are weak stiff materials (chalk) as well as strong ones (steel). A common numerical measure of the stiffness of a material is Young's modulus of elasticity. (See principal stress.)
Strain*	In mechanics, any deformation or change in the shape or dimensions of a structure caused by any kind of load applied to it. It does not mean strength or stress. Compression and tension strains are usually expressed as the change in length per unit of original length and may be given as unit strain (decimal fraction of the unit length) or microstrain (millionths of the unit length). Thus, a compression strain of 0.02% of the original length of some bone or part thereof is exactly the same thing as a 0.0002 unit strain or a 200 microstrain. The emerging importance of strain rate (i.e., how rapidly a strain occurs) and frequency (i.e., how many times an hour or day it occurs) in regard to mechanical effects on biological structures and materials means that in the future some numerical figure of merit or parameter must be devised to express those factors.
Streaming potential*	In physical chemistry an electrical voltage (the Zeta potential) produced when an electrolyte moves rapidly through very small spaces, clefts or tubes. Also associated with

	transient ionic imbalance in the composition of the fluid emerging from those spaces.[267,658]
Strength*	In mechanics, the measure of the magnitude of the external load required to break a structure, or a unit amount of its structural material. Usually expressed in terms of unit or total stress or load, but it can also be expressed as strain.
Stress*	In mechanics, the internal force in matter that resists its deformation by an applied external load of any kind. It results from the elastic resistance of the intermolecular bonds to that deformation. One may speak of the *total* stress in the whole section — say a complete cross-section of a bone or tendon — or of the *unit stress* in a typical unit area of the section such as a square millimeter, a square inch, or a square meter. In physiology, the stress syndrome as described by Selye, which represents a systemic endocrinologic and biochemical response to varied challenges, a response that also can modify normal local tissue activities and local reponses to varied challenges.
Subsigma**	A period of time shorter than the sigma value for a system.
Temporal coherence***	The state in which two or more units of a biological system begin and proceed together through their subsequent internal steps or sequential internal activities.
Temporal incoherence***	The state in which two or more units of a biological system begin and proceed through their internal steps independently in time.
Tension*	In mechanics, pulling apart, referred to a load, a strain, or a stress, any one of which implies the other two.
Toughness*	In mechanics, the property in which a material absorbs a lot of energy before it ruptures or fractures. The opposite of fragility. Nylon and tendon are tough, chalk is fragile.
Turnover	In physiology, removal of older material and its replacement by newer material of the same kind.
Transient*	In cybernetics and control systems analysis, any change caused by a challenge to a system that must arise because of the nature of the system but which then must also disappear before the steady-state response of the system to that challenge can arise. Thus, a subsigma response.
Typical***	In this book, equivalent to the arithmetic mean.
Uniaxial*	In mechanics in referring to a mechanical load, this means that the line of action of that load or force is directed accurately parallel to and coincident with the longitudinal axis of the structure, and applied exactly in the middle of one of its ends or evenly distributed over the surface of that end.
Unit load*	In mechanics, the magnitude of the loading force carried by a typical unit area of the cross-section of a structure (e.g., bone, tendon, tooth) or by a typical unit area of a joint or other surface. In that same sense one may also speak of unit stress and unit strain, see Figure 4.
Unit strain*	The deformation of a unit length of material or structure, expressed as a decimal fraction of its original length. Thus, where L = length, $\Delta L/L$. Multiplied by 1,000,000 it

	becomes microstrain, and by 100, the percent strain. Thus, 0.5% strain = 0.005 unit strain = 5000 microstrain.
Unit stress*	See above.
Units*	The dimensions in which numerical data are expressed. Table 3 provides conversions from English units to those of the metric system that apply to most of the material in this book.
Viscoelasticity*	A kind of elastic behavior in which most of the mechanical work done in deforming a structure is dissipated as heat or in other forms, but not as mechanical work done on its environment as the structure returns to its original dimensions. A good example is a tennis ball with a large leak in its casing. Cartilage is quite viscoelastic.

Table 1
SOME MULTIVALUED OR AMBIGUOUS SKELETAL TERMS (AS OF 1985)

Activation**	Formation surface**	Plexiform bone
Activity	Fragility*	Property***
Appositional rate**	Function*	Precursor cell
Biomechanical competence**	Growth	Regulation***
Bone balance	Homeostasis	Repair
Bone formation	Hyperosteoidosis**	Remodeling***
Bone density	Load*	Resorption
Bone mass	Lamellar bone	Resorption surface**
BMU**	Mean wall thickness**	Sigma**
Bone resorption	Mechanical competence**	Spontaneous fracture
Brittle*	Mesenchymal cell	Stiffness*
BSU**	Microdamage**	Strain*
Bone volume**	Mineralization defect	Strength*
Compliance*	Modeling**	Stress*
Creep*	On-Off**	Toughness*
Crush fracture	Osteoid	Turnover
Fatigue*	Osteomalacia	Woven bone
Fibrosis	Osteopenia	Wolff's law
Force*	Osteoporosis	

Table 2
DEFINITION OF ABBREVIATIONS AND ACRONYMS

ABV:	absolute bone volume	MC:	multiple-chained
ADFR:	activate-depress-free-repeat	MDx:	microdamage
ARF:	activate-resorb-form	mwt:	mean wall thickness
BMU:	basic multicellular unit	OPn:	osteopenia
BSU:	basic structural unit	OPs:	osteoporosis
CGFRC:	chondral growth/force response curve	PTH:	parathyroid hormone
kσ:	constant-rate-sigma	μ:	mu, the activation frequency
LBO:	losing betting odds	RAP:	regional acceleratory phenomenon
LOBO:	lamellar on bone only	σ:	sigma, the lifespan
MCN:	multiple-chained-neutralizing	SATMU:	skeletal adaptation to mechanical usage
		SOS:	"save our souls"

Table 3
CONVERSION OF UNITS OF FORCE AND PRESSURE (ENGLISH TO METRIC)

Force
    1 pound = 4.4 Newtons = 0.45 kilograms
    1 Newton = .22 pounds = 0.102 kilograms
    1 kilogram = 2.2 pounds = 9.8 Newtons
Pressure
    1 pound/square inch = 0.00073 kilograms/square millimeter
                              = 0.073 kilograms/square centimeter
                              = 6896 Newtons/square meter
    1 kilogram/square millimeter = 1375 pounds/square inch
                              = 100 kilograms/square centimeter
                              = 9,800,000 Newtons/square meter
    1 kilogram/square centimeter = 13.7 pounds/square inch
                              = 0.01 kilogram/square millimeter
                              = 980,000 Newtons/square meter
    1 MPa = 1 megaPascal = 10^6 Newtons/square meter
                = 142 pounds/square inch
    10 MPa = 1420 pounds/square inch
    1000 pounds/square inch = 6.9 megaPascal

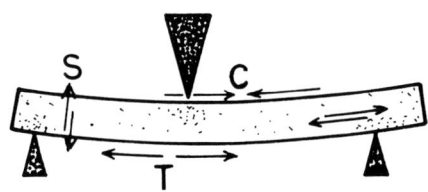

FIGURE 1. When the horizontal beam accepts a direct compression flexual load, it strains as shown. That generates compression strain and stress in its upper fibers, tension in its lower ones, and two planes of maximum shear, one perpendicular to the long axis, the other parallel to it. Those combinations of the three principal stresses are known as flexural moments, and they arise in any structure subjected to flexure, whether by the direct mode or the cantilever one or by a combination of them. It can be seen that the pattern of tension and compression shown here implies or would allow one to predict the pattern of shear. The converse also holds true, and one meaning of that is that the modeling laws for skeletal tissues could also be expressed in the referent of shear if one wished to. However, they would then become harder for most nonengineers to understand.

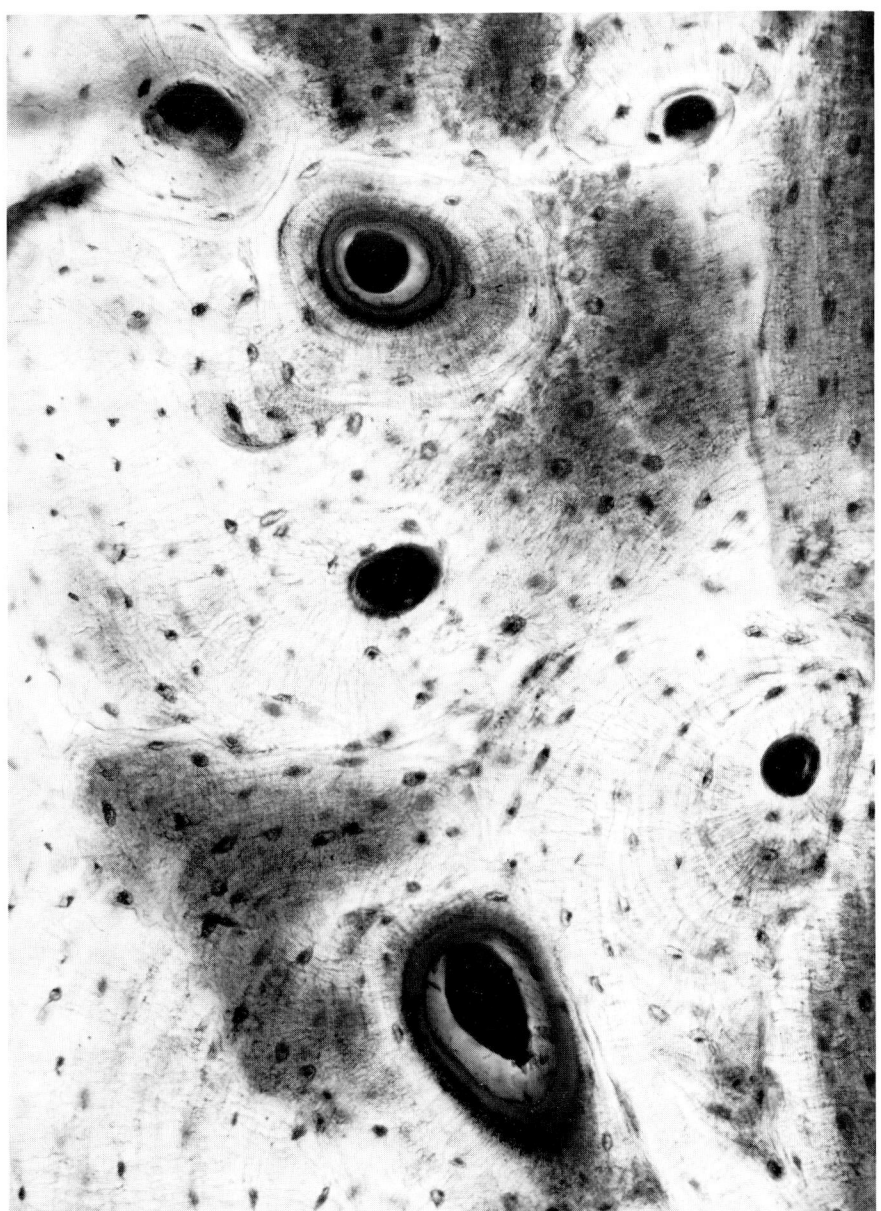

FIGURE 2. Undecalcified cross-section of human compacta, basic fuchsin, 240×. An osteoid seam lies at 12 o'clock and another at 6 o'clock, consisting of the dark hyaline rings. The opaque blobs in their haversian canals are the overstained capillaries, blasts, and supporting cells in this 70-μm thick section. The darker surrounding regions represent feathered semipermeable bone and the small dots scattered over the field, those osteocyte lacunae that were in focus. The field shows six haversian canals. The bone around each canal represents an osteon or secondary haversian system.

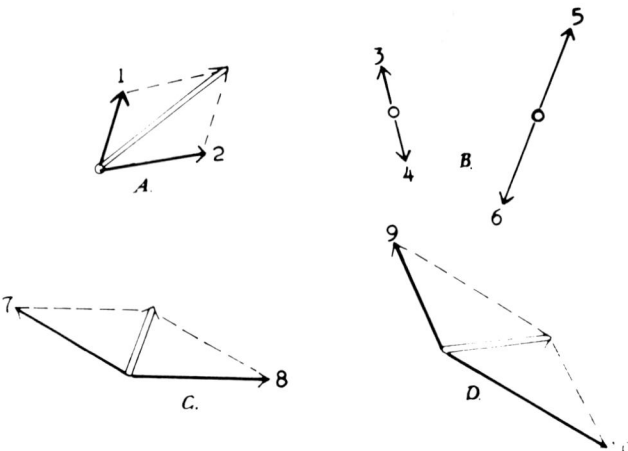

FIGURE 3. In the upper left drawing, two separate tension forces indicated by the solid arrows act at the same time and in the directions shown. Their lengths are proportional to their magnitude and both have the same point of origin. The doubly lined arrow that forms the diagonal of the paralellogram completed by the dotted lines is the resultant of those two forces, meaning that it alone would exert exactly the same action on the origin as the other two forces acting together. Each of those two loads therefore is a component of their resultant. When, as at B, two forces are equal but exactly opposite in their direction, then their resultant equals zero. At C the upwards-acting load on A is shown to be itself a resultant of two other simultaneously acting components, and likewise for D. As the figure is set up then, the resultant shown in A is also the resultant for the eight separate forces or loads shown in B to D. This illustration of the mechanical engineer's use of the paralellogram rule in designing structures omits something of great importance in biological structural design and that is that in the skeleton the above eight forces do not usually all act at the same time. Accordingly, at one moment load #1 may act on, say, the knee joint or the subtrochanteric region of a femur, and at the next moment #4, and at the next #8, and so on. It becomes immediately clear that a structure designed to withstand indefinitely a load represented by the resultant at A could fail from the effects of #1 to 8 acting separately. Hence, the load partition principal and another usually overlooked effect of time in skeletal biomechanics.

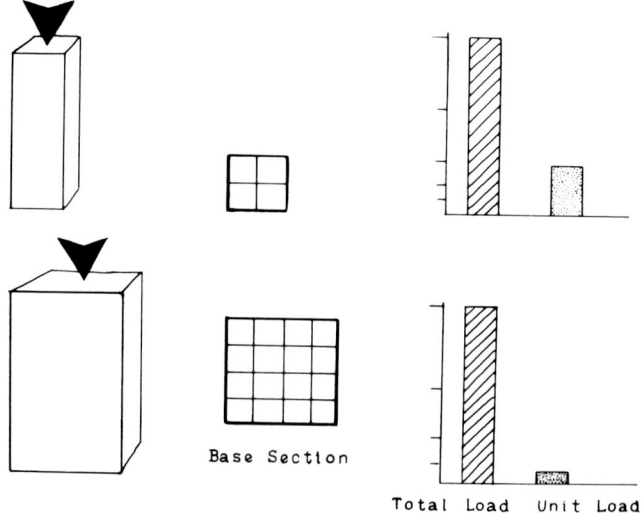

FIGURE 4. This diagram explains the difference between total load and unit load (similar definitions apply to total and unit stress). The total load is the same on both left-hand columns, but the lower one divides it into four times more area units at its base than the upper column, so each of those area units carries one quarter as much of the load as in the upper column.

ACKNOWLEDGMENTS

Only another who has written a tome such as this one can appreciate the labor, frustrations, problems, and anxieties it involves, and also how much an author owes and to how many people for the help and support they render in the doing. It is a pat platitude but true — without such help it cannot be done and an acknowledgement inserted somewhere in a text seems an inadequate recompense, but it is an established one, so this one applies to this book, a link in some chain, the value of which others and time will judge but which could not have been forged without the following.

Mrs. Betty Uhernik and Miss Greta Hanson wore out their fingers typing and retyping the many drafts of the parts of the book as well as the final manuscript. Colleagues in the Southern Colorado Clinic, particularly Drs. C. A. Hanson and J. Chimento accepted with understanding and good humor the many intrusions of this task on the conduct of the medical and business affairs of the author and clinic. David Gavin did most of the drawings, including many revisions thereof, and capably and promptly. Mrs. Doris Oliphint supplied support, warmth, encouragement, and steady supplies of coffee. The Medical Illustration Department of Henry Ford Hospital supplied many of the photomicrographs and X-ray prints.

Less directly, but of no less importance, many colleagues in the clinical and research communities rendered help, support, experience, ideas, and criticisms that weave into the fabric of this text, imperceptably perhaps to the reader thereof but nonetheless there. The clinicians include particularly Drs. F. N. Potts (deceased), J. D. Godfrey, P. A. Casagrande, J. W. O'Meara (deceased), A. Haddad, C. L. Mitchell, J. L. Fleming (deceased), L. Shifrin, P. Shifrin, K. Wu, B. Frame, H. Duncan, B. N. Epker, M. Bartley, J. S. Arnold, A. M. Parfitt, D. Mitchell, S. Stanisavljevic, K. Gitlin, F. Henny (deceased), R. Buerki, C. O. Bechtol, R. Ramsey, H. Pedersen, E. Sedlin, R. S. Hattner, M. R. Urist, H. Takahashi, T. Sakou, P. J. Neunier, P. Coupron, F. Melsen, A. Miyazaki, Z. F. G. Jaworski, J. Albright, D. VanSickle, R. P. Heaney, J. Talbot (deceased), B. E. C. Nordin, H. Copp, G. Bauer, A. Nachemson, H. Schoene, R. A. Robinson, A. R. Arnstein, L. Belanger, J. C. Pickett, H. Boyd, and W. Armstrong (deceased).

Those in the research community (or who wore the laboratory "hat" in some of their interactions with the author) include particularly Drs. W. S. S. Jee, J. S. Arnold, A. M. Parfitt, E. Radin, B. N. Epker, Z. F. G. Jaworski, B. Martin, G. Sumner-Smith, D. Kimmel, W. B. High, H. Black, R. Norrdin, R. R. Recker, D. Enlow, C. C. Johnston, Jr., H. DeLuca, R. V. Talmage, A. R. Villanueva, C. Anderson, S. Teitelbaum, H. Malluche, R. Baron, E. Roberts, W. Peck, E. Slatopolsky, J, Coburn, C. Arnaud, S. Arnaud, F. Glorieux, P. Bordier (deceased), H. Rasmussen, H. Norimatsu, H. E. Meema, J. DeQueker, F. G. Evans, D. Young, D. Burr, I. Clark, E. Collins, L. Matthews, and R. Martin. Much was learned from critics, most of them sincere and motivated by a desire to understand, but several by less admirable desires and they will surely appreciate not being named here.

Institutions also had a constructive part in this for it began at the Yale University School of Medicine in 1955, and with the support of C. O. Bechtol who allowed a young man to try on different faces in search of his own, and it grew and flourished at Henry Ford Hospital where that face was finally found, and under the aegis of Drs. C. L. Mitchell, R. Buerki, and F. Henney, each a wise, accomplished, and dedicated man. To repeat, both colleagues and staff at the Southern Colorado Clinic have accepted the many inconveniences that can arise when one of their clinical orthopedic surgeons spends a significant part of his time and energy in writing, and in traveling, teaching, consulting, and lecturing elsewhere. The University of Utah School of Medicine has also supported much of this work (and many of the people who did it) that this book summarizes via its sponsorship of what have become known in the field as the Annual Sun Valley Bone Workshops, which were conceived and organized by W. S. S. Jee and his staff and fellows at the Radiation Biology Laboratory at the University of Utah. The University of Ottawa in Canada via Prof. Z. F. G. Jaworski, and the Université Claude Bernard at Lyon, France, via the influence of Prof. P. J. Meunier, have also played significant roles in the work and in the development and diffusion of thought and understanding that lies behind it.

On another matter, certain original apostles of what has at times been called the "new bone" should be identified. They were at the right place at the right time and in the right frame of mind to observe and perceive for themselves something new, something higher than cells and lower than organs that represented the missing link that binds the two together in nature to form the whole. More than any other single factor, their own subsequent work, teachings, writings, students, and influence made the IO concept survive its early problems and grow so this book could be written and published here rather than during some future generation of man. Each of professorial status and an authority in his own right, with his own constellation of students (P. J. Meunier calls them his scientific metastases), they include in alphabetical order, J. S. Arnold, P. Coupron, H. Duncan, B. N. Epker, B. Frame, Z. F. G. Jaworski, C. C. Johnston, Jr., W. S. S. Jee, C. L. Matthews, P. J. Meunier, A. M. Parfitt, and H. Takahashi.

REFERENCES

Because of the scope of the main text and the Notes, providing pertinent references posed a problem. To do justice to that material would require over 4000 citations, which would be quite impractical to both author and publisher, so a compromise was made. That still left nearly 800 citations (and wore out one young typist). For much of the material that could be considered standard, few citations have been provided. Hence, very few current standard texts are cited for anatomy, physiology, histology, pharmacology, endocrinology, embryology, orthopedic pathology, biochemistry, physics, internal medicine, orthopedic surgery, neurosurgery, general surgery, plastic surgery, pediatrics, metabolic diseases, genetic diseases, stereology, radiology, statistics, forensic medicine, material science, mechanical engineering, kinesiology, and physical medicine. The organization adopted is this.

The references are provided in three parts, signified as I, II, and III:

Part I: *Compendia.* This lists relevant symposium volumes or volumes of abstracts that contain well over 2000 pertinent reports and discussions. There are 29 such volumes and they represent useful sources of further work, data, and citations.

Part II: *Specific citations.* This part lists particular publications by author, subject, source, and year, and in alphabetical order; there are 798 of them. Their own lists of references provide access to a vast volume of information and thought, past and present.

Part III: *Group citations.* This provides groups of citations from Parts I and II that are too numerous to list conveniently and individually as such in the main text. There are 40 of them.

I. SYMPOSIA AND MULTIAUTHOR TEXTS

 A. Frost, H. M., Ed., *Bone Biodynamics,* Little, Brown, Boston, 1964.
 B. Blackwood, H. J., Ed., *Bone and Tooth,* MacMillan, New York, 1964.
 C. Jaworski, Z. F. G., Ed., *Bone Morphometry,* Proceedings First International Workshop, University of Ottawa Press, Ottawa, Ontario, Canada, 1976.
 D. Meunier, P. J., Ed., *Bone Histomorphometry,* Proceedings Second International Workshop, Armour-Montagu, Paris, 1977.
 E. Jee, W. S. S. and Parfitt, A. M., Eds., *Bone Histomorphometry,* Proceedings Third International Workshop, Armour-Montagu, Levallois, 1981.
 F. Recker, R. R., Ed., *Bone Histomorphometry, Techniques and Interpretation,* CRC Press, Boca Raton, Fla., 1983.
 G. Frame, B., Parfitt, A. M., and Duncan, H., Eds., *Clinical Aspects of Metabolic Bone Disease,* Excerpta Medica, Amsterdam, 1973.
 H. Urist, M. R., Ed., *Fundamental and Clinical Bone Physiology,* J. B. Lippincott, Philadelphia, 1980.
 I. DeLuca, H. F., Frost, H. M., Jee, W. S. S., Johnston, C. C., Jr., and Parfitt, A. M., Eds., *Osteoporosis,* University Park Press, Baltimore, 1981.
 J. Avioli, L., Bordier, P., Fleisch, H., Massry, S., and Slatopolsky, E., Eds., *Phosphate Metabolism: Kidney and Bone,* Armour-Montagu, Paris, 1976.
 K. Norman, A. W., Schaefer, K., Coburn, J. W., DeLuca, H. F., Fraser, D., Grigoleit, H. G., and Herrath, P. V., Eds., *Vitamin D: Biochemical, Chemical and Clinical Aspects Related to Calcium Metabolism,* Walter de Gruyter, Berlin, 1977.
 L. Talmage, R. V. and Belanger, L. F., Eds., *Parathyroid Hormone and Thyrocalcitonin (Calcitonin),* Excerpta Medica, New York, 1968.
 M. Horton, J. G., Tarpley, T. M., and Davis, W. F., Eds., *Mechanisms of Localized Bone Loss,* Information Retrieval, (Suppl. Calcif. Tissue Int.), Arlington, Va., 1978.
 N. Nielsen, S. P. and Hjorting-Hansen, E., Eds., *Calcified Tissues 1975,* Calcif. Tissue Suppl., 21, 1976.
 O. Frost, H. M., Ed., *Orthopedic Clinics of North America,* Vol. 12, W. B. Saunders, Philadelphia, 1981.

P. Frame, B. and Potts, J. T., Eds., *Clinical Disorders of Bone and Mineral Metabolism,* Excerpta Medica, Amsterdam, 1983.
Q. *Orthopaedic Knowledge Update 1,* American Academy of Orthopaedic Surgeons, Chicago, 1984.
R. Transactions, Orthopaedic Research Society, American Academy of Orthopaedic Surgeons, Chicago, 1981.
S. Transactions, Orthopaedic Research Society, American Academy of Orthopaedic Surgeons, Chicago, 1982.
T. Transactions, Orthopaedic Research Society, American Society of Orthopaedic Surgeons, Chicago, 1983.
U. Transactions, Orthopaedic Research Society, American Academy of Orthopaedic Surgeons, Chicago, 1984.
V. Albright, J. A. and Brand, R. A., Eds., *The Scientific Basis of Orthopaedics,* Appleton-Century-Crofts, New York, 1979.
W. Sumner-Smith, G., Ed., *Bone in Clinical Orthopaedics,* W. B. Saunders, Philadelphia, 1982.
X. Freeman, M. A. R., Ed., *Adult Articular Cartilage,* Grune & Stratton, New York, 1973.
Y. Cruess, R. L., Ed., *Musculoskeletal System: Embryology, Biochemistry and Physiology,* Churchill-Livingstone, New York, 1982.
Z. Cohn, D. V., Fujita, T., Potts, J. T., Jr., and Talmage, R. V., Eds., *Endocrine Control of Bone and Calcium Metabolism,* Vol. 8A, International Congress Series, Excerpta Medica, New York, 1984.
AA. Fournier, R. A., Garabedian, M., Sebert, J. L., and Meunier, P. J., Eds., *Vitamin D et Maladies des Os et du Métabolisme Mineral,* Masson et Cie, Paris, 1984.
BB. *American Society for Bone and Mineral Research,* Program Abstracts, Shearer Graphic Arts, Lakeport, Calif., 1984.

II. SPECIFIC CITATIONS

1. Aaro, H., Eerola, F., Aho, A. J., and Penttinent, R., Spinal cord trauma and fracture healing, in *Osteoporosis,* DeLuca, H. F., Frost, H. M., Jee, W. S. S., Johnston, C. C., Jr., and Parfitt, A. M., Eds., University Park Press, Baltimore, 1981, 497.
2. Aaron, J. E., Gallagher, J. C., and Anderson, J., Frequency of osteomalacia and osteoporosis in fracture of the proximal femur, *Lancet,* 1, 229-223, 1974.
3. Aaron, J. E., Autoclasis — a mechanism of bone resorption and an alternative explanation for osteoporosis, *Calcif. Tissue Res., Suppl.,* 22, 247-254, 1977.
4. Adams, P. H. and Jowsey, J., Sodium fluoride in the treatment of osteoporosis and other bone diseases, *Ann. Intern. Med.,* 63, 1151-1155, 1965.
5. Aegerter, E. and Kirkpatrick, J. A., Jr., *Orthopaedic Diseases,* W. B. Saunders, Philadelphia, 1958.
6. Aegerter, E. and Kirkpatrick, J. A., Jr., *Orthopaedic Diseases,* 4th ed., W. B. Saunders, Philadelphia, 1975.
7. Akeson, W. H., Woo, S. L. Y., Coults, R. D., Matthews, J. V., Gonsalves, M., and Amiel, D., Quantative histological evaluation of early fracture healing of cortical bones immobilized by stainless steel and composite plates, *Calcif. Tissue Res.,* 19, 27-38, 1975.
8. Albin, J. and Wu, R., Abnormal hypothalamic-pituitary function in polyostotic fibrous dysplasia, *Clin. Endocrinol.,* 14, 435-443, 1981.
9. Albright, F., Bauer, W., Ropes, M., and Aub, J. C., Studies of calcium and phosphorous metabolism. IV. The effect of parathyroid hormone, *J. Clin. Invest.,* 1, 39-53, 1929.
10. Albright, F., Smith, P. H., and Richardson, A. M., Postmenopausal osteoporosis. Its clinical features, *JAMA,* 116, 2465-2474, 1941.
11. Albright, F. and Reifenstein, E. C., Jr., *The Parathyroid Glands and Metabolic Bone Disease,* Williams & Wilkins, Baltimore, 1948.
12. Albright, J. A., Bone: physical properties, in *The Scientific Basis of Orthopaedics,* Albright, J. A. and Brand, R. A., Eds., Appleton-Century-Crofts, New York, 1979, 135-183.
13. Alexandre, C., Meunier, P. J., Edouard, C., Khair, R. A., and Johnston, C. C., Effects of ethane-1 hydroxy-1, 1-diphosphonate (5mg/kg/day dose) on quantitative bone histology in Paget's disease of bone, *J. Metab. Bone Dis. Relat. Res.,* 4, 309-316, 1981.
14. Aloia, J. F., Roginski, M. S., Jowsey, J., Dombrowski, C. S., Shukla, K. K., and Cohn, S. H., Skeletal metabolism and body composition in acromegaly, *J. Clin. Endocrinol. Metab.,* 35, 543-551, 1972.
15. Aloia, J. F., Zanzi, I., Vaswani, A., Ellis, K., and Cohn, S. H., Combination therapy for osteoporosis, *Metabolism,* 26, 787-792, 1977.
16. Alwens, W., Spätrachitis, Osteomalazie, Senile Osteoporose, Hungerosteopathie, in *Hanbuch der Inneren Medizin,* Bergmann, G. V. and Staehlin, R., Eds., Springer, Berlin, 1926, 584-676.

17. Amiel, D., Akeson, W. H., Harwood, F. L., and Frank, C. B., Stress deprivation effects of metabolic turnover of the medial collateral ligament collagen: a comparison between nine and 12-week immobilization, *Clin. Orthop. Relat. Res.,* 172, 265-270, 1983.
18. Ampére, A.-M. (1775—1836), *Encyclopedia Britannica,* 15th ed., Micropaedia I, 323, 1974.
19. Amprino, R. and Engström, A., Risultati di uno studio sull' assorbimento sulla diffranzione dei raggi Roentgen da parte del tessuto osseo, *Boll. Soc. Ital. Biol. Sper.,* 26, 148-151, 1950.
20. Amprino, R., Rapporti fra processi di ricostruzione e distribuzione dei minerali nelle ossa. II. Ricerche con metodo autoradiografico, *Z. Zellforsch.,* 37, 240-273, 1952.
21. Anderson, C. and Danylchuk, K. D., Age-related variations in cortical bone remodeling measurements in male beagles 10 to 26 months of age, *Am. J. Vet. Res.,* 40, 869-872, 1979.
22. Anderson, C., Danylchuk, K. D., and DeLuca, H. F., An alteration in the plasma levels of 1, 25 dihydroxycholecalciferol is not responsible for the bone lesion in chronic cadmium exposure in dogs, *J. Metab. Bone Dis. Relat. Res.,* 2, 247-250, 1980.
23. Anderson, C., personal communications, 1983.
24. Anderson, C., Cape, R. D. T., Crilly, R. G., Hodsman, A. B., and Wolfe, B. M. J., Preliminary observations of a form of coherence therapy for osteoporosis, *Calcif. Tissue Int.,* 36, 1984.
25. Anderson, W. A. D. and Kissane, J. M., *Pathology,* 7th ed., C. V. Mosby, St. Louis, 1977.
26. Andral, G., *Précis D'Anatomie Pathologique,* Paris, 1827.
27. Andriacchi, T. P., Andersson, G. B. J., Fernier, R. W., Stern, D., and Galante, J. O., A study of lower limb mechanics during stair climbing, *J. Bone Jt. Surg.,* 62A, 740-748, 1980.
28. Angel, J. L., The length of life in Ancient Greece, *J. Gerontol.,* 2, 18-24, 1947.
29. Ariet, J., Ficat, P., Durroux, R., and Girou de Gercourt, R., Histopathologie des lésions osseuses et cartilagineuses dans l'algodystrophie sysmpathique reflèxe du genou, *Rev. Rheum.,* 48, 315-321, 1981.
30. Arkin, A. M. and Katz, J. F., The effects of pressure on epiphyseal growth — the mechanism of plasticity of growing bone, *J. Bone Jt. Surg.,* 38A, 1056-1076, 1956.
31. Arlot, M. E., Bonjean, M., Chavassieux, P. M., and Meunier, P. J., Bone histology in adults with aseptic necrosis, *J. Bone Jt. Surg.,* 65A, 1319-1327, 1983.
32. Armelagos, G. J., Disease in ancient Nubia, *Science,* 163, 255-259, 1969.
33. Arnold, J. S., A method for embedding undecalcified bone for histologic sectioning, and its application to radioautography, *Science,* 114, 178-180, 1951.
34. Arnold, J. S., Jee, W. S. S., and Johnson, K., Observations and quantitative radioautographic studies of calcium 45 deposited *in vivo* in forming haversian systems and old bone of rabbit, *Am. J. Anat.,* 99, 291-313, 1956.
35. Arnold, J. S. and Jee, W. S. S., Bone growth and osteoclastic activity as indicated by radioautographic distribution of plutonium, *Am. J. Anat.,* 101, 367-373, 1957.
36. Arnold, J. S., The quantitation of bone mineralization as an organ and tissue in osteoporosis, in *Dynamic Studies of Metabolic Bone Disease,* Pearson, O. H. and Joplin, G. V., Eds., F. A. Davis, Philadelphia, 1964, 54-70.
37. Arnold, J. S., Bartley, M. H., Tont, S. A., and Jenkins, D. P., Skeletal changes in aging and disease, *Clin. Orthop.,* 49, 17-38, 1966.
38. Arnold, J. S., Focal excessive resporption in aging and senile osteoporosis, in *Osteoporosis,* Barzel, U. S., Ed., Grune & Stratton, New York, 1970, 80-113.
39. Arnold, J. S., External and trabecular changes in lumbar vertebrae in aging, in *Progress in Methods of Bone Mineral Measurement,* Whedon, G. D. and Cameron, J. R., Eds., Department of Health, Education and Welfare, Washington, D.C., 1970.
40. Arnold, J. S., Frost, H. M., and Buss, R. O., The osteocyte as a bone pump, *Clin. Orthop.,* 78, 47-55, 1971.
41. Arnold, J. S., Barnes, W. E., Lihedkar, N., and Nelson, M., Computerized kinetic analysis of two 99^m-Tc-Sn diphosphonates demonstrating different binding characteristics, in *Fourth International Conference on Bone Measurement,* Mazess, R. B., Ed., National Institute of Health, Washington, D.C., 1980, 454-466.
42. Arnold, J. S., personal communications, 1979, 1982.
43. Arnstein, A. R., Regional osteoporosis, *Orthop. Clin. N. Am.,* 585-600, 1972.
44. Asti, P. A., Loutit, J. F., and Townsend, H. N. C., Osteoclasts derive from hematopoietic stem cells according to marker, giant lysosomes of beige mice, *Clin. Orthop. Relat. Res.,* 155, 249-258, 1981.
45. Askanazy, J. and Ruitishauser, E., Die Knochen der Basedow-Kranken. Beitrag zur latenten osteodystrophia fibrosa, *Virchows. Arch. Pathol. Anat. Physiol.,* 291, 653-681, 1933.
46. d'Aubigné, M., Functional results of hip arthroplasty with acrylic prostheses, *J. Bone Jt. Surg.,* 36A, 451-475, 1954.
47. Austin, L. A. and Heath, H., Calcitonin: physiology and pathophysiology, *N. Engl. J. Med.,* 304, 269-277, 1981.

48. Avioli, L. V., Calcitonin therapy for bone disease and hypercalcemia, *Arch. Int. Med.*, 142, 2076-2079, 1982.
49. Axelrod, J. and Reising, T. D., Stress hormones: their interaction and regulation, *Science*, 224, 452-459, 1984.
50. Axhausen, G., Über anämische infarkte am knochensystem und irhre Bedeutung für die Lehre von den primaren Epiphysennekrosen, *Arch. Klin. Chir.*, 151, 72-98, 1928.
51. Baker, L. D., Spontaneous fracture of the femoral neck following irradiation, *J. Bone Jt. Surg.*, 23, 354-366, 1941.
52. Baron, R. and Vignery, A., Behavior of osteoclasts during a rapid change in their number induced by high doses of parathyroid hormone or calcitonin in intact rats, *J. Metab. Bone Dis. Relat. Res.*, 2, 339-346, 1981.
53. Baron, R., Vignery, A., and Tranvan, P., The significance of lacunar erosion without osteoclasts; studies on the reversal phase of the remodeling sequence, in *Osteoporosis*, Deluca, H. F. et al., Eds., University Park Press, Baltimore, 1986.
54. Barondes, S. H., Soluble lectins: a new class of extracellular proteins, *Science*, 223, 1259-1264, 1984.
55. Barry, H. C., *Paget's Disease of Bone*, E. and S. Livingstone, Edinburgh, 1969.
56. Bartels, E. C. and Haggart, G. E., Osteoporosis in hyperthyroidism. Report of two cases with compression fracture of the vertebrae, *N. Engl. J. Med.*, 219, 273-278, 1938.
57. Bartholinus, C., Diaphragmatis structura nova, *Acta Med. Phil. Hafn.*, 4, 14-16, 1676.
58. Bartolucci, A., Casi interessanti di osteite malacia nei bovini, *Med. Zooiatr.*, 23, 194-197, 1912.
59. Barzel, U. S., Osteoporosis in young men, *Arch. Int. Med.*, 142, 2079-2080, 1982.
60. Basle, M., Rebel, A., Pouplard, A., Filmon, R., and Lepatezour, A., L'étiologie virale de la maladie osseuse de Paget, *Nouv. Press Med.*, 10, 1193-1197, 1981.
61. Basle, M. F., Rebel, A., and Renier, J. S., Bone tissue in reflex sympathetic dystrophy syndrome-Sudeks atrophy: structural and ultrastructural studies, *Metab. Bone Dis. Relat. Res.*, 4, 305-311, 1983.
62. Bassett, C. A. L., Pulsing electromagnetic fields: a new method to modify cell behavior in calcified and noncalcified tissues, *Calcif. Tissue Int.*, 34, 1-8, 1982.
63. Baud, C. A., Pouezat, J. A., and Tochon-Danguy, H. J., Quantitative analysis of amorphous and crystalline bone tissue mineral in women with osteoporosis, in *Calcified Tissues 1975*, Nielson, S. P. and Hjorting-Hansen, E., Eds., Fadl, Copenhagen, 1976, 452-456.
64. Bauer, G. C. H., Carlsson, A., and Lindquist, B., Accretion rate of bone salt in osteoporosis studied by means of ^{32}P, *Acta. Med. Scand.*, 158, 139-142, 1957.
65. Bauer, W., Aub, J. C., and Albright, F., Studies of calcium and phosphorus metabolism. V. A study of the bone trabeculae as a readily available reserve supply of calcium, *J. Exp. Med.*, 49, 145-156, 1929.
66. Baylink, D. J. and Liv, C. C., The regulation of endosteal bone volume, *J. Periodontol.*, 50, 43-49, 1979.
67. Baylink, D. J. and Bernstein, D. S., The effect of fluoride therapy on metabolic bone disease: a histological study, *Clin. Orthop.*, 51, 55-67, 1967.
68. Beals, R. K., Hemihypertrophy and hemihypotrophy, *Clin. Orthop. Relat. Res.*, 166, 199-203, 1982.
69. Beals, R. K., Endosteal hyperostosis, *J. Bone Jt. Surg.*, 58A, 1172-1173, 1976.
70. Becker, H. and Diegelmann, R. F., The influence of tension on intrinsic tendon fibroplasia, *Orthop. Rev.*, 13, 153-159, 1984.
71. Beckmann, J., Rodegerdts, V., and Buddecke, ^{14}C-glucose and ^{35}S-metabolism of pig epiphyseal cartilage and its variations after osteotomy, *J. Bone Jt. Surg.*, 57B, 507-510, 1975.
72. Becks, H., Ray, R. D., Simpson, M. E., and Evans, H. M., Effects of thyroxine and the anterior pituitary growth hormone on endochondral ossification, *Arch. Pathol.*, 34, 334-357, 1942.
73. Belchier, J., An account of the bones of animals being changed to a red color by aliment only, *Philos. Trans.*, 39, 287-288, 1736.
74. Berliner, R. W., Outline of renal physiology, in *Diseases of the Kidney*, Struss, M. B. and Welt, L. G., Eds., Little, Brown, Boston, 1971, 31-85.
75. Bernstein, D. S. and Cohen, P., Use of sodium fluoride in the treatment of osteoporosis, *J. Clin. Endocrinol.*, 27, 197-210, 1967.
76. Biening, A. and Iversen, T., Osteogenesis imperfecta with Ehlers-Danlos syndrome, *Acta Pediatr.*, 44, 279-288, 1955.
77. Black, H. E. and Jee, W. S. S., A histomorphometric and biochemical evaluation of the effects of a diphosphonate in corticosteroid-treated rabbits, in *Bone Histomorphometry*, Meunier, P. J., Ed., Armour-Montagu, Paris, 1977, 157-170.
78. Bloch-Michel, H., Milhaud, G., and Waltzing, P., Utilization thérapeutique de la thyrocalcitonine, in *Hormones et Calcium*, Klotz, H. P., Ed., Exp. Scientifique Francais, Paris, 1971, 175-352.

79. Bolvin, G., Edouard, C., Chapuy, M. C., and Meunier, P. J., Iliac bone histomorphometry in skeletal fluorosis. Presented at the Fourth International Workshop on Bone Histomorphometry, Aarhus, Denmark, June 1984.
80. Bonucci, E., New knowledge on the origin, function and fate of osteoclasts, *Clin. Orthop. Relat. Res.*, 158, 252-269, 1981.
81. Bordier, P., de Sèze, S., Miravet, L., and Berbir, N., Physiopathologie de l'ostéoporose de l'adulte jeune, *Sem. Hôp. Paris*, 50, 197-206, 1974.
82. Bordin, S., Page, R. C., and Narayanan, A. S., Heterogeneity of normal human diploid fibroblasts: isolation and characterization of one phenotype, *Science*, 223, 171-173, 1984.
83. Bressot, C., Courpron, P., Edouard, C., and Meunier, P. J., *Histomorphométrie des Ostéopathies Endocriniennes*, Univ. Claude Bernard, Lyon, 1976.
84. Bressot, C., Meunier, P. J., Chapuy, M. D., Lejeune, E., Edouard, C., and Darby, A. J., Histomorphometric profile, pathophysiology and reversibility of corticosteroid induced osteoporosis, *Metab. Bone Dis. Relat. Res.*, 1, 303-311, 1979.
85. Briançon, D. and Meunier, P. J., *Le traitement de l'Ostéoporose par l'Association Fluorure de Sodium, Calcium, Vitamine D*, Univ. Claude Bernard, Lyon, 1979.
86. Brighton, C. T., Lackman, R. D., and Cuckler, J. M., Absence of the glycerol phosphate shuttle in the various zones of the growth plate, *J. Bone Jt. Surg.*, 65A, 663-666, 1983.
87. Brighton, C. T., Pfeffer, G. B., and Pollack, S. R., In vivo growth plate stimulation in various capacitively coupled electrical fields, *J. Orthop. Res.*, 1, 42-49, 1983.
88. Brighton, C. T., The semiinvasive method of treating nonunion with direct current, *Orthop. Clin. N. Am.*, 15, 33-45, 1984.
89. Bromley, R. G., Dockum, N. L., Arnold, J. S., and Jee, W. S. S., Quantitative histological study of human lumbar vertebra, *J. Gerontol.*, 21, 537-543, 1966.
90. Bubenik, G. A., Bubenik, A. B., Stevens, E. D., and Binnington, A. G., The effect of neurogenic stimulation on the development and growth of bony tissues, *J. Exp. Zool.*, 219, 205-216, 1982.
91. Buckwalter, J. A., Proteoglycan structure in calcifying cartilage, *Clin. Orthop. Relat. Res.*, 172, 207-232, 1983.
92. Bulloch, W., *The History of Bacteriology*, Oxford University Press, London, 1938.
93. Bunger, C., Bulow, J., Bach, P., and Solund, K., Blood supply of the juvenile knee in arthritis at rest and during exercise, *Trans. Orthop. Res. Soc.*, 30, 44, 1984.
94. Bunting, C. H. and Eades, C., Effects of mechanical tension on the polarity of growing fibroblasts, *J. Exp. Med.*, 44, 147-149, 1926.
95. Bunch, W. H., Dech, D. J., and Romer, J., The effect of denervation on bony growth after below knee amputation in rats, *Clin. Orthop. Relat. Res.*, 122, 333-339, 1977.
96. Burr, D. B., Martin, R. B., and Radin, E. L., Threshold values for the production of fatigue damage in bone in vivo, *Trans. Orthop. Res. Soc.*, 29, 69, 1983.
97. Burstein, A. H., Reilly, D. T., and Martens, M., Aging of bone tissue: mechanical properties, *J. Bone Jt. Surg.*, 58A, 82-86, 1976.
98. Busogolia, M., d'Espinay, C. L., Movel, B., Ruegg, H., and Vanèche, J., Eds., *Les Critères de Verité dans la Recherche Scientifique*, Maloine, Paris, 1983.
99. Byers, P. and Smith, R., Trephine for full thickness iliac crest biopsy, *Lancet*, 1, 682-683, 1967.
100. Calot, F., *L'Orthopédie Indispensable*, 7th ed., A. Maloine et Fils, Paris, 1917.
101. Cameron, H. V. and Fornasier, V. L., Trabecular stress fractures, *Clin. Orthop. Relat. Res.*, 111, 266-268, 1975.
102. Cameron, J. R. and Sorenson, J., Measurement of bone mineral in vivo. An improved method, *Science*, 142, 230-232, 1963.
103. Canalis, E. M., Dietrich, J. M., Maina, D. M., and Raisz, L. G., Hormonal control of bone collagen synthesis in vitro, *Endocrinology*, 100, 668-674, 1977.
104. Canalis, E. M., Reardon, G. E., Baron, R., and Raisz, L. G., Dynamic bone morphometry and studies on the effects of serum on bone metabolism in vitro in a case of pycnodysostosis, *J. Metab. Bone Dis.*, 2, 99-106, 1980.
105. Cann, C. E., Genant, H. K., Ettinger, B., and Gordan, G. S., Spinal mineral loss in oophorectomized women, *JAMA*, 244, 2056-2059, 1980.
106. Cann, C. E. and Genant, H. K., Precise measurement of vertebral mineral content using computed tomography, *J. Comput. Assist. Tomogr.*, 4, 493-500, 1980.
107. Cantrell, D. A. and Smith, K. A., The interleukin-2 T cell system: a new cell growth model, *Science*, 224, 1312-1316, 1984.
108. Caplin, A. I., Syftestad, G., and Osdoby, P., The development of embryonic bone and cartilage in tissue culture, *Clin. Orthop. Relat. Res.*, 174, 243-263, 1983.
109. Caputo, C. B., Meadows, D., and Raisz, L. G., Failure of estrogens and androgens to inhibit bone resorption in tissue culture, *Endocrinology*, 98, 1065-1069, 1976.

110. Carter, D. R., The relationship between in vivo strains and cortical bone remodeling, *CRC Crit. Rev. Biomechan. Eng.*, 8, 1-28, 1981.
111. Carter, D. R. and Caler, W. E., Uniaxial fatigue of human cortical bone. The influence of tissue physical characteristics, *J. Biomechan.*, 14, 461-470, 1981.
112. Carter, D. R. and Hayes, W. C., Compact bone fatigue damage; a microscopic examination, *Clin. Orthop. Relat. Res.*, 135, 192-205, 1977.
113. Carter, D. R. and Spengler, D. M., Mechanical properties and composition of cortical bone, *Clin. Orthop. Relat. Res.*, 135, 192-217, 1978.
114. Casagrande, P. and Frost, H. M., *Fundamentals of Clinical Orthopedics*, Grune & Stratton, New York, 1953.
115. Chakkalakal, D. A. and Johnson, M. W., Electrical properties of compact bone, *Clin. Orthop. Relat. Res.*, 161, 146-153, 1981.
116. Chalkley, H. W., Method for the quantitative morphologic analysis of tissue, *J. Natl. Cancer. Inst.*, 4, 47-53, 1943.
117. Chalkley, H. W., Cornfield, J., and Park, H., A method for estimating surface to volume ratios, *Science*, 110, 295-297, 1949.
118. Charbon, G. A. and Pisper, E. E. M., Effect of calcitonin on parathyroid hormone-induced vasodilation, *Endocrinology*, 91, 828-831, 1972.
119. Charcot, J. M., Sur quelques arthropathies qui paraissent dépendre d'une lésion due cerveau ou de la moelle épinière, *Arch. Physiol. Norm. Pathol.*, 1, 161-229, 1868.
120. Charon, S., Bavery, E., Malik, M. E., Touraine, J. L., Edouard, C., Arlot, M., Traeger, J., and Meunier, P. J., Ostéonécrose de la transplantation rénale, *Lyon Med.*, 247, 339-347, 1982.
121. Chapuy, M. C., Meunier, P. J., and Alexandre, C., Comparison of the acute effects of human and salmon calcitonins in Pagetic patients: relation with plasma calcitonin levels, *Metab. Bone Dis. Relat. Res.*, 2, 93-97, 1980.
122. Chayes, F., A simple point counter for thin section analysis, *Am. Mineral.*, 34, 1-11, 1949.
123. Churches, A. E. and Howlett, C. R., Functional adaptation of bone in response to sinusoidally varying controlled compressive loading of the ovine metacarpus, *Clin. Orthop. Relat. Res.*, 168, 265-280, 1982.
124. Coindre, C. J. and Meunier, P. J., *Etude Histomorphométrique de l'os NonPagétique Chez le Pagétique*, Faculté Alexis Carrel, Lyon, 1980.
125. Collip, J. B., The extraction of parathyroid hormone which will prevent or control parathyroid tetany and which regulates the level of blood calcium, *J. Biol. Chem.*, 63, 395-438, 1925.
126. Compston, J. E., Vedi, S., Merret, A. L., Clemens, T. L., O'Riordan, J. L. H., and Woodhead, J. S., Privational and malabsorption metabolic bone disease. Plasma vitamin D metabolic concentrations and their relation to quantitative bone histology, *J. Metab. Bone Dis. Relat. Res.*, 3, 165-170, 1981.
127. Connolly, J. D., Selection, evaluation and indications for electrical stimulation of ununited fractures, *Clin. Orthop. Relat. Res.*, 161, 39-53, 1981.
128. Cooke, W. T. Barclay, J. A., Govan, A. D. T., and Nagley, L., Osteoporosis associated with low serum phosphorus and renal glycosuria, *Arch. Int. Med.*, 80, 147-158, 1947.
129. Cook, S. D., Lavernia, C. J., Burke, S. W., Skinner, H. B., and Haddad, R. J., A biochemical analysis of the etiology of tibia vara, *J. Pediatr. Orthop.*, 3, 449-454, 1983.
130. Cook, S. D., Skinner, H., and Haddad, R. J., A quantitative histologic study of osteoporosis produced by secondary hyperparathyroidism, in dogs, *Clin. Orthop. Relat. Res.*, 175, 105-120, 1983.
131. Copp, D. H. and Cheney, B. A., Calcitonin — a hormone from the parathyroid which lowers the calcium level of the blood, *Nature (London)*, 193, 381, 1962.
132. Courpron, P., personal communications, 1979, 1981.
133. Courpron, P., *Données Histologiques Quantitatives sur le Vieillissement Osseux Humain*, Univ. Claude Bernard, Lyon, 1972.
134. Courpron, P., Bone tissue mechanisms underlying osteoporoses, *Orthop. Clin. N. Am.*, 12, 513-545, 1981.
135. Courpron, P., Meunier, P. J., Edouard, C., Bernard, J., Bringuier, J. P., and Vignon, G., Données histologiques quantitative sur le viellissement osseux humain, *Rev. Rheum.*, 40, 469-483, 1973.
136. Courpron, P., Meunier, P. J., Bressot, C., Le volume trabeculaire osseux iliaque: application au vieillissement osseux et aux ostéopathies, *Lyon Med.*, 19, 465-471, 1978.
137. Courpron, P., Lepine, P. M., and Meunier, P. J., Analyse par l'histomorphométrie osseuse de mécanismes de l'ostéopénie du spongieux iliaque humain, Hôpital A. Charial, Francheville, 1982.
138. Coutelier, L., *Recherches sur la Guérison des Fractures*, Arscia, S. A., Ed., Brussels, 1969.
139. Coryn, G., Influence de la thyroide et de la parathyroide sur le squellette, *J. Radiol. Electrol.*, 20, 123-128, 1936.

140. Coutts, R. D., Woo, S. L. Y., Boyer, J., Doty, D., Gonsalves, M., Amiel, D., Ing, D., and Akeson, W. H., The effect of delayed internal fixation on healing of the osteotomized dog radius, *Clin. Orthop. Relat. Res.,* 163, 254-260, 1982.
141. Cowin, S. C., Lanyon, L. E., and Rodan, G., The Kroc Foundation Conference on Functional Adaptation in Bone Tissue, *Calcif. Tissue Int.,* 36, 51-56, 1984.
142. Crowninshield, R. D., Brand, R. A., Johnston, R. C., and Milroy, J. C., An analysis of femoral component stem design in total hip arthroplasty, *J. Bone Jt. Surg.,* 62A, 68-78, 1980.
143. Crowninshield, R. D., Johnston, R. C., Brand, R. A., Pederson, D. R., Wilson, M. A., and Tolbert, J. R., An engineering analysis of total hip component design, *Orthop. Rev.,* 12, 33-45, 1983.
144. Currey, J. D., The adaptations of bones to stress, *J. Theoret. Biol.,* 20, 91-106, 1968.
145. Currey, J. D. and Butler, G., The mechanical properties of bone tissue in children, *J. Bone Jt. Surg.,* 57A, 810-814, 1975.
146. Cushing, H. and Davidoff, L. M., The pathological findings in four autopsied cases of acromegaly with a discussion of their significance, Monographs of the Rockefeller Institute of Medical Research, New York, 1927.
147. Cushing, H., The basophil adenomas of the pituitary body and their clinical manifestations (pituitary basophilism), *Bull. Johns Hopkins Hosp.,* 50, 3-137, 1932.
148. D'Abro, A., *The Evolution of Scientific Thought from Newton to Einstein,* 2nd ed., Dover, New York, 1950.
149. Dallek, M., Jungbluth, K. H., and Holstein, A. F., Studies on the arrangement of the collagenous fibers in infant epiphyseal plates using polarized light and the scanning electron microscope, *Arch. Orthop. Trauma Surg.,* 101, 239-245, 1983.
150. Dalton, John (1766—1844), *Encyclopedia Britannica,* 15th ed., Micropedia III, 358, 1974.
151. Daly, W. R., Mills, E. J., and Hohn, R. B., In vivo strain analysis of canine long bones and its application to internal fixation, *Arch. Am. Coll. Vet. Surg.,* 6, 11-15, 1977.
152. Danielli, J. F., The dynamic function of cellular and intracellular membranes, in *Bone Biodynamics,* Frost, H. M., Ed., Little, Brown, Boston, 1964, 51-57.
153. Danylchuk, K. D., The Effects of Environmental Pollutants on Bone Remodeling, Thesis, Univ. Western Ontario, London, Canada, 1978, 1-209.
154. Darby, A. J. and Meunier, P. J., Mean wall thickness and formation periods of trabecular bone packets in idiopathic osteoporosis, *Calcif. Tissue Int.,* 33, 199-204, 1981.
155. Darwin, Charles (1809—1882), *Encyclopedia Britannica,* 15th ed., Micropedia III, 385, 1974.
156. Daum, W. J., Chang, S.-L., Simmons, D. J., Webster, D., and Shoenecker, P. L., Healing of canine femoral osteotomies, *Clin. Orthop. Relat. Res.,* 180, 291-300, 1983.
157. Deftos, L. J., Calcitonin in clinical medicine, in *Advances in Internal Medicine,* Stolleman, G. H., Ed., Yearbook, New York, 1978, 159-193.
158. Deiss, W. P., Jr., The metabolism of bone, in Dysbarism-Related Osteonecrosis, Beckman, E. L. and Elliott, D. H., Eds., U.S. Department of Health, Education and Welfare, 1974, 61-66.
159. Delesse, M., Procédé mécanique pour déterminer la composition des roches, *C.R. Acad. Sci.,* 25, 544-552, 1847.
160. Dent, C. E., Idiopathic osteoporosis, *Proc. R. Soc. Med.,* 48, 574-588, 1955.
161. Dent, C. E. and Friedman, M., Idiopathic juvenile osteoporosis, *Q. J. Med.,* 34, 177-210, 1965.
162. Dent, C. E. and Harris, H., Hereditary forms of rickets and osteomalacia, *J. Bone Jt. Surg.,* 38B, 204-226, 1956.
163. Dequecker, J., Remans, J., Franssen, R., and Waes, J., Ageing patterns of trabecular and cortical bone and their relationship, *Calcif. Tissue Res.,* 7, 23-30, 1971.
164. Déscartes, René (1596—1650), *Encyclopedia Britannica,* 15th ed., Micropedia III, 484, 1974.
165. Dewey, J., Armelagos, G. J., and Bartley, M., Femoral cortical involution in three Nubian archeological populations, *Hum. Biol.,* 41, 13-21, 1969.
166. Diaz, M., Keleinknecht, C., and Broyer, M., Growth in experimental renal failure. Role of caloric and amino acid intake, *Kidney Int.,* 8, 349-354, 1975.
167. Dickenson, R. P., Hutton, W. C., and Scott, J. R. R., The mechanical properties of bone in osteoporosis, *J. Bone Jt. Surg.,* 63B, 233-238, 1981.
168. Dingle, J. T., The role of cellular interactions in joint erosions, *Clin. Orthop. Relat. Res.,* 182, 24-30, 1984.
169. Doyle, F., Brown, J., and Lachance, C., Relation between bone mass and muscle weight, *Lancet,* 1, 391-393, 1970.
170. Doyle, F., Involutional osteoporosis, *Clin. Endocrinol. Metab.,* 1, 143-167, 1972.
171. Drummond, R. P. and Rose, G. K., A twenty-one year review of a case of congenital indifference to pain, *J. Bone Jt. Surg.,* 57B, 241-243, 1975.
172. Duhamel, H. L., Sur une racine qui a la faculté de teindre en rouge les os des animaux vivants, *Med. Acad. R. Sci.,* 1-13, 1739.

173. Dull, T. A. and Henneman, P. H., Urinary hydroxyproline as an index of collagen turnover in bone, *N. Engl. J. Med.*, 268, 132-134, 1963.
174. Duncan, C. P. and Shim, S., The autonomic nerve supply of bone, *J. Bone Jt. Surg.*, 59B, 323-324, 1977.
175. Duncan, H., Frame, B., Frost, H. M., and Arnstein, A. R., Migratory osteolysis of the lower extremities, *Ann. Int. Med.*, 66, 1165-1173, 1967.
176. Duncan, H., Osteoporosis in rheumatoid arthritis and corticosteroid induced osteoporosis, *Orthop. Clin. N. Am.*, 3, 571-583, 1972.
177. Duncan, H., Frame, B., Arnstein, A. R., and Frost, H. M., Migratory regional osteoporosis, in *Clinical Aspects of Metabolic Bone Disease*, Frame, B., Parfitt, A. M., and Duncan, H., Eds., Excerpta Medica, Amsterdam, 1973, 245-249.
178. Duncan, H. and Parfitt, H. M., Metabolic bone disease of the spine, in *The Spine*, Rothman, R. H. and Simeone, F. A., Eds., W. B. Saunders, Philadelphia, 1975, 599-720.
179. Duncan, H., personal communications, 1973, 1980.
180. Duncan, H. and Jaworski, Z. F. G., Osteoporosis, in *Practice of Medicine*, Harper & Row, Hagerstown, Va., 5, 1-10, 1979.
181. Duriz, J., L'Atteinte osseuse au cours de la mastocytose cutanée, *Rev. Rheum.*, 42, 71, 1975.
182. Von Ebner, V., Über den feineren bau der Knochensubstanz, *Sitz ungsber Dtsch. Akad. Wiss.*, 72, 49-138, 1875.
183. Ehrlich, Paul (1854—1915), *Encyclopedia Britannica*, 15th ed., Micropedia III, 811-812, 1974.
184. Eicher, E. M., Southard, J.-L., Scriver, C. R., and Glorieux, F. H., Hypophosphatemia: mouse model for familial hypophosphatemic (vitamin D-resistant) rickets, *Proc. Natl. Acad. Sci. U.S.A.*, 73, 4667-4671, 1976.
185. Einstein, Albert, *Special Theory of Relativity*, 1905.
186. Ellis, F. R., Holesch, S., and Ellis, J. W., Incidence of osteoporosis in vegetarians and omnivores, *Am. J. Clin. Nutr.*, 25, 555-558, 1972.
187. Engström, A. and Wegstedt, L., Equipment for microradiography with soft roentgen rays, *Acta Radiol.*, 35, 345-355, 1951.
188. Engström, A. and Engfeldt, B., Lamellar structure of osteons demonstrated by microradiography, *Experientia*, 9, 19, 1953.
189. Enlow, D. H., A plastic seal method for mounting sections of ground bone, *Stain Technol.*, 29, 21-22, 1954.
190. Enlow, D. H., *Principles of Bone Remodeling*, Charles C Thomas, Springfield, Ill., 1963.
191. Enlow, D. H. and Brown, S. O., A comparative histological study of fossil and recent bone tissues, *Tex. J. Sci.*, 10, 187-230, 1958.
192. Epker, B. N., The bone loss phenomena and the bone marrow: quantitative light microscope, autoradiographic and tetracycline labeling study in aging rabbits, in *Bone Morphometry*, Jaworski, Z. F. G., Ed., University of Ottawa Press, Ottawa, 1976, 224-245.
193. Epker, B. N. and Frost, H. M., A histological study of remodeling at the periosteal, haversian canal, cortical endosteal and trabecular surfaces in human rib, *Anat. Rec.*, 152, 129-136, 1965.
194. Epker, B. N., A quantitative histological study of the effects of fluoride on resorption and formation in animal and human bone, *Clin. Orthop.*, 49, 77-87, 1966.
195. Epker, B. N., The role of bone loss at the organ level with aging: a review, in *Bone Morphometry*, Jaworski, Z. F. G., Ed., University of Ottawa Press, Ottawa, 1976, 56-62.
196. Epker, B. N. and Frost, H. M., Correlation of patterns of bone resorption and formation with physical behavior of loaded bone, *J. Dent. Res.*, 44, 33-42, 1965.
197. Epker, B. N. and O'Ryan, F., Determinants of Class II dentofacial morphology. I. A biomechanical theory, in *Effects of Surgical Intervention on Craniofacial Growth*, McNamara, J. A., Jr., Carlson, D. S., and Ribbens, K. A., Eds., University of Michigan, Ann Arbor, 1982, 169-205.
198. Eriksen, E. F., Mosekilde, L., and Melsen, F., Trabecular bone resorption depth decreases with age: difference between normal males and females, presented at the Fourth International Workshop on Bone Histomorphometry, Aarhus, Denmark, June 13, 1984.
199. Eriksen, E. F., Gundersen, H. J. G., Melsen, F., and Mosekilde, L., The remodeling site in iliac trabecular bone from 20 normal individuals, presented at the Fourth International Workshop on Bone Histomorphometry, Aarhus, Denmark, June 10, 1984.
200. Eriksson, C., Streaming potentials and other water dependent effects in mineralized tissues, *Ann. N.Y. Acad. Sci.*, 238, 321-329, 1974.
201. Eriksson, C., Bone morphogenesis and surface change, *Clin. Orthop. Relat. Res.*, 121, 295-302, 1976.
202. Evans, F. G., *Stress and Strain in Bones*, Charles C Thomas, Springfield, Ill., 1957.
203. Evans, F. G., *Mechanical Properties of Bone*, Charles C Thomas, Springfield, Ill., 1973.
204. Fell, H. B., Organ culture and the physiology of skeletal tissues, in *Bone Biodynamics*, Frost, H. M., Ed., Little, Brown, Boston, 1964, 311-318.

205. Fisher, L. W., Termine, J. D., Dejter, S. W., Whitson, S. W., Yanagishita, M., Kimura, J. H., Hascall, V. C., Kleinman, H. K., Hassell, J. R., and Nilsson, B., Proteoglycans of developing bone, *J. Biol. Chem.*, 258, 6588-6594, 1983.
206. Flavelli, J. H., On cognitive development, *Child Dev.*, 53, 1-10, 1982.
207. Fleming, A., On the antibacterial action of cultures of a penicillium, *Br. J. Exp. Pathol.*, 10, 226-231, 1929.
208. Flora, L., Hassing, G. S., Villanueva, A. R., Matthews, C., Crouch, M., Duncan, H., and Parfitt, A. M., Comparative effects of disodium ethane hydroxydiphosphonate (EHDP) and disodium dichloromethane diphosphonate (Cl_2MDP) on bone remodeling in adult beagle dogs, *Calcif. Tissue Int. Suppl.*, 27, A12, 1979.
209. Flora, L., Hassing, G. S., Parfitt, A. M., and Villanueva, A. R., Comparative skeletal effects of two diphosphonate drugs, in *Osteoporosis*, DeLuca, H. F., Frost, H. M., Jee, W. S. S., Johnston, C. C., Jr., and Parfitt, A. M., Eds., University Park Press, Baltimore, 1981, 389-407.
210. Frame, B., personal communications, 1979, 1981, 1983.
211. Frame, B. and Nixon, R. K., Bone marrow factors in osteoporosis, in *Osteoporosis*, Barzel, U. S., Ed., Grune & Stratton, New York, 1970, 238-250.
212. Frame, B., Frost, H. M., Pak, C. Y. C., Reynolds, W., and Argen, R. J., Fibrogenesis imperfecta ossium: a collagen defect causing osteomalacia, *N. Engl. J. Med.*, 285, 769-772, 1971.
213. Frame, B. and Parfitt, A. M., Osteomalacia: current concepts, *Ann. Int. Med.*, 89, 966-982, 1978.
214. Frank, C., Schachar, N., and Dittrich, D., Natural history of healing in the repaired medial collateral ligament, *J. Orthop. Res.*, 1, 179-188, 1983.
215. Freeman, M. A. R., Todd, R. C., and Pirie, C. J., The role of fatigue in the pathogenesis of senile femoral neck fracture, *J. Bone Jt. Surg.*, 56B, 698-702, 1974.
216. Friedenberg, Z. B. and Brighton, C. T., Bioelectric potentials in bone, *J. Bone Jt. Surg.*, 48A, 915-923, 1966.
217. Frost, H. M., Preparation of thin undecalcified bone sections by rapid manual method, *Stain Technol.*, 33, 273-277, 1958.
218. Frost, H. M., Staining of fresh, undecalcified thin bone sections, *Stain Technol.*, 34, 135-146, 1959.
219. Frost, H. M., Presence of microscopic cracks *in vivo* in bone, *Henry Ford Hosp. Med. Bull.*, 8, 25-35, 1960.
220. Frost, H. M., Observations on osteoid seams: the existence of a resting state, *Henry Ford Hosp. Med. Bull.*, 8, 220-224, 1960.
221. Frost, H. M., Observations of fibrous and lamellar bone, *Henry Ford Hosp. Med. Bull.*, 8, 199-207, 1960.
222. Frost, H. M., Villanueva, A. R., and Roth, H., Halo volume, *Henry Ford Hosp. Med. Bull.*, 8, 228-238, 1960.
223. Frost, H. M., Villanueva, A. R., and Roth, H., Measurement of bone formation in a 57 year old man by means of tetracycline, *Henry Ford Hosp. Med. Bull.*, 8, 212-219, 1960.
224. Frost, H. M., A new bone affection: feathering, *J. Bone Jt. Surg.*, 42A, 447-456, 1960.
225. Frost, H. M., *In vivo* impermeability of feathered bone to tetracyclines, *Henry Ford Hosp. Med. Bull.*, 8, 225-227, 1960.
226. Frost, H. M., Micropetrosis, *J. Bone Jt. Surg.*, 42A, 144-150, 1960.
227. Frost, H. M., Roth, H., Villanueva, A. R., and Stanisavljevic, S., Experimental multiband tetracycline measurement of lamellar osteoblastic activity, *Henry Ford Hosp. Med. Bull.*, 9, 312-320, 1961.
228. Frost, H. M., Postmenopausal osteoporosis: a disturbance in osteoclasia, *J. Am. Geriatr. Soc.*, 9, 1078-1085, 1961.
229. Frost, H. M., Ed., *Mathematical Elements of Lamellar Bone Remodeling*, Charles C Thomas, Springfield, Ill., 1964.
230. Frost, H. M., Ed., *Bone Dynamics in Osteoporosis and Osteomalacia*, Charles C Thomas, Springfield, Ill., 1966.
231. Frost, H. M., Dynamics of bone remodeling, in *Bone Biodynamics*, Frost, H. M., Ed., Little, Brown, Boston, 1964, 315-333.
232. Frost, H. M., Tetracycline-based histological analysis of bone remodeling, *Calcif. Tissue Res.*, 3, 211-237, 1969.
233. Frost, H. M., Villanueva, A. R., Jaworski, Z. F. G., Meunier, P. J., and Shimizu, A. G., Evaluation of cellular-level haversian bone resorption in human hyperparathyroid states, *Henry Ford Hosp. Med. J.*, 17, 259-266, 1969.
234. Frost, H. M., Managing the skeletal pain and disability of osteoporosis, *Orthoped. Clin. N. Am.*, 3, 561-569, 1972.
235. Frost, H. M., An efficient way to analyze bone affections, in *Radiobiology of Plutonium*, Stover, B. J. and Jee, W. S. S., Eds., J. W. Press, University of Utah, Salt Lake City, 1972, 293-304.
236. Frost, H. M., Ed., *The Physiology of Cartilaginous, Fibrous and Bony Tissue*, Charles C Thomas, Springfield, Ill., 1972.

237. Frost, H. M., Ed., *Bone Remodeling and Its Relation to Metabolic Bone Disease,* Charles C Thomas, Springfield, Ill., 1973.
238. Frost, H. M., Ed., *Bone Modeling and Skeletal Modeling Errors,* Charles C Thomas, Springfield, Ill., 1973.
239. Frost, H. M., The origin and nature of transients in human bone remodeling dynamics, in *Clinical Aspects of Metabolic Bone Disease,* Frame, B., Parfitt, A. M., and Duncan, H., Eds., Excerpta Medica, Amsterdam, 1973, 124-137.
240. Frost, H. M., The spinal osteoporoses: mechanisms of pathogenesis and pathophysiology, *Clin. Endocrinol. Metab.,* 2, 257-275, 1973.
241. Frost, H. M., Ed., *Orthopedic Biomechanics,* Charles C Thomas, Springfield, Ill., 1973, 1-652.
242. Frost, H. M., Jee, W. S. S., Kimmel, D., and Teitelbaum, S., Histomorphometric analysis of trabecular bone in renal dialysis patients treated with 25-hydroxy vitamin D_3: preliminary report, in *Vitamin D: Biochemical, Chemical, and Clinical Aspects Related to Calcium Metabolism,* Norman, A. W., Schaefer, K., Coburn, J. W., DeLuca, H. F., Fraser, D., Grigoleit, H. G., and Herrath, D. V., Eds., Walter de Gruyter, New York, 1977, 885-895.
243. Frost, H. M., A method of analysis of trabecular bone dynamics, in *Bone Histomorphometry,* Meunier, P. J., Ed., Armour-Montagu, Paris, 1977, 445-475.
244. Frost, H. M., A chondral modeling theory, *Calcif. Tissue Int.,* 28, 181-200, 1979.
245. Frost, H. M., Treatment of osteoporoses by manipulation of coherent bone cell populations, *Clin. Orthop. Relat. Res.,* 143, 227-244, 1979.
246. Frost, H. M., Ed., *Bone Biodynamics,* Little, Brown, Boston, 1964.
247. Frost, H. M., Skeletal physiology and bone remodeling: an overview, in *Fundamental and Clinical Bone Physiology,* Urist, M. R., Ed., J. B. Lippincott, Philadelphia, 1980, 242-267.
248. Frost, H. M., A theory of lamellar bone modeling, in *Physiology and Pathology of Bone and Cartilage Metabolism,* Japan Orthopedic Research Society, Tokyo, 1980, 571-628.
249. Frost, H. M., Mechanical determinants of bone modeling, *J. Metab. Bone Dis. Relat. Res.,* 4, 217-230, 1983.
250. Frost, H. M., Griffith, D. L., Jee, W. S. S., Kimmel, D. B., McCandlis, R. P., and Teitelbaum, S. L., Histomorphometric changes in trabecular bone of renal failure patients treated with calcifediol, *J. Metab. Bone Dis.,* 2, 285-295, 1981.
251. Frost, H. M., Resting seams: "On and Off" in lamellar bone forming centers, in *Bone Histomorphometry: Third International Workshop,* Jee, W. S. S. and Parfitt, A. M., Eds., *J. Metab. Bone Dis. Relat. Res., Suppl.,* 167-170, 1981.
252. Frost, H. M., Mechanical microdamage, bone remodeling and osteoporosis. A review, in *Osteoporosis,* DeLuca, H. F., Frost, H. M., Jee, W. S. S., Johnston, C. C., Jr., and Parfitt, A. M., Eds., University Park Press, Baltimore, 1981, 185-190.
253. Frost, H. M., Pharmacology and the osteoporoses: a field in flux, *Method Find. Exp. Clin. Pharmacol.,* 5, 5-16, 1983.
254. Frost, H. M., The osteocyte as a water pump, in *Bone Modeling and Skeletal Modeling Errors,* Charles C Thomas, Springfield, Ill., 1973, 119-150.
255. Frost, H. M., An optimizing strategy for medical research, in *The Physiology of Cartilaginous, Bony and Fibrous Tissue,* Charles C Thomas, Springfield, Ill., 1972, 185-204.
256. Frost, H. M., Frame, B., Ormond, R. S., and Hunter, R. B., Atypical axial osteomalacia. A report of 3 cases, *Clin. Orthop.,* 23, 285-295, 1962.
257. Frost, H. M., Coherence treatment of osteoporoses, *Orthop. Clin. N. Am.,* 12, 649-669, 1981.
258. Frost, H. M., The history of the "BMU controversy", *Orthop. Clin. N. Am.,* 12, 716-721, 1981.
259. Frost, H. M., Appropriate and inappropriate animal models, *Orthop. Clin. N. Am.,* 12, 735-737, 1981.
260. Frost, H. M., The minimum effective strain: a determinant of bone architecture, *Clin. Orthop. Relat. Res.,* 175, 286-292, 1983.
261. Frost, H. M., The regional acceleratory phenomenon, *Henry Ford Hosp. Med. J.,* 31, 3-9, 1983.
262. Frost, H. M., The labelling escape error, in *Bone Histomorphometry: Techniques and Interpretation,* Recker, R. R., Ed., CRC Press, Boca Raton, Fla., 1983, 133-142.
263. Frost, H. M., The skeletal intermediary organization, *J. Metab. Bone Dis. Relat. Res.,* 4, 281-290, 1983.
264. Frost, H. M., Mechanical determinants of skeletal architecture: bone modeling, in *The Scientific Basis of Orthopaedics,* 2nd ed., Albright, J. A. and Brand, R. A., Eds., Appleton-Century-Crofts, New York, in press.
265. Frost, H. M., Mechanical determinants of skeletal architecture: chondral modeling, in *The Scientific Basis of Orthopaedics,* 2nd ed., Albright, J. A. and Brand, R. A., Eds., Appleton-Century-Crofts, New York, in press.

266. Frost, H. M., Mechanical determinants of skeletal architecture: fibrous tissue modeling, in *The Scientific Basis of Orthopaedics,* 2nd ed., Albright, J. A. and Brand, R. A., Eds., Appleton-Century-Crofts, New York, in press.
267. Frost, H. M., An unpublished and/or personal observation, or an original proposal by the author.
268. Fukada, E. and Yasuda, J., On the piezoelectric effect of bone, *J. Physiol. Soc. Jpn.,* 12, 1158-1160, 1957.
269. Gaillard, P. J., Herrmann-Erlee, M. P. M., Hekkelman, J. W., Burger, E. H., and Nijweide, P. J., Skeletal tissue in culture, *Clin. Orthop. Relat. Res.,* 142, 196-214, 1979.
270. Galen (129—199), *Encyclopedia Britannica,* 15th ed., Micropedia IV, 385, 1974.
271. Gallagher, J. C., Aaron, J., Horsman, A., Wilkinson, R., and Nordin, B. E. C., Corticosteroid osteoporosis, *Clin. Endocrinol. Metab.,* 2, 355-368, 1973.
272. Gallavotti, C., *The Elements of Mechanics,* Springer-Verlag, New York, 1983.
273. Garn, S. M., *The Earlier Gain and the Later Loss of Cortical Bone,* Charles C Thomas, Springfield, Ill., 1970.
274. Garn, S. M., The phenomenon of bone formation and bone loss, in *Osteoporosis,* DeLuca, H. F., Frost, H. M., Jee, W. S. S., Johnston, C. C., Jr., and Parfitt, A. M., Eds., University Park Press, Baltimore, 1981, 3-16.
275. Geddes, A. C., The origin of the osteoblast and of the osteoclast, *J. Anat. Physiol.,* 47, 159-171, 1912.
278. Gegenbaur, C., Üeber die Bildung des knochengewebes, *Jena Z. Med. Naturwiss.,* 1, 343-369, 1867.
279. Gilles de la Tourette, F. and Marinesco, G., Note sur l'anatomie pathologique de l'ostéite déformante de Paget, *Bull. Mem. Soc. Méd. Hôp. Paris,* 11, 422-451, 1894.
280. Giroux, J. M., Courpron, P., and Meunier, P. J., *Histomorphométrie de l'Ostéopénie Physiologique Sénile,* Univ. Claude Bernard, Lyon, 1975.
281. Glorieux, F. H., Marie, P. J., Pettifor, J. M., and Delvin, E. E., Bone response to phosphate salts, ergocalciferol and calcitonin in hypophosphatemic vitamin D-resistant rickets, *N. Engl. J. Med.,* 303, 1023-1031, 1980.
282. Glorieux, F. H., personal communications, 1982.
283. Goby, P., Une application nouvelle des rayons x: la microradiographie, *C.R. Acad. Sci.,* 156, 686-688, 1913.
284. Goby, P., La microradiographie stéréoscopique en relief et en pseudo-relief: la stéréomicroradiographie, *C.R. Acad. Sci.,* 180, 735-737, 1925.
285. Golub, L., Stern, B., Glimcher, M., and Goldhaber, P., The inhibition of the maturation of newly synthesized bone collagen by B aminopropionitrile in tissue culture, *Proc. Exp. Biol. Med.,* 129, 465-469, 1968.
286. Goodman, C. R., Osteoporosis as an early complication of hemiplegia, *N.Y. State Med. J.,* 7, 1943-1945, 1971.
287. Goodship, A. F., Lanyon, L. E., McFie, H., Functional adaptation of bone to increased stress, *J. Bone Jt. Surg.,* 61A, 539-546, 1979.
288. Goodsir, J. and Goodsir, H. D. S., *Anatomical and Pathological Observations,* M. MacPhail, Edinburgh, 1845.
289. Goldhaber, P., Stern, B. D., Glimcher, M., and Chao, J., The effects of parathyroid extract and thyrocalcitonin on bone remodeling in tissue culture, in *Parathyroid Hormone and Thyrocalcitonin (Calcitonin),* Talmage, R. V. and Belanger, L. F., Eds., Excerpta Medica, New York, 1968, 182-195.
290. Gordan, G. S., Picchi, J., and Roof, B. S., Antifracture efficacy of long term estrogens for osteoporosis, *Trans. Assoc. Am. Physicians,* 86, 326-332, 1973.
291. Gordan, G. S., Postmenopausal osteoporosis; cause, prevention and treatment, *Clin. Obstetr. Gynecol.,* 4, 169-178, 1977.
292. Gotcher, J. E. and Jee, W. S. S., The progress of the periodontal syndrome in the rat. I. Morphometric and autoradiographic studies, *J. Periodontol. Res.,* 16, 275-291, 1981.
293. Goto, S., Inoue, S., Kurihara, M., Shimo, R., and Ozawa, T., Studies of normal and deformed spine development by morphometrical labelling methods, in *Handbook of Bone Morphometry,* Takahashi, H., Ed., Nipponia, Tokyo, 1983, 154-164.
294. Graves, R. J., Clinical lectures, *London Med. Surg. J.,* Part II, 7, 516-593, 1835.
295. Greenwald, I., Some chemical changes in the blood of dogs after thyroparathyroidectomy, *J. Biol. Chem.,* 61, 649-657, 1924.
296. Griffith, C. C., Nichols, G., Asher, J. D., and Flannagan, B., Heparin osteoporosis, *JAMA,* 193, 91-95, 1965.
297. Groer, G. and Marshall, J. H., Mechanism of calcium exchange at bone surfaces, *Calcif. Tissue Res.,* 12, 175-192, 1973.
298. Grusin, H. and Samuel, E., A syndrome of osteoporosis in Africans and its relationship to scurvy, *Am. J. Clin. Nutr.,* 5, 644-650, 1957.

299. Gunderson, H. J. G., Stereology — or how figures for spatial shape and content are obtained by observation of structures in sections, *Microsc. Acta*, 83, 409-426, 1980.
300. Haba, T., personal communication, 1983.
301. Hales, S., *Vegetable Statics*, W. Innys, London, 1731.
302. Hahn, T. J., Halstead, L. R., Teitelbaum, S. L., and Hahn, B. H., Altered mineral metabolism in glucocorticoid induced osteopenia, *J. Clin. Invest.*, 64, 655-665, 1979.
303. Hall, M. T., *The Architecture of Bone*, Charles C Thomas, Springfield, Ill., 1966.
304. Hangartner, T. J., Overton, T. R., Harley, C. H., Van den Berg, L., and Crockford, P. M., Skeletal challenge: an experimental study of pharmacologically induced changes in bone density in the distal radius, using gamma-ray computed tomography, *Calcif. Tissue Int.*, in press.
305. Hanson, C. A., Sagrin, J. W., and Duncan, H., The osteoporosis of ankylosing spondylitis, *Clin. Orthop.*, 74, 59-66, 1971.
306. Hansson, T., *The Bone Mineral Content and Biomechanical Properties of Lumbar Vertebrae*, Uno Lundgren Tryckeri, A. B., Göteborg, Sweden, 1977.
307. Hardrup, T. and Tehag, H., Mitosis of chondrocytes in normal adult joint cartilage, *Clin. Orthop. Relat. Res.*, 153, 248-252, 1980.
308. Harrington, I. J., Static and dynamic loading patterns in knee joints with deformities, *J. Bone Jt. Surg.*, 65A, 247-259, 1983.
309. Harris, E. D., Pathogenesis of rheumatoid arthritis, *Clin. Orthop. Relat. Res.*, 182, 14-23, 1984.
310. Harris, W. H. and Heaney, R. P., Skeletal renewal and metabolic bone disease, *N. Engl. J. Med.*, 280, 193-202, 253-259, 303-311, 1969.
311. Harris, W. H., Heaney, R. P., and Jowsey, J., Growth hormone: the effects on skeletal renewal in the adult dog, *Calcif. Tissue Res.*, 10, 1-13, 1978.
312. Harrison, M. and Fraser, R., Bone metabolism in rats studied with stable strontium, *J. Endocrinol.*, 2, 191-196, 1960.
313. Harvey, W. M. (1528—1657), *Encyclopedia Britannica*, 15th ed., Micropedia IV, 935-936, 1974.
314. Hassall, A. H., *Microscopic Anatomy of the Human Body*, Pratt, Woodward, New York, 1851.
315. Hattner, R., Epker, B. N., and Frost, H. M., Suggested sequential mode of control of changes in cell behavior in adult bone remodeling, *Nature (London)*, 206, 489-490, 1965.
316. Haussler, M. R. and McCain, T. A., Basic and clinical concepts related to vitamin D metabolism and action, *N. Engl. J. Med.*, 297, 974-1041, 1977.
317. Havers, C., *Osteologia Nova*, London, 1691.
318. Heaney, R. P., A unified concept of osteoporoses. A second look, in *Osteoporosis*, Barzel, U. S., Ed., Grune & Stratton, New York, 1970, 257-265.
319. Heller-Steinberg, M., Ground substance, bone salts and cellular activity in bone formation and destruction, *Am. J. Anat.*, 89, 347-379, 1951.
320. Hench, P. S., The reversiblity of certain rheumatic and non-rheumatic conditions by the use of cortisone or of the pituitary adrenocorticotrophic hormone, *Ann. Int. Med.*, 36, 1-19, 1950.
321. Henkin, W. A., Collapse of vertebral bodies in sickle cell anemia, *Am. J. Roentgenol.*, 62, 395-402, 1949.
322. Hennig, A., A critical survey of volume and surface measurement in microscopy, *Zeiss Werkzeitschr.*, 30, 78-87, 1958.
323. Hérissant, M., Eclaircissements sur l'ossification, *Mem. Acad. R. Sci.*, 322-336, 1758.
324. Hesp, R., Hulme, P., Williams, D., and Reeve, J., The relationship between changes in femoral bone density and calcium balance in patients with involutional osteoporosis treated with human parathyroid hormone fragment (APTH 1-34), *J. Metab. Bone Dis. Relat. Res.*, 2, 331-334, 1981.
325. Hess, A. F. and Weinstock, M., Antirachitic properties imparted to inert fluids and to green vegetables by ultraviolet irradiation, *J. Biol. Chem.*, 62, 301-313, 1924.
326. Heys, P. M., Osteogenesis imperfecta and odontogenesis imperfecta, *J. Pediatr.*, 56, 334-339, 1960.
327. High, W. B., Capen, C. C., and Black, H. E., Histomorphometric evaluations of the effects of intermittent 1,25-dihydroxy cholecalciferol administration on cortical bone remodeling in adult dogs, *Am. J. Pathol.*, 104, 41-49, 1981.
328. High, W. B., Capen, C. C., and Black, H. E., Effects of low dose parathyroid hormone administration on cortical bone remodeling in adult dogs, *Lab. Invest.*, 44, 449-454, 1981.
329. High, W. B., personal communications, 1980, 1981, 1982, 1983.
330. High, W. B., Capen, C. C., and Black, H. E., Histomorphometric effects of thyroxine on cortical bone remodeling in adult dogs, *Am. J. Pathol.*, 102, 438-446, 1981.
331. High, W. B., Capen, C. C., and Black, H. E., Histomorphometric Evaluation of the Effects of 1,25 Dihydroxycholecalciferol, Parathormone and Thyroxine on Trabecular Bone Remodeling in Adult Dogs, Thesis, Ohio State Univ., Columbus, 1980.
332. Hioco, D., Miravet, L., and Bordier, P., Physiopathology and treatment of osteoporosis in younger men, in *Osteoporosis*, Hioco, D., Ed., Masson et Cie, Paris, 1964, 365-398.

333. Hitt, O., Jaworski, Z. F. G., Shimizu, A. G., and Frost, H. M., Tissue level bone formation rates in chronic renal failure measured by means of tetracycline bone labeling, *Can. J. Physiol. Pharmacol.*, 48, 824-828, 1970.
334. Hjelmstedt, A., Asplund, S., and Rauschning, W., Cryodissection and cryosectioning in biomechanical studies on congenital dislocation of the hip, *Acta Clin.*, 4, 13-21, 1982.
335. Hjelmstedt, A. and Asplund, S., Congenital dislocation of the hip: a biomechanical study in autopsy specimens, *J. Pediatr. Orthop.*, 3, 491-497, 1983.
336. Hollander, J. L. and McCarty, D. J., Jr., *Arthritis and Allied Conditions,* Lea & Febiger, Philadelphia, 1972.
337. Holtrop, M. E., Cox, K. A., Holick, M. F., and Arnast, C. S., Effects of hormones on the activity of osteoclasts, in *Osteoporosis,* DeLuca, H. F., Frost, H. M., Jee, W. S. S., Johnston, C. C., Jr., and Parfitt, A. M., Eds., University Park Press, Baltimore, 1981, 59-66.
338. Hooke, R. (1635—1703), *Encyclopedia Britannica,* 15th ed., Micropedia V, 120, 1974.
339. Hooke, R., *Micrographia,* Royal Society, London, 1665, 1-270.
340. Hoover, G. H. and Frost, H. M., Dynamic correction of spastic rocker — bottom foot, *Clin. Orthop.*, 65, 175-182, 1969.
341. Hori, M., Takahashi, H., Konno, T., and Haba, T., A classification of double labels in iliac trabecular bone in beagles, in *Handbook of Bone Morphometry,* Takahashi, H., Ed., Nipponia, Tokyo, 1983, 147-153.
342. Horton, J. E., Tarpley, T. M., and Davis, W. F., *Proceedings Mechanisms of Localized Bone Loss,* Spec. Suppl., Calcif. Tissue Abstr., 1-454, 1978.
343. Howship, J., Experiments and observations in order to determine the means employed by the animal economy in the formation of bone, *Trans. Med. Chir. Soc. (London),* 6, 263-301, 1817.
344. Hughes, S., Davies, R., Kahn, R., and Kelley, P., Fluid space in bone, *Clin. Orthop. Relat. Res.*, 134, 232-341, 1978.
345. Hughes, S., Kahn, R., Davies, R., and Lavender, P., The uptake by the canine tibia of the bone scanning agent 99^m Tc-MDP before and after an osteotomy, *J. Bone Jt. Surg.,* 60B, 579-582, 1978.
346. Huldschinsky, K., Cure of rickets my means of artificial heliotherapy, *Dtsch. Med. Wochenschr.,* 45, 712-713, 1919.
347. Hunter, D., The significance to clinical medicine of studies in calcium and phosphorus metabolism, 1, 947-957, 1930.
348. Hunter, J., Experiments and observations on the growth of bones (1798), in *Hunter's Works,* Palmer, D. F., Ed., Longman, London, 1837.
349. Hunter, W., A case of acromegaly. Hypertrophy of the pituitary body and thyroid; changes in the bone marrow, *Trans. Pathol. Soc. London,* 49, 246-278, 1897.
350. Iannaccone, A., Gabrilove, J. L., Brahms, S. A., and Soffer, L. J., Osteoporosis in Cushing's Syndrome, *Ann. Int. Med.,* 52, 570-586, 1960.
351. Imhotep (ca.2700 B.C.), *Encyclopedia Britannica,* 15th ed., Micropedia V, 310, 1974.
352. Indong, O. and Harris, W. H., Proximal strain distribution in the loaded femur, *J. Bone Jt. Surg.,* 60A, 75-85, 1978.
353. Inman, V. T., Functional aspects of the abductor muscles of the hip, *J. Bone Jt. Surg.,* 28, 607-617, 1947.
354. Inoue, T., personal communications, 1983.
355. Irons, B. and Shrive, N., *Finite Element Primer,* John Wiley & Sons, New York, 1983.
356. Ishikawa, H. and Hirohata, K., An immunoelectron-microscopic study of the immunoglobulin synthesizing cells of the rheumatoid synovial membrane, *Clin. Orthop. Relat. Res.,* 185, 280-289, 1984.
357. Iskrant, A. P. and Surita, R. W., Osteoporoses in women 45 years and over related to subsequent fractures, *Public Health Rep.,* 84, 33-38, 1969.
358. Ivey, J. L. and Baylink, D. J., Postmenopausal osteoporosis: proposed roles of defective coupling and estrogen deficiency, *J. Metab. Bone Dis. Relat. Res.,* 3, 3-7, 1981.
359. Jackson, A. S., *Goiter and Other Diseases Of The Thyroid Gland,* Paul B. Hoeber, New York, 1926.
360. Jacob, F. and Monod, J., Genetic regulatory mechanisms in the synthesis of proteins, *J. Mol. Biol.,* 3, 318-356, 1961.
361. Jaffe, J. L., The resorption of bone; a consideration of the underlying processes, particularly in pathological conditions, *Arch. Surg.,* 20, 355-369, 1930.
362. Jaffe, H. L., *Metabolic, Degenerative and Inflammatory Disease of Bones and Joints,* Lea & Febiger, Philadelphia, 1972.
363. Jaffe, M. D. and Willis, P. W., Multiple fractures associated with long term sodium heparin therapy, *JAMA,* 193, 158-160, 1965.
364. Jande, S. S., Heterogeneity of osteocyte population, in *Bone Morphometry,* Jaworski, Z. F. G., Ed., University of Ottawa Press, Ottawa, 1976, 274-278.
365. Janssen, M., *On Bone Formation: Its Relations to Tension and Pressure,* Longmans, London, 1920.

366. Jastrup, B., Mosekilde, L., Melsen, F., Lund, B. J., and Sørensen, O. H., Serum levels of vitamin D metabolites and bone remodeling in hyperthyroidism, *Metabolism,* 31, 126-132, 1982.
367. Jaworski, Z. F. G. and Lok, E., The rate of osteoclastic bone erosion in haversian remodeling sites of adult dog rib, *Calcif. Tissue Res.,* 10, 103-112, 1972.
368. Jaworski, Z. F. G., personal communications, 1977—1983.
369. Jaworski, Z. F. G. and Wellington, J. L., Impaired osteoclastic function and linear bone erosion rate in secondary hyperparathyroidism associated with renal failure, *Clin. Orthop. Relat. Res.,* 107, 297-310, 1975.
370. Jaworski, Z. F. G., The quantum concept in bone remodeling in adults, in *Vitamin D: Biochemical, Chemical and Clinical Aspects Related to Calcium Metabolism,* Norman, A. M., Schaefer, K., Coburn, J. W., DeLuca, H. F., Fraser, D., Grigoleit, H. G., and Herrath, P. V., Eds., W. De-Gruyter, Berlin, 1981, 103-113.
371. Jaworski, Z. F. G., Ed., *Bone Morphometry,* University of Ottawa Press, Ottawa, 1976.
372. Jaworski, Z. F. G., Liskova-Kiar, M., and Uhthoff, H., Effect of long term immobilization on the pattern of bone loss in older dogs, *J. Bone Jt. Surg.,* 62B, 104-110, 1980.
373. Jaworski, Z. F. G. and Hooper, C., Study of cell kinetics within evolving secondary Haversian systems, *J. Anat.,* 131, 91-102, 1980.
374. Jaworski, Z. F. G., Duck, B., and Sekaly, G., Kinetics of osteoclasts and their nuclei in evolving secondary haversian systems, *J. Anat.,* 133, 397-405, 1981.
375. Jaworski, Z. F. G., Medical treatment of Paget's disease of bone: current status, in *Current Concepts of Diagnosis and Treatment of Bone and Soft Tissue Tumors,* Springer-Verlag, Heidelberg, 1984, 395-403.
376. Jaworski, Z. F. G., Lamellar bone turnover system and its effector organ, *Calcif. Tissue Int.,* 36, 546-555, 1984.
377. Jee, W. S. S., and Arnold, J. S., Rate of individual haversian system formation, *Anat. Rec.,* 118, 315, 1954.
378. Jee, W. S. S., Black, H. E., and Gotcher, J. E., Effect of dichloromethane diphosphonate on cortisol-induced bone loss in young adult rabbits, *Clin. Orthop. Relat. Res.,* 156, 39-51, 1981.
379. Jee, W. S. S., The skeletal tissues, in *Histology,* 5th ed., Weiss, L., Ed., Elsevier/North-Holland, Amsterdam, 1983, 200-255.
380. Jee, W. S. S. and Nolan, P. D., Origin of osteoclasts from the fusion of phagocytes, *Nature (London),* 200, 325, 1963.
381. Jenkins, D. H. R., Cheng, D. H. F., and Hodgson, A. R., Stimulation of bone growth by periosteal stripping, *J. Bone Jt. Surg.,* 57B, 482-484, 1975.
382. Jenner, Edward (1749—1823), *Encyclopedia Britannica,* 15th ed., Micropedia V, 542, 1974.
384. Jensen, A. and Chenowith, H. H., *Statics and Strength of Materials,* 3rd ed., McGraw-Hill, New York, 1975.
385. Jett, S., Wu, K., and Frost, H. M., Tetracycline-based histological measurement of cortical endosteal bone formation in normal and osteoporotic rib, *Henry Ford Hosp. Med. J.,* 15, 325-344, 1967.
386. Jett, S., Duncan, H., and Frost, H. M., Adrenalcorticoid and salicylate actions on human and canine haversian bone remodeling, *Clin. Orthop.,* 68, 301-315, 1970.
387. Johnson, F., Leitl, S., and Waugh, W., The distribution of load across the knee, *J. Bone Jt. Surg.,* 62B, 346-349, 1980.
388. Johnson, L. C., The kinetics of skeletal remodeling, in *Structural Organization of the Skeleton,* Milch, R. A. and Robinson, R. A., Eds., National Foundation March of Dimes, New York, 1966, 66-142.
389. Johnston, C. C., Osteoporosis: an overview, in *Clinical Disorders of Bone and Mineral Metabolism,* Frame, B. and Potts, J. T., Jr., Eds., Excerpta Medica, Amsterdam, 1983, 317-322.
390. Johnston, C. C. and Norton, J. A., How can effectiveness of treatment be determined?, in *Osteoporosis,* DeLuca, H. F., Frost, H. M., Jee, W. S. S., Johnston, C. C., Jr., and Parfitt, A. M., Eds., University Park Press, Baltimore, 1981, 375-382.
391. Johnston, C. C., Norton, J. A., Khair, M. R. A., and Longcope, C., Age-related bone loss, in *Osteoporosis II,* Barzel, U. S., Ed., Grune & Stratton, New York, 1979, 91-100.
392. Johnston, C. C., Smith, D. M., Yu, P. L., and Deiss, W. P., In vivo measurement of bone mass in the radius, *Metabolism,* 17, 1140-1153, 1968.
393. Jones, H. H., Priest, J. D., Hayes, W. C., Tichenor, C. C., and Nagel, D. A., Humeral hypertrophy in response to exercise, *J. Bone Jt. Surg.,* 59A, 204-208, 1977.
394. Jones, J. P., Jr., Alcholism, hypercortisonism, fat embolism and osseous vascular necrosis, in *Idiopathic Ischemic Necrosis of the Femoral Head in Adults,* Zinn, W. M., Ed., George Thieme, Stuttgart, 1971, 112-132.
395. Jones, S. J., Boyde, A., and Shapiro, M., The response of osteoblasts to parathyroid hormone (PTH 1-34) in vitro, *J. Metab. Bone Dis. Relat. Res.,* 2, 335-338, 1981.

396. Joubin, J., Pattrone, C. F., and Pettrone, F. A., Cornelia de Lange's syndrome, *Clin. Orthop. Relat. Res.*, 171, 186-195, 1982.
397. Jowsey, J., Variations in bone mineralization with age and disease, in *bone Biodynamics,* Frost, H. M., Ed., Little, Brown, Boston, 1964, 461-479.
398. Jowsey, J., Riggs, B. L., Kelly, P. J., and Hoffman, D. L., Effect of combined therapy with sodium fluoride, vitamin D and calcium in osteoporosis, *Am. J. Med.*, 53, 43, 1972.
399. Jowsey, J. and Offord, K. P., Osteoporosis: juvenile, idiopathic and postmenopausal, in *Mechanisms of Localized Bone Loss,* Horton, J. E., Tarpley, T. M., and Davis, W. F., *Calcif. Tissue Int. (Suppl.),* 345-364, 1978.
400. Karaharju, E., Ryoppy, S. A., and Makinen, R. J., Remodeling by asymmetrical epiphyseal growth, *J. Bone Jt. Surg.*, 58B, 122-126, 1976.
401. Kasabach, H. H. and Dyke, C. G., Osteoporosis circumscripta of the skull as a form of osteitis deformans, *Am. J. Roentgenol.*, 28, 192-199, 1932.
402. Kasai, R. K., Ichisaka, A., Okumura, J., and Yamamuro, T., Minicomputer system for bone morphometry, in *Handbook of Bone Morphometry,* Takahashi, H., Ed., Nipponia, Tokyo, 1983, 197-206.
403. Kassarian, L., Cann, L., Parfitt, A. M., Simmons, D., More-Holton, E., A 14-day ground-based hypokinesia study in nonhuman primates. A compilation of results. N.A.S.A. T.M. #41268, 1981.
404. Katayama, H., Suruga, K., Kurashige, T., and Kimoto, T., Bone changes in congenital biliary atresia, *J. Clin. Invest.*, 42, 1471-1475, 1963.
405. Kellgren, J. H., Ball, J., and Tutton, G. K., The articular and other limb changes in acromegaly, *Q. J. Med.*, 21, 405-424, 1952.
406. Kember, N. F., Cell population kinetics of bone growth: the first ten years of autoradiographic studies with tritiated thymidine, *Clin. Orthop.*, 76, 213-220, 1971.
407. Kember, N. F. and Sissons, H. A., Quantitative histology of the human growth plate, *J. Bone Jt. Surg.*, 58B, 426-435, 1976.
408. Khairi, M. R. A. and Johnston, C. C., Jr., Treatment of Paget's disease of bone (osteitis deformans) with sodium etidronate (EHDP), *Clin. Orthop. Relat. Res.*, 127, 94-105, 1977.
409. Khairi, M. R. A., Cronin, J. H., Robb, J. A., Smith, D. M., Yu, P. L., and Johnston, C. C., Jr., Femoral trabecular pattern index and bone mineral content measurement by photon absorption in senile osteoporosis, *J. Bone Jt. Surg.*, 58A, 221-226, 1976.
410. Khoury, S., Silberman, F. S., and Cabrini, R. L., Stimulation of the longitudinal growth of long bones by peristeal stripping, *J. Bone Jt. Surg.*, 45A, 1679-1684, 1963.
411. Kimmel, D. B., personal communications, 1981, 1983.
412. Kimmel, D. B. and Jee, W. S. S., Bone cell kinetics during longitudinal growth in the rat, *Calcif. Tissue Int.*, 32, 123-133, 1980.
413. Kimmel, D. B. and Jee, W. S. S., A quantitative histologic analysis of the growing long bone metaphysis, *Calcif. Tissue Int.*, 32, 113-122, 1980.
414. Kimmel, D. B., Cellular basis of bone accumulation during growth: implications for metabolic bone disease, in *Proceedings, Bone Histomorphometry Third International Workshop,* Jee, W. S. S. and Parfitt, A. M., Eds., Armour-Montagu, Levallois, 1981, 87-95.
415. Kimmel, D. B., Ritz, E., Krempien, B., and Mehls, O., Bone cell kinetics and function during experimental uremia, *J. Metab. Bone Dis. Relat. Res.*, 3, 191-198, 1981.
416. Kimmel, D. B., A computer simulation of the mature skeleton, presented at the Fourth International Workshop on Bone Histomorphometry, Aarhus, Denmark, June 10, 1984.
417. Klein, L., Dawson, M. H., and Heiple, K. G., Turnover of collagen in the adult rat after denervation, *J. Bone Jt. Surg.*, 59A, 1065-1067, 1977.
418. Klein, L., Player, J. S., Heiple, K. G., Bahniuk, E., and Goldberg, V. M., Isotopic evidence for resorption of soft tissues and bone in immobilized dogs, *J. Bone Jt. Surg.*, 64A, 225-230, 1982.
419. Koch, J. C., The laws of bone architecture, *Am. J. Anat.*, 21, 177-293, 1917.
420. Koch, R., Die aetiologie der tuberculose, *Klin. Wochenschr.*, 19, 221-260, 1882.
421. Koch, Robert (1843—1910), *Encyclopedia Britannica,* 15th ed., Micropedia V, 864, 1974.
422. von Koelliker, R., Die normale Resorption des knochengewebes, und ihre Bedeutung für die Entstenhung der typischen knochenformen, F.C.W. Vogel, Leipzig, 1873.
423. Kolář, J., Babický, A., and Vrabec, R., *The Physical Agents and Bone,* Czech Academy of Science, Prague, 1965.
424. Konno, T. and Takahashi, H., Bone histomorphometry: semiautomated image analysis, in *Handbook of Bone Morphometry,* Takahashi, H., Ed., Nipponia, Tokyo, 1983, 87-89.
425. Kragstrup, J., Richards, A., and Fejerskov, O., Experimental osteofluorosis in the domestic pig. A stereologic study of vertebral trabecular bone, presented at the Fourth International Workshop on Bone Histomorphometry, Aarhus, Denmark, June 10, 1984.
426. Kragstrup, J. and Melsen, F., Three-dimensional morphology of trabecular bone osteons reconstructed from serial sections, *J. Metab. Bone Dis. Relat. Res.*, 5, 127-130, 1984.

427. Krane, S. M., Assessment of mineral and matrix turnover, in *Clinical Disorders of Bone and Mineral Metabolism,* Frame, B. and Potts, J. T., Jr., Amsterdam, 1983, 95-98.
428. Krempien, B., Ritz, E., Ditzen, K., and Hudelmeir, G., Über den einfluss der Niereninsuffizienz auf knochenbildung und knochen resorption, *Virchows Arch.,* A335, 354-366, 1972.
429. Kruse, H. P. and Kuhlencordt, F., Osteopetrosis Albers-Schönberg mit gesteigertem knochenumbau, *Klin. Wochenschr.,* 51, 762-766, 1973.
430. Ksiazek, T., Bone induction by calcified cartilage transplant, *Clin. Orthop. Relat. Res.,* 172, 243-250, 1983.
431. Kuhlencordt, F., Bauditz, W., Lanzano-Tonkin, C., Kruse, H. P., Rehpenning, W., and Bartelheimer, H., Osteopathien und Calcium-Phosphatstoffwechsel bei chronischer Hämodialyse, *Klin. Werkzeitshr.,* 49, 134-155, 1971.
432. Kuhlencordt, F. and Bartelhiemer, H., *Klinische Osteologie,* 6th ed., Springer-Verlag, Heidelberg, 1980.
433. Kummer, R. H., Recherches sur le métabolisme minéral dans la maladie de Basedow, *Rev. Med. Suisse Rom.,* 37, 439-450, 1917.
434. Labat, M. L., Tzehoral, E., Moricard, Y., Feldman, M., and Milhaud, G., Lack of a T-cell dependent subpopulation of macrophages in (dichloromethylene) disphosphonate treated mice, *Biomed. Pharmacoth.,* 37, 270-276, 1983.
435. Lacapère, J. H., Drieux, H., and Dalaville, G., Etude sur l'ostéophytose expérimentale, *Ann. Med.,* 56, 603-627, 1955.
436. Lacroix, P., *The Organization of Bones,* McGraw-Hill, New York, 1951.
437. Lacroix, P., Remarques sur l'ostéoporose post-traumatique, in *L'Ostéoporose,* Hioco, D. J., Ed., Masson et Cie, Paris, 1964, 34-41.
438. Lafferty, J. F., Winter, W. G., and Gambaro, S. A., Fatigue characteristics of posterior elements of vertebrae, *J. Bone Jt. Surg.,* 57A, 154-158, 1977.
439. Landeros, O. and Frost, H. M., A cell system in which rate and amount of protein synthesis are separately controlled, *Science,* 145, 1323-1324, 1964.
440. Langloh, N. D., Hunder, G. G., Riggs, B. L., and Kelley, P. J., Transient painful osteoporoses of the lower extremities, *J. Bone Jt. Surg.,* 55A, 1188-1196, 1973.
441. Lanyon, L. E. and Hartman, W., Strain related electrical potentials recorded in vitro and in vivo, *Calcif. Tissue Res.,* 22, 315-327, 1977.
442. Lanyon, L. E., Paul, I. L., Rubin, C. T., Thrasher, E. L., DeLaura, R., Rose, R. M., and Radin, E. L., In vivo strain measurements from bone and prosthesis following total hip replacement, *J. Bone Jt. Surg.,* 63A, 989-1000, 1981.
443. Lanyon, L. E., Goodship, A. E., Pye, C. J., and MacFie, J. H., Mechanically adaptive bone remodeling, *J. Biomech.,* 15, 141-154, 1982.
444. Lanyon, L. E., Functional strain as a determinant for bone remodeling, *Calcif. Tissue Int.,* 36, S56-S61, 1984.
445. Larsson, S. E., Lortentzon, R., and Boquist, L., Low doses of 1,25 dihydrocholecalciferol increase mature bone mass in adult normal rats, *Clin. Orthop. Relat. Res.,* 127, 228-235, 1977.
446. Leblond, C. P., Wilkinson, G. W., Belanger, L. F., and Robichon, J., Radioautographic visualization of bone formation in the rat, *Am. J. Anat.,* 86, 289-342, 1950.
447. van Leeuwenhoeck, A., Microscopical observations concerning blood, milk, bones, the brain, spittle and cuticula, *Philos. Trans.,* 106, 125-144, 1674.
448. van Leeuwenhoeck, A., Several observations on the texture of the bones of animals compared with that of wood, *Phil. Trans. R. Soc. London,* 17, 838-843, 1693.
449. Leman, J., Jr., Litzow, J. R., and Lennon, E. J., The effects of chronic acid loads in normal man. Further evidence for the participation of bone mineral in the defense against chronic metabolic acidosis, *J. Clin. Invest.,* 45, 1608-1614, 1966.
450. Lemnius, L., De miraculis occultis nature, *Colonia,* 1567.
451. Leriche, R., Désequilibres rasomoteurs posttraumatiques primitifs des extrêmités, *Lyon Chir.,* 20, 746-753, 1923.
452. Leriche, R. and Policarde, A., *Les Problèmes de la Physiologie Normal et Pathologique de l'Os,* Masson et Cie, Paris, 1926.
453. Levine, M., Hypothesis behavior by humans during discrimination learning, *J. Exp. Psychol.,* 71, 331-338, 1966.
454. Lewis, F. T., *Stohr's Histology,* P. Blakiston's Son & Co., Philadelphia, 1906.
455. Lewis, A., Harootunian, A., Huang, J., and Smith, B., Time-resolved x-ray absorption spectroscopy of carbon monoxide-myoglobin recombination after laser photolysis, *Science,* 223, 811-813, 1984.
456. Lewis, J. L., Askew, M. J., Wixson, R. L., Kramer, G. M., and Tarr, R. R., The influence of prosthetic stem stiffness and of a calcar collar on stresses in the proximal end of the femur with a cemented femoral component, *J. Bone Jt. Surg.,* 66A, 280-286, 1984.

457. Lindahl, O. and Lindgren, A. G. H., Grading of osteoporosis in autopsy specimens, *Acta Orthop. Scand.*, 32, 85-100, 1962.
458. Lindsay, R., Aitken, J. M., Anderson, J. B., Hart, D. M., MacDonald, E. B., and Clarke, A. C., Long-term prevention of postmenopausal osteoporosis by oestrogen, *Lancet*, i, 1038-1042, 1976.
459. Lindsay, R., McLean, A., Kraszewsik, A., Hart, D. M., Clarke, A. C., and Garwood, J., Bone response to termination of oestrogen treatment, *Lancet*, i, 1325-1328, 1978.
460. Liskova, M. and Hert, J., Reaction of bone to mechanical stimuli. II. Periosteal and endosteal reaction of tibial diaphysis to intermittent loading, *Folia Morphol.*, 19, 301, 1971.
461. Lister, J., On the antiseptic principle in the practice of surgery, *Br. Med. J.*, 2, 246-269, 1867.
462. Little, K., Osteoporotic mechanisms, *J. Int. Med. Res.*, 1, 509-529, 1973.
463. Looser, E., Über Spätrachitis und Osteomalacia: Klinische, roentgenologische und pathologisch anatomische untersuchungen, *Dtsch. Z. Chir.*, 152, 210-357, 1920.
464. Louyot, P., Mathieu, J., and Gaucher, A., L'ostéose raréfiante des gastrectomisés (étude de 50 sujets gastrectomisés avec anastomose gastro-jejunale), *Arch. Fr. Mal. Appar.*, 50, 20-38, 1961.
465. Luck, J. V., *Bone and Joint Diseases*, Charles C Thomas, Springfield, Ill., 1950.
466. MacCallum, W. G. and Voegtlin, C., On the relation of tetany to the parathyroid glands and to calcium metabolism, *J. Exp. Med.*, 2, 118-151, 1908.
467. Mack, P. B. and LaChance, P. A., Effects of recumbency and space flight on bone density, *Am. J. Clin. Nutr.*, 20, 1194-1205, 1967.
468. Malluche, H. H., Ritz, E., Lange, H. P., Kutschera, J., Hodgson, M., Sieffert, U., and Schoeppe, W., Bone histology in incipient and advanced renal failure, *Kidney Int.*, 9, 355-362, 1976.
469. Malm, O., *Calcium Requirement and Adaptation in Adult Men*, University of Oslo Press, Norway, 1958.
470. Malpighi, M., *De Ossium Structura*, Ex. Op. Posth., Venice, 1743.
471. Mankin, H. J. and Lipiello, L., Turnover of adult rabbit articular cartilage, *J. Bone Jt. Surg.*, 51A, 862-874, 1969.
472. Manske, P. R. and Lesker, P. A., Nutritional pathways to extensor tendons within the extensor retinacular components, *Clin. Orthop. Relat. Res.*, 181, 234-237, 1983.
473. Manske, P. R. and Lesker, P. A., Comparative nutrient pathways to the flexor profundis tendon in Zone II of various experimental animals, *J. Surg. Res.*, 34, 83-92, 1982.
474. Maquet, P. G. J., *Biomechanics of the Knee*, 2nd ed., Springer-Verlag, New York, 1984.
475. Marie, P. and Marinesco, G., Sur l'anatomie pathologique de l'acromégalie, *Arch. Med. Exp. Anat. Pathol.*, 3, 539-560, 1981.
476. Marks, S. C., Jr., Tooth eruption depends on bone resorption: experimental evidence from osteopetrotic (ia) rats, *J. Metab. Bone Dis. Relat. Res.*, 3, 107-115, 1981.
477. Maroteaux, P., Fontaine, G., Scharfman, W., and Farriaux, J. P., L'hyperostose corticale généralisée a transmission dominante (Type Worth), *Arch. Fr. Pediatr.*, 28, 685-698, 1971.
478. Marotti, G., Quantitative studies on bone reconstruction, *Acta Anat.*, 52, 291-333, 1963.
479. Matthews, J. L. and Talmage, R. V., Influence of parathyroid hormone on bone cell ultrastructure, *Clin. Orthop. Relat. Res.*, 156, 27-38, 1981.
480. Marx, J. L., Organizing the cytoplasm, *Science*, 222, 1109-1111, 1983.
481. Matkovic, U., Ciganovic, M., Tominac, C., and Kostial, K., Osteoporosis and epidemiology of fractures in Croatia, *Henry Ford Hosp. Med. J.*, 28, 116-126, 1980.
482. Matthews, J. L., Talmage, R. V., and Doppelt, R., Responses of osteocyte lining cell complex — the bone cell unit — to calcitonin, *J. Metab. Bone Dis. Relat. Res.*, 2, 113-122, 1980.
483. Mattsson, S., The reversibility of disuse osteoporosis, *Acta Orthop. Scand., Suppl.*, 144, 5-135, 1972.
484. Maxwell, James C. (1831-1879), *Encyclopedia Britannica*, 15th ed., Micropedia VI, 717-718, 1974.
485. Mayer, E., Cause and effect in biology, *Science*, 134, 1501-1509, 1961.
486. Mayer, R. E., *Thinking, Problem Solving, Cognition*, W. B. Freeman, New York, 1983.
487. Mazess, R. B., On aging bone loss, *Clin. Orthop. Relat. Res.*, 165, 239-252, 1982.
488. Mazess, R. B. and Vetler, J., Measurement of marrow fat in trabecular bone using dual-energy computed tomography, presented at the Fourth International Workshop on Bone Histomorphometry, Aarhus, Denmark, June 13, 1984.
489. McBeath, A. A., Schopley, S. A., and Narechania, R. G., Circumferential and axial strain in the proximal femur, *Clin. Orthop. Relat. Res.*, 150, 301-305, 1980.
490. McLean, F. C. and Urist, M. R., *Bone: An Introduction to the Physiology of Skeletal Tissue*, Univ. Chicago Press, Chicago, 1955.
491. McRae, W. M. and Sweet, E. M., Diagnosis of osteoporosis in childhood, *Br. J. Radiol.*, 40, 104-107, 1967.
492. Meema, H. E., Rabinovich, S., Pierratos, A., Katirtzoglu, A., Murray, T. M., and Oreopoulos, D. G., The healing effect of 1,25 dihydroxy vitamin D_3 on periosteal and intracortical resorption in renal osteodystrophy, *Metab. Bone Dis. Relat. Res.*, 2, 223-231, 1980.

493. Meema, H. E. and Meema, S., Microradioscopic bone structure of the hand in thyrotoxicosis, renal osteodystrophy and acromegaly, in *Clinical Aspects of Metabolic Bone Disease,* Frame, B., Parfitt, A. M., and Duncan, H., Eds., Excerpta Medica, Amsterdam, 1973, 10-19.
494. Melsen, F., personal communications, 1980, 1983.
495. Melsen, F. and Nielson, H. E., Osteonecrosis following renal allotransplantation, *Acta Pathol. Microbiol. Scand. Sect. A,* 85, 99-104, 1977.
496. Melsen, F. and Mosekilde, L., Morphometric and dynamic studies of bone changes in hyperthyroidism, *Acta Pathol. Microbiol. Scand.,* 85, 141-150, 1977.
497. Melsen, F., *Histomorphometric Analysis of Iliac Bone in Normal and Pathological Conditions,* Aarhus Amtssygehus, Aarhus, Denmark, 1978, 1-83.
498. Melsen, F. and Mosekilde, L., Dynamic studies of trabecular bone formation and osteoid maturation in normal and certain pathological conditions, *Metab. Bone Dis. Relat. Res.,* 1, 45-55, 1978.
499. Melsen, F. and Mosekilde, L., The role of bone biopsy in the diagnosis of metabolic bone disease, *Orthop. Clin. N. Am.,* 12, 571-602, 1981.
500. Melsen, F. and Mosekilde, L., Tetracycline double labelling of iliac trabecular bone in 41 normal adults, *Calcif. Tissue Res.,* 26, 99-102, 1978.
501. Melsen, F., Mosekilde, L., and Kragstrup, J., Metabolic bone diseases as evaluated by bone histomorphometry, in *Bone Histomorphometry: Techniques and Interpretations,* Recker, R. R., Ed., CRC Press, Boca Raton, Fla., 1983, 265-284.
502. Mendel, Gregor J. (1822-1884), *Encyclopedia Britannica,* 15th ed., Micropedia VI, 782-783, 1974.
503. Menon, T. J., Thjellesen, D., and Wroblewski, B. M., Charnley low-friction arthroplasty in diabetic patients, *J. Bone Jt. Surg.,* 65B, 580-581, 1983.
504. Meunier, P. J., Bianchi, G. G. S., Edouard, C. M., Bernard, J. C., Courpron, P., and Vignon, G. E., Bony manifestations of thyrotoxicosis, *Orthop. Clin. N. Am.,* 3, 745-774, 1972.
505. Meunier, P. J., Courpron, P., Edouard, C., Bernard, J., Bringuier, J., and Vignon, G., Physiological senile involution and pathological rarefaction of bone. Quantitative and comparative histological data, *Clin. Endocrinol. Metab.,* 2, 239-256, 1973.
506. Meunier, P. J., Ed., *Bone Histomorphometry,* Armour-Montagu, Paris, 1977.
507. Meunier, P. J., Courpron, C., Edouard, C., Alexandre, C., Bressot, C., Lips, P., and Boyce, B. F., Bone histomorphometry in osteoporotic states, in *Osteoporosis II,* Barzel, U. S., Ed., Grune & Stratton, New York, 1979, 27-47.
508. Meunier, P. J., personal communications, 1979, 1983.
509. Merz, W. A. and Schenk, R. K., A quantitative histological study on bone formation in human cancellous bone, *Acta Anat.,* 76, 1-11, 1970.
510. Michelsson, J. E. and Rauschning, W., Pathogenesis of experimental heterotopic bone formation following temporary forcible exercising of immobilized limbs, *Clin. Orthop. Relat. Res.,* 176, 265-272, 1983.
511. Mielke, J. H., Armelagos, G. J., and Van Gerven, D. P., Trabecular involution in femoral heads of a prehistoric (x group) population from Sundanes Nubia, *Am. J. Phys. Anthropol.,* 36, 39-44, 1972.
512. Milch, R. A. and Robinson, R. A., Eds., *Structural Organization Of The Skeleton,* National Foundation March of Dimes, New York, 1966.
513. Milch, R. A., Rall, D. P., and Tobie, J. F., Fluorescence of tetracycline antibiotics in bone, *J. Bone Jt. Surg.,* 40A, 897-910, 1958.
514. Milhaud, G. and Aubert, J.-P., Quelques remarques concernant l'analyse cinétique a l'aide de calcium radio-actif dans l'ostéoporose, in *L'Ostéoporose,* Hioco, D. J., Ed., Masson et Cie, Paris, 1964, 60-62.
515. Milkman, L. A., Multiple spontaneous idiopathic symmetrical fractures, *Am. J. Roentgenol.,* 32, 622-634, 1934.
516. Miller, M. R. and Kasahara, M., Observations on the innervation of human long bones, *Anat. Rec.,* 145, 13-23, 1963.
517. Miller, S. C. and Jee, W. S. S., The microvascular bed of fatty bone marrow in the adult beagle, *J. Metab. Bone Dis.,* 2, 239-246, 1980.
518. Mills, B. G., Singer, F. R., Weiner, L. P., and Hulst, P. A., Cell cultures from bone affected by Paget's disease, *Arth. Rheum.,* 23, 1115, 1980.
519. Minaire, P., Meunier, P. J., Edouard, C., Bernard, J., Courpron, P., and Bournet, H., Quantitative histological data on disuse osteoporosis, *Calcif. Tissue Res.,* 17, 57-73, 1974.
520. Minaire, P., *L'Ostéoporose D'Immobilization. Données Biologiques et Histologiques Quantitatives,* Thesis, Ediprim, Lyon, 1973, 1-122.
521. Minkoff, E. C., *Evolutionary Biology,* Addison-Wesley, N. Reading, Mass., 1983.
522. Miravet, L. and Landron, A., Etudes cinétique avec le Ca^{45} dans l'ostéoporose, in *L'Ostéoporose,* Hioco, D. J., Ed., Masson et Cie, Paris, 1964, 74-91.
523. Miyakawa, G. and Stearns, G., Severe osteoporosis (or osteomalacia) associated with long continued low-grade steatorrhea, *J. Bone Jt. Surg.,* 24, 429-435, 1942.

524. Mizaldus, A., Memorabilium utilium et jucundorum centurie, Lutetiae, 1566.
525. Monro, A., *The Anatomy of the Human Bones,* 7th ed., E. Hamilton and J. Balfour, Edinburgh, 1763.
526. Moreland, M. S., Morphological aspects of tension applied to growing bone, *J. Bone Jt. Surg.,* 62B, 230-237, 1980.
527. Morgan, B., *Osteomalacia, Renal Osteodystrophy and Osteoporosis,* Charles C Thomas, Springfield, Ill., 1973.
528. Morrisey, R. T., Steele, R. W., and Gerdes, M. H., Localized immune complexes and slipped upper femoral epiphysis, *J. Bone Jt. Surg.,* 65B, 574-579, 1983.
529. Mosekilde, L., Melsen, F., Bagger, O., Jensen, M., and Sørensen, N. S., Bone changes in hyperthyroidism, *Acta Endocrinol.,* 85, 515-520, 1977.
530. Mosekilde, L. and Melsen, F., Morphometric and dynamic studies of bone changes in hypothyroidism, *Acta Pathol. Microbiol. Scand.,* 86, 56-62, 1978.
531. Mosekilde, L. and Melsen, F., A tetracycline based histomorphometric evaluation of the maturation period of osteoid in normal and certain pathological conditions, *Calcif. Tissue Int.,* 27, A30, 117, 1979.
532. Mosekilde, L. and Melsen, F., A tetracycline based histomorphometric evaluation of bone resorption and bone turnover in hyperthyroidism and hyperparathyroidism, *Acta Med. Scand.,* 204, 97-102, 1978.
533. Mosekilde, L. and Melsen, F., Dynamic differences in trabecular bone remodeling between patients after jejuno-ileal bypass for obesity and epileptic patients receiving anticonvulsant therapy, *J. Metab. Bone Dis.,* 2, 77-82, 1980.
534. Moyen, B. J. L., Lahey, P. J., Winberg, E. H., and Harris, W. H., Effects on intact femurs of dogs of the application and removal of plates, *J. Bone Jt. Surg.,* 61A, 866-872, 1979.
535. Muller, H., Über die Entwicklung der knochensubstanz nebst Bemerkungen ueber den Bau rachitischer knochen, *Z. Zool.,* 9, 147-233, 1858.
536. Nakamura, T. and Kanda, S., The role of microphages in osteoclast neogenesis in vivo: quantitative analysis of osteoclasts and macrophages in endosteal bone development in rats treated with hydrocortisone or calcitonin, *J. Metab. Bone Dis. Relat. Res.,* 5, 139-143, 1984.
537. Nester, E. W., Roberts, C. E., Lidstrom, M. E., Pearsall, N. N., and Nester, M. T., *Microbiology,* 3rd ed., W. B. Saunders, Philadelphia, 1983.
538. Newton, I., *Philosophiae Naturalis Principia Mathematica,* 1687.
539. Nathan, P. A. and Fowler, A., Remodeling of a metacarpal bone graft in a child, *J. Bone Jt. Surg.,* 58A, 719-722, 1976.
540. Nilsson, B. E., Spinal osteoporosis and femoral neck fracture, *Clin. Orthop.,* 68, 93-95, 1970.
541. Nilsson, B. E. and Westlin, N. E., The fracture incidence after gastrectomy, *Acta. Surg. Scand.,* 137, 533-534, 1971.
542. Nordin, B. E. C., The application of basic science to osteoporosis, in *Bone Biodynamics,* Frost, H. M., Ed., Little, Brown, Boston, 1964, 521-542.
543. Nordin, B. E. C., MacGregor, J., and Smith, D. A., The incidence of osteoporosis in normal women; its relation to age and the menopause, *Q. J. Med.,* 35, 25-38, 1966.
544. Norimatsu, H., Vander Wiel, C. J., and Talmage, R. V., Morphological support of a role for cells lining bone surfaces in maintenance of plasma calcium concentration, *Clin. Orthop. Relat. Res.,* 138, 254-262, 1979.
545. Norimatsu, H., Weil, C. J. V., and Talmage, R. V., Electron microscopic study of the effects of calcitonin on bone cells and their extracellular milieu, *Clin. Orthop. Relat. Res.,* 139, 250-258, 1979.
546. Norman, A. W., Schaefer, K., Coburn, J. W., DeLuca, H. F., Fraser, D., Grigoleit, H. G., and Herrath, D. V., *Vitamin D. Biochemical, Chemical and Clinical Aspects Related to Calcium Metabolism,* Walter de Gruyter, New York, 1977.
547. Norrdin, R. W. and Shih, M. S., Profiles of cortical remodeling sites in longitudinal rib sections of beagles with renal failure and parathyroid hyperplasia, *J. Metab. Bone Dis. Relat. Res.,* 5, 353-359, 1983.
548. Norrdin, R. W., Villafane, F. A., and Lopresti, C. A., Haversian remodeling activity in various bones of beagles and the effect of chronic renal failure, in *Bone Histomorphometry,* Meunier, P. J., Ed., Armour-Montagu, Paris, 1977, 171-178.
549. O'Reagan, S., Mehoom, D. K., and Newman, A. J., Methotrexate-induced bone pain in childhood leukemia, *Am. J. Dis. Child.,* 126, 489-490, 1974.
550. Obrant, K. J. and Nilsson, B. E., Histomorphologic changes in the tibial epiphysis after diaphyseal fracture, *Clin. Orthop. Relat. Res.,* 185, 270-275, 1984.
551. Olah, A. J., Histomorphometric analysis of bone formation in postmenopausal osteoporosis and its reactions to sodium fluoride, presented at the Fourth International Workshop on Bone Histomorphometry, Aarhus, Denmark, June 11, 1984.
552. Ossemman, E. F., Plasma cell myeloma, *N. Engl. J. Med.,* 26, 952-960, 1959.

553. Osterberg, E. E. and Mills, R. G., The absence of osteoporosis demonstrated chemically in clinical hyperthyroidism, *Am. J. Med. Sci.*, 184, 399-412, 1932.
554. Overstall, P. W., Exton-Smith, A. N., Imms, F. J., and Johnson, A. L., Falls in the elderly related to postural imbalance, *Br. Med. J.*, 1, 261-264, 1977.
555. Overton, R. R., Silverberg, D. S., Grace, M., Rigal, W. R., Higgins, M., Bettcher, K. B., Dossetor, J. B., Harley, F., and DeLuca, H. F., Bone demineralization in renal failure: a longitudinal study of the distal femur using photon absorptiometry, *Br. J. Radiol.*, 49, 921-925, 1976.
556. Overton, T. R., Whitmore, K., Heath, R., Menon, D., Hangartner, T. N., and Elsasser, U., A multiple detector gamma-ray CT system: implementation and application, *J. Comput. Assist. Tomogr.*, 3, 1979.
557. Owen, M., The origin of bone cells, *Int. Rev. Cytol.*, 28, 213-238, 1970.
558. Owen, M., Studies on cell population kinetics in bone, in *Bone Morphometry*, Jaworski, Z. F. G., Ed., University of Ottawa Press, Ottawa, 1976, 303-309.
559. Owen, M., The histogenesis of bone cells, *Calcif. Tissue Res.*, 25, 205-207, 1978
560. Pak, C. Y. C., Zisman, E., Evans, R., Jowsey, J., DeLea, C., and Bartter, F. C., The treatment of osteoporosis with calcium infusions: clinical studies, *Am. J. Med.*, 47, 7-22, 1969.
561. Pak, C. Y. C. and Bartter, F. C., Treatment of osteoporosis with calcium infusions, *Semin. Drug. Treat.*, 2, 39-46, 1972.
562. Pappas, A. M. and Miller, J. T., Cogenital ball-and socket ankle joints and related lower-extremity malformations, *J. Bone Jt. Surg.*, 64A, 672-679, 1982.
563. Pardis, G. R. and Kelley, P. J., Blood flow and mineral deposition in canine tibial fractures, *J. Bone Jt. Surg.*, 57A, 220-226, 1975.
564. Parlier, R. and Hioco, D. J., Traitment de l'ostéoporose, in *L'Ostéoporose,* Hioco, D. J., Ed., Masson et Cie, Paris, 1964, 234-249.
565. Parfitt, A. M., Morphologic basis of bone mineral measurements: transient and steady state effects of treatment in osteoporosis, *Min. Elect. Metab.*, 4, 273-287, 1980.
566. Parfitt, A. M. and Duncan, H., Metabolic bone disease affecting the spine, in *The Spine*, Rothman, R. H. and Simeone, F. A., Eds., W. B. Saunders, Philadelphia, 1975, 599-720.
567. Parfitt, A. M., Oliver, I., and Villanueva, A. R., Bone histology in metabolic bone disease: the diagnostic value of bone biopsy, *Orthop. Clin. N. Am.*, 10, 329-345, 1979.
568. Parfitt, A. M., Matthews, C. H. G., Villanueva, A. R., Rao, D. S., Rogers, M., Kleerekoper, M., and Frame, B., Microstructural and cellular basis of age-related bone loss and osteoporosis, in *Clinical Disorders of Bone and Mineral Metabolism,* Frame, B. and Potts, J. T., Jr., Eds., Excerpta Medica, Amsterdam, 1983, 328-332.
569. Parfitt, A. M., Assessment of trabecular bone status — introduction, in *Clinical Disorders of Bone and Mineral Metabolism,* Frame, B. and Potts, J. T., Jr., Eds., Excerpta Medica, Amsterdam, 1983, 27-29.
570. Parfitt, A. M., Recent developments in bone physiology — introduction, in *Clinical Disorders of Bone and Mineral Metabolism,* Frame, B. and Potts, J. T., Jr., Eds., Excerpta Medica, Amsterdam, 1983, 171-172.
571. Parfitt, A. M., Equilibrium and disequilibrium hypercalcemia: new light on an old concept, *Metab. Bone Dis. Relat. Res.*, 1, 279-293, 1979.
572. Parfitt, A. M., The cellular basis of bone remodeling: the quantum concept reexamined in the light of recent advances in the cell biology of bone, *Calcif. Tissue Int. Suppl.*, 36, S37-S45, 1984.
573. Parfitt, A. M. and Kleerekoper, M., The divalent ion homeostatic system: physiology and metabolism of calcium, phosphorous, magnesium and bone, in *Clinical Disorders of Fluid and Electrolyte Metabolism,* 3rd ed., Maxwell, M. and Kleeman, C. R., Eds., McGraw-Hill, New York, 1979, 269-398.
574. Parfitt, A. M., The quantum concept of bone remodeling and turnover: implications for the pathogenesis of osteoporosis, *Calcif. Tissue Int.*, 28, 1-7, 1979.
575. Parfitt, A. M. and Chir, B., Richmond Smith as a clinical investigator, *Henry Ford Hosp. Med. J.*, 28, 95-107, 1980.
576. Parson, J. A., Parathyroid physiology and the skeleton, in *Biochemistry and Physiology of Bone,* Bourne, G. H., Ed., Academic Press, New York, 1976, 159-182.
577. Parsons, J. A. and Zanelli, J. M., Physiological role of the parathyroid glands, in *Klinische Osteologie,* 6th ed., Kuhlencourt, F. and Bartelheimer, H., Eds., Springer-Verlag, Heidelberg, 1980, 135-172.
578. Pasteur, Louis (1822-1895), *Encyclopedia Britannica,* 15th ed., Micropedia VII, 791-792, 1974.
579. Paterson, C. R., *Metabolic Disorders of Bone,* Blackwell, Oxford, 1974.
580. Pauwels, F., Kurzer überlick über die mechanische Beanspruchung des knochens und Bedeutung für die funktionelle Anpassung, *Z. Orthop.*, 3, 671-695, 1973.
581. Pechet, M. M., Day, L. C., Carrol, E., and Howard, D. E., The metabolic effects of neutral phosphate and studies of the prevention of corticosteroid induced osteoporosis, in *Phosphate Metabolism; Kidney and Bone,* Avioli, L., Bordier, P., Fleish, H., Massry, S., and Slatopolsky, E., Eds., Armour-Montagu, Paris, 1976, 47-62.

582. Peck, W. A., Shen, V., and Rifas, L., Locally elaborated factors in the coordination of bone cell activity, in *Clinical Disorders of Bone and Mineral Metabolism,* Frame, B. and Potts, J. T., Jr., Eds., Excerpta Medica, Amsterdam, 1983, 179-182.
583. Peck, W. A., Kohler, G., and Rifas, L., Bone cell function studied *in vitro:* the importance of local chemical signals, in *Osteoporosis,* DeLuca, H. F., Frost, H. M., Jee, W. S. S., Johnston, C. C., Jr., and Parfitt, A. M., Eds., University Park Press, Baltimore, 1981, 67-76.
584. Perry, H. M., III, Drake, D. M., and Avioli, L. V., Alternate calcitonin and etidronate disodium therapy for Paget's bone disease, *Arch. Int. Med.,* 144, 929-933, 1984.
585. Petersen, H., Die organe des Skelett Systems, in *Handbuch Mikroskopischen Anatomie,* Mollendorf, H., Ed., Springer, Berlin, 1930.
586. Petrovic, A., Oudet, C., and Stutzman, J., L'Hormone somatotrope peut-elle stimuler la résorption osseuse?, *C.R. Acad. Aci.,* 276, 2961-2966, 1973.
587. Phillip, E. T., Five cases of femoral neck injury in women after x-ray treatment for carcinoma of the uterus, *Strahlentherapie,* 44, 363-376, 1932.
588. Pienkowski, D. and Pollack, S. R., The origin of stress-generated potentials in fluid saturated bone, *J. Orthop. Res.,* 1, 30-41, 1983.
589. Plummer, W. and Kunlap, H. F., Cases showing osteoporosis due to decalcification in exophthalmic goiter, *Proc. Staff Meet. Mayo Clin.,* 3, 119-131, 1928.
590. Pogrund, H., Makin, M., Robin, C., Menczel, J., and Steinbert, R., Osteoporosis in patients with fractured femoral neck in Jerusalem, *Clin. Orthop.,* 124, 165-172, 1976.
591. Polanyi, M., Life's irreducible structure, *Science,* 160, 1308-1312, 1968.
592. Policard, A., *L'Appareil de Croissance des os longs. Ses Méchanismes a l'Etat Normale et Pathologique,* Masson et Cie, Paris, 1941.
593. Pollack, J. A., Schiller, A. L., and Crawford, J. D., Rickets myopathy cured by removal of a non-ossifying fibroma of bone, *Pediatrics,* 52, 364-371, 1973.
594. Pollack, S. R., Bioelectrical properties of bone, *Orthop. Clin. N. Am.,* 15, 3-14, 1984.
595. Pommer, G., Über die Osteoklastentheorie, *Arch. Pathol. Anat.,* 92, 296-316, 1883.
596. Pommer, G., Untersuchengen über Osteomalacie und Rachitis, nebst betragen zur kenninis der durchbochrenden gefasse, Voegel, Leipzig, 1885.
597. Pommer, G., Über osteoporose, ihren Ursprung und ihre differential-diagnostische Bedeutung, *Arch. Klin. Chir.,* 136, 1-68, 1925.
598. Ponlot, R., *Le Radiocalcium (Ca^{45}) dans l'Etude des Os,* Arscia, S. A., Ed., Brussels, 1959.
599. Popov, E. P., *The Mechanics of Materials,* Prentice-Hall, Englewood Cliffs, N.J., 1952.
600. Popovtzer, M. M., Stjernholm, M., and Hutfer, W. E., Treatment of osteoporosis with alternating phosphorus and calcium infusions, in *Phosphate Metabolism: Kidney and Bone,* Avioli, L., Bordier, P., Fleish, H., Massry, S., and Slatopolsky, E., Eds., Armour-Montagu, Paris, 1976, 213-222.
601. Porter, R. W., The effect of tension across a growing epiphysis, *J. Bone Jt. Surg.,* 60B, 242-255, 1978.
602. Ponseti, I. V. and Baird, W. A., Scoliosis and dissecting aneurysm of the aorta in rats fed with *Lathyrus odoratus* seed, *Am. J. Pathol.,* 28, 1059-1067, 1952.
603. Potter, C. and Frost, H. M., Determining alignment of the knee, *Clin. Orthop. Relat. Res.,* 103, 32, 1974.
604. Potts, J. T. and Deftos, L. J., Parathyroid hormone, calcitonin, vitamin D, bone and bone mineral metabolism, in *Diseases of Metabolism,* 7th ed., Bondy, P. K. and Rosenberg, L. E., Eds., W. B. Saunders, Philadelphia, 1974, 1225-1430.
605. Prevate, J. J., *Human Physiology,* McGraw-Hill, New York, 1983.
606. Price, P. A., Williamson, M. K., Haba, T., Dell, R. B., and Jee, W. S. S., Excessive mineralization with growth plate closure in rats on chronic warfarin treatment, *Proc. Natl. Acad. Sci. U.S.A.,* 79, 7734-7738, 1982.
607. Prokop, D. J. and Sjoerdsma, A., Significance of urinary hydroxyproline in man, *J. Clin. Invest.,* 40, 843-849, 1961.
608. Putschar, W. G. J., General pathology of the musculoskeletal system, in *Handbuch der Allgemeinen Pathologie,* Buchner, F., Letterer, E., and Roulet, F., Eds., Springer-Verlag, Berlin, 1960, 361-488.
609. Rabischong, P., Bomel, F., Oonishi, A., Asaada, P., and Micaleff, J. P., Comportement bioméchanique du bassin a l'état normal et avec prosthèses totale de la hanche, *Rev. Chir. Orthop. Suppl. II,* 63, 95, 1977.
610. Radin, E. L., Trabecular microfractures in response to stress: the possible mechanism of Wolff's Law, in *Orthopaedic Surgery and Traumatology,* Excerpta Medica, Amsterdam, 1972, 59-65.
611. Radin, E. and Tolkoff, J., Subchondral bone changes in patients with early degenerative joint diseases, *Arth. Rheum.,* 13, 400-409, 1970.
612. Rafii, M., Firooznia, H., Galimbu, C., and Balthazar, E., Pathologic fracture in systemic mastocytosis, *Clin. Orthop. Relat. Res.,* 180, 260-267, 1983.

613. Rahn, B. A., Bone healing: histologic and physiologic concepts, in *Bone in Clinical Orthopaedics*, Sumner-Smith, G., Ed., W. B. Saunders, Philadelphia, 1982, 335-386.
614. Raisz, L. G., Trummel, C. L., Holick, M., and DeLuca, H. F.,1,25 dihyroxycholecalciferol: a potent stimulator of bone resorption in tissue culture, *Science*, 175, 768-769, 1972.
615. Raisz, L. G., Dietrich, J. W., and Mainz, D., Effects of phosphate on bone formation and resorption in tissue culture, in *Phosphate Metabolism; Kidney and Bone*, Bordier, P., Fleisch, H., Massry, S., and Slatopolsky, E., Eds., Armour-Montagu, Paris, 1976, 213-222.
616. Raisz, L. G., What marrow does to bone, *N. Engl. J. Med.*, 304, 1485-1486, 1981.
617. Ramser, J. R., Villanueva, A. R., Pirok, O. J., and Frost, H. M., Tetracycline based measurement of bone dynamics in three women with osteogenesis imperfecta, *Clin. Orthop.*, 49, 151-162, 1966.
618. Rasmussen, H. and Bordier, P., *The Physiological and Cellular Basis of Metabolic Bone Disease*, Williams & Wilkins, Baltimore, 1974.
619. Rasmussen, H., Theoretical considerations in the treatment of osteoporosis, in *Osteoporosis*, DeLuca, H. F., Frost, H. M., Jee, W. S. S., Johnston, C. C., Jr., and Parfitt, A. M., Eds., University Park Press, Baltimore, 1981, 383-392.
620. Rathburn, J. C., Hypophosphatasia: new development abnormality, *Am. J. Dis. Child.*, 75, 822-831, 1948.
621. Rao, S. D., Matkovic, V., and Duncan, H., Transiliac bone biopsy, *Henry Ford Hosp. Med. J.*, 28, 112-115, 1980.
622. Ray, R. D., Uptake of calcium[45] and carbon[14] labelled proline by dead and living bone, in *Radioisotopes and Bone*, McLean, F. C., Lacroix, P., and Budy, A. M., Eds., Davis, Philadelphia, 1962, 63-83.
623. Ranvier, L., Quelques faits relatif au dévelopment du tissu osseux, *C. R. Acad. Sci.*, 77, 1105-1129, 1873.
624. Recker, R. R., Saville, P. O., and Heaney, R. P., Effect of estrogens and calcium carbonate on bone loss in postmenopausal women, *Ann. Intern. Med.*, 87, 649-655, 1977.
625. Recker, R. R. and Heaney, R. P., Age-related bone loss and osteoporosis due to a primary defect in bone remodeling?, in *Osteoporosis*, DeLuca, H., Frost, H. M., Jee, W. S. S., Johnston, C. C., Jr., and Parfitt, A. M., Eds., University Park Press, Baltimore, 1981, 407-409.
626. Recker, R. R., Continuous treatment of osteoporosis. Current status, *Orthop. Clin. N. Am.*, 12, 611-627, 1981.
627. Recker, R. R., personal communications, 1979, 1983.
628. Recker, R. R., Ed., *Bone Histomorphometry: Techniques and Interpretation*, CRC Press, Boca Raton, Fla. 1983.
629. Reifenstein, E. C., Jr., Albright, F., and Wells, S. L., Accumulation, interpretation and presentation of data pertaining to metabolic balances, notably those of calcium, phosphorus and nitrogen, *J. Clin. Endocrinol. Metab.*, 5, 367-395, 1945.
630. Reifenstein, E. C., Jr., Osteoporosis, in *Principles of Internal Medicine*, Harrison, T. R., Adams, R. D., Bennett, I. L., Resnik, W. H., Thorn, C. W., and Wintrobe, M. M., Eds., McGraw-Hill, New York, 1962, 703-708.
631. Reilly, D., Walker, P. S., Ben-Dov, M., and Ewald, F. C., Effects of tibial components on load transfer in the upper tibia, *Clin. Orthop. Relat. Res.*, 165, 273-282, 1982.
632. Reilly, D. T. and Burstein, A. H., The mechanical properties of cortical bone, *J. Bone Jt. Surg.*, 36A, 1001-1022, 1974.
633. Resnik, L. B., Mathematics and science learning: a new conception, *Science*, 220, 447-478, 1983.
634. Reynolds, J. J., Pavlovitch, H., and Balsaw, S., 1, 25 dihydroxycholecaliferol increases bone resorption in thyroparathyroidectomized mice, *Calcif. Tissue Res.*, 21, 207-212, 1976.
635. Rhoades, C. E., Neff, J. R., Rengachary, S. S., Bratnitzky, S., Ketcherside, J., Price, H. I., and Jacobs, R. R., Diagnosis of post-traumatic syringohydromyelia presenting as neuropathic joints, *Clin. Orthop. Relat. Res.*, 180, 182-187, 1983.
636. Riggs, B. L., Jowsey, J., Kelley, P. J., Hoffman, D. L., and Arnaud, C. D., Studies on pathogenesis and treatment in postmenopausal and senile osteoporosis, *Clin. Endocrinol. Metab.*, 22, 317-332, 1973.
637. Riggs, B. L., Jowsey, J., and Kelley, P. J., Effects of oral therapy with calcium and vitamin D in primary osteoporosis, *J. Clin. Endocrinol. Metab.*, 42, 1139-1144, 1976.
638. Rindfleisch, E., *A Manual of Pathological Histology*, The New Syndenham Society, London, VII, 271-272, 1873.
639. Ring, P. A. and Ward, B. C. H., Paralytic bone lengthening following poliomyelitis, *Lancet*, 2, 551-553, 1958.
640. Ring, P. A., The influence of the nervous system on the growth of bones, *J. Bone Jt. Surg.*, 43B, 121-140, 1961.
641. Roberts, W. E. and Jee, W. S. S., Cell kinetics of orthodontically stimulated and nonstimulated rat periodontal ligament, *Arch. Oral Biol.*, 19, 17-26, 1974.

642. Robin, C., Note sur les éléments anatomiques appelés myeloplaxes, *J. Anat. Physiol. Norm.*, 1, 88-109, 1864.
643. Robinson, R. A., Observations regarding compartments for tracer calcium in the body, in *Bone Biodynamics*, Frost, H. M., Ed., Little, Brown, Boston, 1964, 423-439.
644. Roelfsema, F., Over let bot on de Calciumstofwisseling bij acromegalie, Drukkerus de kempenaer, Oegstgeest, 1972.
645. Rokitansky, C., *A Manual of Pathological Anatomy*, Vol. III, Blanchard and Lea, Philadelphia, 1855.
646. Romer, A. S., Bone in early vertebrates, in *Bone Biodynamics*, Frost, H. M., Ed., Williams & Wilkins, Boston, 1964, 13-40.
647. Romer, A. S., *Vertebrate Paleontology*, 3rd ed., University of Chicago Press, Chicago, 1966.
648. Rose, G. A., The art and science of calcium balance studies, *Proceedings Fourth European Symposium on Calcified Tissues*, Excerpta Medica, London, 1966, 91-93.
649. Roos, B. O. and Sköldborn, H., Dual photon absorptiometry in lumbar vertebrae, *Acta Radiol.*, 13, 266-280, 1974.
650. Rosendal, R. H. L., *Some Aspects of the Development In Vitro of Skeletal Elements of the Chick Embryo*, G. Van Soest, Amsterdam, 1964.
651. Rothschild, B. M. and Hanissian, A. S., Severe generalized (Charcot-like) joint destruction in juvenile arthritis, *Clin. Orthop. Relat. Res.*, 155, 75-80, 1981.
652. Rowland, R. E., Resorption and bone physiology, in *Bone Biodynamics*, Frost, H. M., Ed., Little, Brown, Boston, 1964, 335-355.
653. Rubin, P., *Dynamic Classification of Bone Dysplasias*, Yearbook, Chicago, 1964.
654. Rubens-Dural, A., Villaumey, J., Kaplan, G., Cerf, M., Duchiey, J., Gouin, B., and Marchie, C., Etude de perturbations du métabolism du lactose au cours de l'ostéoporose primitive, *Rev. Rheum.*, 37, 639-644, 1970.
655. Rüegsegger, P. and Elsasser, V., Computerassistierte photonenabsorptionmessung zur quantifizierung der Spondiosadicte, in *Proceedings First CEMO Symposium*, La Chaux-de-Fonds, 1975.
656. Rüegsegger, P., Anliker, M., and Dambacher, M., Quantification of trabecular bone with low dose computed tomography, *J. Comp. Assist. Tomogr.*, 5, 384-390, 1980.
657. Rutishauser, E. and Majno, G., Lésions osseuses par surcharge dans le squelette normale et pathologique, *Bull. Schweiz. Akad. Med. Wiss.*, 5, 333-342, 1950.
658. Sally into theory, An original prediction made from theory by the author.
659. Saunders, J. B. de C. M. and O'Malley, C. D., *Vesalius*, World, New York, 1950.
660. Saville, P. D., Changes in bone mass with age and alcoholism, *J. Bone Jt. Surg.*, 47A, 492-499, 1965.
661. Saville, P. D., Changes in skeletal mass and fragility with castration in the rat. A model of osteoporosis, *J. Am. Geriatr. Soc.*, 17, 155-160, 1969.
662. Schafer, M. F. and Dias, L. S., *Myelomeningocele: Orthopedic Treatment*, Williams & Wilkins, Baltimore, 1983.
663. Schatzker, J., Manley, P. A., and Sumner-Smith, G., In vivo strain guage study of bone response to loading with and without internal fixation, in *Current Concepts of Internal Fixation of Fractures*, Uhthoff, H. K., Ed., Springer-Verlag, Berlin, 1980, 306-314.
664. Schenk, R. K., Endosteal formation surface estimated by histological technique in iliac bone, in *Bone Morphometry*, Jaworski, Z. F. G., Ed., University of Ottawa Press, Ottawa, 1976, 185-188.
665. Schenk, R. K., Basic sterological principles, in *Bone Morphometry*, Jaworski, Z. F. G., Ed., University of Ottawa Press, Ottawa, 1976, 21-23.
666. Schenk, R. K., Die Histologie der Primären knochenheilung in Lichte neuer konzeptionen Über den knochenumbau, *Unfallheilkunde*, 81, 219-227, 1978.
667. Schenk, R. K. and Olah, A. J., Histomorphometrie, in *Klinische Osteologie*, 6th ed., Kuhlencordt, F. and Bartelheimer, H., Eds., Springer-Verlag, Heidelberg, 1980, 437-494.
668. Schultz, A., Anderson, G., Örtengren, R., Haderspeck, K., and Nachemson, A., Loads on the lumbar spine, *J. Bone Jt. Surg.*, 64A, 713-720, 1982.
669. Sedlin, E. D., Frost, H. M., and Villanueva, A. R., Variations in cross section area of rib cortex with age, *J. Gerontol.*, 18, 9-13, 1963.
670. Sedlin, E. D., Uses of bone as a model system in the study of aging, in *Bone Biodynamics*, Frost, H. M., Ed., Little, Brown, Boston, 1964, 655-666.
671. Seeforf, K. S., *Osteogenesis Imperfecta*, University Aarhus Press, Aarhus, 1949.
672. Semmelweiss, I. P., Die Aetiologie, der Begriff und die Prophylaxis des kindbettfiebers, Pest, Vienna, 1861.
673. Shapiro, J. R., Triche, T., Rowe, D. W., Monabi, A., Catell, H., and Schlesinger, S., Osteogenesis imperfecta and Paget's disease of bone, *Arch. Int. Med.*, 143, 2250-2257, 1983.
674. Shannon, C. E. and Weaver, W., *The Mathematical Theory of Communication*, Univ. Illinois Press, Urbana, 1962.

675. Sharnoff, M., Karcher, T. H., and Brehm, L. P., Microdifferential holography and the polysarcomic unit of activation of skeletal muscle, *Science*, 223, 822-825, 1984.
676. Sharpey, W., *Human Anatomy*, Lea and Blanchard, Philadelphia, 1856.
677. Sheldon, H., On the natural history of falls in old age, *Br. Med. J.*, 2, 1685-1690, 1960.
678. Shih, M. S. and Nordin, R. W., Regional acceleration of remodeling during healing of bone defects in Beagles of various ages, presented at the Fourth International Workshop on Bone Histomorphometry, Aarhus, Denmark, June 13, 1984.
679. Shimomur, Y. and Suzuki, F., Cultured growth cartilage cells, *Clin. Orthop. Relat. Res.*, 184, 93-105, 1984.
680. Shing, Y., Folkman, J., Sullivan, R., Butterfield, C., Murray, J., and Klagsbrun, M., Heparin affinity: purification of a tumor derived capillary endothelial cell growth factor, *Science*, 223, 1296-1298, 1984.
681. Schiano, A., Elsinger, J., and Acquaviva, P. C., *Les Algodystrophies*, Armour-Montagu, Paris, 1976.
682. Shipley, P. G., Park, F. A., McCollum, E. V., and Simmonds, N., Is there more than one kind of rickets?, *Am. J. Dis. Child.*, 23, 91-100, 1922.
683. Siefert, M. F., The biology of macrophages in osteopetrosis, *Clin. Orthop. Relat. Res.*, 182, 270-277, 1984.
684. Siegler, R. A., How knowledge influences learning, *Am. Sci.*, 71, 631-638, 1983.
685. Silberman, M., Mirsky, N., Levitan, S., and Weisman, Y., The effect of 1,25 dihydroxy vitamin D_3 on cartilage growth in neonatal mice, *J. Metab. Bone Dis. Relat. Res.*, 4, 337-345, 1983.
686. Sillence, D. O., Senn, A., and Danks, D. M., Genetic heterogeneity in osteogenesis imperfecta, *J. Med. Genet.*, 16, 101-116, 1979.
687. Sillence, D. O., Osteogenesis imperfecta: an expanding panorama of variants, *Clin. Orthop. Relat. Res.*, 159, 11-25, 1981.
688. Sirc, R., *L'Historadiographie, Nouvelle Méthode D'Investigation Histologique*, Thesis, Montpellier, 1938.
689. Singer, F. R., Paget's disease of bone. A slow virus infection, *Calcif. Tissue Int.*, 31, 185-187, 1980.
690. Singer, F. R. and Ahrne-Collier, I., Salmon calcitonin therapy: antibodies and clinical resistance in patients with Paget's disease of bone, in *Molecular Endocrinology*, MacIntyre, I. and Szelke, M., Eds., Elsevier/North-Holland, New York, 1977, 207-212.
691. Simmons, D. J. and Cohen, M., Postfracture linear growth in rats, *Clin. Orthop. Relat. Res.*, 149, 240-248, 1980.
692. Sissons, H. A., Osteoporosis of Cushing's Syndrome, in *Bone as a Tissue*, Rodahl, K., Nicholson, J. T., and Brown, E. M., Eds., McGraw-Hill, New York, 1960, 3-17.
693. Skeels, R. F., Reversibility of osteoporosis in Cushing's Disease, *J. Clin. Endocrinol.*, 18, 61-69, 1958.
694. Skinner, H. B., Barrack, R. L., and Cook, S. D., Age-related decline in proprioception, *Clin. Orthop. Relat. Res.*, 184, 208-211, 1984.
695. Sledge, C. B., Growth hormone and articular cartilage, *Fed. Proc., Fed. Am. Soc. Exp. Biol.*, 32, 1053-1057, 1973.
696. Smith, R. R., Malcolm, A. J., Madeley, C. R., and Gregg, P. J., Pathologic effects of vaccinia virus on rabbit bone and marrow, *Clin. Orthop. Relat. Res.*, 180, 278-286, 1983.
697. Smith, R. W., Eyler, W. R., and Mellinger, R. C., On the incidence of senile osteoporosis, *Ann. Int. Med.*, 52, 773-881, 1960.
698. Smith, R. W. and Walker, R. R., Femoral expansion in aging women: implications for osteoporosis and fractures, *Science*, 145, 156-157, 1964.
699. Smith, R. W. and Rizek, J., Epidemiologic studies of osteoporosis in women of Puerto Rico and Southeastern Michigan, *Clin. Orthop.*, 45, 31-48, 1966.
700. Snapper, I., *Bone Diseases in Medical Practice*, Grune & Stratton, New York, 1957.
701. Somerman, M. J., Hotchkiss, R. N., Bowers, M. R., and Termine, J., Comparison of fetal and adult human bone: identification of a chemotactic factor in fetal bone, *J. Metab. Bone Dis. Relat. Res.*, 5, 75-80, 1984.
702. Spengler, D. M., Baylink, D. J., and Rosenquist, J. B., Effect of β-aminopropionitrile on bone mechanical properties, *J. Bone Jt. Surg.*, 59A, 670-672, 1977.
703. Spranger, J. W., Langer, L. O., and Wiedemann, H. R., *Bone Dysplasias*, W. B. Saunders, Philadelphia, 1974.
704. Stanisavljevic, S., *Diagnosis and Treatment of Congenital Hip Pathology in the Newborn*, Williams & Wilkins, Baltimore, 1964.
705. Stanley, C. and Gultman, L., Measurement of internal boundaries in three dimensional structures by random sectioning, *Trans. A.I.M.E.*, January, 81-87, 1953.

706. Ste-Marie, L. G., Charhon, S. A., Edouard, C., and Meunier, P. J., Iliac bone histomorphometry in adults and children with osteogenesis imperfecta, presented at the Fourth International Workshop on Bone Histomorphometry, Aarhus, Denmark, June 12, 1984.
707. Steinberg, J. J. and Sledge, C. B., Co-cultivation models of joint destruction: early production of synovial inductive factors that direct cartilage breakdown, *Trans. Orthop. Res. Soc.*, 29, 198, 1983.
708. Steinberg, M. E. and Trueta, J., Effects of activity on bone growth and development in the rat, *Clin. Orthop. Relat. Res.*, 156, 52-60, 1981.
709. Storey, E., Osteosclerosis after intermittent administration of large doses of vitamin D in the rat, *J. Bone Jt. Surg.*, 42B, 606-625, 1960.
710. Stout, S. D., Histological structure and its preservation in ancient bone, *Curr. Anthropol.*, 19, 601-603, 1978.
711. Stout, S. D., The effects of long term immobilization on the histomorphology of human cortical bone, *Calcif. Tissue Int.*, 34, 337-342, 1982.
712. Strobino, L. J., French, G. O., and Colonna, P. C., The effect of increasing tensions on the growth of epiphyseal bone, *Surg. Gynecol. Obstetr.*, 95, 694-700, 1952.
713. Sudek, P., Die kollateralen Entzündungreaktionen an der Glied-massen, *Arch. Klin. Chir.*, 191, 710-753, 1938.
714. Takahashi, H., personal communications, 1976, 1979, 1982, 1983.
715. Takahashi, H., Epker, B. N., Hattner, R., and Frost, H. M., Evidence that bone resorption precedes formation at the cellular level, *Henry Ford Hosp. Med. Bull.*, 12, 359-364, 1964.
716. Takahashi, H. and Frost, H. M., A tetracycline-based evaluation of the relative prevalence and incidence of formation of secondary osteons in human cortical bone, *Can. J. Physiol. Pharmacol.*, 43, 785-791, 1965.
717. Takahashi, H. and Frost, H. M., Age and sex related changes in the amount of cortex in human ribs, *Acta Orthop. Scand.*, 37, 122-130, 1966.
718. Takahashi, H., Epker, B. N., and Frost, H. M., The relation between age and the size of osteons in man, *Henry Ford Hosp. Med. Bull.*, 13, 25-31, 1965.
719. Takahashi, H., Norimatsu, H., Konno, T., Inoue, H., and Yanagi, K., Different patterns of tetracycline uptake in varying forms of rickets and osteomalacia, *J. Metab. Bone Dis. Relat. Res.*, 2, 87-93, 1980.
720. Takahashi, H., Norimatsu, H., Watanabe, G., Konno, T., Inoue, J., and Fukuda, M., The remodeling period (sigma) in canine and human cortical bone, in *Calcium Endocrinology*, Yoshitoshi, Y. and Fujita, T., Eds., Chugai Igaku, Tokyo, 1980, 13-31.
721. Takahashi, H., Watanabe, G., Togawa, Y., Hanzoka, T., Konno, T., Sarto, Y., and Suzuki, H., The effect of various types of microelectrical current on internal remodeling of bone in dogs, *Trans. Orthop. Res. Soc.*, 7, 369, 1982.
722. Tashjian, A. H., Wright, D. R., Ivy, J. L., and Pont, A., Calcitonin binding sites in bone. Relationships to biological response and "escape", *Recent Prog. Horm. Res.*, 34, 285-334, 1978.
723. Teitelbaum, S. L. and Bates, M., Relationship of static and kinetic histomorphometric features of bone, *Clin. Orthop. Relat. Res.*, 146, 239-245, 1980.
724. Termine, J., Osteonectin and other newly described proteins of developing bone, in *Bone and Mineral Research*, Peck, W., Ed., Excerpta Medica, Amsterdam, 1983, 144-156.
725. Thomson, D. L. and Frame, B., Involutional osteopenia: current concepts, *Ann. Int. Med.*, 85, 789-803, 1976.
726. Thorngren, J. I. and Hansson, C., Cell kinetics and morphology of the growth plate in the normal and hypophysectomized rat, *Calcif. Tissue Res.*, 13, 113-129, 1973.
727. Todd, R. B. and Bowmann, W., *The Physiological Anatomy and Physiology of Man*, Blanchard and Lea, Philadelphia, 1845.
728. Todd, R. C., Freeman, M. A. R., and Pirie, C. J., Isolated trabecular fatigue fractures in the femoral head, *J. Bone Jt. Surg.*, 54B, 723-728, 1972.
729. Tomes, J. and DeMorgan, C., Observations on the structure and development of bone, *Phil. Trans. R. Soc.*, 143, 109-139, 1853.
730. Tonino, A. J., Davidson, C. L., Klopper, P. J., and Linclau, L. A., Protection from stress in bone and its effects. Experiments with stainless steel and plastic plates, *J. Bone Jt. Surg.*, 58B, 107-113, 1976.
731. Totot, G. and Sarronat, F., Ostéomalacie et goître exophthalmique, *Rev. Med. Paris*, 26, 445-461, 1906.
732. Tougard, L., Melsen, F., and Mosekilde, L., Bone phosphorus-hydroxyproline ratio and bone histomorphometry in normals, *Scand. J. Clin. Lab. Invest.*, 38, 89-95, 1978.
733. Triffitt, T. J., The organic matrix of bone tissue, in *Fundamental and Clinical Bone Physiology*, Urist, M. R., Ed., J. B. Lippincott, Philadelphia, 1980, 45-82.
734. Trotter, M., Broman, G. E., and Petersen, R. R., Densities of bones of whites and negro skeletons, *J. Bone Jt. Surg.*, 42A, 50-58, 1960.

735. Trotter, M., Hixon, B. B., and Deaton, S. S., Sequential changes in weight of the skeleton and in length of long limb bones of *Macaca mulatta, Am. J. Phys. Anthropol.*, 43, 79-89, 1975.
736. Trotter, M. and Petersen, R. R., The relationship of ash weight and organic weight in human skeletons, *J. Bone Jt. Surg.*, 44A, 669-681, 1962.
737. Turner, A. S., Mill, E. J., and Gabel, A. A., In vivo measurement of bone strain in the horse, *Am. J. Vet. Res.*, 36, 1573-1580, 1975.
738. Uhthoff, H. K. and Jaworski, Z. F. G., Bone loss in response to long term immobilization, *J. Bone Jt. Surg.*, 60B, 420-429, 1978.
739. Urist, M. R., Gurvey, M. S., and Fareed, D. O., Long term observations on aged women with pathological osteoporosis, in *Osteoporosis*, Barzel, U. S., Ed., Grune & Stratton, New York, 1970.
740. Urist, M. R., Orthopedic management of osteoporosis in postmenopausal women, *Clin. Endocrinol. Metab.*, 2, 159-171, 1973.
741. Urist, M. R., Ed., *Fundamental and Clinical Bone Physiology*, J. B. Lippincott, Philadelphia, 1980.
742. Urist, M. R., DeLange, R. J., and Finerman, G. A. M., Bone cell differentiation and growth factors, *Science*, 220, 680-686, 1983.
743. Venable, C. S. and Stuck, W. G., *The Internal Fixation of Fractures*, Charles C Thomas, Springfield, Ill., 1947.
744. Vesalius, A., *De Humani Corporis Fabrica*, Vols. 1-7, Padua, Italy, 1543.
745. Vignon, G. and Meunier, P. J., *Les Ostéoses Décalcifiantes Diffuses de L'Adulte*, Ciba-Geigy, Basel, 1973.
746. Villafane, F., Norrdin, R. W., Lopresti, C. A., and Kimmel, D., Bone remodeling in chronic renal failure in perinatally irradiated dogs, *Calcif. Tissue Res.*, 23, 171-178, 1977.
747. Villanueva, A. R., Ilnicki, L., Duncan, H., and Frost, H. M., Bone and cell dynamics in the osteoporoses: a review of measurements by tetracycline bone labelling, *Clin. Orthop.*, 49, 163-168, 1966.
748. Villanueva, A. R. and Frost, H. M., Bone formation in human osteogenesis imperfecta measured by tetracycline bone labelling, *Acta. Orthop. Scand.*, 41, 531-538, 1970.
749. Villanueva, A. R., Jaworski, Z. F. G., Hitt, O., Sarnsethsiri, P., and Frost, H. M., Cellular level bone resorption in chronic renal failure and primary hyperparathyroidism, *Calcif. Tissue Res.*, 5, 289-304, 1970.
750. Villanueva, A. R. and Frost, H. M., Evaluation of factors determining the tissue-level haversian bone formation rate in man, *J. Dent. Res.*, 49, 836-846, 1970.
751. Vincent, J., *Recherches Sur La Constitution de L'Os Adulte*, Thesis, Arscia, S. A., Ed., Brussels, 1955.
752. Virchow, R., Knochen und Knorpelkorpenchen, *Verh. Physik. Med. Ges.*, 1, 193-197, 1850.
753. Virchow, R., Ueber Drei nach Form zusammengesetzte skoliotische Rümpfe, *Z. Orthop. Chir.*, 29, 263-273, 1911.
754. Vose, G. P., Quantitative microradiography of osteoporotic compact bone, *Clin. Orthop.*, 24, 206-211, 1962.
755. Vose, G. P. and Baylink, D. J., Effect of fibrillar structure of pericanalicular and intercanalicular bone on x-ray absorption, *Anat. Rec.*, 166, 239-246, 1970.
756. Wakamatsu, E. and Sissons, H. A., The cancellous bone of the iliac crest, *Calcif. Tissue Res.*, 4, 147-161, 1969.
757. Wallach, S., Cohn, S. H., Atkins, H. L., Ellis, J. K., Kohberger, R., Aloia, J. D., and Zanzi, I., Effect of salmon calcitonin on skeletal mass in osteoporosis, *Curr. Ther. Res.*, 22, 556-569, 1977.
758. Wallach, S., Hormonal factors in osteoporosis, *Clin. Orthop. Relat. Res.*, 144, 284-292, 1979.
759. Walker, D. G., Control of bone resorption by hematopoietic tissue, *J. Exp. Med.*, 142, 651-663, 1975.
760. Walton, M. and Rothwell, A. G., Reactions of thigh tissues of sheep to blunt trauma, *Clin. Orthop. Relat. Res.*, 176, 273-281, 1983.
761. Wasserman, F. and Yaeger, J. A., Fine structure of the osteocyte capsule and the wall of the lacunae in bones, *Z. Zellforsch.*, 67, 636-640, 1965.
762. Watson, M., Microfractures in the head of the femur, *J. Bone Jt. Surg.*, 57A, 696-698, 1975.
763. Weibel, E. R., Principles and methods for the morphometric study of the lung and other organs, *Lab. Invest.*, 12, 131-142, 1963.
764. Wehrenberg, W. B., Baird, A., and Ling, N., Potent interaction between glucocorticoids and growth hormone-releasing factor in vivo, *Science*, 221, 556-557, 1983.
765. Weinmann, J. P. and Sicher, H., *Bone and Bones*, C. V. Mosby, St. Louis, 1955.
766. Weinstein, R. L. and Lutcher, C. L., Erythroid hyperplasia and bone formation, *J. Med. Bone Dis. Relat. Res.*, 5, 7-12, 1983.
767. Weiss, P., Introduction to the living cell, in *Bone Biodynamics*, Frost, H. M., Ed., Little, Brown, Boston, 1964, 51-57.
768. Weijs, W. A. and DeJongh, H. J., Strain in mandibular alveolar bone during mastication in the rabbit, *Arch. Oral Biol.*, 22, 667-675, 1977.

769. Weinstein, J. N., Koo, K. N., and Millar, E. A., Congenital coxa vara. A retrospective review, *J. Pediatr. Orthop.*, 4, 70-77, 1984.
770. Wener, J. A., Gorton, S. J., and Raisz, L. G., Escape from inhibition of resorption in cultures of fetal bone treated with calcitonin and parathyroid hormone, *Endocrinology*, 90, 752-759, 1972.
771. Wenger, D. R., Mickelson, M., and Maynard, J. A., The evolution and history of adolescent tibia vara, *J. Pediatr. Orthop.*, 4, 78-88, 1984.
772. Wheaton, B. R., *The Tiger and the Shark*, Cambridge University Press, New York, 1983.
773. Whedon, G. D., Effects of high calcium intake on bones, blood and soft tissue: relationships of calcium intake to balance in osteoporosis, *Fed. Proc., Fed. Am. Soc. Exp. Biol.*, 18, 1112-1118, 1959.
774. Wheeler, M., Osteoporosis, *Med. Clin. N. Am.*, 60, 1213-1224, 1976.
775. Whyte, M. P., Fallon, M. D., Murphy, W. A., and Teitelbaum, S. L., Axial osteomalacia. Clinical, laboratory and genetic investigation of an affected mother and son, *Am. J. Med.*, 71, 1041-1049, 1981.
776. Williams, B., Orthopaedic features in the presentation of syringomyelia, *J. Bone Jt. Surg.*, 61B, 314-318, 1979.
777. Williams, W. J., Beutler, E., Ersler, A., and Lichtman, M. A., *Hematology*, 3rd ed., McGraw-Hill, New York, 1983.
778. Wiener, N., *Cybernetics*, MIT Press, Cambridge, Mass., 1948.
779. Wilde, C. D., Jaworski, Z. F. G., Villanueva, A. R., and Frost, H. M., Quantitative histological measurements of bone turnover in primary hyperparathyroidism, *Calcif. Tissue Res.*, 12, 137-142, 1973.
780. Wiltse, L. L., Widell, H., and Jackson, D. W., Fatigue fracture: the basic lesion in isthmic spondylolisthesis, *J. Bone Jt. Surg.*, 57A, 17-22, 1975.
781. Wirth, C. R., Kay, J., and Bourke, R., Diaphyseal dysplasia (Engelmann's Syndrome), *Clin. Orthop. Relat. Res.*, 171, 186-195, 1982.
782. Wolff, J., *Des Gesetz der Transformation der Knochen*, A. Hirschwald, Berlin, 1892.
783. Woo, S. L.-Y., Kuei, S. C., Amiel, D., Gomez, M. A., Hayes, W. C., White, F. C., and Akeson, W. H., The effect of prolonged physical training on the properties of long bone. A study of Wolff's Law, *J. Bone Jt. Surg.*, 63A, 780-787, 1981.
784. Worth, H. M. and Wollin, D. G., Hyperostosis corticalis generalisata congenita, *J. Can. Assoc. Radiol.*, 17, 67-74, 1966.
785. Wronski, T. J., Smith, J. M., and Jee, W. S. S., The microdistribution of injected ^{239}Pu on trabecular bone surfaces of the beagle: implications for the induction of osteosarcoma, *Radiat. Res.*, 83, 74-89, 1980.
786. Wronski, T. J. and Morey, E. R., Inhibition of cortical and trabecular bone formation in the long bones of immobilized monkeys, *Clin. Orthop. Relat. Res.*, 181, 269-276, 1983.
787. Wronski, T. J., Smith, J. M., and Jee, W. S. S., Variation in mineral appositional rate of trabecular bone in the beagle skeleton, *Calcif. Tissue Int.*, 33, 583-586, 1981.
788. Wu, K., Jett, S., and Frost, H. M., Bone resorption rates in physiological, senile and postmenopausal osteoporoses, *J. Lab. Clin. Med.*, 69, 810-818, 1967.
789. Wu, K. and Frost, H. M., Bone formation in osteoporosis: appositional rate measured by tetracycline labelling, *Arch. Pathol.*, 88, 508-510, 1969.
790. Wu, K., Schubeck, K. E., Frost, H. M., and Villanueva, A. R., Haversian bone formation rates determined by a new method in a mastodon, and in human diabetes mellitus and osteoporosis, *Calcif. Tissue Res.*, 6, 204-219, 1970.
791. Wu, Y. K. and Miltner, L. A., A procedure for stimulation of longitudinal growth of bone, *J. Bone Jt. Surg.*, 18, 909-921, 1937.
792. Wyman, A. L., Paradinas, F. J., and Daly, J. R., Hypophosphatemic osteomalacia associated with a malignant tumor of the tibia. Report of a case, *J. Clin. Pathol.*, 30, 328-335, 1977.
793. Yamada, H., *Strength of Biological Materials*, Williams & Wilkins, Baltimore, 1970.
794. Yamada, K., Yamamoto, H., Nakagawa, Y., Tezuka, A., Tamura, T., and Kawata, S., Etiology of idiopathic scoliosis, *Clin. Orthop. Relat., Res.*, 184, 50-57, 1984.
795. Yasuda, I., Electrical callus and callus formation by electret, *Clin. Orthop. Relat. Res.*, 124, 53-56, 1977.
796. Yasuda, N., Ono, K., Konomi, H., and Nagai, Y., Transitions in collagen types during endochondral ossification in human growth cartilage, *Clin. Orthop. Relat. Res.*, 183, 215-218, 1984.
797. Zanzi, I., Roginsky, M. S., Ellis, K. J., Blau, S., and Cohn, S. H., Skeletal mass in rheumatoid arthritis; a comparison with forearm bone mineral content, *Am. J. Roentgenol.*, 726, 1305-1306, 1976.
798. Zinn, W. M., Ed., *Idiopathic Ischemic Necrosis of the Femoral Head in Adults*, Georg Thieme, Stuttgart, 1971.

III. GROUP CITATIONS

799. The following references document the point in question, and they are listed in the sense of documentation rather than of criticism. The list cites two of the author's early publications: 1—6, 9—11, 15—29, 47—49, 59—62, 64—67, 71, 75, 78, 81, 90, 91, 103, 104, 108, 109, 114, 126, 127, 131, 138, 152, 157, 158, 160—163, 173, 181, 186, 204, 214, 228, 269, 271, 285, 284, 291, 296, 316, 318, 332, 336, 337, 342, 361, 362, 381, 393, 395, 398, 399, 403, 417, 418, 428, 429, 434, 443, 458—460, 467, 471, 490, 492, 493, 527, 528, 543, 546, 560—562, 576, 677, 581, 584, 586, 600, 604, 614—616, 634, 636, 637, 654, 660, 661, 683, 685, 689, 695, 701, 722, 724, 757—761, 766, 767, 770, 773, 774.

800. The following citations provide examples of trial treatments extrapolated from lower systems to the intact human: 4, 10, 11, 15, 44, 47, 48, 49, 62, 65, 66, 67, 85, 87, 121, 157, 208, 209, 284, 312, 332, 358, 381, 394, 408, 410, 458, 459, 551, 560, 564, 576, 577, 581, 600, 604, 614—616, 619, 630, 634, 637, 773, 774.

801. In evidence of the present discrepancy between the depth and detail of knowledge of the cell and the man, and the IO, one might skim the entries, evidence, and methodologies found in post-1980 references such as P to U and the more than 3000 abstracts of the ABMR meetings for 1982 to 1984, of the International Calcified Tissue Meetings for 1982 to 1984, of the Calcium Regulating Hormones meetings for the same years, of the Orthopaedic Research Society for 1981 to 1984, and the Kroc Foundation Symposium (in a *Calcif. Tissue Int. Suppl.*) in 1984.

802. Citations that relate to predictions of IO theory that have been verified either after the fact or by perceiving the previously obscure meaning of already cataloged facts, would include the following. In some cases the phenomena in the intact subject were already known but only after 1964 did growing IO knowledge provide satisfactory explanations of them. 0, 24, 38, 40, 52, 53, 77, 83—85, 96, 104, 124, 130, 133—137, 153, 175—178, 180, 192—197, 208, 211, 213, 229, 230, 236—239, 244—246, 248, 249, 251—254, 257, 259—262, 264—266, 280, 304, 315, 327—331, 366—376, 378, 385, 386, 410, 412—416, 425, 426, 458, 459, 495—501, 504—508, 529—533, 544, 545, 547, 548, 550, 551, 560—563, 565—575, 582, 583, 606, 610, 611, 616, 617, 624, 626, 666, 669, 670, 678, 691, 710, 711, 715—721, 723, 725, 738, 746—750, 770, 779, 786—791. Also the author's *The Laws of Bone Structure*, Charles C Thomas, Springfield, Ill., 1964.

803. The following citations can provide an idea of how rudimentary and flawed was our knowledge of physiology and disease and the methods available for studying them prior to 1926, and especially before 1850. We have come a long way — 26, 58, 73, 94, 100, 119, 125, 148, 150, 164, 172, 182, 183, 189 (the first chapter provides a fascinating review of the development of skeletal knowledge), 246, 270, 275, 279, 283, 284, 288, 294, 295, 301, 313, 314, 317, 323, 325, 338, 339, 343, 346, 348, 349, 351, 359, 365, 382, 419—422, 433, 447, 448, 450—452, 454, 461, 466, 470, 475, 484, 502, 524, 525, 535, 538, 578, 596, 597, 623, 638, 642, 645, 672, 727, 729, 731, 744, 752, 753, 782.

804. This citation identifies phenomena that are truly general knowledge and therefore should not require the support of specific citations. The relevant material appears in up-to-date standard textbooks and reviews of human physiology, embryology, histology, anatomy, pediatrics, medicine, and genetics. However, the terminology used in such texts may differ from that used in this text to signify particular phenomena, and while the phenomena themselves are described in those texts, often they are not singled out for specific comment as they are in the present text.

805. The following citations concern parathyroid hormone effects on the IO: J, K, L, N, P, R—U, 11, 52, 66, 80, 103, 130, 139, 257, 269, 284, 309, 324, 328, 333, 337, 361, 362, 369, 395, 415, 431, 432, 479, 492, 493, 497, 498, 531, 532, 547, 571, 573, 576, 577, 604, 607, 745, 746, 749, 770, 779.

806. The following citations concern calcitonin effects on the IO: 48, 66, 78, 80, 118, 121, 131, 257, 269, 284, 337, 375, 408, 431, 432, 536, 546, 584, 604, 618, 690, 722, 757, 758, 770.

807. The following citations deal with the development of the statistical, geometrical, and other mathematical bases for the methodology of dynamic histomorphometry: 116, 117, 159, 187, 188, 232, 243, 262, 283, 284, 299, 322, 416, 509, 664—667, 756, 763, C—F (the histomorphometry symposia contain numerous articles and sources that pertain to the present subject). In addition, the solutions to the section orientation problem for the trabecular bone and for the labeling escape error began with manuscripts circulated to the σ bone group.

808. The original and subsequent citations for the game rules that apply to the mechanical control of the macromodeling of bone, cartilage, and fibrous tissue include the following. The minimum effective strain: 238, 260, and originally in *An Introduction to Biomechanics,* Charles C Thomas, Springfield, Ill., 1963. The flexure-drift relation: 238, 260, and originally in *The Laws of Bone Structure,* Charles C Thomas, Springfield, Ill., 1964. For dynamic strain — meaning strain rate or change — as opposed to stress, static, or otherwise: the above *Laws,* plus the above *Introduction,* plus 238, 241, 244, 247, 248, 249, 252—254, and 260. For the chondral drift barrier or closed gate effect: the above *Laws* plus 241, 248, 249, 252, and 253. For the flexural neutralization mechanism: originally the above *Laws,* then 238, 241, 247—249, and 265—266. For the chondral growth/force response curve: 236, 244, and 265. For the stretch-hypertrophy relation in fibrous tissue: 237, 241, and 266. For the way in which bone fluid streaming effects can provide, amplify bone strain, and polarize biological signals that control bone surface drifts: in a discussion, in response to a question from W. Hayes at the Orthopedic Research Society Meeting in Augusta, Ga. 1983; mentioned but not detailed in citations 248 and 249; described in detail in 264 submitted in June 1983 to J. A. Albright.

809. Citations concerning the status of BMU-related lamellar bone remodeling parameters include the following, most of them histomorphometric studies: 13, 21, 22, 24, 35—39, 52, 53, 67, 77, 79, 83—85, 118, 120, 121, 124, 126, 130, 133—137, 153, 154, 176, 177, 192—196, 198, 199, 208, 209, 237, 240, 242, 250, 252, 263, 280, 281, 310, 311, 327—331, 333, 366—378, 385, 386, 412—416, 424—426, 439, 468, 495—501, 504—507, 517, 519, 520, 529—533, 57, 548, 550, 551, 565, 574, 617, 664—667, 669, 670, 678, 706, 711, 716—721, 723, 746—750, 775, 779, 786—790.

810. Citations dealing with studies of bone envelope behavior during normal growth and aging include the following: 21, 32, 89, 163, 170, 193, 273, 274, 372, 391, 475, 577, 669, 670, 717, 718, 734—736.

811. Citations concerning envelope-specific bone changes in disease or in response to drugs or physical agents include the following: 68, 69, 104, 130, 181, 212, 404, 429, 445, 460, 464, 477, 492, 493, 552, 606, 609, 616, 620, 660, 661, 681, 698, 706, 709, 711, 725, 738, 748, 749, 783, 784, 786, 789, 797.

812. Examples of envelope-specific effects of hormones, drugs, and other agents occur in the following citations: P, T, U, 43, 47, 77, 84, 95, 120, 127, 140, 176, 193, 363, 372, 403, 437, 443, 458, 459, 483, 504, 549, 582, 583, 606, 678, 702, 738, 745, 759, 766, 798.

813. Citations relating to microdamage occurring in vivo include the following: 97, 101, 145, 167, 215, 219, 252, 622, 677, 694, 728, 762.

814. Citations dealing with fragilization of the skeleton by concurrent disease include: 16, 45, 51, 56, 59, 96, 128, 139, 160, 161, 169, 170, 176—178, 181, 186, 201, 271, 280, 286, 291, 296, 298, 306, 321, 350, 357, 481, 586, 587, 590, 654, 660, 662, 699, 725, 731, 745, 747, 758, 766, 786.

815. Citations that deal with fluoride effects on bone should include the bibliography of a monograph by Briançon and Meunier (*Le Traitement de l'Ostéoporose par l'Association Fluorure de Sodium, Calcium, Vitamine D,* Université-Claude Bernard), 1939, plus the following citations: 4, 58, 67, 75, 79, 194, 398, 399, 425, 551.

816. Citations concerning clinical and biochemical features of osteomalacias include the following: A, B, G, J, K, L, N, P, 2, 16, 58, 130, 162, 173, 178, 184, 213, 230, 237, 256, 281, 310, 342, 431, 432, 452, 463, 490, 492, 523, 527, 535, 541, 546, 566, 579, 593, 596, 604, 607, 618, 629, 682, 700, 731, 732, 745, 775, 792.

817. Citations dealing with the histology and histomorphometry of osteomalacias include the following: C, D, E, F: 2, 6, 126, 130, 184, 194, 209, 232, 242, 243, 250, 251, 333, 362, 424, 462, 468, 499, 500, 501, 531, 548, 567, 596, 608, 667, 719, 723, 732, 745, 746, 749, 765, 779.

818. Citations dealing with the origins of osteoclasts include these: 44, 80, 275, 373, 374, 376, 380, 434, 536, 558, 683, 701, 707, 759.

819. Strain, and particularly dynamic strain, meaning changing strain and thus strain rate, and particularly some minimum or threshold value of it were first proposed as the governing materials properties phenomena for the control of the biological processes that determine bone architecture by the writer in *An Introduction to Biomechanics,* Charles C Thomas, Springfield, Ill., 1963, 86, and in *The Laws of Bone Structure,* Charles C Thomas, Springfield, Ill., 1964. A related article by Epker and the author contained the same proposal (Reference 196). Several later publications from the author's hand repeated it, including these citations: 236, 237, 238, 241, 244, 247, 248, 249, 252, 260, and 264.

820. Citations containing evidence compatible with the idea that some diseases require multifactorial abnormalities include these: 82, 107, 108, 168, 199, 292, 309, 360, 364, 386, 405, 434, 492, 493, 540, 541, 593, 671, 686, 687, 689, 701, 704, 707, 760, 764, 792.

821. The citations for the original and subsequent descriptions of the principles of the responses of chondral macromodeling to mechanical forces are as follows: 236, 244, 247, and 265. Comments also appear in *Orthopaedic Surgery in Spasticity,* Charles C Thomas, Springfield, Ill., 1973.

822. The following articles were published in *Calcif. Tissue Int.*, Suppl. 2V-36, 1984; Cowin, S. C., et al., S1-S6; Currey, J. D., S7-S10; Rubin, C. T., S11-18; Carter, D. R., S19-S22; Meade, J. D., et al., S25-S30; Katz, J. L., et al., S31-S36; Lanyon, L. E., S56-S61; Roberts, W. E., et al., S62-S65; Yeh, C. L. and Rodan, G. A., S67-S71; Johnson, M. W., S72-S76; Pollack, S. R., et al., S77-S81; Binderman, I., et al., S82-S85; Davidovich, Z., et al., S86-S97; Cowin, S. C., S98-S103; Huiskes, R. and Nunamaker, D., S110-S117; Hart, R. T., et al., S104-S109; Durrey, J. D., S118-S122; Smith, E. L., et al., S129-138; Mohan, S., et al., S139-S145; Whedon, G. D., S146-S150; Schneider, V. S. and McDonald, J., S151-S154; Poss, R., S155-S161; also, Ogden, J. A., in *op. cit.*[v] pp. 41—103; Albright, J. A., in *op. cit.*[v] p. 174; Lanyon, L. E., in [w] pp. 273—304; Marie, P. J., in *op. cit.* pp. 109—170;[y] Creuss, R. L., in op. cit. pp. 191—217;[y] 62, 86—88, 594, 588.
823. Frost, H. M., *The Laws of Bone Structure,* Charles C Thomas, Springfield, Ill., 1964.
824. Frost, H. M., Bone Remodeling Dynamics; An Introduction to Biomechanics, *Charles C Thomas,* both published in 1963.
825. Good current descriptions of the chemical composition of bone, cartilage, and fibrous tissues are provided by various authors in the cited basic science textbooks[H,P,V] so they need not be given in detail here. Also, all current standard texts of general human histology contain good descriptions of the histologic structure of those tissues and of the dental complex.
826. Some current texts that reflect that dogma would include References H, I, J, K, M, P, Q, V, and W. See also any recent edition of a textbook of general medicine and physiology.
827. Early awareness of the role of the parathyroid gland in bone physiology is sampled in the following citations: 9, 45, 65, 128, 139, 295, 347, 466, and 597.
828. A sampling of citations that deal with the exclusive role of unassociated osteoclasts in maintaining calcium homeostasis would include: A, G, H, I, J, K, L, M, N, P, R-W, 52, 53, 130, 204, 269, 284, 316, 318, 337, 342, 399, 418, 427, 514, 522, 546, 604, 605, 614, 615, 616, 634, 766, 770, 774.
829. Citations dealing with the composition, histology, and other properties of bone include the following: H, V, and W; L. Weiss' 1984 edition of *Histology* (Elsevier Biomedical); Bloom and Fawcett's 10th ed. of Histology (Saunders); O, P, 6, 25, 36, 37, 80, 145, 174, 201, 232, 237, 241, 242, 243, 252, 261, 263, 297, 306, 344, 362, 379, 412, 430, 438, 516, 572, 588, 594, 608, 632, 793.
830. Citations dealing with the composition, histology, and other properties of cartilage include the following: the texts listed above in Reference 829; Freeman, M. A. R., *Adult Articular Cartilage,* Grune & Stratton, New York, 1973; 336, 6, 25, 41, 91, 149, 236, 241, 244, 261, 263, 334, 335, 336, 362, 379, 407, 471, 606, 608, 668, 695, 704, 707, 726, 769, 771, 796.
831. Citations dealing with the composition, histology, and other properties of fibrous tissues include the following: the texts listed in Reference 829; 6, 25, 70, 214, 236, 241, 261, 263, 417, 418, 608, 641, 668.
832. These citations reveal some of the thinking that prevailed in this field from 1960 to 1970 concerning skeletal health and disease, and that when reviewed with the benefit of 1986 knowledge and insight are to varying degrees in error and/or inadequate and/or humorous. Some of the author's are included too, and all citations are offered in the spirit that they offer another yardstick against which to measure how far the field has progressed in the past 2 decades alone: A, B, L, Q, 5, 55, 78, 131, 158, 190, 202, 203, 229, 230, 227, 228, 231, 284, 312, 318, 337, 394, 457, 460, 490, 512, 622, 630, 700, 751, 754, 757, 773, 774, 798, 823, 824.
833. Citations dealing with the bone balance properties of the $\Delta B \cdot BMU$ parameter in health and disease and under the influence of drugs include the following: 134, 154, 198, 199, 232, 327—330, 376, 385, 425, 426, 468, 496—501, 522, 533, 547—548, 718—720, 746, 745—750.
834. Histomorphometric and other works other than the author's that accept the modeling-remodeling distinction would include the following partial list of citations: C—I, O, P, T, U, V, 13, 21, 38, 41, 45, 53, 77, 79, 83, 85, 96, 120, 121, 133—137, 153, 180, 197, 198, 199, 208, 209, 213, 281, 292, 293, 305, 327—321, 366, 367—376, 378—380, 412—416, 418, 425, 426, 432, 468, 495—501, 505—508, 519, 520, 529—533, 544, 547, 548, 550, 551, 565—575, 606, 613, 626, 666, 678, 706, 711, 719—721, 723, 746, 766, 775, 785—787.
835. The concepts listed in Table 6, Note 3, Chapter 7 that prevailed during 1963 to 1973 in the field at large concerning the mechanics, determinants of bone architecture, and the conceptually important properties of bone and other structural tissues appear in these citations: A, B, 114, 158, 190, 202, 203, 388, 423, 608, and 765; *Orthopaedic Biomechanics,* Frankel and Burstein, Lea and Febiger, 1970; Galante, J. O.; *Acta Orthop. Scand.,* 100, 1962; Rothman, H., *Engineering in the Practice of Medicine,* Williams & Wilkins, Baltimore, 1967; Kenidi, R. M., *Biomechanics and Related Bioengineering Topics,* Pergamon Press, Elmsford, N.Y., 1965.

836. References for Note 2, Chapter 4 include these: Courpron, thesis, Univ. Claude Bernard, 1972; Courpron et al., *Rev. Rheum.*, 40, 469, 1972; Epker and Frost, *Anat. Rec.*, 152, 129, 1965; Frost and Villanueva, *Henry Ford Hosp. Med. Bull.*, 10, 229, 1962; Hattner et al., *Nature (London)*, 206, 489, 1965; Hitt et al., *Can. J. Physiol. Pharmacol.*, 48, 824, 1970; Jett et al., *Henry Ford Hosp. Med. Bull.*, 15, 325, 1967; Meunier, thesis, Univ. Claude Bernard, 1967; Minaire, thesis, Univ. Claude Bernard, 1973; Sedlin et al., *Anat. Rec.*, 146, 201, 1963.
837. Further discussions of Wolff's Law and of the contributions of some of his contemporary analysts include the following: Roesler, H., *Mechanical Properties of Bone,* Vol. 45, ASME, New York, 1982, 27—42; Roux, W., *Gesammelte Abhandlungen über die Entwicklungs mechanik der Organismen,* W. Engelmann, Leipzig, 1885; Rauber, A., *Elastizität and Festigkeit der Knochen,* W. Engelmann, Leipzig, 1876.
838. Additional studies of "On-Off" in bone remodeling were provided by the Niigata histomorphometrists. See Hori, H., et al., A classification of *in vivo* bone labels after double labelling in canine bones, Sun Valley Bone Workshop, 1982; Takahashi, et al., Various bone labels after double labeling in ilium of 1-34 hPTH-treated beagle dogs, Fourth International Workshop of Bone Histomorphometry, Aarhus, Denmark, 1984.

INDEX

A

Abbreviations, 282
Abductor muscle, 34
Abscopic involvement, 114
Acetabulum, 4, 34
Acondroplasia, 5, 27, 30, 185, 205
Acid-base balance, 74, 83
Acidosis, 83, 87, 184
Acromegaly, 180, 184, 263
Acronyms, 282
Acrylic plastic, 223
Activate-depress-free-repeat (ADFR) sequence, 173—175, 269
Activation event, 64, 173, 268
Activation frequency, 269
Activation-resorption-formation (ARF) sequence, 49—50
Activity, 268
Addison's disease, 155
Adductor muscle, 34
Adenoma, parathyroid, 78
ADFR sequence, see Activate-depress-free-repeat sequence
Adrenal cortex, 128
Adrenalcorticalsteroid, 6, 57, 102—103, 132, 159
Adventitia, 44
Age
 macromodeling and, 169
 remodeling and, 173
Albers-Schonberg disease, see Osteopetrosis
Algodystrophy, 117
Alkalosis, 83, 97
Aluminum, 155
Alveolar ridge recession, 180
β-Aminopropionitrile, 155
Amputation, 135
Amyloid, 111
Andistributive property, 171—172
Angiogenic factor, 127, 158
Angulation, 145, 161
Anholistic concepts, 158, 269
Animal model bone systems, 191
Ankle, 15, 20—25, 58, 120
 anterior-unstable, 58
 artificial, 40, 242
Ankle mortise, 20
Ankle valgus, 13
Antidistributive property, 269
Anti-inflammatory agent, 154
Apophyseal plate, 4, 12
Apophysis, traction, 12
Appositional rate, 69—70, 78, 183, 269
Arcane knowledge, 269
A response, abnormal, 155
ARF sequence, see Activation-resorption-formation sequence
Arrest line, 166
Arteriolar embolism, 219
Arteriosclerosis, 220
Arteriotomy, 110
Arthritis, 205, 222—223, 262
Arthrodesis, 1, 117, 136—137
Arthrofibrosis, 118
Arthroplasty, 220
 total hip replacement, 221—249
Arthrotomy, 110
Articular cartilage, 4—5, 20—25, 33—36, 40
 femoral, 24—25
 growth of, 7, 12
 increased thickness of, 29
 modeling errors in, 13
 tibial, 24—25
Asperite, 234, 241
Association, 269
Astragalus, 120
Athetosis, 263
Atlanto-occipital joint, 24
Attuned perception, 63—70
Aves mirabilis ansapiens, 212, 214
Axon flow products, 127

B

Back-knee, spastic, 13
Back pain, 132
Baker's disease, 155, 184
Ball and socket joint, 20—22
Ball-socket ankle, 13
Basement membrane, 44
Base plate, of hip prosthesis, see Hip replacement prosthesis
Basic fuchsin stain, 87, 90, 92—93, 283
Basic multicellular unit (BMU), 64, 67, 269
Basic multicellular unit (BMU) activation, 64—65, 68, 78, 81, 174, 204
Basic multicellular unit (BMU)-based remodeling, 76, 80—84, 95, 100—101, 132, 137, 140, 166—167, 173, 188—190, 231—234, 258, 261, 268
 fibrous tissue, 49—51
Basic structural unit (BSU), 106, 173, 204, 270
Battering ram effect, 224, 238, 241
Beef bone, implant of, 123
Bicarbonate, 41
Biliary stenosis, 180, 184
Bimalleolar sprain, 58
Biological stimulation, direct, 159
Biomechanical competence, 57—59, 132, 145—147, 269
Biopsy, bone, 75, 90, 123, 204
Birth rate, 70, 105
Blast, 269
Blood supply, impaired local, 118
Blount's disease, 13, 30, 185, 263

BMP, 158—159
BMU, see Basic multicellular unit
Bone, see also specific bones; specific kinds of bones, 39—40
Bone balance, 64, 270
Bone-blood interactions, 73—107
Bone disease
    classification of, 192
    envelope-specific, 186
    function and, 200
    growth modeling-dependent, 178—181
    intraskeletal mechanism of, 178—191
    remodeling-dependent, 183—184
Bone graft, 137, 149—150, 158—159
Bone marrow, 80, 167
    juxtaosseous, 74, 186
    transfusion of, 181
    transplantation of, 181
Bone marrow cavity, 140
Bone mass, 270
Bone scan, 113
    hot region in, 111
Bone-seeking isotope, 85—86, 94, 111
Bone surface-lining cell, 79, 84—85, 95, 189—191
Bone transplantation, 137
Bone volume, 172
    absolute, 268
Bow leg, 10, 14
Brace, 1—3
Brachial plexus injury, 115
Brittle material, 270
BSU, see Basic structural unit
Buerger's disease, 220
Buffering capacity, 79, 95
Bunionectomy, 122
Burn, 110, 138
Bursting load, 231—232

C

Calcar, 224, 228, 234, 238—241
Calcitonin, 80, 82, 85, 103—104, 202
Calcium, 73, 80, 100—103, 179
Calcium polyphosphate, 82
Callus, 137, 140, 148, 244
    excessive, 155
    inadequate, 149—152, 155, 158—159
Callus phase, by-passing of, 158—159
Canaliculae, 39, 79, 85—91
Capillary, 45, 48, 50, 67, 101, 139, 154, 169, 185, 268
    afferent, 80, 83
    dilation of, 119
    efferent, 80, 83
    synovial, 7
Capsule, growth of, 46—48
Caput, epiphyseal plate of, 4
Carbonate, 39, 80, 100
Carpal tunnel compression, of median nerve, 54
Cartilage, see also specific kinds of cartilage
    abnormally thick, 29
    composition of, 40—41
    dominance of skeletal architecture by, 26—27
    growth of, 41
    mineralized, 189, 206—207
    properties of, 40—41
    in skeletal growth, 3
Cartilage column, disorganized, 30, 32
Cartilage space, 207—208
Catastrophic wear, on prosthesis, 223
C defect, 156
Cell biologist, 197
Cell culture, 184—186, 195—196
Cell level, 5, 59, 200
Cell proliferation, 45, 168
Cement, 222—224, 244
Cement line, 63, 143, 204
Cementoclast, 258
Central nervous system, 2—3, 113, 118
Cerebral palsy, 2, 263
CGFRC, see Chondral growth/force response curve
Charcot joint, 117—120
Chemical injury, 138
Chondral bar, mineralized, 188
Chondral cell column, 176—177
Chondral conductor, 182
Chondral creep, 27—32
Chondral dominance, 26—27
Chondral growth, 6—7, 41, 126, 177
    abnormal, 185
    accelerated regional, 112
    errors in, 5
Chondral growth/force response curve (CGFRC), 10—12, 20, 23—25, 37, 270
Chondral matrix, 40—41
Chondral modeling, 1—42, 126, 132
    abnormal, 185
    control of, 9
    dynamic averaging property of, 10
    errors in, 13, 27—29
    in utero, 9
    L_o-level, clinical examples of, 13—26
    mechanically controlled, 8
    minimum effective signal in, 9—10
    neuromotor system and, 25—26
    occurence during growth, 10
    rate of, relative growth and, 26—27
Chondral modeling axioms, 9—12
Chondral modeling state equations, 12
Chondral structures, by level of organization, 5
Chondroblast, 5, 139—140
Chondroclast, 188, 258
Chondrocyte, 5, 30—32, 40—41
Chondrodystrophy, 32
Chondroosseoid, 41
Cicatrization, 111
Circumferential lamellae, 62, 168, 180, 188
Cleft, ultramicroscopic, 39, 88—89, 92, 94
Clinic, communication with laboratory, 247—249
Clinician, 199, 212—214, 247—249

Clone, 176
Clone elongation, 18
Clone enlargement, 45
Clone multiplication, 6, 17—18, 45
Clone size, 6
Closed mind, 67—68
Closed system, 69
Club foot, 13, 185, 263
Club hand, 13
Coherence, 270
Coherence research, 125, 131—132, 157, 159
Coherence treatment, 83, 132, 157, 261, 270
Collagen, 44—46, 270
   aging of, 45
   cross-links in, 44—46
   orientation of, 30—32
   in scar, 57
   turnover of, 51
   Type I, 39, 41, 44
   Type II, 40, 44
   Type III, 41—42, 44
Collagen bundle, 44, 46, 50, 62
Collagen fibril, 41—42, 44, 46
Collateral ligament
   ankle, 58
   ulnar, 28
Colle's fracture, 125
Communication, between laboratory and clinic, 247—
Compacta, 113, 143—144, 228, 283
Compact bone processes, 189
Competent foil, 212
Compliance, 270
Component, 270
Compression, 271
Compression plate, 117, 150
Concave-tending facing, 161, 241
Concave-tending surface, 194
Concept, 43—44
Cone-piston effect, 235—237, 239
Connective tissue, 45, 48
Continuum, 151—152, 159, 271
Contracture, joint, 111
Control, 271
Contusion, 110—111
Convex-tending surface, 161
Cooley's anemia, 180
Cortical-endosteal bone formation, 194
Cortical-endosteal envelope, 169
Cortical-endosteal modeling, 189
Cortical-endosteal surface, 167, 235
Corticosteroid, 118, 177, 219
Cosmology, 73
Costal cartilage, 5
Counterproductive reaction, 212
Coupling, 100, 101, 146, 258—259, 268, 271—272
Coxa valga, 13, 185
Coxa vara, 13, 30—31, 185
Creep, 10, 40, 56, 132, 207, 272
   bulk, 182
   chondral, 27—32, 177, 179

   total, 29
   unit, 29—30
Creep compensation mechanism, 42, 52, 132
Cruciate ligament, 53
Crumbling, 234, 239—240
Crushing injury, 110
Crutches, 220
Cubitus valgum, 13
Cubitus varum, 13
Cures, ways to find, 199
Cushing's syndrome, 184
Cyte, 272

D

Dead bone, 85, 88, 93
Death rate, 105
Default judgment, 59
Defensive tissue reaction, 111
Deformity
   of neuromotor origin, 26
   postpolio, 1—2
   spastic, 2
ΔB.BMU, 65, 83, 106—107, 174—179, 186—190, 204, 258
Dendritic property, 203
Denervation, 110, 117—120
Dental caries, 132
Dentinogenesis imperfecta, 263
Dependent variable, 105
Dermatomyositis, 110
Diabetes mellitus, 118, 154
Diaphysis, 26—27, 272
Didronil, 155
Differentiation, 268
Digit
   artificial, 242
   supernumerary, 263
1,25-Dihydroxyvitamin D, 132
Diphosphonate, 104—105, 131, 154, 231
DIP joint, 25
Discrete property, 272
Disc space degeneration, 64—66
Disease, 198, 272
Displacement, progressive, 145
Distraction, 152
Distributive property, 171—174, 192
Disuse, acute, 173
Domain, 272
Dorsiflexor, ankle, paralyzed, 1
Downwards search, 198—199
D response, abnormal, 155
Drift, see also Formation drift; Resorption drift, 272
Drop foot, see Equinus
Drugs, 131—133
Dura, 44
Dwarfism, 5, 185, 262—263
Dynamic averaging property, 10
Dynamic peak strain axiom, 52

E

Ear, 40
Edema, 111, 118—119
Effector cell, 185, 272
Ehlers-Danlos syndrome, 52
Ehrenfried's disease, 13, 27—28, 30
Either-or assumption, 258
Elastic cartilage, 5, 9, 40
Elasticity, 272
   Hookean, 40
Elbow, 15, 23
   artificial, 242
Electrical potential, 10
Electrical stimulation, 117, 128, 149, 157—158
Electron microscopy, 88—89
Embolism, 219—220
Enabling function, 159, 262—263, 272
Enamel, 103
End-load, 194
Endocrine system, 6, 131—133
Endoprosthesis, 132, 205, 272
   hip, see Hip replacement prosthesis
Endosteal envelope, 181
Endosteum, 272
Endothelial cell, 5
Endothelial thickening, 112
Energy dissipation, 20
Envelope, 63, 272
Envelope-specific disease, 186
Epiphyseal height, 25
Epiphyseal pad, 20
Epiphyseal plate, 5, 13—20, 40, 174, 177, 242, 273
   capital, 4, 112
   creep in, 29
   diameter of, 17—20
   distal femoral, 16
   distal radial, 11
   growth of, 6—7, 12
   in knee alignment, 14—17
   modeling errors in, 13
   proximal humeral, 11
   proximal tibial, 16
   ulnar, 207
   upper tibia, 13—15
Epiphyseal plate cartilage, 4
Epiphysis, 273
   distal femoral, 22
   distal radial, 16, 28
   expanded, 182
   proximal tibial, 22
Equilibrium, 273
Equinus
   postpolio, 2
   spastic, 2
Error, 212
Erythema, 111, 118
Escherichia coli, 65
Estrogen, 6, 65, 132

Etidronate, 132
Evolution, 55
Exercise, against maximal resistance, 54
Exostoses, congenital hereditary, 13, 27—28, 30
Experience, 43—44
Extracellular fluid, 41—42, 90, 95
Ex vivo system, 186

F

Fail in shear, 230—231
Fascia, 12, 40—41, 44—45
   bulk strength of, 51—52
   growth of, 46—48
   transplant of, 57
Fat, 48
Fatigue, 232, 259, 273
   in fibrous tissue, 56—57
Fatigue failure, 56, 60, 147, 153, 231
Fatigue life, 225
Feathered bone, 87, 89, 273, 283
Feedback control, 60, 169, 171, 273
   negative, 14—16, 276
   positive, 16—17, 31, 241, 277
Femoral articular surface, 23
Femoral head, 26—27, 219, 222
Femoral neck, 4, 112, 206, 222
Femoral neck-shaft angle, 31
Femoral shaft, 170
Femoral torsion, 13
Femur, 142, 209
   fracture of, 115
   motion on tibia, 53
Fetus, 3, 9, 173
Fiber bone, see Woven bone
Fibroblast, 41—42, 46—50, 55, 139
Fibrocartilage, 5, 9, 33, 40
Fibrocyte, 42, 46—51, 54
Fibrogenesis imperfecta ossium, see Baker's disease
Fibronectin, 158—159
Fibrosis, 119
Fibrous dysplasia, 139, 180
Fibrous tissue, 5, 258
   aging of, 42
   anatomy of, 59—60
   biomechanical competence of, 57—59
   composition of, 41—42
   fatigue, 56—57
   functions of, 47, 59
   growth of, 45—49
   microdamage in, 56—57
   properties of, 41—42
   regional accelaratory phenomenon in, 114
   structure of, 47—48
   turnover of, 49—51, 111
Fibrous tissue modeling, 28, 43—71, 132
   dynamic peak strain axiom, 52
   minimum effective strain, 52—54
   rules of thumb for, 55—56

state equations for, 55
stretch hypertrophy axiom, 53—55
tension-creep compensation, 52
time-averaging axiom, 52
Fibula, 113, 115
Fixation, 67, 116
  biological, 244
  internal, 145, 150
  rigid mechanical, 158
Flexor pollicis longus tendon, 53
Flexural neutralization, 161, 261
Flexure, 231, 273
  cantilever, 14, 194, 270
  concave-tending, 141
  convex-tending, 141
  direct, 272
  dynamic, 161
  neutral axis for, 228
  outflaring, 236—237, 239, 241
Flexure-drift relation, 161, 261
Fluid space, bone, 39
Fluoride, 132, 154—156, 204
Foot, rheumatoid, 121—122
Force, 273
  units of, 282
Foreign body response, 124
Formation drift, 27, 66, 141, 167—172, 183, 188, 194, 235
Fracture, see also specific bones, 110—111, 116, 147, 219
  crush, 267
  fatigue, 231
  fully healed, 144
  greenstick, 145, 147
  march, 147, 263
  open, 109, 135
  overriding of, 145
  spontaneous, 117, 184, 263
  stress, 205, 220
Fracture gap, 152
Fragility, 273
Free bone surface, 79
Frost criterion, 159
Frost-Heaney observation, 174
Frostian balance, 71, 99, 104—106
Function, 198, 273
  upwards evolution of, 200—201
Function/disease relation, 200
Function/property ratio, 200

G

Gain, net, 100
Gallstone, 65
Gating, 6—7, 48, 261, 274
  parallel, 12, 159, 186, 276
  primary, 55
  serial, 159, 279
Gaucher's disease, 181, 184
Genu valgum, 16, 30

Genu varum, 13, 17, 30
Giant cell tumor, 139
Gigantism, 5, 184—185, 263
Ginglymus, 20
Gla protein, 100
Gluteus maximus muscle, 34
Gluteus minimus muscle, 34
Grain, 50, 140, 228—231
Granulation tissue, 111, 137, 139, 150—151, 154
Greater loading situation, 227
Greater trochanter, 4, 12, 206
Growth, 111, 168, 172, 182, 259
  bone, longitudinal, 123—124
  chondral, see Chondral growth
  chondral modeling rate and, 26—27
  of fibrous tissue, 45—49
  L_1-mode longitudinal, 18
Growth factor, 127, 158—159
Growth hormone, 65, 159, 168
Growth plate, 177
Growth rate, 69
Growth retardation, 5, 30, 124
Gunshot wound, 115

H

Hagfish, 90—91
Hallux varus, 13
Halo volume, 89, 93, 274
Hamstring, 34
Haversian system, 62, 143, 274, 283
Healing process, see also Repair process, 111, 116—117
  bone, 39
  delayed, 117
  primary phase of, 57, 145
  seconday phase of, 57, 138, 145
  wound, see Wound healing
Helper cell, 40, 65
Hematoma, 139
Hematoxylin and eosin stain, 114, 176—177
Hemiatrophy, 5
Hemiplegia, 113
Hemosiderin, 112
Hernia, inguinal, 52
Hindfoot varus, 13
Hinge joint, 20
Hip, 112
  artificial, 40
  child's, 4
  congenital dysplasia of, 20, 185
  dislocation of, 13, 20
  load on, 226—227
  normal, 225—232
  rotation of, 34
  support in, 226
Hip replacement prosthesis, 221—249
  acetabular component of, 222, 224
  Austin-Moore, 241
  base plate of, 232—235

battering ram effect in, 224, 238, 241
biological fixation and, 244
biology of, 244—245
cone-piston effect in, 235—237, 239
differential longitudinal stiffness, 235
femoral component of, 222, 238, 242
inwaisting in, 243—244
Judet, 223
load focusing in, 241
load multiplication in, 232—235
mechanical fit of, 244
outflaring cortical flexure in, 241
shear-shunt mechanism in, 237—240, 242—243
skin mechanism in, 237—240, 242—243
toggling in, 243—244
tolerances in, 244
transmission of vertical load in, 242
Histamine, 127
Histogenesis, 151—152, 274
Histological system, 79—84
Histomorphometry, 63, 274
quantitative, 95
regional acceleratory phenomenon in, 123
History, patient, 135
Holistic concepts, 274
Homeostasis
disorders of, 89
intermediary organization and, 73—107
skeletal role in, 79—95, 132, 258
Hormones, 146
Howship's lacunae, 84, 166
Humerus, head of, 15
Hyaline cartilage, 4, 40, 139—140
biomechanical functions of, 26
chondral growth/force response curve for, 10—12
IO-intrinsic disease of, 28—29
modeling of, 1—42
regional acceleratory phenomenon in, 114
Hyaluronic acid, 42
Hydrogen ions, 80
Hydronium ions, 80
Hydrostatic pressure gradient, 9
Hydroxyapatite, 39, 82
Hydroxyproline, 100
Hypercalcemia, 74, 107
Hypercalciuria, 115
Hypercompliant bone, 144
Hypernatremia, 107
Hyperostosis frontalis interna, 153
Hyperparathyroidism, 74, 184, 263
primary, 155
secondary, 29, 87, 155, 182, 202
Hyperphosphatasia, 183
Hypertrophic zone, 176
Hypertrophy, 111
Hyponatremia, 107
Hypophosphatasia, 184
Hypothyroidism, 29—30

I

Iliotibial band, 45
Immobilization, 102
Immune reaction, 112
Immunosuppressive agent, 147, 154, 159, 219, 231
Implant, 158
Independent variable, 105
Infarction, 110, 116, 138, 147, 219, 231
Infection, 39, 110—116, 138, 153
Infiltrate
eosinophilic, 111
lymphocytic, 111—112
monocytic, 111
Inflammation, 110—112, 118—119
Innervation, 40—41, 119, 146
Innovation, 212
Insertion, 274
Intermediary organization (IO), 274
basic concepts of, 252
chronology of perception of, 252
conclusions about, 254—256
future directions in understanding of, 256
pre-IO concepts of skeletal physiology, 257—265
Intermediary organization (IO)-extrinsic disorder, 3, 5, 31, 155, 166, 180—181
Intermediary organization (IO)-intrinsic disorder, 3, 5, 35—36, 155, 166, 181—184
Intermuscular septae, 45
Interosseous membrane, 44
Interstitial bone fluid, 86, 91, 94
Interstitial fluid, 9, 39—41
Interstitial lamellae, 86, 188
Interstitial water, 86
Intervertebral disc, 5, 11, 40
Intramedullary nailing, 117
Investigator, 199, 247—249
In vitro system, IO-oriented properties of, 184—186
Inwaisting, 228, 241, 243—244
IO, see Intermediary organization
Ions, diffusion in wet bone, 86—93
IP joint, 15, 24
Ivory peg, implant of, 123

J

Jansen's disease, 5, 30, 185
Joint
alignment problems in spasticity, 3
artificial, 67
degenerative disease of, 132
false new, 148
stiffness of, 118
Joint capsule, 44
fibrosis of, 118
Joint surface, 31

Juxtaosseous tissue, 169

K

Knee, 24, 35—36, 40, 208
    adolescent's, 23
    artificial, 40, 242, 244
    child's, 14—16, 23, 25
    synovitis of, 127
Knock knee, 10, 16
Krσ property, 48, 68—71, 258, 274

L

Labeling escape error, 67
Laboratory, communication with clinic, 247—249
Lactate, 41
Lacunae, 40
Ladder graph, 65
Lamellar bone, 5, 39—40, 86, 132, 137, 140, 153, 168, 189
    macromodeling of, 168—173
    regional accelleratory phenomenon in, 114
    remodeling of, 173—176
    turnover of, 111
Lamellar-on-bone-only (LOBO) property, 39—40, 132, 258, 274
Langhan's giant cell, 111
Larynx, 5
Lathyrism, 52
LBO property, see Losing betting odds property
L_2-domain remodeling, 238
Lead, 132
Leg, 141
    artificial, 135
Leprosy, 117
Lesser loading situation, 227
Lesser trochanter, 4, 12, 224
Leukine, 158
Leukotriene, 127, 158
Lifespan, 70, 105—106, 171
Ligament, 5, 12, 41, 44—45
    attachment to bone, 40
    bulk strength of, 51—52
    graft of, 57
    growth of, 46—48
Limb
    alignment problems in spasticity, 3
    deficiency of, congenital, 263
    denervated, 118
Limb misalignment, 13
Limb overgrowth, 5
Limb shortening, 5
Limb torsion, 29
Linea aspera, 170
Lining cell, see Bone surface-lining cell
Lithiasis, 115
Load, 274, 284
    compression, 227—231, 282
    continuous shearing, 32
    flexural side, 194
    on hip, 226—227
    instantaneous, 23
    mechanical, 168—169
    multiplication of, 232, 238—241
    shearing, 228—231, 237
    tangential hoop, 231
    tension, 50
    time-averaged, 23—25
    uniaxial, 14—15, 280
    vertical, 14—15, 227—229, 242
Load bearing mechanism, 226—227
Load defocusing, 13—15, 20, 25, 132, 228—230, 242
Load focusing, 14—15, 20, 132, 224, 241
Load orientation, 231
Load partition principle, 53
Load resultant, average, 227
Load-transfer situation, 227—228
LOBO property, see Lamellar-on-bone-only property
Longitudinal tunneling, 113, 121, 150
Looser's zone, 117, 136—137, 153, 155
Losing betting odds (LBO) property, 132, 154, 195, 258, 274
Loss, net, 100
Lower intermediary organization, 5, 59, 200
L_1-space-time, 185
L_o-space-time domain, 103
L_2-space-time domain, 103
Lues, 114, 118, 153
Lupus, 110
Lysozyme, 127

M

Macromodeling, see also Modeling, 215, 275
    age incidence of, 169
    control of, 168—169, 171
    distribution of, 169, 171
    distributive properties of, 171—172
    fibrous tissue, 49
    functions of, 172—173
    of lamellar bone, 168—173
    operational features of, 169—171
    purposes of, 60
    sublimal, 54
Macromolecular level, 200
Macrorepair, 136
Madelung's deformity, 30
Magnesium, 39, 80, 100, 133
Malfunction
    causative, control of, 203—205
    identification of, 201—202
Malleolus, posteriorinferior lateral, groove in, 8
Malnutrition, 30, 154, 184
Malunion, 16
Marfan's syndrome, 5, 184—185, 263
Marginal erosion, 208

Marrow, see Bone marrow
Master blueprint, 59, 260
Materials, 268
Maturation period, 70
Mature bone, 89, 144
Maxillofacial problem, 64
MCN effect, see Multiple, chained, neutralizing (MCN) effect
M defect, 156
Mean tissue age, 29—32, 167
Mechanical abuse, 116
Mechanical deloading, 124
Mechanical disuse, 169
Mechanical incompetence, 275
Mechanical load
    L mode, 8
    S mode, 8
Mechanical loading experiments, 124—125
Mechanical usage, 183
Median nerve, 54
Meningomyelocele, 17
Meniscectomy, 115
Meniscus, 5, 40
    knee, 33
MES, see Minimum effective signal
Mesenchymal cell, 275
Metabolic balance study, 104
Metabolic bone disease, 64
Metabolism, anaerobic, 202
Metacarpal, 15, 207
Metamorphosis, 251—254
Metaphysis, 275
    flared bony, 182
Metastasis, bony, 112, 136—139, 153
Matatarsal, 15
Metatarsophalangeal joint, 122
Metatarsus adductus, 13, 185
Microdamage, 40, 42, 104, 136, 207, 219—220, 231, 234, 249, 259—260, 263, 275
    to fibrous tissue, 56—57
Micromodeling, see also Modeling, 275
    fibrous tissue, 49—50
Micropetrosis, 86, 88, 275
Middle intermediary organization, 5, 59, 200
Mineral, solubilization of, 80—81
Mineral ions, 83
Mineralization, 40, 67, 82, 86, 89, 142, 233, 257
    abnormal, 97
    depression of, 179
    impaired, 152—153
Mineralization lag time, 268
Minimum effective signal (MES), 52—54, 132, 138, 161, 194, 209, 229—231, 258, 275
    chondral, 9—10
    set-point of, 120—121, 233
Minimum effective signal (MES) curve, 37
Modeling, 215, 275
    chondral, see Chondral modeling
    compared to remodeling, 180
    defects in, 139
    distributive properties of, 192
    fibrous tissue, see Fibrous tissue modeling
    L_2-level properties of, 172
    malfunctions in, 132
    pathologically reactivated, 180
Modeling-only system, 170, 193
Model systems, for skeletal research, 165—196
Molecular biologist, 197
Molecular-level turnover, in fibrous tissue, 51
Molecular sieve effect, 6, 88, 92—93
Moment, 275
Monoplegia, 113
Morquio's disease, 5, 27, 29—30, 182, 185
Mosaic pattern, 114
Motor innervation, 119
Motor neuron lesion, 113
Move, 198
MP joint, 24—25
Multiple, chained, neutralizing (MCN) effect, 65, 78, 132, 156, 159, 195, 258, 274
Muscle, 48
    anchoring to bone, 62
    neurologic control of, 171
    peripheral, communication with central nervous system, 2—3
    transfer of, 1
Muscular dystrophy, 37, 180, 184
Musculoskeletal injury, 62
MWT, 275
Myelodysplasia, 183—184
Myelofibrosis, 137, 153
Myositis ossificans, 137, 139, 153
Myositis ossificans progressiva, 137—139
Myositis ossificans traumatica, 137, 153, 155

N

Nasal plate, 40
Natural unit, 105—106
Nearthrosis, 120
Necrosis, 111—112
    aseptic, 156
Negative work, 20
Nervous system, in spasticity, 2
Neurofibromatosis, 154—155
Neuromotor system, chondral modeling and, 25—26
Neuropathy, peripheral, 118
Neutron activation, 104
Nonunion, 111, 132, 136, 263
    atrophic, 116, 147
    biological failure, 116
    hypertrophic, 147
    oligotrophic, 116
    threatened, 152
Noxious stimulus, see also Regional acceleratory phenomenon, 109
Nutrition, 146

O

OAF, 127
Observation, accidental, 211
Obturator muscle, 34
O defect, 156
Olecranon, 12
Ollier's disease, 5
On-off phenomenon, 66—67, 82, 91—94, 121—123, 158, 170, 276
Operation, bone, 110
Organ capsule, 44
Organ culture, 184—186, 195—196
Organelle level, 200
Organic matrix
 bone, 39
 cartilage, 40—41
Organization, 268
Organization ladder, 3
Organ level, 5, 59, 200
Origin, 276
Ossification, heterotropic, 136
Ossification center, 129, 179
 chondral, 39
 endochondral, 27, 126
 epiphyseal, 14, 20, 176
 of femur, 23
 of tibia, 23
Osteoarthritis, 263
Osteoarthropathy, pulmonary hypertrophic, 153, 155, 180
Osteoblast, 66—67, 74, 79—82, 95, 139—140, 171, 173, 257—260
 brush border region of, 80
Osteoblast jumping distance, 152
Osteochondritis dissecans, 33
Osteochondroma, 139
Osteoclast, 63, 74, 79—83, 95, 171, 173, 188, 258, 260
 number of nuclei per, 104—105
Osteoclastic drift, 188
Osteocyte, 79, 85, 87, 91—92
Osteocyte lacunae, 39, 85—92, 97
Osteocyte syncytium, 40
Osteodystrophy
 experimental canine, 105
 renal, 182
Osteogenesis imperfecta, 53, 155—156, 182—184, 263
Osteogenesis imperfecta tarda, 180
Osteoid, 39, 66, 249
 mineralization of, 67
Osteoid seam, 66, 166, 258, 276, 283
Osteoma, 153
 lamellar periosteal, 139
 osteoid, 116, 124, 139, 180
Osteomalacia, 74, 132, 153—154, 183, 231, 257—262, 276
 axial, 181
 low-turnover, 263
 vitamin D-deficiency, 155
Osteomyelitis, 111, 124, 139, 153, 180
Osteon, 202, 276, 283
 primary, 62
 secondary, 62, 86—87, 92, 113, 117, 144, 166—168, 188
 waltzing, 169
Osteonectin, 127
Osteopathy, 263
Osteopenia, 141, 174, 180, 187, 276
 corticosteroid-induced, 186
 postmenopausal, 186
 RAP-induced, 103
 reversible, 115
 senile, 186
 treatment of, 211
Osteopetrosis, 181—184, 206
Osteophyte, 10, 126, 180
Osteoporosis, 83, 131—132, 204, 257, 260—262, 276
 corticosteroid-induced, 184
 juvenile, 184
 low-turnover, 263
 migratory, 117, 184
 postmenopausal, 66, 184, 205
 senile, 184, 202
 spinal, 201
Osteosarcoma, 39—40, 114, 139, 153
Osteotomy, 16, 137
 Chiari, 34
 experimental, 142
 subtrochanteric valgus, 31
 surgical, 148
 tibial, 35
Outwaisting, 228, 241
Overload, 228, 232

P

Packet, 276
Pagetoid bone, 114
Paget's disease, 136, 139, 153, 180, 184, 263
Pain
 absence of, 120
 preoperative hip, 224
Panhypopituitary state, 155
Paralysis, 110, 115, 123—124, 180, 184—185
Paraplegia, 113
Parathyroid, 74, 202
Parathyroid hormone, 65, 74—75, 78, 81—85, 103, 131—132, 159, 258
Partition property, 276
Patella, subluxing, 13
Patellofemoral joint, 25
Percolation system, 79, 85—95, 189—191
Perfusion, 111
Perichondral ring, 4—5, 40, 277
 growth of, 6—7, 17—20
Perilacunar bone, 97
Periodontal disease, 132, 205, 262

Periodontal ligament, 41, 44
Periosteal bone formation, 194
Periosteal envelope, 169
Periosteal modeling, 189
Periosteal stripping, 123, 125
Periosteal surface, 167
Periosteum, 167, 277
Permanganate stain, 87, 93
Peroneal muscle, 1, 20
Peroneal tendon, 8
Perspective, 251
Pes planus, 13
Phase contrast effect, 97
Phase lag pool, 268
Phosphate, 80, 100, 103
Photon absorption study, 111, 113
Physical examination, 135
Physical science, evolution of, 211
Physical therapy, 1
Physiological balances
   fifth axiom of, 106
   first axiom of, 99
   fourth axiom of, 103
   nature of, 99—101
   population logistics in, 104—106
   properties of, 99—107
   push-pull phenomena in, 102—104
   reutilization phenomena in, 101—102
   second axiom of, 101
   seventh axiom of, 107
   sixth axiom of, 107
   third axiom of, 103
Piezoelectric potential, 39
PIP joint, 25
Piston-wedge effect, 244
Plexiform bone, 40, 124
Poliomyelitis, anterior, 1—2, 119
Polyethylene socket, 222
Polymethylmethacrylate, 222
Pool, 99, 107
Population logistics, 104—106
Postinjury phenomena, 110
Postulates, 191—193
Precursor cell, 5—6, 55, 140, 155, 181, 262, 268, 277
Pre-IO knowledge, 257—265
P response, abnormal, 155
Pressure, units of, 282
Pressure pad, 13—14, 33
Primitive bone, see Woven bone
Progenitor cell, see Precursor cell
Progeria, 185
Progress, 213
Property, 198, 277
Propositions, 191—193
Prostaglandins, 65, 118, 127, 158—159, 187
Prostatectomy, 64
Prosthesis
   hip, see Hip replacement prosthesis
   load-bearing, 40
Proteoglycan, 32, 39—42, 51, 100

Protons, 80, 83
Pseudoachondroplasia, 5
Pseudoarthrosis, 148, 155
Pseudogout, 115
Pseudohyperparathyroidism, 184
Pseudohypoparathyroidism, 263
Psoas muscle, 34
Pubic symphysis, 40
Pus, 111
Push-pull phenomenon, 102—104, 188
Pyle's disease, 184

Q

Quantum, 277

R

Radial closure rate, 70
Radial head subluxation, 13
Radiation damage, 155, 219, 231
Radiocarpal joint, 23—24
Radiohumeral joint, 25
Radius, 16, 28
Random choice, 198—200
RAP, see Regional acceleratory phenomenon
Rat calvarium, fetal, 184
Rat femur, fetal, 185
Reactive bone, see Woven bone
Recipe, 265—266
Reduction, open, 150
Referent, 99, 277
Regional acceleratory phenomenon (RAP), 57, 67, 109—133, 136—138, 141, 143, 184, 232—233, 277
   abnormal, 117, 154
   absent, 117
   anatomical distribution of, 112—113
   causes of, 110, 127—128
   characteristics of, 110—116
   clinical examples of, 116—123
   compact bone and, 113
   duration of, 115—116
   electrical stimulation and, 158
   experimental design and, 125—127
   hyperactive, 154
   inadequate, 116, 149—150
   nature of, 110—112
   obtunded, 109, 117, 121—123, 154
   osseous features of, 121
   pathological, 117—118
   positive, 116, 123
   prolonged, 154
   research on, 123—125
   runaway, 117
   signs in skeletal tissue, 114
   SOS role of, 116
   system properties of, 125
   tissue and activities affected by, 111

ubiquity of, 128
Regulation, 278
  cell population property of, 81
Reinfarction, 219
Reiter's disease, 110
Remodeling, see also BMU-based remodeling, 165—167, 215, 278
  age incidence of, 173
  compared to modeling, 180
  control of, 173
  disorders of, 139, 153, 193
  distribution of, 173
  distributive properties of, 173—174, 192
  functions of, 175—176
  L_2-level, 180
  operational composition of, 173
  secondary, 63
  subepiphyseal, 175
  temporal properties of, 64
  variants of, in spongiosa of growing animals, 174—175
Remodeling space, 76—78, 141, 150, 268
Renal dialysis, 66, 209
Renal failure, 66, 180, 182, 207, 209
Repair process, 20, 135—164, 231, 245, 278
  biologic, 137—147
  biological failures in, 147—148
  clinical problems related to, 147—153
  in disease, 153
  levels of, 136
  malfunctions of, 137
  nontraumatic clinical situations causing, 139
  normal variations of, 147
  problems of quantity and kind, 150—153
  rate problems in, 148—150
  technical failures in, 147—148
Research, 63—68, 125—127, 163—164, 247—249
  algorithm, 198—205
  awards for, 163—164
  coherence, 125, 131—132, 157, 159
  funding of, 163
  goals of, 198—199
  history of, 205
  model systems for, 165—196
  optimizing strategy for, 197—220
  program time, 163
  publication of, 163
Resilence, 278
Resorption, 78, 83, 139, 187, 278
  chemistry of, 80
  connection to formation, 63
  defect in, 206
Resorption drift, 101, 141, 161, 167—172, 194, 206, 235
Resultant, 278
Reutilization
  local, 101—102
  systemic, 102
Revascularization, 219
Reversal line, 166
Rheumatic fever, 110

Rheumatoid arthritis, 52, 57, 110, 116, 121, 132, 139, 202, 208, 263
  juvenile, see Still's disease
Rheumatoid phenomenon, 121—122
Rib, 167, 183
Ribbing's disease, 180, 184
Rickets, 27, 30, 32, 155, 258, 263
  congenital, 5
  nutritional, 5
  renal, 182, 207
  vitamin D-deficiency, 153—154, 180, 185
  vitamin D-resistant, 29, 184—185
Rigidity, 278
Rockerbottom foot, 13
Roots, 200—203
Rules of thumb, for fibrous tissue modeling, 55—56

S

Saturation effect, 57, 132, 159, 279
Scar, 45, 50—51, 57, 60, 112, 155
  anarchic, 46, 57
  fibrous tissue, 151
Scientific posture, investment in, 217—218
Sclerae, 41
Scoliosis, 13, 30
Sculpting, 165—166
Second injury phenomenon, 158
S effect, 155, 159
Sensory innervation, 119, 127
Septicemia, 65
Sequestrae, 88
Serendipity, 198—199, 205
Sex hormone, 159
Sharpey's fibers, 62
Shear, 279
Shear-locking mechanism, 62, 228—231, 243
Shear-shunt mechanism, 237—243
Shear skid, 224
Shoulder, 23, 25
  artificial, 242
  subluxing, 13
Sigma value, 64, 70, 105, 116, 186, 190, 279
Singularity assumption, 258
Sink-reservoir capacity, 79
Size, individual, 71
Skeletal balance phenomena, 104
Skeletal research, see Research
Skid effect, 242—243
Skid mechanism, 237—240
Skin, epithelialization of, 111
Skin graft, 150
Slipped carpal femoral epiphysis, 30, 263
Smearing, 144—145
Sodium, 39
Soft tissue
  devitalized, 139
  neuropathic problems of, 118
Somatomedin, 6, 159

Somatotrophic hormone, 6
SOS response, 52, 116—117, 124, 132, 136
Space-time domain, 123
Spasticity, 2, 3, 33
Spine
  crush fracture of, 267
  injury to, 120
  lumbosacral, 126
Spine fusion, 137
Spongiosa, 228
  collapse of, 220
  metaphyseal, 15
  primary, 132, 174—177, 188—190, 207
  secondary, 132, 175, 179, 188—190
State equations, for fibrous tissue modeling, 55
Steady-state phenomenon, 64, 70, 105, 168, 185, 279
Sternoclavicular joint, 40
Stickler's disease, 5, 35—36
Stiffness, 279
  differential longitudinal, 235
  hoop, 33
Stiffness lag, 42, 53—54, 57, 144, 233
  prolonged, 153
Still's disease, 112, 124, 129
Strain, 60, 279
  compression, 228, 231, 282
  flexural, 167, 194
  outflaring, 235, 239
  principal, 277
  shearing, 229—231, 236
  tangential hoop, 235—236
  time-averaged, 23
  torsion, 50
Strain-averaging mechanism, 132
Strain gauge study, 235
Strain gradient, 9
Strain history, dynamic, 23
Strain rate, 52
Streaming potential, 10, 39, 42, 55, 144, 279
Strength, 280
  bulk, 231
Strength in shear, 230
Stress, 102, 128, 280
  compression, 228—229, 259, 282
  principal, 277
  shearing, 228—230
  tension, 259
Stress shielding effect, 67
Stretch-hypertrophy axiom, 132
Student, 251
Subchondral bone, 5
Subsigma, 280
Subtalar joint, 20—22
Sudek's atrophy, 111, 117
Summation, 251—266
Sun Valley Bone Workshop, 123, 187, 197—198
Surface-to-volume ratio, 184
Symbols, definition of, 268
Sympathetic innervation, 127
Sympathetic nerve block, 117

Syndactyly, 263
Synovia, 5, 7, 48
  regional acceleratory phenomenon in, 114
Synovial cell, 5
Synovial fluid, 7, 100, 111
Synovitis, 127
Syrinx, 118
Systemic factors, 268
  in repair, 146

T

Talus, 115, 129
Tarsus, 129
Teacher, 251
Temporal coherence, 103, 159—160, 258, 280
Temporal incoherence, 156—157, 280
Temporal overlap, 144—145
Tendon, 12, 41, 44—45
  attachment to bone, 40
  biomechanically adapted, 54
  bulk strength of, 51—52
  child's, 47
  growth of, 46—48
  length of, control of, 51
  rupture of, 53, 57, 117, 263
  transplant of, 1, 57
Tension, 280
Tension-creep compensation, 52
Tension force, 284
Terminology, skeletal, 281—282
Test confusion, 135
Tethering, 268
Tetracycline bone label, 66, 75, 86, 90, 142, 167, 170, 182—183
Theory, value and uses of, 211—217
Thrombosis, 220
Thumb, 11
Thyroid hormone, 65
Thyroxine, 132
Tibia, 113, 179, 187, 190, 209
  articular surface of, 8
  child's, 188
  fracture of, 115, 149
  pseudoarthrosis of, 155
Tibial joint surface, 23
Tibial torsion, 13, 185
Tidemark, 31—32
Time-averaging property, 22, 52
Tissue culture, 184—186, 195—196
Tissue debris, 119, 139
Tissue time marker, 67, 166
Toggling, 240, 243—244
Torque, 231
Torus, 147
Toughness, 280
Trabeculum, 81, 168
Trachea, 5
Transient phenomenon, 64, 103, 107, 185, 280
Trauma, 112, 147—153, 220

Trivial property, 201—202
Tropocollagen, 41, 46
Tumor, bone, 39, 139
Turner's syndrome, 5, 184
Turnover, 100, 111, 132, 166, 187—191, 280
   in adult skeleton, 190—191
   annual bone, 69
   of fibrous tissue, 49—51
   in growing skeleton, 189—190

U

Ulna, 28
Ulnohumeral joint, 25
Union, delayed, 117, 121, 132, 136—137, 149, 154
Unit, 281
   conversion of, 282
Unit load, 10—11, 15, 19—20, 280
Unit strain, 280—281
Unit stress, 281
Unwitting assumption, 22
Upper intermediary organization, 5, 59, 200
Up time-down time, 66
Upwards search, 198, 202

V

Varus deformity, 14, 16

Vascular channel, 39
Vascular disease, 118, 154
Vascular embolism, 219
Vertebral body, 237
Vertebral centrae, 15
Viscera, 48, 128
Viscoelastic deformability, 29
Viscoelasticity, 281
Vitamin D, 132, 205
Volumetric sterotypism, 175

W

Warp-and-woof pattern, 39, 144
Water, interstitial, 86
Weld, biological, 138, 143
Wound healing, 42, 45, 117
Wound infection, 150
Woven bone, 5, 39, 46, 111, 119, 139—144, 152, 157—159, 188—189, 244
   histogenesis of, 139
   turnover of, 111

X

X-ray, 111

Z

Zero loading, 11